Large Cardinals, Determinacy and Other Topics: The Cabal Seminar, Volume IV

The proceedings of the Los Angeles Caltech-UCLA "Cabal Seminar" were originally published in the 1970s and 1980s. *Large Cardinals, Determinacy and Other Topics* is the final volume in a series of four books collecting the seminal papers from the original volumes together with extensive unpublished material, new papers on related topics and discussion of research developments since the publication of the original volumes.

This final volume contains Parts VII and VIII of the series. Part VII focuses on "Extensions of AD, models with choice", while Part VIII ("Other topics") collects material important to the Cabal that does not fit neatly into one of its main themes. These four volumes will be a necessary part of the book collection of every set theorist.

ALEXANDER S. KECHRIS is Professor of Mathematics at the California Institute of Technology. He is the recipient of numerous honors, including the J. S. Guggenheim Memorial Foundation Fellowship and the Carol Karp Prize of the Association for Symbolic Logic. He is also a member of the Scientific Research Board of the American Institute of Mathematics.

BENEDIKT LÖWE is Universitair Hoofddocent at the Universiteit van Amsterdam, Professor of Mathematics at the Universität Hamburg and Fellow of Churchill College at the University of Cambridge. He is currently the president of the Deutsche Vereinigung für Mathematische Logik und für Grundlagenforschung der Exakten Wissenschaften (DVMLG) and the Secretary General of the Division for Logic, Methodology and Philosophy of Science and Technology (DLMPST).

JOHN R. STEEL is Professor of Mathematics at the University of California, Berkeley. Prior to that, he was a professor in the mathematics department at the University of California, Los Angeles. He is a recipient of the Carol Karp Prize of the Association for Symbolic Logic and of a Humboldt Prize. Steel is also a former Fellow at the Wissenschaftskolleg zu Berlin and the Sloan Foundation.

LECTURE NOTES IN LOGIC

A Publication of The Association for Symbolic Logic

This series serves researchers, teachers, and students in the field of symbolic logic, broadly interpreted. The aim of the series is to bring publications to the logic community with the least possible delay and to provide rapid dissemination of the latest research. Scientific quality is the overriding criterion by which submissions are evaluated.

More information, including a list of the books in the series, can be found at http://aslonline.org/books/lecture-notes-in-logic/

LECTURE NOTES IN LOGIC 49

Large Cardinals, Determinacy and Other Topics: The Cabal Seminar, Volume IV

Edited by

ALEXANDER S. KECHRIS
California Institute of Technology

BENEDIKT LÖWE
Universiteit van Amsterdam, Universität Hamburg and University of Cambridge

JOHN R. STEEL
University of California, Berkeley

ASSOCIATION FOR SYMBOLIC LOGIC

CAMBRIDGE
UNIVERSITY PRESS

CAMBRIDGE
UNIVERSITY PRESS

University Printing House, Cambridge CB2 8BS, United Kingdom

One Liberty Plaza, 20th Floor, New York, NY 10006, USA

477 Williamstown Road, Port Melbourne, VIC 3207, Australia

314–321, 3rd Floor, Plot 3, Splendor Forum, Jasola District Centre, New Delhi – 110025, India

79 Anson Road, #06–04/06, Singapore 079906

Cambridge University Press is part of the University of Cambridge.

It furthers the University's mission by disseminating knowledge in the pursuit of education, learning, and research at the highest international levels of excellence.

www.cambridge.org
Information on this title: www.cambridge.org/9781107182998
DOI: 10.1017/9781316863534

Association for Symbolic Logic
Richard A. Shore, Publisher
Department of Mathematics, Cornell University, Ithaca, NY 14853
http://aslonline.org

First published 2021

A catalogue record for this publication is available from the British Library.

ISBN – 4 Volume Set 978-1-108-92022-3 Hardback
ISBN – Volume I 978-0-521-89951-2 Hardback
ISBN – Volume II 978-0-521-76203-8 Hardback
ISBN – Volume III 978-1-107-03340-5 Hardback
ISBN – Volume IV 978-1-107-18299-8 Hardback

CONTENTS

PREFACE

This book is the last of the series of volumes containing reprints of the papers in the original Cabal Seminar volumes of the Springer *Lecture Notes in Mathematics* series [CABAL i, CABAL ii, CABAL iii, CABAL iv], unpublished material, and new papers. The first volume, [CABAL I], contained papers on games, scales and Suslin cardinals. The second volume, [CABAL II], contained papers on Wadge degrees and pointclasses and projective ordinals. The third volume, [CABAL III], contained papers on ordinal definability and recursion theory. In this volume, we conclude with Parts VII and VIII of the project: *Extensions of* AD*, models with choice* and *Other topics*.

In addition to the reprinted papers, this volume contains new papers by Larson (*A brief history of determinacy*), Martin (*Games of countable length*), Farah (*The extender algebra and* Σ^2_1*-absoluteness*) and Caicedo & Löwe (*The fourteen Victoria Delfino problems and their status in 2020*). Table 1 gives an overview of the papers in this volume with their original references.

As emphasized in our first three volumes, our project is not to be understood as an historical edition of old papers. In the retyping process, we uniformized and modernized notation and numbering of sections and theorems. As a consequence, references to papers in the old Cabal volumes will not always agree with references to their reprinted versions. In this volume, references to papers that already appeared in reprinted form will use the new numbering. In order to help the reader to easily cross-reference old and new numberings, we provide a list of changes after the preface.

The typing and design were partially funded by the *Marie Curie Research Training Site* GLoRiClass (MEST-CT-2005-020841) of the European Commission. Infrastructure was provided by the Institute for Logic, Language and Computation (ILLC) of the *Universiteit van Amsterdam*. Many people were involved in typing, laying out, and proofreading the papers. We should like to thank (in alphabetic order) Can Baskent, Hanne Berg, Pablo Cubides Kovacsics, Jined Elpitiya, Thomas Göbel, Leona Kershaw, Anston Klev, Alexandru Marcoci, Kian Mintz-Woo, Antonio Negro, Maurice Pico de los Cobos, Sudeep Regmi, Philipp Rohde, Cesar Sainz de Vicuña, Stephan

	PART VII	
Larson	*A brief history of determinacy*	NEW
Kechris	"AD *plus Uniformization" is equivalent to "Half* $\mathsf{AD}_{\mathbb{R}}$"	[CABAL iv, pp. 98–102]
Solovay	*The independence of* DC *from* AD	[CABAL i, pp. 171–184]
Martin	*Games of countable length*	NEW
Woodin	*Some consistency results in* ZFC *using* AD	[CABAL iii, pp. 172–198]
Kechris	*Subsets of* $\aleph_1$ *constructible from a real*	[CABAL iv, pp. 110–116]
Woodin	AD *and the uniqueness of the supercompact measures on* $\wp_{\omega_1}(\lambda)$	[CABAL iii, 67–71]
Farah	*The extender algebra and* Σ^2_1*-absoluteness*	NEW
	PART VIII	
Steel	*On Vaught's conjecture*	[CABAL i, pp. 193–208]
Dellacherie	*Capacities and analytic sets*	[CABAL ii, pp. 1–31]
Foreman	*More saturated ideals*	[CABAL iii, pp. 1–27]
Caicedo & Löwe	*The fourteen Victoria Delfino problems and their status in the year 2020*	NEW

TABLE 1

Schroevers, Sam van Gool, and Daniel Velkov for their important contribution as typists and diligent proofreaders. We should like to mention that the original LaTeX stylefile for the retyping was designed by Dr. Samson de Jager. Very special thanks are due to Dr. Joel Uckelman, who took over the typesetting coordination from de Jager in 2007.

The Editors
Alexander S. Kechris, *Pasadena, CA*
Benedikt Löwe, *Amsterdam, Hamburg, & Cambridge*
John R. Steel, *Berkeley, CA*

REFERENCES

ALEXANDER S. KECHRIS, BENEDIKT LÖWE, AND JOHN R. STEEL
[CABAL I] *Games, Scales, and Suslin cardinals: the Cabal Seminar, volume I*, Lecture Notes in Logic, vol. 31, Cambridge University Press, 2008.

[CABAL II] *Wadge Degrees and Projective Ordinals: the Cabal Seminar, volume II*, Lecture Notes in Logic, vol. 37, Cambridge University Press, 2012.

[CABAL III] *Ordinal Definability and Recursion Theory: the Cabal Seminar, volume III*, Lecture Notes in Logic, vol. 43, Cambridge University Press, 2016.

ALEXANDER S. KECHRIS, DONALD A. MARTIN, AND YIANNIS N. MOSCHOVAKIS

[CABAL ii] *Cabal Seminar* 77–79, Lecture Notes in Mathematics, vol. 839, Berlin, Springer-Verlag, 1981.

[CABAL iii] *Cabal Seminar* 79–81, Lecture Notes in Mathematics, vol. 1019, Berlin, Springer-Verlag, 1983.

ALEXANDER S. KECHRIS, DONALD A. MARTIN, AND JOHN R. STEEL

[CABAL iv] *Cabal Seminar* 81–85, Lecture Notes in Mathematics, vol. 1333, Berlin, Springer-Verlag, 1988.

ALEXANDER S. KECHRIS AND YIANNIS N. MOSCHOVAKIS

[CABAL i] *Cabal Seminar* 76–77, Lecture Notes in Mathematics, vol. 689, Berlin, Springer-Verlag, 1978.

ORIGINAL NUMBERING

Numbering in the reprints may differ from the original numbering. Where numbering differs, the original designation is listed on the left, with the corresponding number in the reprint listed on the right. In cases where an item numbered in the reprint had neither a number nor a name in the original, we have indicated that with a '—'.

"AD plus Uniformization" is equivalent to "Half $\mathsf{AD}_{\mathbb{R}}$*"*, Kechris, [CABAL iv, pp. 98–102]

Theorem	Theorem 1.1	Theorem	Theorem 1.3
Corollary (of the proof)	Corollary 1.2	Lemma	Lemma 1.4

The independence of DC *from* AD, Solovay, [CABAL i, pp. 171–183]

§0	§1	Lemma 2.2	Lemma 3.2
Definition 0.1	Definition 1.1	Definition 2.3	Definition 3.3
Lemma 0.2	Lemma 1.2	Corollary 2.4	Corollary 3.4
§1	§2	Theorem 2.5	Theorem 3.5
Lemma 1.1	Lemma 2.1	§3	§4
Theorem 1.2	Theorem 2.2	Lemma 3.1	Lemma 4.1
Lemma	Lemma 2.3	—	Definition 4.2
Theorem 1.3	Theorem 2.4	—	Definition 4.3
Lemma 1.4	Lemma 2.5	—	Definition 4.4
Lemma 1.5	Lemma 2.6	—	Lemma 4.5
Claim	Claim 2.7	—	Lemma 4.6
Lemma 1.6	Lemma 2.8	—	Lemma 4.7
Claim	Claim 2.9	—	Lemma 4.8
§2	§3	§4	§5
Lemma 2.1	Lemma 3.1	—	§5.1

—	§5.2	—	Lemma 9.3
—	§5.3	—	§9.10
—	§5.4	—	Lemma 9.4
—	§5.5	—	§10
—	§5.6	—	§10.1
—	§5.7	—	§10.2
—	Proposition 5.1	—	§10.3
—	§5.8	—	§10.4
—	§6	—	§10.5
—	§6.1	—	§10.6
—	§6.2	—	§10.7
—	§6.3	—	§11
—	§7	—	§11.1
—	§7.1	—	§11.2
—	§7.2	—	§12
—	§7.3	—	§12.1
—	§7.4	—	§12.2
—	§7.5	—	§12.3
—	§7.6	—	Theorem 12.1
—	§8	—	Theorem 12.2
—	§8.1	—	§12.4
—	§8.2	—	Lemma 12.3
—	Proposition 8.1	—	Lemma 12.4
—	§8.3	—	Theorem 12.5
—	§9	—	Corollary 12.6
—	§9.1	—	Corollary 12.7
—	Proposition 9.1	—	Theorem 12.8
—	§9.2	—	§12.5
—	§9.3	—	Theorem 12.9
—	§9.4	—	§13
—	§9.5	—	§13.1
—	Lemma 9.2	§6	§13.2
—	§9.6	—	§13.3
—	§9.7	—	§13.3.1
—	§9.8	—	§13.3.2
—	§9.9	—	§13.3.3

Some consistency results in ZFC *using* AD, Woodin, [Cabal iii, pp. 172–198]

(1)	Claim 2.3	(2)	Claim 2.8
(2)	Claim 2.4	(3)	Claim 2.9
(3)	Claim 2.5	(4)	Claim 2.10
Theorem 2.3	Theorem 2.6	Claim	Claim 3.8
(1)	Claim 2.7	Theorem 3.8	Theorem 3.9

Subsets of $\aleph_1$ constructible from a real, Kechris, [Cabal iv, pp. 110–116]

Theorem	Theorem 1.1		

AD *and the uniqueness of the supercompact measures on* $\wp_{\omega_1}(\lambda)$, Woodin, [Cabal iii, pp. 67–71]

Definition	Definition 1.1	Theorem 3	Theorem 1.4
Theorem 1	Theorem 1.2	Theorem 4	Theorem 1.5
Theorem 2	Theorem 1.3	Conjecture	Conjecture 2.1

On Vaught's conjecture, Steel, [Cabal i, pp. 193–208]

Theorem 1.2.1	Theorem 1.1	Definition 1.5.6	Definition 1.17
Corollary 1.2.2	Corollary 1.2	Definition 1.5.7	Definition 1.18
Corollary 1.2.3	Corollary 1.4	Definition 1.5.8	Definition 1.19
Corollary 1.2.4	Corollary 1.5	Theorem 1.5.9	Theorem 1.20
Theorem 1.3.1	Theorem 1.6	Theorem 1.5.10	Theorem 1.21
Remark	Remark 1.7	Theorem 1.5.11	Theorem 1.22
1.4.1	Example 1.8	Theorem 2.1.1	Theorem 2.1
1.4.2	Example 1.9	Lemma 2.1.2	Lemma 2.2
1.4.3	Example 1.10	Lemma 2.1.3	Lemma 2.3
A	§1.5.1	Subcase A	Subcase 1
Theorem 1.5.1	Theorem 1.11	Subcase B	Subcase 2
Remark	Remark 1.12	Corollary 2.3.1	Corollary 2.4
Theorem 1.5.2	Theorem 1.13	Corollary 2.3.2	Corollary 2.5
Theorem 1.5.3	Theorem 1.14	Lemma 2.3.3	Lemma 2.6
Remark 1.5.4	Remark 1.15	Corollary 2.3.4	Corollary 2.7
Remark 1.5.5	Remark 1.16	Corollary 2.3.5	Corollary 2.8
B	§1.5.2		

Capacities and analytic sets, Dellacherie, [Cabal ii, pp. 1–31]

Chapter 1	§1	Corollary	Corollary 1.10
§1	§1.1	§2	§1.2
Definition	Definition 1.1	Example 2.1	Example 1.11
Definition	Definition 1.3	Theorem 2.1	Theorem 1.12
Theorem 1.1	Theorem 1.5	Example 2.2	Example 1.14
Remark 1	Remark 1.6	Theorem 2.2	Theorem 1.15
Remark 2	Remark 1.7	Definition	Definition 1.16
Definition	Definition 1.8	Theorem 2.3	Theorem 1.17
Theorem 1.2	Theorem 1.9	Example 2.3	Example 1.19

Theorem 2.4	Theorem 1.20
Example 2.4	Example 1.21
Theorem 2.5	Theorem 1.22
Example 2.5	Example 1.23
§3	§1.3
Theorem 3.1	Theorem 1.24
Appendix	Appendix 1.3
Lemma	Lemma 1.26
Theorem	Theorem 1.27
Chapter 2	§2
§1	§2.1
Definition	Definition 2.1
Example 1.1	Example 2.2
Example 1.2	Example 2.3
Example 1.3	Example 2.4
Definition	Definition 2.5
Theorem 1.1	Theorem 2.6
§2	§2.2
Definition	Definition 2.7
Theorem 2.1	Theorem 2.7
Example 2.1	Example 2.9
Example 2.2	Example 2.10
Example 2.3	Example 2.11
Example 2.4	Example 2.12
Example 2.5	Example 2.13
Example 2.6	Example 2.14
§3	§2.3
Definition	Definition 2.15
Theorem 3.1	Theorem 2.16
Corollary 1	Corollary 2.17
Corollary 2	Corollary 2.18
Corollary 3	Corollary 2.19
Theorem 3.2	Theorem 2.20
§4	§2.4
Theorem 4.1	Theorem 2.23
Definition	Definition 2.24
Theorem 4.2	Theorem 2.25
Corollary 1	Corollary 2.26
Corollary 2	Corollary 2.27
Definition	Definition 2.28
Corollary 3	Corollary 2.29
Corollary 4	Corollary 2.30

§5	§2.5
Proposition A	Proposition 2.31
Theorem 5.1	Theorem 2.32
Chapter 3	§3
§1	§3.1
Definition	Definition 3.2
Example 1.1	Example 3.3
Example 1.2	Example 3.4
Example 1.3	Example 3.5
Example 1.4	Example 3.7
Example 1.5	Example 3.8
§2	§3.2
Definition	Definition 3.9
Example 2.1	Example 3.10
Example 2.2	Example 3.11
Example 2.3	Example 3.13
Theorem 2.1	Theorem 3.14
Theorem 2.2	Theorem 3.15
Theorem 2.3	Theorem 3.16
Theorem 2.4	Theorem 3.17
Theorem 2.5	Theorem 3.18
§3	§3.3
Definition	Definition 3.19
Theorem 3.1	Theorem 3.20
Corollary	Corollary 3.21
Chapter 4	§4
§1	§4.1
Definition	Definition 4.1
Proposition 1.1	Proposition 4.2
Example 1.1	Example 4.3
Example 1.2	Example 4.4
Example 1.3	Example 4.5
Example 1.4	Example 4.6
§2	§4.2
Proposition 2.1	Proposition 4.7
Lemma 2.1	Lemma 4.8
Proposition 2.2	Proposition 4.9
Theorem 2.1	Theorem 4.10
Corollary	Corollary 4.11
Theorem 2.2	Theorem 4.12
Theorem 2.3	Theorem 4.13

More saturated ideals, Foreman, [CABAL iii, pp. 1–27]

Theorem	Theorem 1	Lemma 6	Lemma 2.9
Theorem	Theorem 1	Lemma 7	Lemma 2.10
Theorem	Theorem 1	Lemma 8	Lemma 2.11
Proposition 1	Proposition 1.1	Lemma 9	Lemma 2.12
Theorem 1	Theorem 1.2	Problem	Open Problem 2.13
Theorem 2	Theorem 1.3	Problem	Open Problem 2.14
Theorem 3	Theorem 1.4	Problem	Open Problem 2.15
§1	§2	Problem	Open Problem 2.16
Theorem	Theorem 2	Proposition 2	Proposition 3.1
Lemma 1	Lemma 2.1	Lemma 10	Lemma 3.2
Lemma 2	Lemma 2.2	Proposition 3	Proposition 3.3
Lemma 3	Lemma 2.3	Lemma 8	Lemma 3.4
Lemma 4	Lemma 2.4	Lemma 9	Lemma 3.5
Lemma 5	Lemma 2.6	Corollary 1	Corollary 3.6
Definition	Definition 2.7	Claim	Claim 3.7
Example	Example 2.8	Claim	Claim 3.8

PART VII: EXTENSIONS OF AD, MODELS WITH CHOICE

A BRIEF HISTORY OF DETERMINACY

PAUL B. LARSON

§1. Introduction. Determinacy axioms are statements to the effect that certain games are *determined*, in that each player in the game has an optimal strategy. The commonly accepted axioms for mathematics, the Zermelo-Fraenkel axioms with the Axiom of Choice (ZFC; *cf.* [Jec03, Kun11]), imply the determinacy of many games that people actually play. This applies in particular to many **games of perfect information**, games in which the players alternate moves which are known to both players, and the outcome of the game depends only on this list of moves, and not on chance or other external factors. Games of perfect information which must end in finitely many moves are determined. This follows from the work of Ernst Zermelo [Zer13], Dénes Kőnig [Kőn27] and László Kálmar [Kal28], and also from the independent work of John von Neumann and Oskar Morgenstern (in their 1944 book, reprinted as [vNM04]).

As pointed out by Stanisław Ulam [Ula60], determinacy for games of perfect information of a fixed finite length is essentially a theorem of logic. If we let $x_1,y_1,x_2,y_2,\dots,x_n,y_n$ be variables standing for the moves made by players player I (who plays $x_1,\dots,x_n$) and player II (who plays $y_1,\dots,y_n$), and A (consisting of sequences of length $2n$) is the set of runs of the game for which player I wins, the statement

$$\exists x_1 \forall y_1 \dots \exists x_n \forall y_n \langle x_1, y_1, \dots, x_n, y_n\rangle \in A$$

essentially asserts that the first player has a winning strategy in the game, and its negation,

$$\forall x_1 \exists y_1 \dots \forall x_n \exists y_n \langle x_1, y_1, \dots, x_n, y_n\rangle \notin A$$

essentially asserts that the second player has a winning strategy.[1]

The author was supported in part by NSF grants DMS-0801009, DMS-1201494 and DMS-1764320. This paper is a revised version of [Lar12].

[1]If there exists a way of choosing a member from each nonempty set of moves of the game, then these statements are actually equivalent to the assertions that the corresponding strategies exist. Otherwise, in the absence of the Axiom of Choice the statements above can hold without the corresponding strategy existing.

Large Cardinals, Determinacy, and Other Topics: The Cabal Seminar, Volume IV
Edited by A. S. Kechris, B. Löwe, J. R. Steel
Lecture Notes in Logic, 49

We let ω denote the set of natural numbers $0, 1, 2, \ldots$; for brevity we will often refer to the members of this set as "integers". Given sets X and Y, ${}^{X}Y$ denotes the set of functions from X to Y. The **Baire space** is the space ${}^{\omega}\omega$, with the product topology. The Baire space is homeomorphic to the space of irrational real numbers (*cf.*, *e.g.*, [Mos09, p. 9]), and we will often refer to its members as "reals" (though in various contexts the Cantor space ${}^{\omega}2$, the set of subsets of ω ($\wp(\omega)$) and the set of infinite subsets of ω ($[\omega]^{\omega}$) are all referred to as "the reals").

Given $A \subseteq {}^{\omega}\omega$, we let $\mathrm{G}_{\omega}(A)$ denote the game of perfect information of length ω in which the two players collaborate to define an element f of ${}^{\omega}\omega$ (with player I choosing $f(0)$, player II choosing $f(1)$, player I choosing $f(2)$, and so on), with player I winning a run of the game if and only if f is an element of A. A game of this type is called an **integer game**, and the set A is called the **payoff set**. A **strategy** in such a game for player I (player II) is a function Σ with domain the set of sequences of integers of even (odd) length such that for each $a \in \mathrm{dom}(\Sigma)$, $\Sigma(a)$ is in ω. A run of the game (partial or complete) is said to be **according to** a strategy Σ for player I (player II) if every initial segment of the run of odd (nonzero even) length is of the form $a^\frown\langle\Sigma(a)\rangle$ for some sequence a. A strategy Σ for player I (player II) is a **winning strategy** if every complete run of the game according to Σ is in (out of) A. We say that a set $A \subseteq {}^{\omega}\omega$ is **determined** (or the corresponding game $\mathrm{G}_{\omega}(A)$ is determined) if there exists a winning strategy for one of the players. These notions generalize naturally for games in which players play objects other than integers (*e.g.*, **real games**, in which they play elements of ${}^{\omega}\omega$) or games which run for more than ω many rounds (in which case player I typically plays at limit stages).

The study of determinacy axioms concerns games whose determinacy is neither proved nor refuted by the Zermelo-Fraenkel axioms ZF (without the Axiom of Choice). Typically such games are infinite. Axioms stating that infinite games of various types are determined were studied by Stanisław Mazur, Stefan Banach and Ulam in the late 1920s and early 1930s; were reintroduced by David Gale and Frank Stewart [GS53] in the 1950s and again by Jan Mycielski and Hugo Steinhaus [MS62] in the early 1960s; gained interest with the work of David Blackwell [Bla67] and Robert Solovay in the late 1960s; and attained increasing importance in the 1970s and 1980s, finally coming to a central position in contemporary set theory.

Mycielski and Steinhaus introduced the Axiom of Determinacy (AD), which asserts the determinacy of $\mathrm{G}_{\omega}(A)$ for all $A \subseteq {}^{\omega}\omega$. Work of Banach in the 1930s shows that AD implies that all sets of reals satisfy the property of Baire. In the 1960s, Mycielski and Stanisław Świerczkowski proved that AD implies that all sets of reals are Lebesgue measurable, and Mycielski showed that AD implies countable choice for reals. Together, these results show that determinacy provides a natural context for certain areas of mathematics, notably analysis, free of the paradoxes induced by the Axiom of Choice.

Unaware of the work of Banach, Gale and Stewart [GS53] had shown that AD contradicts ZFC. However, their proof used a wellordering of the reals given by the Axiom of Choice, and therefore did not give a nondetermined game of this type with definable payoff set. Starting with Banach's work, many simply definable payoff sets were shown to induce determined games, culminating in D. Anthony Martin's celebrated 1974 result [Mar75] that all games with Borel payoff set are determined. This result came after Martin had used measurable cardinals to prove the determinacy of games whose payoff set is an analytic sets of reals.

The study of determinacy gained interest from two theorems in 1967, the first due to Solovay and the second to Blackwell. Solovay proved that under AD, ω_1 (the least uncountable ordinal) is a measurable cardinal, setting off a study of strong Ramsey properties on the ordinals implied by determinacy axioms. Blackwell used open determinacy (proved by Gale and Stewart) to reprove a classical theorem of Kazimierz Kuratowski. This also led to the application, by John Addison, Martin, Yiannis Moschovakis and others, of stronger determinacy axioms to produce structural properties for definable sets of reals. These axioms included the determinacy of $\utilde{\mathbf{\Delta}}^1_n$ sets of reals, for $n \geq 2$, statements which would not be proved consistent relative to large cardinals until the 1980s.

The large cardinal hierarchy was developed over the same period, and came to be seen as a method for calibrating consistency strength. In the 1970s, various special cases of $\utilde{\mathbf{\Delta}}^1_2$ determinacy were located on this scale, in terms of the large cardinals needed to prove them. Determining the consistency (relative to large cardinals) of forms of determinacy at the level of $\utilde{\mathbf{\Delta}}^1_2$ and beyond would take the introduction of new large cardinal concepts. Martin (in 1978) and W. Hugh Woodin (in 1984) would prove $\utilde{\mathbf{\Pi}}^1_2$-determinacy and $\mathsf{AD}^{\mathbf{L}(\mathbb{R})}$ respectively, using hypotheses near the very top of the large cardinal hierarchy. In a dramatic development, the hypotheses for these results would be significantly reduced through work of Woodin, Martin and John Steel. The initial impetus for this development was a seminal result of Matthew Foreman, Menachem Magidor and Saharon Shelah which showed, assuming the existence of a supercompact cardinal, that there exists a generic elementary embedding with well-founded range and critical point ω_1. Combined with work of Woodin, this yielded the Lebesgue measurability of all sets in the inner model $\mathbf{L}(\mathbb{R})$ from this hypothesis. Shelah and Woodin would reduce the hypothesis for this result further, to the assumption that there exist infinitely many Woodin cardinals below a measurable cardinal.

Woodin cardinals would turn out to be the central large cardinal concept for the study of determinacy. Through the study of tree representations for sets of reals, Martin and Steel would show that $\utilde{\mathbf{\Pi}}^1_{n+1}$-determinacy follows from the existence of n Woodin cardinals below a measurable cardinal, and that this hypothesis is not sufficient to prove stronger determinacy results for the projective hierarchy. Woodin would then show that the existence of infinitely many Woodin cardinals below a measurable cardinal implies $\mathsf{AD}^{\mathbf{L}(\mathbb{R})}$, and he would locate the exact consistency strengths of $\boldsymbol{\Delta}^1_2$-determinacy and $\mathsf{AD}^{\mathbf{L}(\mathbb{R})}$ at one Woodin cardinal and ω Woodin cardinals respectively.

In the aftermath of these results, many new directions were developed, and we give only the briefest indication here. Using techniques from inner model theory, the exact consistency strengths of many determinacy hypotheses were established. Using similar techniques, it has been shown that almost every natural statement (*i.e.*, not invented specifically to be a counterexample) implies directly those determinacy hypotheses of lesser consistency strength. For instance, by Kurt Gödel's Second Incompleteness Theorem, ZFC cannot prove that the AD holds in $\mathbf{L}(\mathbb{R})$, as the latter implies the consistency of the former. Empirically, however, every natural extension T of ZFC of sufficient consistency strength (*i.e.*, such that Peano Arithmetic does not prove the consistency of of T from the consistency of ZF+ AD) does appear to imply that AD holds in $\mathbf{L}(\mathbb{R})$. This sort of phenomenon is taken by some as evidence that the statement that AD holds in $\mathbf{L}(\mathbb{R})$, and other determinacy axioms, should be counted among the true statements extending ZFC (*cf.*, *e.g.*, [KW]).

The history presented here relies heavily on those given by Jackson [Jac10], Kanamori [Kan95, Kan03], Moschovakis [Mos09], Neeman [Nee04] and Steel [Ste08B]. As the title suggests, this is a selective and abbreviated account of the history of determinacy. We have omitted many interesting topics, including, *e.g.*, Blackwell games [Bla69, Mar98, MNV03] and proving determinacy in second-order arithmetic [LSR87, LSR88, KW10].

§2. Early developments. The first published paper in mathematical game theory appears to be Zermelo's paper [Zer13] on chess. Although he noted that his arguments apply to all games of reason not involving chance, Zermelo worked under two additional chess-specific assumptions. The first was that the game in question has only finitely many states, and the second was that an infinite run of the game was to be considered a draw. Zermelo specified a condition which is equivalent to the existence of a strategy in such a game guaranteeing a win within a fixed number of moves, as well as another condition equivalent to the existence of a strategy guaranteeing that one will not lose within a given fixed number of moves. His analysis implicitly introduced the notions of **game tree**, **subtree** of a game tree, and **quasi-strategy**.[2]

The paper states indirectly, but does not quite prove, or even define, the statement that in any game of perfect information with finitely many possible positions such that infinite runs of the game are draws, either one player

[2]As defined above, a strategy for a given player specifies a move in each relevant position; a quasi-strategy merely specifies a set of acceptable moves. The distinction is important when the Axiom of Choice fails, but is less important in the context of Zermelo's paper.

has a strategy that guarantees a win, or both players have strategies that guarantee at least a draw. A special case of this fact is determinacy for games of perfect information of a fixed finite length, which is sometimes called Zermelo's Theorem.

Kőnig [Kőn27] applied the fundamental fact now known as **Kőnig's Lemma** to the study of games, among other topics. While Kőnig's formulation was somewhat different, his lemma is equivalent to the assertion that every infinite finitely branching tree with a single root has an infinite path (a path can be found by iteratively choosing any successor node such that the tree above that node is infinite). Extending Zermelo's analysis to games in which infinitely many positions are possible while retaining the condition that each player has only finitely many options at each point, Kőnig used the statement above to prove that in such a game, if one player has a strategy (from a given point in the game) guaranteeing a win, then he can guarantee victory within a fixed number of moves. The application of Kőnig's Lemma to the study of games was suggested by von Neumann.

Kálmar [Kal28] took the analysis a step further by proving Zermelo's Theorem for games with infinitely many possible moves in each round. His arguments proceeded by assigning transfinite ordinals to nodes in the game tree, a method which remains an important tool in modern set theory. Kálmar explicitly introduced the notion of a winning strategy for a game, although his strategies were also quasi-strategies as above. In his analysis, Kálmar introduced a number of other important technical notions, including the notion of a **subgame** (essentially a subtree of the original game tree), and classifying strategies into those which depend only on the current position in the game and those which use the history of the game so far.[3]

Games of perfect information for which the set of infinite runs is divided into winning sets for each player appear in a question by Mazur in the Scottish Book, answered by Banach in an entry dated 4 August 1935 (*cf.* [Mau81, p. 113]). Following up on Mazur's question (still in the Book), Ulam asked about games where two players collaborate to build an infinite sequence of 0's and 1's by alternately deciding each member of the sequence, with the winner determined by whether the infinite sequence constructed falls inside some predetermined set E. Essentially raising the issue of determinacy for arbitrary $G_\omega(E)$, Ulam asked: for which sets E does the first player (alternately, the second player) have a winning strategy? (§ 2.1 below has more on the Banach-Mazur game.)

Games of perfect information were formally defined in 1944 by von Neumann and Morgenstern [vNM04]. Their book also contains a proof that games of perfect information of a fixed finite length are determined (p. 123).

Infinite games of perfect information were reintroduced by Gale and Stewart [GS53], who were unaware of the work of Mazur, Banach and Ulam (Gale,

[3] *Cf.* [SW01] for much more on these papers of Zermelo, Kőnig and Kálmar.

personal communication). They showed that a nondetermined game can be constructed using the Axiom of Choice (more specifically, from a wellordering of the set of real numbers).[4] They also noted that the proof from the Axiom of Choice does not give a definable undetermined game, and raised the issue of whether determinacy might hold for all games with a suitably definable payoff set. Toward this end, they introduced a topological classification of infinite games of perfect information, defining a game (or the set of runs of the game which are winning for the first player) to be **open** if all winning runs for the first player are won at some finite stage (*i.e.*, if, whenever $\langle x_0, x_1, x_2, \ldots \rangle$ is a winning run of the game for the first player, there is some n such that the first player wins all runs of the game extending $\langle x_0, \ldots, x_n \rangle$). Using this framework, they proved a number of fundamental facts, including the determinacy of all games whose payoff set is a Boolean combination of open sets (*i.e.*, in the class generated from the open sets by the operations of finite union, finite intersection and complementation). The determinacy of open games would become the basis for proofs of many of the strongest determinacy hypotheses. Gale and Stewart also asked a number of important questions, including the question of whether all Borel games are determined (to be answered positively by Martin [Mar75] in 1974).[5] Classifying games by the definability of their payoff sets would be an essential tool in the study of determinacy.

2.1. Regularity properties. Early motivation for the study of determinacy was given by its implications for regularity properties for sets of reals. In particular, determinacy of certain games of perfect information was shown to imply that every set of reals has the property of Baire and the perfect set property, and is Lebesgue measurable.[6] These three facts themselves each contradict the Axiom of Choice. We will refer to Lebesgue measurability, the property of Baire and the perfect set property as the **regularity properties**, the fact that there are other regularity properties notwithstanding.

---

[4]Given a set Y, we let AC_Y denote the statement that whenever $\{X_a : a \in Y\}$ is a collection of nonempty sets, there is a function f with domain Y such that $f(a) \in X_a$ for all $a \in Y$. Zermelo's **Axiom of Choice** (AC) [Zer04] is equivalent to the statement that AC_Y holds for all sets Y. A linear ordering $\leq$ of a set X is a **wellordering** if every nonempty subset of X has a $\leq$-least element. The Axiom of Choice is equivalent to the statement that there exist wellorderings of every set.

Kőnig's Lemma is a weak form of the Axiom of Choice and cannot be proved in ZF (*cf.* [Lév79, Exercise IX.2.18]).

[5]The **Borel** sets are the members of the smallest class containing the open sets and closed under the operations of complementation and countable union. The collection of Borel sets is generated in ω_1 many stages from these two operations. A natural process assigns a measure to each Borel set (*cf.*, *e.g.*, [Hal50]).

[6]A set of reals X has the **property of Baire** if $X \triangle O$ is meager for some open set O, where the **symmetric difference** $A \triangle B$ of two sets A and B is the set $(A \setminus B) \cup (B \setminus A)$, where $A \setminus B = \{x \in A : x \notin B\}$. A set of reals X has the **perfect set property** if it is countable or contains a perfect set (an uncountable closed set without isolated points). A set of reals X is **Lebesgue measurable** if there is a Borel set B such that $X \triangle B$ is a subset of a Borel measure 0 set. *Cf.* [Oxt80].

Question 43 of the Scottish Book, posed by Mazur, asks about games where two players alternately select the members of a shrinking sequence of intervals of real numbers, with the first player the winner if the intersection of the sequence intersects a set given in advance. Banach posted an answer in 1935, showing that such games are determined if and only if the given set is either meager (in which case the second player wins) or comeager relative to some interval (in which case the first player wins). The determinacy of the restriction of this game to each interval implies then that the given set has the Baire property (*cf.* [Oxt80, pp. 27–30] and [Kan03, pp. 373–374]). The game has come to be known as the Banach-Mazur game. Using an enumeration of the rationals, one can code intervals with rational endpoints with integers, getting a game on integers.

Morton Davis [Dav64] studied a game, suggested by Lester Dubins, where the first player plays arbitrarily long finite strings of 0's and 1's and the second player plays individual 0's and 1's, with the payoff set a subset of the set of infinite binary sequences as before. Davis proved that the first player has a winning strategy in such a game if and only if the payoff set contains a perfect set, and the second player has a winning strategy if and only if the payoff set is finite or countably infinite. The determinacy of all such games then implies that every uncountable set of reals contains a perfect set (asymmetric games of this type can be coded by integer games of perfect information). It follows that under AD there is no set of reals whose cardinality falls strictly between $\aleph_0$ and $2^{\aleph_0}$.[7]

Mycielski and Świerczkowski showed that the determinacy of certain integer games of perfect information implies that every subset of the real line is Lebesgue measurable [MŚ64]. Simpler proofs of this fact were later given by Leo Harrington (*cf.* [Kan03, pp. 375–377]) and Martin [Mar03].

By way of contrast, an argument of Vitali [Vit05] shows that under ZFC there are sets of reals which are not Lebesgue measurable. Banach and Tarski (*cf.* [BT24]; *cf.* also [Wag93] for a modern exposition), building on work of Hausdorff [Hau14], showed that under ZFC the unit ball can be partitioned into five pieces which can be rearranged to make two copies of the same sphere, again violating Lebesgue measurability as well as physical intuition. As with the undetermined game given by Gale and Stewart, the constructions of Vitali and Banach-Tarski use the Axiom of Choice and do not give definable examples of nonmeasurable sets. Via the Mycielski-Świerczkowski theorem, determinacy results would rule out the existence of definable examples, for various notions of definability.

2.2. Definability. As discussed above, ZFC implies that open sets are determined, and implies also that there exists a nondetermined set. The study of determinacy merges naturally with the study of sets of reals in terms of their

[7] *I.e.*, for every set X, if there exist injections $f : \omega \to X$ and $g : X \to 2^\omega$, then either X is countable or there exists a bijection between X and 2^ω.

definability (*i.e.*, descriptive set theory), which can be taken as a measure of their complexity. In this section we briefly introduce some important definability classes for sets of reals. Standard references include [Mos80, Kec95]. While we do mention some important results in this section, much of the section can be skipped on a first reading and used for later reference.

A **Polish space** is a topological space which is separable and completely metrizable. Common examples include the integers ω, the reals $\mathbb{R}$, the open interval $(0, 1)$, the Baire space ${}^{\omega}\omega$, the Cantor space ${}^{\omega}2$ and their finite and countable products. Uncountable Polish spaces without isolated points are a natural setting for studying definable sets of reals. For the most part we will concentrate on the Baire space and its finite powers.

Following notation introduced by Addison [Add58B],[8] open subsets of a Polish space are called $\mathbf{\Sigma}^0_1$, complements of $\mathbf{\Sigma}^0_n$ sets are $\mathbf{\Pi}^0_n$, and countable unions of $\mathbf{\Pi}^0_n$ sets are $\mathbf{\Sigma}^0_{n+1}$. More generally, given a positive $\alpha < \omega_1$, $\mathbf{\Sigma}^0_\alpha$ consists of all countable unions of members of $\bigcup_{\beta<\alpha} \mathbf{\Pi}^0_\beta$, and $\mathbf{\Pi}^0_\alpha$ consists of all complements of members of $\mathbf{\Sigma}^0_\alpha$. The Borel sets are the members of $\bigcup_{\alpha<\omega_1} \mathbf{\Sigma}^0_\alpha$.

A **pointclass** is a collection of subsets of Polish spaces. Given a pointclass $\Gamma \subseteq \wp({}^{\omega}\omega)$, we let $\mathsf{Det}(\Gamma)$ and Γ**-determinacy** each denote the statement that $G_\omega(A)$ is determined for all $A \in \Gamma$. Philip Wolfe proved $\mathbf{\Sigma}^0_2$-determinacy in ZFC[Wol55]. Davis followed by proving $\mathbf{\Pi}^0_3$-determinacy [Dav64]. Jeffrey Paris would prove $\mathbf{\Sigma}^0_4$-determinacy [Par72]. However, this result was proved after Martin had used a measurable cardinal to prove analytic determinacy (*cf.* § 5.2).

Continuous images of $\mathbf{\Pi}^0_1$ sets are said to be $\mathbf{\Sigma}^1_1$, complements of $\mathbf{\Sigma}^1_n$ sets are $\mathbf{\Pi}^1_n$, and continuous images of $\mathbf{\Pi}^1_n$ sets are $\mathbf{\Sigma}^1_{n+1}$. For each $i \in \{0, 1\}$ and $n \in \omega$, the pointclass $\mathbf{\Delta}^i_n$ is the intersection of $\mathbf{\Sigma}^i_n$ and $\mathbf{\Pi}^i_n$. The **boldface projective pointclasses** are the sets $\mathbf{\Sigma}^1_n$, $\mathbf{\Pi}^1_n$, and $\mathbf{\Delta}^1_n$ for positive $n \in \omega$. These classes were implicit in work of Lebesgue as early as [Leb18]. They were made explicit in independent work by Nikolai Luzin [Luz25C, Luz25B, Luz25A] and Wacław Sierpiński [Sie25]. The notion of a boldface pointclass in general (*i.e.*, possibly non-projective) is used in various ways in the literature. We will say that a pointclass Γ is **boldface** (or **closed under continuous preimages** or **continuously closed**) if $f^{-1}[A] \in \Gamma$ for all $A \in \Gamma$ and all continuous functions f between Polish spaces (where A is a subset of the codomain). The classes $\mathbf{\Sigma}^0_\alpha$, $\mathbf{\Pi}^0_\alpha$, $\mathbf{\Delta}^0_\alpha$ are also boldface in this sense.

The pointclass $\mathbf{\Sigma}^1_1$ is also known as the class of **analytic sets**, and was given an independent characterization by Mikhail Suslin [Sus17]: A set of reals A is analytic if and only if there exists a family of closed sets D_s (for each finite

[8]The papers [Add58B] and [Add58A] appear in the same volume of ***Fundamenta Mathematicae***. The front page of the volume gives the date 1958–1959. The individual papers have the dates 1958 and 1959 on them, respectively.

sequence s consisting of integers) such that A is the set of reals x for which there is an ω-sequence S of integers such that $x \in \bigcap_{n \in \omega} D_{S \restriction n}$.[9] Suslin showed that there exist non-Borel analytic sets, and that the Borel sets are exactly the $\utilde{\mathbf{\Delta}}^1_1$ sets.

We let $\exists^0$ and $\exists^1$ denote existential quantification over the integers and reals, respectively, and $\forall^0$ and $\forall^1$ the analogous forms of universal quantification. Given a set $A \subseteq ({}^\omega\omega)^{k+1}$, for some positive integer k, $\exists^1 A$ is the set of $(x_1, \ldots, x_k) \in ({}^\omega\omega)^k$ such that for some $x \in {}^\omega\omega$, $(x, x_1, \ldots, x_k) \in A$, and $\forall^1 A$ is the set of $(x_1, \ldots, x_k) \in ({}^\omega\omega)^k$ such that for all $x \in {}^\omega\omega$, $(x, x_1, \ldots, x_k) \in A$. Given a pointclass Γ, $\exists^1\Gamma$ consists of $\exists^1 A$ for all $A \in \Gamma$, and $\forall^1\Gamma$ consists of $\forall^1 A$ for all $A \in \Gamma$. It follows easily that for each positive integer n, $\exists^1\utilde{\mathbf{\Pi}}^1_n = \utilde{\mathbf{\Sigma}}^1_{n+1}$ and $\forall^1\utilde{\mathbf{\Sigma}}^1_n = \utilde{\mathbf{\Pi}}^1_{n+1}$.

Given a pointclass Γ, $\check{\Gamma}$ is the set of complements of members of Γ, and Δ_Γ is the pointclass $\Gamma \cap \check{\Gamma}$; Γ is said to be **selfdual** if $\Delta_\Gamma = \Gamma$. A set $A \in \Gamma$ is said to be **Γ-complete** if every member of Γ is a continuous preimage of A. If Γ is closed under continuous preimages and Γ-determinacy holds, then $\check{\Gamma}$-determinacy holds. Each of the regularity properties for a set of reals A is given by the determinacy of games with payoff set simply definable from A (indeed, continuous preimages of A), but not necessarily with payoff A itself. It follows that when Γ is a boldface pointclass, Γ-determinacy implies the regularity properties for sets of reals in Γ.

A simple application of Fubini's theorem shows that if Γ is a boldface pointclass and there exists in Γ a wellordering of a set of reals of positive Lebesgue measure, then there is a non-Lebesgue measurable set in Γ. Skipping ahead for a moment, in the early 1970s Alexander Kechris and Martin, using a technique of Solovay called **unfolding**, proved that for each integer n, $\utilde{\mathbf{\Pi}}^1_n$-determinacy plus **countable choice for sets of reals**[10] implies that all $\utilde{\mathbf{\Sigma}}^1_{n+1}$ sets of reals are Lebesgue measurable, have the Baire property and have the perfect set property (*cf.* [Kan03, pp. 380–381]).

As developed by Stephen Kleene, the **effective** (or **lightface**) pointclasses Σ^0_n, Π^0_n, Δ^0_n [Kle43] and Σ^1_n, Π^1_n, Δ^1_n [Kle55C, Kle55B, Kle55A] are formed in the same way as their boldface counterparts, starting instead from Σ^0_1, the collection of open sets O such that the set of indices for basic open sets contained in O (under a certain natural enumeration of the basic open sets) is recursive (*cf.*, *e.g.*, [Mos09]). Sets in Σ^0_1 are called **semirecursive**, and sets in Δ^0_1 are called **recursive**. Given $a \in {}^\omega\omega$, $\Sigma^0_1(a)$ is the collection of open sets O such that the set of indices for basic open sets contained in O is recursive in a, and the **relativized lightface projective pointclasses** $\Sigma^0_n(a)$, $\Pi^0_n(a)$, $\Delta^0_n(a)$,

[9] For a function S with domain ω, and $n \in \omega$, $S \restriction n = \langle S(0), \ldots, S(n-1) \rangle$.

[10] The statement that whenever X_n ($n \in \omega$) are nonempty sets of reals, there is a function $f : \omega \to \mathbb{R}$ such that $f(n) \in X_n$ for each n. Countable choice for sets of reals is a consequence of AD, as shown by Mycielski [Myc64] (*cf.* § 2.3).

$\Sigma^1_n(a)$, $\Pi^1_n(a)$, $\Delta^1_n(a)$ are built from Σ^0_n in the manner above. It follows that each boldface pointclass is the union of the corresponding relativized lightface classes (relativizing over each member of ${}^{\omega}\omega$).

Following [Mos09], a pointclass is **adequate** if it contains all recursive sets and is closed under finite unions and intersections, bounded universal and existential integer quantification (*cf.* [Mos09, p. 119]) and preimages by recursive functions.[11] The relativized lightface projective pointclasses are adequate (*cf.* [Mos09, pp. 118–120]).

Given a Polish space $\mathfrak{X}$, an integer k, a set $A \subseteq \mathfrak{X}^{k+1}$ and $x \in \mathfrak{X}$, A_x is the set of $(x_1, \ldots, x_k)$ such that $(x, x_1, \ldots, x_k) \in A$. A set $A \subseteq \mathfrak{X}^{k+1}$ in a pointclass Γ is said to be **universal** for Γ if each subset of $\mathfrak{X}^k$ in Γ has the form A_x for some $x \in \mathfrak{X}$. Pointclasses of the form Σ^1_n, Π^1_n have universal members. Those of the form Δ^1_n do not. Each member of each boldface pointclass is of the form A_x for A a member of the corresponding effective class. Similarly, as each member of each lightface projective pointclass listed above is definable, each member of each corresponding boldface pointclass is definable from a real number as a parameter.

A set of reals is said to be Σ^2_1 (Π^2_1) if is definable by a formula of the form $\exists X \subseteq \mathbb{R}\varphi$ ($\forall X \subseteq \mathbb{R}\varphi$), where all quantifiers in φ range over the reals or the integers.

In the **Lévy hierarchy** [Lév65B], a formula φ in the language of set theory is Δ_0 (equivalently Σ_0, Π_0) if all quantifiers appearing in φ are bounded (*cf.* [Jec03, Chapter 13]); Σ_{n+1} if it has the form $\exists x\psi$ for some Π_n formula ψ; and Π_{n+1} if it has the form $\forall x\psi$ for some Σ_n formula ψ. A set is Σ_n-*definable* if it can be defined by a Σ_n formula (and similarly for Π_n). We say that a model M is **Γ-correct**, for a class of formulas Γ, if for all φ in Γ and $x \in M$, $M \models \varphi(x)$ if and only if $\mathbf{V} \models \varphi(x)$. If M is a model of ZF, we say that a set in M is Σ^M_n if it is definable by a Σ_n formula relativized to M (and similarly for other classes of formulas).

Gödel's inner model $\mathbf{L}$ is the smallest transitive model of ZFC containing the ordinals. For any set A, $\mathbf{L}$ generalizes to two inner models $\mathbf{L}(A)$ and $\mathbf{L}[A]$, developed respectively by András Hajnal [Haj56, Haj61] and Azriel Lévy [Lév57, Lév60] (*cf.* [Jec03, Chapter 13] or [Kan03, p. 34]). Given a set A, $\mathbf{L}(A)$ is the smallest transitive model of ZF containing the transitive closure of $\{A\}$ and the ordinals,[12] and $\mathbf{L}[A]$ is the smallest transitive model of ZF containing the ordinals and closed under the function $X \mapsto A \cap X$. Alternately, $\mathbf{L}(A)$ is constructed in the same manner as $\mathbf{L}$, but introducing the members of the transitive closure of the set $\{A\}$ at the first level, and $\mathbf{L}[A]$ is constructed as $\mathbf{L}$,

[11] A function f from a Polish space $\mathfrak{X}$ to a Polish space $\mathfrak{Y}$ is said to be **recursive** if the set of pairs $x \in \mathfrak{X}$, $n \in \omega$ such that $f(x)$ is in the nth basic open neighborhood of $\mathfrak{Y}$ is semi-recursive.

[12] A set x is **transitive** if $z \in x$ whenever $y \in x$ and $z \in y$. The **transitive closure** of a set x is the smallest transitive set containing x.

but by adding a predicate for membership in A to the language. When A is contained in $\mathbf{L}$, $\mathbf{L}(A)$ and $\mathbf{L}[A]$ are the same. While $\mathbf{L}[A]$ is always a model of AC, $\mathbf{L}(A)$ need not be. Indeed, $\mathbf{L}(\mathbb{R})$ is a model of AD in the presence of suitably large cardinals, and is thus a natural example of a "smaller universum" as described in the quote from [MS62] in § 2.3.

Although it can be formulated in other ways, we will view the set $0^\#$ ("zero sharp") as the theory of a certain class of ordinals which are indiscernibles over the inner model $\mathbf{L}$. This notion was independently isolated by Solovay [Sol67A] and by Jack Silver in his 1966 Berkeley Ph.D. thesis (*cf.* [Sil71]). The existence of $0^\#$ cannot be proved in ZFC, as it serves as a sort of transcendence principle over $\mathbf{L}$. For instance, if $0^\#$ exists then every uncountable cardinal of $\mathbf{V}$ is a strongly inaccessible cardinal in $\mathbf{L}$.[13] For any set X there is an analogous notion of $X^\#$ ("X sharp") serving as a transcendence principle over $\mathbf{L}(X)$ (*cf.* [Kan03]).

2.3. The Axiom of Determinacy. The Axiom of Determinacy, the statement that all length ω integer games of perfect information are determined, was proposed by Mycielski and Steinhaus [MS62].[14] In a passage that anticipated a commonly accepted view of determinacy, they wrote

> It is not the purpose of this paper to depreciate the classical mathematics with its fundamental "absolute" intuitions on the universum of sets (to which belongs the axiom of choice), but only to propose another theory which seems very interesting although its consistency is problematic. Our axiom can be considered as a restriction of the classical notion of a set leading to a smaller universum, say of determined sets, which reflect some physical intuitions which are not fulfilled by the classical sets ... Our axiom could be considered as an axiom added to the classical set theory claiming the existence of a class of sets satisfying (A) and the classical axioms (without the axiom of choice).

Mycielski and Steinhaus summarized the state of knowledge of determinacy at that time, including the fact that AD implies that all sets of reals are Lebesgue measurable and have the Baire property. They also noted that by results of Gödel and Addison [Add58B], assuming that $\mathbf{V} = \mathbf{L}$ there exists a Δ^1_2 wellordering of the reals, and thus a Δ^1_2 set which is not determined.

In his [Myc64], Mycielski proved several fundamental facts about determinacy, including the fact that AD implies countable choice for set of reals (he

[13] A cardinal κ is **strongly inaccessible** if it is uncountable, regular and a strong limit (*i.e.*, $2^\gamma < \kappa$ for all $\gamma < \kappa$). If κ is a strongly inaccessible cardinal, then $\mathbf{V}_\kappa$ is a model of ZFC. It follows that the existence of strongly inaccessible cardinals cannot be proved in ZFC. *Cf.* [Jec03] for the definition of $\mathbf{V}_\alpha$, for an ordinal α.

[14] We continue to use the now-standard abbreviation AD for the Axiom of Determinacy; it was called (A) in [MS62].

credits this result to Świerczkowski, Dana Scott and himself, independently). Thus, while AD contradicts the Axiom of Choice, it implies a form of Choice which suffices for many of its most important applications, including the countable additivity of Lebesgue measure. Via countable choice for sets of reals, AD implies that ω_1 is regular.[15] Mycielski also showed that AD implies that there is no uncountable wellordered sequence of reals. In conjunction with the perfect set property, this implies that under determinacy, $\omega_1^{\mathbf{V}}$ is a strongly inaccessible cardinal in $\mathbf{L}$ (and even in $\mathbf{L}[a]$ for any real number a), a fact which was to be greatly extended by Solovay, Martin and Woodin. Harrington [Har78] would show that Π^1_1-determinacy implies that $0^\#$ exists, and thus that Π^1_1-determinacy is not provable in ZFC.

In the same paper, Mycielski showed that ZF implies the existence of an undetermined game of perfect information of length ω in which the first move by player I is a countable ordinal instead of an integer (such a game can also be coded by a game of length ω_1 where both players play integers). An interesting aspect of the proof is that it does not give a specific undetermined game. As a slight variant on Mycielski's argument, consider the game in which the first player plays a countable ordinal α (and then makes no other moves for the rest of the game) and the second player plays a sequence of integers coding a linear order of ordertype α, under some fixed coding of binary relations on ω by reals. Since the first player cannot have a winning strategy in this game, determinacy for the game implies the existence of an injection from ω_1 into $\mathbb{R}$, which contradicts AD but is certainly by itself consistent with ZF, as it follows from ZFC. Later results of Woodin (building on work of Neeman [Nee04]) would show that, assuming the consistency of certain large cardinal hypotheses, ZFC is consistent with the statement that every integer game of length ω_1 with payoff set definable from real and ordinal parameters is determined (*cf.* § 6.3, and [Nee04, p. 298]). Mycielski noted that under AD there are no nonprincipal ultrafilters[16] on ω (this follows from Lebesgue measurability for all sets of reals plus a result of Sierpiński [Sie38] showing that nonprincipal ultrafilters on ω give rise to nonmeasurable sets of reals), which implies that every ultrafilter (on any set) is countably complete (*i.e.*, closed under countable intersections). Finally, in a footnote on the first page of the paper, Mycielski reiterated a point

---

[15]A cardinal κ is **regular** if, for every $\gamma < \kappa$, every function $f : \gamma \to \kappa$ has range bounded in κ. Under ZFC, every successor cardinal is regular. Solomon Feferman and Azriel Lévy [FL63] (*cf.* also [HR98, pp. 153–154]) showed that the singularity of ω_1 is consistent with ZF. Moti Gitik showed that it is consistent with ZF (relative to large cardinals) that ω is the largest regular cardinal [Git80].

[16]An **ultrafilter** on a set X is a collection U of nonempty subsets of X which is closed under supersets and finite intersections, and which has the property that for every $A \subseteq X$, exactly one of A and $X \backslash A$ is in U. An ultrafilter is **nonprincipal** if it contains no finite sets. The existence of nonprincipal ultrafilters on ω follows from ZFC, but (as this result shows) requires the Axiom of Choice.

made in the passage quoted above from his paper with Steinhaus, suggesting that an inner model containing the reals could satisfy AD. In a followup paper, Mycielski [Myc66] presented a number of additional results, including the fact that there is a game in which the players play real numbers whose determinacy implies *uniformization* (*cf.* § 3.2) for subsets of the plane, another weak form of the Axiom of Choice.

In 1964, one year after Paul Cohen's invention of forcing, Solovay proved that if there exists a strongly inaccessible cardinal, then in a forcing extension there exists an inner model containing the reals in which every set of reals satisfies the regularity properties from § 2.1 [Sol70]. Shelah later showed that a strongly inaccessible cardinal is necessary, in the sense that the Lebesgue measurability of all sets of reals (and even the perfect set property for $\utilde{\Pi}^1_1$ sets) implies that ω_1 is strongly inaccessible in all models of the form $\mathbf{L}[a]$, for $a \subseteq \omega$ [She84]. In the introduction to his paper, Solovay conjectured (correctly, as it turned out) that large cardinals would imply that AD holds in $\mathbf{L}(\mathbb{R})$.

The year 1967 saw two major results in the study of determinacy, one by Blackwell [Bla67] and the other by Solovay. Reversing chronological order by a few months, we discuss Blackwell's result and its consequences in the next section, and Solovay's in § 4.

§3. Reduction and scales. Blackwell [Bla67] used open determinacy to reprove a theorem of Kuratowski [Kur36] stating that the intersection of any two analytic sets A, B in a Polish space $\mathfrak{Y}$ is also the intersection of two analytic sets A' and B' such that $A \subseteq A'$, $B \subseteq B'$, and $A' \cup B' = \mathfrak{Y}$.[17] Briefly, the argument is as follows. Since A and B are analytic, there exist continuous surjections $f : {}^\omega\omega \to A$ and $g : {}^\omega\omega \to B$. For each finite sequence $\langle n_0, \dots, n_k \rangle$, let $\Omega(\langle n_0, \dots, n_k \rangle)$ be the set of $x \in {}^\omega\omega$ with $\langle n_0, \dots, n_k \rangle$ as an initial segment; let $R(\langle n_0, \dots, n_k \rangle)$ be the closure (in $\mathfrak{Y}$) of the f-image of $\Omega(\langle n_0, \dots, n_k \rangle)$; and let $S(\langle n_0, \dots, n_k \rangle)$ be the closure of the g-image of $\Omega(\langle n_0, \dots, n_k \rangle)$. Then for each $z \in \mathfrak{Y}$, let $G(z)$ be the game where players player I and player II build x and y in ${}^\omega\omega$, with player I winning if for some integer k, $z \in R(x \restriction k) \setminus S(y \restriction k)$, player II winning if for some integer k, $z \in S(y \restriction k) \setminus R(x \restriction (k+1))$, and the run of the game being a draw if neither of these happens. Roughly, each player is creating a real (x or y) to feed into his function, and trying to maintain for as long as possible that the corresponding output can be made arbitrarily close to the target real z; the loser is the first player to fail to maintain this condition. Let A' be the set of z for which player I has a strategy guaranteeing at least a draw, and let B' be the set of z for which player II has such a strategy. Then the determinacy of open games implies that ${}^\omega\omega = A' \cup B'$, and $A \subseteq A'$, $B \subseteq B'$ and $A' \cap B' = A \cap B$ follow from the fact that A is the range of f and B is the range of g. The sets A' and B' are analytic, as A' is a projection of

[17] Blackwell describes the discovery of his proof in [AA85, p. 26].

the set of pairs (φ, z) such that φ is (a code for) a strategy for player I in $G(z)$ guaranteeing at least a draw, which is Borel, and similarly for B'.[18]

3.1. Reduction, separation, norms and prewellorderings. In his [Kur36], Kuratowski defined the **reduction theorem** (now called the **reduction property**) for a pointclass Γ to be the statement that for any A, B in Γ there exist disjoint A', B' in Γ with $A' \subseteq A$, $B' \subseteq B$ and $A' \cup B' = A \cup B$. He showed in this paper that $\boldsymbol{\Pi}^1_1$ and $\boldsymbol{\Sigma}^1_2$ have the reduction property; Addison showed this for $\Pi^1_1(a)$ and $\Sigma^1_2(a)$, for each real number a [Add58B]. Blackwell's argument proves the reduction property for $\boldsymbol{\Pi}^1_1$, working with the corresponding $\boldsymbol{\Sigma}^1_1$ complements.

Kuratowski also defined the **first separation theorem** (now called the **separation property**) for a pointclass Γ to be the statement that for any disjoint A, B in Γ there exists C in Δ_Γ with $A \subseteq C$ and $B \cap C = \varnothing$. This property had been studied by Sierpiński [Sie24] and Luzin [Luz30A] for initial segments of the Borel hierarchy. Kuratowski also noted that the reduction property for a pointclass Γ implies the separation property for $\check{\Gamma}$. Luzin proved that the pointclass $\boldsymbol{\Sigma}^1_1$ satisfies the separation property, by showing that disjoint $\boldsymbol{\Sigma}^1_1$ sets are contained in disjoint Borel sets [Luz27, pp. 51–55]. Petr Novikov showed that $\boldsymbol{\Pi}^1_2$ satisfies the separation property and $\boldsymbol{\Sigma}^1_2$ does not [Nov35]. Novikov (in the case of $\boldsymbol{\Sigma}^1_2$ sets) and Addison showed that if Γ satisfies the reduction property and has a so-called **doubly universal** member, and Δ_Γ has no universal member, then Γ does not have the separation property, so $\check{\Gamma}$ does not have the reduction property [Nov35, Add58B].[19] Addison showed that if all real numbers are constructible, then the reduction property holds for $\boldsymbol{\Sigma}^1_k$, for all $k \geq 2$ [Add58A, Add58B].

Inspired by Blackwell's argument, Addison and Martin independently proved that $\boldsymbol{\Delta}^1_2$-determinacy implies that $\boldsymbol{\Pi}^1_3$ has the reduction property. Since the pointclass $\boldsymbol{\Sigma}^1_3$ has a doubly universal member, this shows that $\boldsymbol{\Delta}^1_2$-determinacy implies the existence of a nonconstructible real. This fact also follows from Gödel's result (discussed in [Add58A]) that the Lebesgue measurability of all Δ^1_2 sets implies the existence of a nonconstructible real. Determinacy would soon be shown to imply stronger structural properties for the projective pointclasses.

The key technical idea behind the (pre-determinacy) results listed above on separation and reduction for the first two levels of the projective hierarchy was the notion of *sieve* (in French, *crible*). This construction first appeared in a paper of Lebesgue [Leb05], in which he proved the existence of Lebesgue-measurable sets which are not Borel. In Lebesgue's presentation, a **sieve** is an association of a closed subset F_r of the unit interval [0, 1] to each rational

[18]A **projection** of a set $A \subseteq ({}^\omega\omega)^k$ (for some integer $k \geq 2$) is a set of the form $\{(x_0, \ldots, x_{i-i}, x_{i+1}, \ldots, x_{k-1}) \mid \exists x_i (x_0, \ldots, x_{k-1}) \in A\}$, for some $i < k$.

[19]Members U and V of a pointclass Γ are **doubly universal** for Γ if for each pair A,B of members of Γ there exist an $x \in {}^\omega\omega$ such that $U_x = A$ and $V_x = B$. The non-selfdual projective pointclasses (*e.g.*, $\Sigma^1_1(a)$, $\Pi^1_1(a)$, $\Sigma^1_2(a)$, $\Pi^1_2(a)$, …) all have doubly universal members.

number r in this interval. The sieve then represents the set of $x \in [0, 1]$ such that $\{r \mid x \in F_r\}$ is wellordered, under the usual ordering of the rationals. Using this approach, Luzin and Sierpiński showed that $\mathbf{\Sigma}^1_1$ sets and $\mathbf{\Pi}^1_1$ sets are unions of $\aleph_1$ many Borel sets [LS18, LS23].

Much of the classical work of Luzin, Sierpiński, Kuratowski and Novikov mentioned here was redeveloped in the lightface context by Kleene [Kle43, Kle55C, Kle55B, Kle55A], who was unaware of their previous work. The two theories were unified primarily by Addison (*e.g.*, [Add58A]). While Blackwell's argument generalizes throughout the projective hierarchy, Moschovakis (*cf.* [Mos67, Mos69A, Mos69B, Mos70, Mos71]; *cf.* also [Mos09, pp. 202–206]) developed via the effective theory a generalization of the Luzin-Sierpiński approach (decomposing a set of reals into a wellordered sequence of simpler sets) which could be similarly propagated. Moschovakis's goal was to find a uniform approach to the theory of Π^1_1 and Σ^1_2; he was unaware of either Kuratowski's work or determinacy (personal communication). He extracted the following notions, for a given pointclass Γ: a **Γ-norm** for a set A is a function $\rho\colon A \to \mathrm{On}$ for which there exist relations $R^+ \in \Gamma$ and $R^- \in \check{\Gamma}$ such that for any $y \in A$,

$$x \in A \wedge \rho(x) \leq \rho(y) \leftrightarrow R^+(x, y) \leftrightarrow R^-(x, y);$$

a pointclass Γ is said to have the **prewellordering property** if every $A \in \Gamma$ has a Γ-norm.[20] The prewellordering property was first explicitly formulated by Moschovakis in 1964; the definition just given is a reformulation due to Kechris. Kuratowski [Kur36] and Addison [Add58B] had shown that a variant of the property implies the reduction property; the same holds for the prewellordering property as defined by Moschovakis. Moschovakis applied Novikov's arguments to show that if Γ is a projective pointclass such that $\forall^1\Gamma \subseteq \Gamma$, and Γ has the prewellordering property, then so does the pointclass $\exists^1\Gamma$. Martin and Moschovakis independently completed the picture in 1968, proving what is now known as the First Periodicity Theorem (*cf.* [AM68, Mar68]).

THEOREM 3.1 (First Periodicity Theorem). Let Γ be an adequate pointclass and suppose that Δ_Γ-determinacy holds. Then for all $A \in \Gamma$, if A admits a Γ-norm, then $\forall^1 A$ admits a $\forall^1\exists^1\Gamma$-norm.

COROLLARY 3.2. Let Γ be an adequate pointclass closed under existential quantification over reals, and suppose that Δ_Γ-determinacy holds. If Γ satisfies the prewellordering property, then so does $\forall^1\Gamma$.

[20] A **prewellordering** is a binary relation which is wellfounded, transitive and total. A function ρ from a set X to the ordinals induces a prewellording $\preceq$ on X by setting $a \preceq b$ if and only if $\rho(a) \leq \rho(b)$. Conversely, a prewellordering $\preceq$ on a set X induces a function ρ from X to the ordinals, where for each $a \in X$, $\rho(a)$ (the **$\preceq$-rank** of a) is the least ordinal α such that $\rho(b) < \alpha$ for all $b \in X$ such that $b \preceq a$ and $a \not\preceq b$. The range of ρ is called the **length** of $\preceq$.

Projective Determinacy (PD) is the statement that all projective sets of reals are determined. By the First Periodicity Theorem, under Projective Determinacy the following pointclasses have the prewellordering property, for any real a:

$$\Pi^1_1(a), \Sigma^1_2(a), \Pi^1_3(a), \Sigma^1_4(a), \Pi^1_5(a), \Sigma^1_6(a), \ldots$$

By contrast (*cf.* [Kan03, pp. 409–410]), in **L** the pointclasses with the prewellordering property are

$$\Pi^1_1(a), \Sigma^1_2(a), \Sigma^1_3(a), \Sigma^1_4(a), \Sigma^1_5(a), \Sigma^1_6(a), \ldots$$

3.2. Scales. As noted above, the Axiom of Determinacy contradicts the Axiom of Choice, but it is consistent with, and even implies, certain weak forms of Choice. If X and Y are nonempty sets and A is a subset of the product $X \times Y$, a function f **uniformizes** A if the domain of f is the set of $x \in X$ such that there exists a $y \in Y$ with $(x, y) \in A$, and such that for each x in the domain of f, $(x, f(x)) \in A$. A consequence of the Axiom of Choice, **uniformization** is the statement that for every $A \subseteq \mathbb{R} \times \mathbb{R}$ there is a function f which uniformizes A.

Uniformization is not implied by AD, as it fails in $\mathbf{L}(\mathbb{R})$ whenever there are no uncountable wellordered sets of reals (*cf.* [Sol78B]; *cf.* also § 3.3).

Uniformization was implicitly introduced by Jacques Hadamard [Had05], when he pointed out that the Axiom of Choice should imply the existence of functions on the reals which disagree everywhere with every algebraic function over the integers. Luzin [Luz30B] explicitly introduced the notion of uniformization and showed that such functions exist. He also announced several results on uniformization, including the fact that all Borel sets (but not all $\boldsymbol{\Sigma}^1_1$ sets) can be uniformized by $\boldsymbol{\Pi}^1_1$ functions. The result on Borel sets was proved independently by Sierpiński. Novikov showed that every $\boldsymbol{\Sigma}^1_1$ set of pairs has a $\boldsymbol{\Sigma}^1_2$ uniformization [LN35].

A pointclass Γ is said to have the **uniformization property** if every set of pairs in Γ is uniformized by a function in Γ. Motokiti Kondô showed that the pointclasses $\boldsymbol{\Pi}^1_1$ and $\boldsymbol{\Sigma}^1_2$ have the uniformization property [Kon38]. The effective version of this result (*i.e.*, for Π^1_1 and Σ^1_2) was proved by Addison. In some sense this is as far as one can go in ZFC: Lévy [Lév65A] would show that consistently there exist Π^1_2 sets that cannot be uniformized by any projective function. Remarkably, Luzin [Luz25B] had predicted that the question of whether the projective sets are Lebesgue measurable and satisfy the perfect set property would never be solved.

After studying Kondô's proof, Moschovakis in 1971 isolated a property for sets of reals which induces uniformizations. Given a set A and an ordinal γ, a **scale** (or a γ-scale) on A into γ is a sequence of functions $\rho_n \colon A \to \gamma$ ($n \in \omega$) such that whenever $\{x_i : i \in \omega\} \subseteq A$ and $\lim_{i \to \omega} x_i = x$, and the sequence $\langle \rho_n(x_i) : i \in \omega \rangle$ is eventually constant for each $n \in \omega$, then $x \in A$

and, for every $n \in \omega$, $\rho_n(x)$ is less than or equal to the eventual value of $\langle \rho_n(x_i) : i \in \omega \rangle$. The scale is a **Γ-scale** if there exist $R^+ \in \Gamma$ and $R^- \in \check{\Gamma}$ such that for all $y \in A$ and all $n \in \omega$,

$$x \in A \wedge \rho_n(x) \leq \rho_n(y) \leftrightarrow R^+(n, x, y) \leftrightarrow R^-(n, x, y).$$

A pointclass Γ has the **scale property** if every A in Γ has a Γ-scale. Moschovakis proved the following three theorems about the scale property [Mos71].

THEOREM 3.3. If Γ is an adequate pointclass, $A \in \Gamma$, and A admits a Γ-scale, then $\exists^1 A$ admits a $\exists^1\forall^1\Gamma$-scale.

THEOREM 3.4 (Second Periodicity Theorem). Suppose that Γ is an adequate pointclass such that Δ_Γ-determinacy holds. Then for all $A \in \Gamma$, if A admits a Γ-scale, then $\forall^1 A$ admits a $\forall^1\exists^1\Gamma$-scale.

THEOREM 3.5. Suppose that Γ is an adequate pointclass which is closed under integer quantification. Suppose that Γ has the scale property, and that Δ_Γ-determinacy holds. Then Γ has the uniformization property.

Kondô's proof of uniformization for $\boldsymbol{\Pi}^1_1$ shows that $\Pi^1_1(a)$ has the scale property for every real a (*cf.* [Kan03, p. 419]). It follows that under $\boldsymbol{\Delta}^1_{2n}$-determinacy, $\boldsymbol{\Pi}^1_{2n+1}$ and $\boldsymbol{\Sigma}^1_{2n+2}$ have the scale property, and every $\boldsymbol{\Pi}^1_{2n+1}$ relation on the reals can be uniformized by a $\boldsymbol{\Pi}^1_{2n+1}$ relation (and similarly for $\boldsymbol{\Sigma}^1_{2n+2}$). Furthermore, under Projective Determinacy, for any real a, the projective pointclasses with the scale property are the same as those with the prewellordering property: $\Pi^1_1(a), \Sigma^1_2(a), \Pi^1_3(a), \Sigma^1_4(a), \Pi^1_5(a), \Sigma^1_6(a)$, *etc.*

A **tree** on a set X is a collection of finite sequences from X closed under initial segments. Given sets X and Z, a positive integer k and a tree T on $X^k \times Z$, the **projection** of T, $\mathrm{p}[T]$, is the set of $x \in (X^\omega)^k$ such that for some $z \in Z^\omega$, $(x{\restriction}n, z{\restriction}n) \in T$ for all $n \in \omega$ (strictly speaking, this definition involves the identification of finite sequences of k-tuples with k-tuples of finite sequences). If one substitutes the Baire space ${}^\omega\omega$ for $\mathbb{R}$, Suslin's construction for analytic sets (*cf.* § 2.2) essentially presents them as projections of trees on $\omega \times \omega$, modulo the representation of closed intervals. Many descriptive set theorists, starting perhaps with Luzin and Sierpiński [LS23], used trees to represent sets of reals, except that they converted these trees to linear orders via what is now known as the Kleene-Brouwer ordering (after [Bro24] and [Kle55C]). The explicit use of projections of trees as we have presented them here is due to Richard Mansfield [Man70]. As pointed out in [KM78B], given an ordinal γ, a γ-scale for a subset A of the Baire space naturally gives rise to a tree on $\omega \times \gamma$ such that $\mathrm{p}[T] = A$. Given a set Z, a subset of the Baire space is said to be Z-**Suslin** if it is the projection of a tree on $\omega \times Z$. Suslin's representation of analytic sets shows that a set is analytic if and only if it is ω-Suslin. Some authors use "Suslin" to mean "analytic". We will follow a different usage, however, and say that a subset of the Baire space is **Suslin** if is γ-Suslin for some ordinal γ.

Given a tree T on $\omega \times Z$ and a wellordering of Z, a member of $\mathrm{p}[T]$ can be found by following the so-called **leftmost** infinite branch through T (similar to the proof of Kőnig's Lemma, one picks a path through the tree by taking the least next step which is the initial segment of an infinite path through the tree). In a similar manner, a tree on $(\omega \times \omega) \times \gamma$, for some ordinal γ, induces a uniformization of the projection of the tree.

3.3. The game quantifier. Given a Polish space $\mathfrak{X}$ and a set $B \subseteq \mathfrak{X} \times {}^{\omega}\omega$, we let $\Game B$ denote the set of $x \in \mathfrak{X}$ such that player I has a winning strategy in $\mathrm{G}_{\omega}(B_x)$. If Γ is a pointclass, $\Game\Gamma$ is the class $\{\Game B \mid B \in \Gamma\}$. The following facts appear in [Mos09, pp. 245–246].

THEOREM 3.6. If Γ is an adequate pointclass then the following hold.

1. $\Game\Gamma$ is adequate and closed under $\exists^0$ and $\forall^0$.
2. $\exists^1\Gamma \subseteq \Game\Gamma$ and $\forall^1\Gamma \subseteq \Game\Gamma$.
3. If $\mathsf{Det}(\Gamma)$ holds, then $\Game\Gamma \subseteq \forall^1\exists^1\Gamma$.

The First Periodicity Theorem can be stated more generally as the fact that if an adequate pointclass Γ has the prewellordering property, then so does $\Game\Gamma$, and the Second Periodicity Theorem can be similarly stated as saying that if an adequate pointclass Γ has the scale property, then so does $\Game\Gamma$ (*cf.* [Mos09, pp. 246,267]). The propagation of these properties through the projective pointclasses then follows from Theorem 3.6, given that they hold for $\utilde{\Pi}^1_1$ (and its variants).

Modifying the notion of Γ-scale by dropping the requirement that $\rho_n(x)$ is less than or equal to the eventual value of $\langle \rho_n(x_i) : i \in \omega\rangle$, one gets the notion of **Γ-semiscale**. Moschovakis's Third Periodicity Theorem [Mos73] concerns the definability of winning strategies and is stated using the game quantifier and the notion of semiscale.

THEOREM 3.7 (Third Periodicity Theorem). Suppose that Γ is an adequate pointclass, and that $\mathsf{Det}(\Gamma)$ holds. Fix $A \subseteq {}^{\omega}\omega$ in Γ, and suppose that A admits a Γ-semiscale and that player I has a winning strategy in the game $\mathrm{G}_{\omega}(A)$. Then player I has a winning strategy coded by a subset of ω in $\Game\Gamma$.

One consequence of the Third Periodicity Theorem in conjunction with Theorem 3.6 is the following [Mos73]: for any $n \in \omega$, if $\utilde{\Sigma}^1_{2n}$-determinacy holds, $A \subseteq \omega^{\omega}$ is $\Sigma^1_{2n}(a)$ for some real a and player I has a winning strategy in the game with payoff A, then player I has a winning strategy coded by a subset of ω in $\Delta^1_{2n+1}(a)$.

Let $\Game^1$ denote the game quantifier for **real games**, games of length ω where the players alternate playing real numbers. Then $\Game^1\utilde{\Sigma}^0_1$ defines the **inductive** sets of reals.[21] Moschovakis showed that the inductive sets have the scale property [Mos78]. He later showed that, assuming the determinacy of all games with payoff in the class built from the inductive sets by the operations of projection and complementation, coinductive sets have scales in this class [Mos83]. Building on this work, Martin and Steel showed that the pointclass $\underset{\sim}{\Sigma}^2_1$ has the scale property in $\mathbf{L}(\mathbb{R})$ [MS83]. Kechris and Solovay had shown that if there is no wellordering of the reals in $\mathbf{L}(\mathbb{R})$, then there exists in $\mathbf{L}(\mathbb{R})$ a set of reals that does not have a scale since it cannot be uniformized: the set of pairs (x, y) such that y is not ordinal definable from x (*i.e.*, definable from x and some ordinals). This set is Π^2_1 in $\mathbf{L}(\mathbb{R})$.

The **Solovay Basis Theorem** says that if $P(A)$ is a Σ^2_1 relation on subsets of ${}^\omega\omega$ and there exists a witness to $P(A)$ in $\mathbf{L}(\mathbb{R})$, then there is a Δ^2_1 witness. This reflection result, along with the Martin-Steel theorem on scales in $\mathbf{L}(\mathbb{R})$, compensates in many circumstances for the fact that not every set of reals has a scale in $\mathbf{L}(\mathbb{R})$.

Steel [Ste83A] applied Jensen's fine structure theory [Jen72] to the study of scales in $\mathbf{L}(\mathbb{R})$, refining and unifying a great deal of work on scales and Suslin cardinals. Extending [MS83], he showed that for each positive ordinal α, determinacy for all sets of reals in $\mathbf{J}_\alpha(\mathbb{R})$ implies that the pointclass $\Sigma_1^{\mathbf{J}_\alpha(\mathbb{R})}$ has the scale property.

Martin showed how to propagate the scale property using the game quantifier for integer games of fixed countable length (this subsumes propagation by the quantifier $\Game^1$; *cf.* [Mar83]), and Steel did the same for certain games of length ω_1 [Ste88, Ste08C].

3.4. Partially playful universes. The periodicity theorems showed that determinacy axioms imply structural properties for sets of reals beyond the classical regularity properties. It remained to show that these hypotheses were necessary. Towards this end, Moschovakis (*cf.* [Bec78]) identified for each integer n (under the assumption of $\underset{\sim}{\Delta}^1_k$-determinacy, where k is the greatest even integer less than n) the smallest transitive Σ^1_n-correct model of ZF + DC which contains all the ordinals (Joseph Shoenfield [Sho61] had shown that $\mathbf{L}$ is Σ^1_2-correct).[22] This model satisfies AC and $\underset{\sim}{\Delta}^1_k$-determinacy and has a Σ^1_{n+1} wellordering of

[21] Formally, this definition requires a definable association of ω-sequences of reals to individual reals. Alternately, a set of reals is inductive if it is in $\Sigma_1^{\mathbf{J}_{\kappa_\mathbb{R}}(\mathbb{R})}$, where $\mathbf{J}$ refers to Ronald Jensen's constructibility hierarchy and $\kappa_\mathbb{R}$ is the least κ such that $\mathbf{J}_\kappa(\mathbb{R})$ is a model of Kripke-Platek set theory.

[22] The Axiom of Dependent Choices (DC) is the statement that if R is a binary relation on a nonempty set X, and if for each $x \in X$ there is a $y \in X$ such that xRy, then there exists an infinite sequence $\langle x_i : i < \omega \rangle$ such that $x_i R x_{i+1}$ for all $i \in \omega$. This statement is a weakening of the Axiom of Choice, sufficient to prove Kőnig's Lemma, the regularity of ω_1 and the wellfoundedness of ultrapowers by countably complete ultrafilters. *Cf.* [Jec03].

the reals. In this model, Π^1_i has the scale property for all odd $i \leq n$, and Σ^1_i has the scale property for all other positive integers i.

Kechris and Moschovakis [KM78B] introduced the models $\mathbf{L}[T_{2n+1}]$, where T_{2n+1} denotes the tree for a Π^1_{2n+1}-scale for a complete Π^1_{2n+1} set. Moschovakis showed that $\mathbf{L}[T_1] = \mathbf{L}$, and conjectured that $\mathbf{L}[T_{2n+1}]$ is independent of the choice of complete set and scale for all n. This conjecture became the third Victoria Delfino problem [KM78A, KMM81A] and was proved by Howard Becker and Kechris in [BK84] (*cf.* also [KMM83A]).

Solovay showed that if $\mathbf{L} \cap \mathbb{R}$ is countable, then it is the largest countable Σ^1_2 set of reals (*i.e.*, a countable Σ^1_2 set which contains all other such sets; *cf.* [Sol66]). Kechris and Moschovakis showed that for each positive integer n, if $\mathsf{Det}(\Delta^1_{2n})$ holds then there exists a largest countable Σ^1_{2n+2} set [KM72]. The largest countable Σ^1_{2n} set came to be called C_{2n}. Kechris showed that under Projective Determinacy there is for each integer n a largest countable Π^1_{2n+1} set, which he also called C_{2n+1} [Kec75]. The case $n = 0$ follows from $\mathsf{ZF}+\mathsf{DC}$ and was shown independently by David Guaspari, Kechris and Gerald Sacks [Sac76]. Kechris also showed that under Projective Determinacy there are no largest countable Σ^1_{2n+1} or Π^1_{2n} sets. It follows that under Projective Determinacy the lightface projective pointclasses with a largest countable set are the same as those in the pattern above for the prewellording property and the scale property. Harrington and Kechris showed (under the assumption that AD holds in $\mathbf{L}(\mathbb{R})$) that the reals of each $\mathbf{L}[T_{2n+1}]$ are exactly C_{2n+2}, for all integers n [HK81] (the case $n = 1$ was due to Kechris and Martin).

Kechris showed (assuming Projective Determinacy) that each model $\mathbf{L}[\mathrm{C}_{2n}]$ satisfies $\mathsf{Det}(\boldsymbol{\Delta}^1_{2n-1})$ but not $\mathsf{Det}(\boldsymbol{\Sigma}^1_{2n-1})$, and has a Δ^1_{2n} wellordering of its reals. Martin would show that $\mathsf{Det}(\boldsymbol{\Delta}^1_{2n})$ implies $\mathsf{Det}(\boldsymbol{\Sigma}^1_{2n})$ for each positive integer n.

3.5. Wadge degrees. In 1968, William Wadge considered the following game, given two sets of reals A and B: player I builds a real x, player II builds a real y, and player II wins if $x \in A \leftrightarrow y \in B$. Determinacy for this class of games is known as **Wadge determinacy**. Given two sets of reals A, B, we say that $A \leq_{\mathrm{W}} B$ (A has **Wadge rank** less than or equal to B, or is **Wadge reducible** to B) if there is a continuous function f such that for all reals x, $x \in A$ if and only if $f(x) \in B$ (*i.e.*, such that $A = f^{-1}[B]$). Wadge determinacy implies that for any two sets of reals A, B, either $A \leq_{\mathrm{W}} B$ (in the case that player II has a winning strategy) or $\omega^\omega \backslash B \leq_{\mathrm{W}} A$ (in the case that player I does), from which it follows that for any two pointclasses closed under continuous preimages, either the two classes are dual (*i.e.*, a pair of the form Γ, $\check{\Gamma}$) or one is contained in the other. Wadge showed that $\leq_{\mathrm{W}}$ is wellfounded on the Borel sets, and Martin, using an idea of Leonard Monk, extended this to all sets of reals under $\mathsf{AD} + \mathsf{DC}$ (*cf.* [Van78B]).

Wadge determinacy and the wellfoundedness of the Wadge hierarchy divide $\wp(\omega^\omega)$ into equivalence classes by Wadge reducibility and order these classes into a wellfounded hierarchy, where each level consists either of one selfdual equivalence class, or two non-selfdual classes, one consisting of all the complements of the members of the other. Wadge determinacy also implies that every non-selfdual adequate pointclass has a universal set (*cf.* [Van78B, p. 162]).

The discovery of Wadge determinacy led to further progress on separation and reduction. Robert Van Wesep proved that under AD, if Γ is a non-selfdual pointclass which is closed under continuous preimages, then Γ and $\check{\Gamma}$ cannot both have the separation property [Van78A]. Kechris, Solovay and Steel showed that under AD + DC, if $\Gamma \subseteq \mathbf{L}(\mathbb{R})$ is nonselfdual boldface pointclass and Γ is closed under countable intersections and unions and either $\exists^1$ or $\forall^1$, but not complements, then either Γ or $\check{\Gamma}$ has the prewellordering property [KSS81]. In 1981, Steel showed that under AD, if Γ is a nonselfdual pointclass closed under continuous preimages, then either Γ or $\check{\Gamma}$ has the separation property, and if one assumes in addition that Δ_Γ is closed under finite unions, then either Γ or $\check{\Gamma}$ has the reduction property [Ste81B].

§4. Partition properties and the projective ordinals. A cardinal κ is **measurable** if there is a nonprincipal κ-complete ultrafilter on κ, where κ**-completeness** means closure under intersections of fewer than κ many elements. In ZFC measurable cardinals are strongly inaccessible. In 1967, Solovay (*cf.* [Jec03, p. 633] or [Kan03, p. 348]) showed that AD implies that the club filter on ω_1 is an ultrafilter, which implies that ω_1 is a measurable cardinal.[23] Ulam had shown that under ZFC there are stationary, co-stationary subsets of ω_1; Solovay's result shows the opposite under AD. Solovay also showed that under AD every subset of ω_1 is constructible from a real (*i.e.*, exists in $\mathbf{L}[a]$ for some real number a). Since the measurability of ω_1 implies that the sharp of each real exists, this gives another proof that the club filter on ω_1 is an ultrafilter, since for any real a, if $a^\#$ exists, then every subset of ω_1 in $\mathbf{L}[a]$ either contains or is disjoint from a tail of the a-indiscernibles below ω_1, which is a club set.

A **Turing degree** is a nonempty subset of $\wp(\omega)$ closed under equicomputability. A **cone** of Turing degrees is the set of all degrees above (or computing) a given degree.[24] Martin showed that under AD the cone measure on Turing degrees is an ultrafilter, *i.e.*, that every set of Turing degrees either contains or is disjoint from a cone [Mar68]. This important fact has a relatively short and simple proof: the two players collaborate to build a real, with the winner decided by whether the Turing degree of the real falls inside the payoff set; the cone above the degree of any real coding a winning strategy must contain or be disjoint from the payoff set. Martin used this result to find a simpler proof of the measurability of ω_1.

---

[23]A subset of an ordinal is **closed unbounded** (or **club**) if it is unbounded and closed in the order topology on the ordinals, and **stationary** if it intersects every club set. The **club filter** on an ordinal γ consists of all subsets of γ containing a club set.

[24]*Cf.* [Soa87, Coo04] for more on the Turing degrees, including a more precise statement of their definition.

Solovay followed by showing that ω_2 is measurable as well. **Turing determinacy** is the restriction of AD to payoff sets closed under Turing equivalence. This form of determinacy is easily seen to suffice for Martin's result. In the early 1980s, Woodin would show that, in $\mathbf{L}(\mathbb{R})$, AD and Turing determinacy are equivalent.

Given an ordered set X and an ordinal β, $[X]^\beta$ denotes the set of subsets of X of ordertype β. Given ordinals α, β, δ, and γ, the expression $\alpha \to (\beta)^\gamma_\delta$ denotes the statement that for every function $f : [\alpha]^\gamma \to \delta$, there exists an $X \in [\alpha]^\beta$ such that f is constant on $[X]^\gamma$. Frank Ramsey proved that $\omega \to (\omega)^n_2$ holds for each positive $n \in \omega$ (this fact is known as **Ramsey's Theorem**; *cf.* [Ram30]). For infinitary partitions, Paul Erdős and András Hajnal showed (in ZFC) that for any infinite cardinal κ there is a function $f : [\kappa]^\omega \to \kappa$ such that for every $X \in [\kappa]^\kappa$, the range of $f \restriction X$ is all of κ [EH66].

In 1968, Adrian Mathias showed that $\omega \to (\omega)^\omega_2$ holds in Solovay's model from [Sol70], in which all sets of reals satisfy the regularity properties [Mat68, Mat77]. A set $Y \subseteq [\omega]^\omega$ is said to be **Ramsey** if there exists an $X \in [\omega]^\omega$ such that either $[X]^\omega \subseteq Y$ or $[X]^\omega \cap Y = \varnothing$. The statement $\omega \to (\omega)^\omega_2$ is equivalent to the statement that every subset of $[\omega]^\omega$ is Ramsey. Karel Příkrý showed that under $\mathsf{AD}_\mathbb{R}$ (determinacy for games of perfect information of length ω for which the players play real numbers) every subset of $[\omega]^\omega$ is Ramsey [Pří76]. It follows from the main theorem of [MS83] that AD + $\mathbf{V}$=$\mathbf{L}(\mathbb{R})$ implies that every such set is Ramsey. Whether AD alone suffices is still an open question.

In late 1968, Martin (*cf.* [Kan03, p. 392]) showed that AD implies $\omega_1 \to (\omega_1)^\omega_2$ (this implies, *e.g.*, that the club filter on ω_1 is an ultrafilter). Kenneth Kunen then showed that AD implies that ω_1 satisfies the weak partition property, where a cardinal κ satisfies the **weak partition property** if $\kappa \to (\kappa)^\alpha_2$ holds for every $\alpha < \kappa$. Martin followed by showing that $\omega_1 \to (\omega_1)^{\omega_1}_2$, again under AD. The proof actually shows $\omega_1 \to (\omega_1)^{\omega_1}_{2^\omega}$ and $\omega_1 \to (\omega_1)^{\omega_1}_\alpha$ for every countable ordinal α. Martin and Paris (in an unpublished note [MP71], *cf.* [Kec78]) showed that under AD + DC, ω_2 has the weak partition property.

Before continuing with this line of results, we briefly discuss the Coding Lemma and the projective ordinals.

4.1. The ordinal Θ, the Coding Lemma and the projective ordinals. Following convention, we let Θ denote the least ordinal that is not a surjective image of $\mathbb{R}$. Under ZFC, $\Theta = \mathfrak{c}^+$, but under AD, Θ is a limit cardinal, as noted by Harvey Friedman (*cf.* [Kan03, p. 398]). This fact follows from a theorem known as the *Coding Lemma*, due to Moschovakis [Mos70], extending earlier work of Friedman and Solovay.

Given a subset P of some Polish space, let $\utilde{\Sigma}^1_1(P)$ denote the pointclass of sets which are Σ^1_1-definable using P and individual reals as parameters.

THEOREM 4.1 (Coding Lemma). Assume ZF+AD. Let $\preceq$ be a prewellordering of a set of reals X. Let ξ be the length of $\preceq$ and let A be a subset of ξ. Then there exists a $Y \subseteq X$ in $\utilde{\Sigma}^1_1(\preceq)$ such that A is the set of $\preceq$-ranks of elements of Y.

As an immediate consequence, under AD, if $\xi < \Theta$, then there is a surjection from $\mathbb{R}$ onto $\wp(\xi)$ (furthermore, if $\xi < \Theta^M$ for some wellfounded model M of ZF containing the reals, then such a surjection can be found in M). The proof of the Coding Lemma uses a version of Kleene's Recursion Theorem (first proved in [Kle38] for partial recursive functions on the integers), which can be stated as saying that given a suitable coding under which each real x codes a continuous partial function $\hat{x}$ (our notation) on the reals, for each two-variable continuous partial function g on the reals there is a real x such that $\hat{x}(w) = g(x, w)$ for all reals w.

If Γ is a pointclass, δ_Γ denotes the supremum of the lengths of the prewellorderings of the reals in Δ_Γ. The notation $\underset{\sim}{\boldsymbol{\delta}}^1_n$ is used to denote $\delta_{\underset{\sim}{\boldsymbol{\Sigma}}^1_n}$ (which is the same as $\delta_{\underset{\sim}{\boldsymbol{\Pi}}^1_n}$). The **projective ordinals** are the ordinals $\underset{\sim}{\boldsymbol{\delta}}^1_n$, for $n \in \omega\backslash\{0\}$. It follows from the results of [LS23] that $\underset{\sim}{\boldsymbol{\Sigma}}^1_1$ prewellorderings of the reals have countable length, and therefore that the ordinal $\underset{\sim}{\boldsymbol{\delta}}^1_1$ is equal to ω_1. Moschovakis showed (under AD, using the Coding Lemma) that for each $n \in \omega$, $\underset{\sim}{\boldsymbol{\delta}}^1_{n+1}$ is a cardinal, and that $\underset{\sim}{\boldsymbol{\delta}}^1_{2n+1}$ is regular and (using just PD) strictly less than $\underset{\sim}{\boldsymbol{\delta}}^1_{2n+2}$ [Mos70]. Martin showed (without AD) that $\underset{\sim}{\boldsymbol{\delta}}^1_2 \leq \omega_2$ (*cf.* [KM78B]); together these results show that under AD, $\underset{\sim}{\boldsymbol{\delta}}^1_2 = \omega_2$.

Kunen and Martin [KM78B] independently established from ZF + DC that every wellfounded κ-Suslin prewellordering has length less than κ^+ (this fact is sometimes called the **Kunen-Martin Theorem**). Moschovakis (*cf.* [Mos70]; *cf.* also [Mos09, 4C.14]) showed (from PD) that any $\underset{\sim}{\boldsymbol{\Pi}}^1_{2n+1}$-norm on a complete $\underset{\sim}{\boldsymbol{\Pi}}^1_{2n+1}$ set has length $\underset{\sim}{\boldsymbol{\delta}}^1_{2n+1}$ (this result also uses Kleene's Recursion Theorem). By the scale property for $\underset{\sim}{\boldsymbol{\Pi}}^1_{2n+1}$ sets (under the assumption of DC + $\underset{\sim}{\boldsymbol{\Delta}}^1_{2n}$-determinacy, given $n \in \omega$ [Mos71]), every $\underset{\sim}{\boldsymbol{\Pi}}^1_{2n+1}$ set (and thus every $\underset{\sim}{\boldsymbol{\Sigma}}^1_{2n+2}$ set) is $\underset{\sim}{\boldsymbol{\delta}}^1_{2n+1}$-Suslin, and, since $\underset{\sim}{\boldsymbol{\delta}}^1_{2n+1}$ is regular, every $\underset{\sim}{\boldsymbol{\Sigma}}^1_{2n+1}$ set is λ-Suslin for some $\lambda < \underset{\sim}{\boldsymbol{\delta}}^1_{2n+1}$. It follows that under the same hypothesis, $\underset{\sim}{\boldsymbol{\delta}}^1_{2n+2} \leq (\underset{\sim}{\boldsymbol{\delta}}^1_{2n+1})^+$, and under AD that $\underset{\sim}{\boldsymbol{\delta}}^1_{2n+2} = (\underset{\sim}{\boldsymbol{\delta}}^1_{2n+1})^+$ for each $n \in \omega$.

Kechris proved (assuming AD) that $\underset{\sim}{\boldsymbol{\delta}}^1_{2n+1}$ is a successor cardinal [Kec74] (its predecessor is called λ_{2n+1}). It follows from his arguments, and those of the previous paragraph, that the pointclasses $\underset{\sim}{\boldsymbol{\Sigma}}^1_{2n+2}$ and $\underset{\sim}{\boldsymbol{\Sigma}}^1_{2n+1}$ are exactly the $\underset{\sim}{\boldsymbol{\delta}}^1_{2n+1}$-Suslin and λ_{2n+1}-Suslin sets respectively.

Given an ordinal λ, the λ-**Borel** sets of reals are those in the smallest class containing the open sets and closed under complements and well-ordered unions of length less than λ. Martin showed that if κ is a cardinal of uncountable cofinality, then all κ-Suslin sets are κ^+-Borel. He also showed (using AD + DC, the Coding Lemma and Wadge determinacy) that the $\underset{\sim}{\boldsymbol{\delta}}^1_{2n+1}$-Borel sets are $\underset{\sim}{\boldsymbol{\Delta}}^1_{2n+1}$, for each $n \in \omega$ (the reverse inclusion follows from the results of Moschovakis [Mos71] mentioned above). Using this fact, Kechris proved (again, under AD) that λ_{2n+1} has cofinality ω. It follows (under AD) that $\underset{\sim}{\boldsymbol{\delta}}^1_{2n} < \underset{\sim}{\boldsymbol{\delta}}^1_{2n+1}$ for each $n \in \omega$, so that under AD the sequence $\langle \underset{\sim}{\boldsymbol{\delta}}^1_{n+1} : n \in \omega \rangle$

is a strictly increasing sequence of successor cardinals. Kunen showed that $\underset{\sim}{\delta}^1_n$ is regular for each positive $n \in \omega$ [Kun71B].

Solovay noted that under AD, Θ is the Θth cardinal, and that under the further assumption of $\mathbf{V} = \mathbf{L}(\mathbb{R})$, Θ is regular (*cf.* [Kan03, p. 398]). He showed [Sol78B] that under DC, Θ has uncountable cofinality, and also that ZFC + $\mathsf{AD}_{\mathbb{R}}$ + cf(Θ)>ω proves the consistency of ZF + $\mathsf{AD}_{\mathbb{R}}$, so that by Gödel's Second Incompleteness Theorem, if ZF + $\mathsf{AD}_{\mathbb{R}}$ is consistent, then so is ZFC + $\mathsf{AD}_{\mathbb{R}}$ + cf(Θ)=ω.[25] Kechris [Kec84], using the proof of the Third Periodicity Theorem and work of Martin, Moschovakis and Steel on scales [MMS82], showed that DC follows from AD + $\mathbf{V}$=$\mathbf{L}(\mathbb{R})$. Woodin (*cf.* [Kec84]) strengthened Solovay's result that DC does not follow from AD by showing that, assuming AD + $\mathbf{V}$=$\mathbf{L}(\mathbb{R})$ there is an inner model of a forcing extension satisfying ZF + AD + $\neg\mathsf{AC}_\omega$ (DC directly implies AC_ω). Whether AD implies DC($^\omega\omega$) (DC for relations on $^\omega\omega$) is still open.

4.2. Partition properties and ultrafilters. Kunen in an unpublished note proved that $\underset{\sim}{\delta}^1_{2n} \to (\underset{\sim}{\delta}^1_{2n})^\lambda_2$ for all positive $n \in \omega$ and $\lambda < \omega_1$, under AD[Kun71C]. He also showed (under the same hypothesis) that $\underset{\sim}{\delta}^1_{2n} \to (\underset{\sim}{\delta}^1_{2n})^{\underset{\sim}{\delta}^1_{2n}}_2$ is false [Kun71D]. Martin, in another unpublished note from 1971, showed that $\underset{\sim}{\delta}^1_{2n+1} \to (\underset{\sim}{\delta}^1_{2n+1})^\lambda_2$ for all positive $n \in \omega$ and $\lambda < \omega_1$, under AD.

While Erdős and Hajnal [EH58] had shown how to derive partition properties from measurable cardinals, Eugene Kleinberg proved the following result in the other direction, which shows (via $\lambda = \omega$) that $\underset{\sim}{\delta}^1_n$ is measurable for each positive $n \in \omega$.[26]

THEOREM 4.2 (Kleinberg; [Kle70]). If $\lambda < \kappa$, λ is regular, and $\kappa \to (\kappa)^{\lambda+\lambda}_2$ holds, then $\mathcal{C}^\lambda_\kappa$ is a normal ultrafilter over κ.

In 1970, Kunen proved, using Martin's result on the cone measure on the Turing degrees, that under AD, any ω_1-complete filter on an ordinal $\lambda < \Theta$ can be extended to an ω_1-complete ultrafilter, and that every ultrafilter on an ordinal less than Θ is definable from ordinal parameters (*cf.* [Kan03, pp. 399–400]). Solovay proved that under $\mathsf{AD}_{\mathbb{R}}$, there is a normal ultrafilter on $\wp_{\aleph_1}(\mathbb{R})$ [Sol78B]: for each $A \subseteq \wp_{\aleph_1}(\mathbb{R})$, consider the game where player I and player II collaborate to build a sequence $\langle s_i : i < \omega\rangle$ consisting of finite sets of reals, and player I wins if and only if $\bigcup\{s_i : i \in \omega\} \in A$.[27] This implies (again, under $\mathsf{AD}_{\mathbb{R}}$) that for each ordinal $\gamma < \Theta$ there is a normal ultrafilter on $\wp_{\aleph_1}(\gamma)$ (*i.e.*, that ω_1 is γ-**supercompact**). It is not known whether AD suffices for this

[25]The end of § 6.2 continues this line of results.

[26]We let $\mathcal{C}^\lambda_\kappa$ denote the filter generated by the set of λ-closed unbounded subsets of κ. A filter is **normal** if every regressive function on a set in the filter is constant on a set in the filter.

[27]Given a cardinal κ and a set X, $\wp_\kappa(X)$ denotes the collection of subsets of X of cardinality less than κ. An ultrafilter U on $\wp_\kappa(X)$ is **normal** if for each $Y \in U$, if f is a regressive function on Y (*i.e.*, if dom(f) = Y and $f(A) \in A$ for all nonempty $A \in Y$) then f is constant on a set in U.

result, though Harrington and Kechris showed that if AD holds and γ is less than a Suslin cardinal, then there is a normal ultrafilter on $\wp_{\aleph_1}(\gamma)$ [HK81].[28] Extending work of Becker [Bec81] (who proved it in the case that γ is a Suslin cardinal), Woodin showed that there is just one such ultrafilter for each $\gamma < \Theta$, if either $\mathsf{AD}_{\mathbb{R}}$ holds or AD holds and γ is below a Suslin cardinal [Woo83]. The end of § 6.4 mentions more recent progress on these topics.

A cardinal κ is said to have the **strong partition property** if $\kappa \to (\kappa)^{\kappa}_{\mu}$ holds for every $\mu < \kappa$. As mentioned above, Martin showed that under AD, ω_1 has the strong partition property. In late 1977, Kechris adapted Martin's argument to show that under AD there exists a cardinal κ with the strong partition property such that the set of $\lambda < \kappa$ with the strong partition property is stationary below κ (*cf.* [Kan03, p. 432]). Pushing this further, Kechris, Kleinberg, Moschovakis and Woodin showed (using a uniform version of the Coding Lemma) that AD implies that unboundedly many cardinals below Θ have the strong partition property and are stationary limits of cardinals with the strong partition property [KKMW81]. They also showed that whenever λ is an ordinal below a cardinal with the strong partition property, all λ-Suslin sets are determined. Using work of Steel [Ste83A] and Martin [Mar83], Kechris and Woodin showed that in $\mathbf{L}(\mathbb{R})$, AD is equivalent to the assertion that Θ is a limit of cardinals with the strong partition property, and also to the statement that all Suslin sets are determined [KW83]. James Henle, Mathias and Woodin later showed that the first equivalence does not follow from ZF + DC, as the existence of a nonprincipal ultrafilter on ω is consistent with Θ being a limit of cardinals with the strong partition property [HMW85].

A key step in the proof of the Kechris-Woodin theorem was a transfer theorem extending results of Harrington and Martin (discussed in § 5.3). Harrington and Martin had shown from ZF + DC that, for each real a, $\Pi^1_1(a)$-determinacy is equivalent to determinacy for the larger class $\bigcup_{\beta<\omega^2}\beta$-$\Pi^1_1(a)$ (see Section 5.3 for an explanation of this notation). Kechris and Woodin showed, from the same hypothesis, that for all positive integers k, $\utilde{\mathbf{\Delta}}^1_{2k}$-determinacy is equivalent to $\Game^{(2k-1)}\bigcup_{\beta<\omega^2}\beta$-$\utilde{\mathbf{\Pi}}^1_1$-determinacy, where $\Game^{(2k-1)}$ indicates an application of $2k-1$ many instances of the game quantifier $\Game$. By Theorem 3.6, this means that $\utilde{\mathbf{\Delta}}^1_{2k}$-determinacy implies $\utilde{\mathbf{\Pi}}^1_{2k}$-determinacy. Martin had proved the lightface version in 1973 (*cf.* [KS85]). Later results of Woodin and Itay Neeman [Nee95] would show that $\utilde{\mathbf{\Pi}}^1_{n+1}$-determinacy is equivalent to $\Game^{(n)}\bigcup_{\beta<\omega^2}\beta$-$\utilde{\mathbf{\Pi}}^1_1$-determinacy for all $n \in \omega$.

4.3. Cardinals, uniform indiscernibles and the projective ordinals. A cardinal κ is **Ramsey** if for every function $f\colon [\kappa]^{<\omega} \to \{0,1\}$ (where $[\kappa]^{<\omega}$ denotes the set of finite subsets of κ) there exists $A \in [\kappa]^{\kappa}$ such that for each $n \in \omega$,

[28] An ordinal (necessarily a cardinal) κ is said to be **Suslin** if there is a set of reals which is κ-Suslin but not λ-Suslin for any $\lambda < \kappa$.

$f \restriction [\kappa]^n$ is constant. Measurable cardinals are Ramsey, and if there exists a Ramsey cardinal then the sharp of each real number exists. Assuming the existence of a Ramsey cardinal, Martin and Solovay showed that nonempty $\utilde{\mathbf{\Sigma}}^1_3$ subsets of the plane have $\utilde{\mathbf{\Delta}}^1_4$ uniformizations [MS69]. As mentioned above, Lévy [Lév65A] had shown that ZFC does not suffice for this result. Martin and Solovay used an analysis of sharps for reals, and modeled their argument after the proof of the Kondô-Addison theorem. Mansfield [Man71] extended the Martin-Solovay analysis to show (using a measurable cardinal) that nonempty $\utilde{\mathbf{\Pi}}^1_2$ sets are uniformized by $\utilde{\mathbf{\Pi}}^1_3$ functions.

Given a positive ordinal α, u_α denotes the αth **uniform indiscernible**, the αth ordinal which is a Silver indiscernible for each real number. As bijections between ω and countable ordinals can be coded by reals, the first uniform indiscernible, u_1, is ω_1. It follows from the basic analysis of sharps that all uncountable cardinals are uniform indiscernibles, so $u_2 \leq \omega_2$. By applying the Kunen-Martin theorem inside models of the form $\mathbf{L}[a]$, for a a real number, and applying the basic analysis of sharps, Martin showed that $\utilde{\boldsymbol{\delta}}^1_2 = u_2$ if the sharp of every real exists (*cf.* [Kec78]). Recall that by the results of § 4.1, $\utilde{\boldsymbol{\delta}}^1_2 = \omega_2$, under AD.

Martin showed from ZF plus the assumption that the sharp of each real exists that every $\utilde{\mathbf{\Sigma}}^1_3$ set is u_ω-Suslin, and from AD that $u_\omega = \omega_\omega$ (*cf.* [Kan03, pp.203–204]). By the Kunen-Martin Theorem, then, AD implies that $\utilde{\boldsymbol{\delta}}^1_3 \leq \omega_{\omega+1}$. Solovay had shown that if the sharp of every real exists, then $u_{\xi+1}$ has the same cofinality as u_2, for every positive ordinal ξ (*cf.* [Kec78]). Since $u_\omega = \omega_\omega$, it follows that each ω_n $(n \geq 2)$ is of the form u_{k+1} for some positive integer k, and thus that each such ω_n has cofinality ω_2. It follows that under AD + DC, $\utilde{\boldsymbol{\delta}}^1_3 = \omega_{\omega+1}$, since $\utilde{\boldsymbol{\delta}}^1_3$ is a regular cardinal, and therefore that $\utilde{\boldsymbol{\delta}}^1_4 = \omega_{\omega+2}$. Kunen and Solovay would then show that $u_n = \omega_n$ for all n satisfying $1 \leq n \leq \omega$.

In 1971, Kunen reduced the computation of $\utilde{\boldsymbol{\delta}}^1_5$ to the analysis of certain ultrapowers of $\utilde{\boldsymbol{\delta}}^1_3$ (*cf.* [Kec78]; as part of his analysis, Kunen showed that $\utilde{\boldsymbol{\delta}}^1_3$ has the weak partition property, *cf.* [Sol78A]). The completion of this project was to take another decade. In the early 1980s, Martin proved new results analyzing these ultrapowers, and Steve Jackson, using joint work with Martin, computed $\utilde{\boldsymbol{\delta}}^1_5$. The following theorem [Jac88, Jac99] completes the calculation of the $\utilde{\boldsymbol{\delta}}^1_n$'s.

THEOREM 4.3 (Jackson). Assume AD. Then for $n \geq 1$, $\utilde{\boldsymbol{\delta}}^1_{2n+1}$ has the strong partition property and is equal to $\omega_{w(2n-1)+1}$, where $w(1) = \omega$ and $w(m+1) = \omega^{w(m)}$ in the sense of ordinal exponentiation.

Jackson's proof of this theorem was over 100 pages long. Elements of his argument (as presented in [Jac99]) include the Kunen-Martin theorem, Kunen's $\utilde{\mathbf{\Delta}}^1_3$ coding for subsets of ω_ω [Sol78A], Martin's theorem that $\utilde{\mathbf{\Delta}}^1_{2n+1}$ is closed under intersections and unions of sequences of sets indexed by ordinals less than $\utilde{\boldsymbol{\delta}}^1_3$, and so-called homogeneous trees, a notion which traces back to [MS69] and a result of Martin discussed in the next section.

§5. **Determinacy and large cardinals.** As discussed above, a strongly inaccessible cardinal is an uncountable regular cardinal which is closed under cardinal exponentiation. If κ is strongly inaccessible, then $\mathbf{V}_\kappa$ is a model of ZFC, so that the existence of strongly inaccessible cardinals is not a consequence of ZFC. While there is no technical definition of **large cardinal**, a typical large cardinal notion (in the context of the Axiom of Choice) specifies a type of strongly inaccessible cardinal. Examples of this type include Ramsey cardinals, measurable cardinals, Woodin cardinals and supercompact cardinals. The **large cardinal hierarchy** orders large cardinals by **consistency strength**. That is, large cardinal notion A is below large cardinal notion B in the hierarchy if the existence of cardinals of type B implies the consistency of cardinals of type A. It is a striking empirical fact that the large cardinal hierarchy is linear, modulo open questions (*e.g.*, the examples just given were listed in increasing order). Even more striking is the fact that many set-theoretic statements having no ostensible relationship to large cardinals are equiconsistent with some large cardinal notion.[29]

By results of Mycielski (discussed in § 2.3), AD implies that ω_1 is strongly inaccessible in $\mathbf{L}$, which means that the consistency of AD cannot be proved in ZFC. Moreover, Solovay's result that AD implies the measurability of ω_1 implies that under AD, ω_1 (as computed in the full universe) is a measurable cardinal in certain inner models of AC, such as **HOD**.[30] As we shall see in this section, the relationship between large cardinals and determinacy runs in both directions: Various forms of determinacy imply the existence of models of ZFC containing large cardinals, and the existence of large cardinals can be used to prove the determinacy of certain definable sets of reals.

5.1. Measurable cardinals. Solovay showed in 1965 that if there exists a measurable cardinal then every uncountable $\mathbf{\Sigma}^1_2$ set of reals contains a perfect set [Sol66]. This result was proved independently by Mansfield (*cf.* [Sol66]). Martin showed that in fact analytic determinacy follows from the existence of a Ramsey cardinal [Mar70].

Roughly, the idea behind Martin's proof is that if A is the projection of a tree T on $\omega \times \omega$ and χ is a Ramsey cardinal, one can modify the original game for A to require the second player to play, in addition to his usual moves, a function $G^* : \omega^{<\omega} \to \chi$ witnessing (via the wellfoundedness of the ordinal χ) that the fragment of T corresponding to the real produced by the two players in their moves from the original game has no infinite branches, and thus that this real is not in the projection of T. This modified game is closed, and thus determined, by Gale-Stewart. If the second player has a winning strategy in the modified game, then he has a winning strategy in the original game by

[29] [Kan03] is the standard reference for the large cardinal hierarchy.

[30] The inner model **HOD** (a model of ZFC) consists of all sets x such that every member of the transitive closure of $\{x\}$ is ordinal-definable (*cf.* [Jec03, Chapter 13]).

ignoring his extra moves. In general there is no reason that a winning strategy for the first player in the modified game will induce a winning strategy for the original game. However, if χ is a Ramsey cardinal, then there is uncountable $X \subseteq \chi$ such that, as long as the range of G^* is contained in X, the first player's strategy does not depend on the extra moves for the second player. Using this fact, the first player can convert his winning strategy in the modified game into a winning strategy in the original game. The notion of a determined (often closed) auxiliary game and a method for transferring strategies from the auxiliary game to the original game is the basis of many determinacy proofs.

Martin later proved the following refinement.

THEOREM 5.1. If the sharp of every real exists, then $\underset{\sim}{\Pi}^1_1$-determinacy holds.

In the 1970s, Kunen and Martin independently developed the notion of a **homogeneous** tree, following a line of ideas deriving from Martin's proof of $\underset{\sim}{\Pi}^1_1$-determinacy (*cf.* [Kec81]). Given a set Z and a cardinal κ, a tree on $\omega \times Z$ is said to be **κ-homogeneous** if for each $\sigma \in \omega^{<\omega}$ there is a κ-complete ultrafilter μ_σ on $Z^{|\sigma|}$ such that

1. for each $\sigma \in \omega^{<\omega}$, $\{z : (\sigma, z) \in T\} \in \mu_\sigma$;
2. p$[T]$ is the set of $x \in \omega^\omega$ such that the sequence $\langle \mu_{x \restriction i} : i \in \omega \rangle$ is countably complete.[31]

A tree is said to be **homogeneous** if it is $\aleph_1$-homogeneous. A set of reals is said to be **homogeneously Suslin** if it is the projection of a homogeneous tree. There are related notions of **weakly homogeneous tree** and **weakly homogeneously Suslin set** of reals, involving a more involved relationship with a set of ultrafilters. Though it is not the original definition, let us just say that a tree on a set of the form $\omega \times (\omega \times Z)$ is weakly homogeneous if and only if the corresponding tree on $(\omega \times \omega) \times Z$ is homogeneous, and note that a set of reals is weakly homogeneously Suslin if and only if it is the projection of a homogeneously Suslin set of pairs.

Martin's proof then shows the following.

THEOREM 5.2 (Martin). Homogeneously Suslin sets are determined.

The unfolding argument mentioned in § 2.2 then shows that weakly homogeneously Suslin sets satisfy the regularity properties.

In retrospect, Martin's proof of analytic determinacy can be broken into two parts, the fact that homogeneously Suslin sets are determined, and the fact that if there is a measurable cardinal then $\underset{\sim}{\Pi}^1_1$ sets are homogeneously Suslin.

The results of [MS69] can similarly be reinterpreted. If $\underset{\sim}{\Pi}^1_1$ sets are homogeneously Suslin, then $\underset{\sim}{\Sigma}^1_2$ sets are weakly homogeneously Suslin. The Martin-Solovay construction can be seen as a method for taking a γ-weakly

[31] *I.e.*, such that for each sequence $\langle A_i : i \in \omega \rangle$ such that each $A_i \in \mu_{x \restriction i}$ there exists a $t \in Z^\omega$ such that $t \restriction i \in A_i$ for each i.

homogeneous tree T (for some cardinal γ) and producing a tree S on $\omega \times \gamma'$, for some ordinal γ', projecting to the complement of the projection of T. From this it follows that all $\mathbf{\Pi}^1_2$ sets, and thus all $\mathbf{\Sigma}^1_3$ sets, are projections of trees on the product of ω with some ordinal. More sophisticated arguments can be carried out from the existence of sharps, using the fact that sharps give ultrafilters over certain inner models.

5.2. Borel determinacy. In 1968, Friedman showed that the Replacement axiom is necessary to prove Borel determinacy, even for sets invariant under Turing degrees [Fri71B] (he also showed that analytic determinacy cannot hold in a forcing extension of **L**). As refined by Martin, his results show (for each $\alpha < \omega_1$) that $\mathsf{ZFC} - \mathsf{PowerSet} - \mathsf{Replacement} +$ "the αth iteration of the power set of ${}^\omega\omega$ exists" does not prove the determinacy of all $\mathbf{\Sigma}^0_{1+\alpha+3}$ sets.

James Baumgartner mixed the method of Martin's $\mathbf{\Pi}^1_1$-determinacy proof with Davis's $\mathbf{\Sigma}^0_3$-determinacy proof to give a new proof of $\mathbf{\Sigma}^0_3$-determinacy in ZFC. Using a similar approach, Martin proved $\mathsf{Det}(\mathbf{\Sigma}^0_4)$ from the existence of a weakly compact cardinal,[32] and then Paris proved it in ZFC [Par72]. Paris noted at the end of his paper that his argument could be carried out without the power set axiom, assuming instead only that the ordinal ω_1 exists.

Andreas Blass [Bla75] and Mycielski (1967, unpublished) independently proved that $\mathsf{AD}_\mathbb{R}$ is equivalent to determinacy for integer games of length ω^2. The key idea in Blass's proof was to reduce determinacy in the given game to determinacy in another, auxiliary, game in such a way that one player's moves in the auxiliary game correspond to fragments of his strategy in the original game. Martin [Mar75] used this basic idea to prove Borel determinacy in 1974 (the auxiliary game was in fact an open game). In his [Mar85], Martin gave a short, inductive, proof of Borel determinacy, and introduced the notion of **unraveling** a set of reals—roughly, finding an association of the set to a clopen set in a larger domain with a map sending strategies in one game to strategies in the other. In his [Mar90], Martin extended this method to games of length ω played on any (possibly uncountable) set, with Borel payoff (in the corresponding sense). Neeman [Nee00, Nee06B] would unravel $\mathbf{\Pi}^1_1$ sets from the assumption of a measurable cardinal κ of Mitchell rank κ^{++} (proved to be an optimal hypothesis by Steel [Ste82B]; *cf.* [Jec03, pp. 357–360] for the definition of Mitchell rank). Complementing Friedman's theorem, Martin proved that for each $\alpha < \omega_1$, the determinacy of each Boolean combination of $\mathbf{\Sigma}^0_{\alpha+2}$ sets follows from $\mathsf{ZF} - \mathsf{PowerSet} - \mathsf{Replacement} + \Sigma_1\text{-}\mathsf{Replacement} +$ "the αth iteration of the power set of ${}^\omega\omega$ exists".

[32] A cardinal κ is **weakly compact** if $\kappa \to (\kappa)^2_2$. Weakly compact cardinals are below the existence of $0^\#$ and above strongly inaccessible cardinals in the consistency strength hierarchy (*cf.* [Kan03, pp. 76, 472]).

5.3. The difference hierarchy. Given a countable ordinal α and a real a, a set of reals X is said to be α-$\Pi^1_1(a)$ if there is wellordering of ω of length α recursive in a with corresponding rank function $R\colon \omega \to \alpha$ and a $\Pi^1_1(a)$ subset A of $\omega \times {}^\omega\omega$ such that for all $n, m \in \omega$, if $R(n) < R(m)$ then $\{x : (m,x) \in A\} \subseteq \{x : (n,x) \in A\}$ and X is the set of reals x for which the least ξ such that either $\xi = \alpha$ or $\xi < \alpha$ and $(R^{-1}(\xi), x) \notin A$ is odd.

This notation has its roots in [Hau08]. When a is itself recursive one writes α-Π^1_1. The union of the sets α-$\Pi^1_1(a)$ for all reals a is denoted α-$\undertilde{\Pi}^1_1$. The union of the sets α-$\undertilde{\Pi}^1_1$ for all $\alpha < \omega_1$ is denoted $\mathrm{Diff}(\undertilde{\Pi}^1_1)$. Note that $\mathrm{Diff}(\undertilde{\Pi}^1_1)$ is a proper subclass of $\undertilde{\Delta}^1_2$.

Friedman [Fri71A] extended Theorem 5.1 to show that $\mathsf{Det}(3\text{-}\undertilde{\Pi}^1_1)$ follows from the existence of the sharp of every real. Martin in 1975 then extended this result to show that the existence of $0^\#$ is equivalent to $\mathsf{Det}(\bigcup_{\beta<\omega^2} \beta\text{-}\Pi^1_1)$ (*cf.* [DuB90]). Harrington [Har78] then proved the converse to Theorem 5.1 by showing that $\mathsf{Det}(\Pi^1_1(a))$ implies the existence of $a^\#$, for each real a.

For the purposes of the next theorem, say that a model has α measurable cardinals and indiscernibles if there exists a set of ordertype α consisting of measurable cardinals of the model, and there exist uncountably many ordinal indiscernibles of the model above the supremum of these measurable cardinals. Martin proved the following theorem after Harrington's result.

THEOREM 5.3. For any real a and any ordinal α recursive in a, the following are equivalent.

(i) $\mathsf{Det}(\bigcup_{\beta<\omega^2} (\omega^2 \cdot \alpha + \beta)\text{-}\Pi^1_1(a))$.
(ii) $\mathsf{Det}((\omega^2 \cdot \alpha + 1)\text{-}\Pi^1_1(a))$.
(iii) There is an inner model of ZFC containing a and having α many measurable cardinals and indiscernibles.

Still, a large-cardinal consistency proof of $\mathsf{Det}(\undertilde{\Delta}^1_2)$, the hypothesis used by Addison and Martin in their extension of Blackwell's argument, remained beyond reach. John Green showed that $\mathsf{Det}(\Delta^1_2)$ implies the existence of an inner model with a measurable cardinal of Mitchell rank 1 [Gre78].

5.4. Larger cardinals. In § 4 we defined a measurable cardinal to be a cardinal κ such that there exists a nonprincipal κ-complete ultrafilter on κ. Equivalently, under the Axiom of Choice, κ is measurable if and only if there is a nontrivial elementary embedding j from the full universe $\mathbf{V}$ into some inner model M whose critical point is κ, *i.e.*, such that κ is the least ordinal not mapped to itself by j. Many large cardinal notions can be expressed both in terms of ultrafilters and in terms of embeddings, although in the Choiceless context (without the corresponding form of Łoś's Theorem, *cf.* [Jec03, p. 159]) it is the definition in terms of ultrafilters which is relevant. For instance, a cardinal κ is **supercompact** if for each $\lambda > \kappa$ there exists a normal fine ultrafilter on $\wp_\kappa(\lambda)$.[33] Under the Axiom of Choice, κ is supercompact if and only if for every $\lambda > \kappa$ there is an elementary embedding j from $\mathbf{V}$ into an inner model M such that the critical point of j is κ and M is closed under sequences of length λ. Every supercompact cardinal is a limit of measurable cardinals. An even larger large cardinal notion is the huge cardinal, where an uncountable cardinal κ is **huge** if for some cardinal $\lambda > \kappa$ there is a κ-complete normal fine ultrafilter on $[\lambda]^\kappa$ (where "normal" and "fine" are defined in analogy with the supercompact case, *cf.* [Kan03, p. 331]). Under AC, κ is huge if and only if there is an elementary embedding $j : \mathbf{V} \to M$ with critical point κ such that M is closed under sequences of length $j(\kappa)$. The existence of huge cardinals does not imply the existence of supercompact cardinals, but it does imply their consistency.

Kunen [Kun71A] put a limit on the large cardinality hierarchy, showing in ZFC that there is no nontrivial elementary embedding from $\mathbf{V}$ into itself. A corollary of the proof is that for any elementary embedding j of $\mathbf{V}$ into any inner model M, if δ is the least ordinal above the critical point of j sent to itself by j, then $\mathbf{V}_{\delta+2} \not\subseteq M$. In 1978, Martin proved $\undertilde{\Pi}^1_2$-determinacy from the hypothesis I2, which states that for some ordinal δ there is a nontrivial elementary embedding of $\mathbf{V}$ into an inner model M with critical point less than δ such that $\mathbf{V}_\delta \subseteq M$ and $j(\delta) = \delta$ [Mar80].

In 1979, Woodin proved that for each $n \in \omega$, $\undertilde{\Pi}^1_{n+1}$ follows (in ZF) from the existence of an *n-fold strong rank-to-rank embedding*.[34] For $n = 1$, this is essentially the theorem of Martin just mentioned. For $n > 1$, these axioms are incompatible with the Axiom of Choice, by Kunen's theorem, though they are not known to be inconsistent with ZF.

In 1984, Woodin proved $\mathsf{AD}^{\mathbf{L}(\mathbb{R})}$ from I0, the statement that for some ordinal δ there is a nontrivial elementary embedding from $\mathbf{L}(\mathbf{V}_{\delta+1})$ into itself with critical point below δ, thus verifying Solovay's conjecture that $\mathsf{AD}^{\mathbf{L}(\mathbb{R})}$ would follow from large cardinals. The axiom I0 is one of the strongest large cardinal hypotheses not known to be inconsistent. The inner model program at the time had produced models for many measurable cardinals, hypotheses far short of I2, and so there was little hope of showing that I2 and I0 were necessary for these results.

New large cardinal concepts would prove to be the missing ingredient. Given an ideal I on a set X, forcing with the Boolean algebra given by the power set of X modulo I gives a $\mathbf{V}$-ultrafilter on the power set of X.[35] The ideal

---

[33]Given a cardinal κ and a set X, a collection U of subsets of $\wp_\kappa(X)$ is **fine** if it contains the collection of supersets of each element of $\wp_\kappa(X)$.

[34]For positive $n \in \omega$, an n-**fold strong rank-to-rank embedding** is a sequence of elementary embeddings $j_1, \ldots, j_n$ such that for some cardinal λ, $j_i : \mathbf{V}_{\lambda+1} \to \mathbf{V}_{\lambda+1}$ whenever $1 \leq i \leq n$ and $\kappa_\omega(j_i) < \kappa_\omega(j_{i+1})$ for all $i < n$, where $\kappa_\omega(j)$ denotes the first fixed point of an elementary embedding j above the critical point.

[35]An **ideal** is a collection of sets closed under subsets and finite unions. Given a model M and a set X in M, an M-**ultrafilter** is a subset of $\wp(X) \cap M$ closed under supersets and finite

I is said to be **precipitous** if the ultrapower of $\mathbf{V}$ by this generic ultrafilter is wellfounded in all generic extensions. If the underlying set X is a cardinal κ, the ideal I is said to be **saturated** if the Boolean algebra $\wp(\kappa)/I$ has no antichains of cardinality κ^+.[36] If κ is a regular cardinal, saturation of I implies precipitousness. Huge cardinals were invented by Kunen [Kun78], who used them to prove the consistency of a saturated ideal on ω_1.

In early 1984, Foreman, Magidor and Shelah showed that if there exists a supercompact cardinal—a hypothesis much weaker than I0 or I2—then there is an ω_1-preserving forcing making the nonstationary ideal on ω_1, denoted by NS_{ω_1}, precipitous [FMS88].

Foreman and Magidor [For86, Mag80] had earlier made a connection between generic elementary embeddings[37] and regularity properties for reals. Magidor [Mag80] in particular had shown that the Lebesgue measurability of Σ^1_3 sets followed from the existence of a generic elementary embedding with critical point ω_1 and wellfounded image model (the existence of such an embedding follows from the Foreman-Magidor-Shelah result mentioned above). Woodin noted that these arguments plus earlier work of his (*cf.* [Woo86]) could be used to extend this to Lebesgue measurability for all projective sets. Woodin also noted that arguments from [FMS88] could be used to prove the Lebesgue measurability of all sets of reals in $\mathbf{L}(\mathbb{R})$, if one could force to produce a saturated ideal on ω_1 without adding reals. Shelah then noted that techniques from [She98] could be used to do just that. It follows then that the existence of a supercompact cardinal implies that all sets of reals in $\mathbf{L}(\mathbb{R})$ are Lebesgue measurable.

Woodin and Shelah then addressed the problem of weakening the hypotheses needed for the Lebesgue measurability of all projective sets of reals.[38] Woodin showed that a superstrong cardinal sufficed. Shelah then isolated a weaker notion now known as a **Shelah cardinal**, and showed that the existence of $n+1$ Shelah cardinals implies that $\undertilde{\Sigma}^1_{n+2}$ sets are Lebesgue measurable.

DEFINITION 5.4. A cardinal κ is a **Shelah cardinal** if for every $f\colon \kappa \to \kappa$ there is an elementary embedding $j\colon \mathbf{V} \to N$ with critical point κ such that $\mathbf{V}_{j(f)(\kappa)} \subseteq N$.

Woodin noted that by modifying Shelah's definition one obtained a weaker, still sufficient, hypothesis, now known as a Woodin cardinal.

---

intersections such that for every $A \subseteq X$ in M, exactly one of A and $X \setminus A$ is in U. Note that U does not need to be an element of M.

[36] An **antichain** in a partial order (or a Boolean algebra) is a set of pairwise incompatible elements. In the case of a Boolean algebra of the form $\wp(\kappa)/I$, an antichain is a collection of subsets of κ not in I which pairwise have intersection in I.

[37] A **generic elementary embedding** is an elementary embedding of the universe $\mathbf{V}$ into some class model M which is definable in a forcing extension of $\mathbf{V}$.

[38] We follow the account in [Nee04].

DEFINITION 5.5. A cardinal δ is a **Woodin cardinal** if for each function $f : \delta \to \delta$ there exists an elementary embedding $j : \mathbf{V} \to M$ with critical point $\kappa < \delta$ closed under f such that $\mathbf{V}_{j(f)(\kappa)} \subseteq M$.

Woodin proved that the existence of n Woodin cardinals below a measurable cardinal implies the Lebesgue measurability of $\underset{\sim}{\mathbf{\Sigma}}^1_{n+2}$ sets, the same amount of measurability that would follow from $\underset{\sim}{\mathbf{\Pi}}^1_{n+1}$-determinacy. All of this work was done within a few weeks of the Foreman-Magidor-Shelah result on the precipitousness of NS_{ω_1}. In [SW90] the hypothesis for the statement that all sets of reals in $\mathbf{L}(\mathbb{R})$ are Lebesgue measurable and have the property of Baire was reduced to the existence of ordertype $\omega + 1$ many Woodin cardinals. The hypothesis was to be reduced even further.

Woodin extracted from the Foreman-Magidor-Shelah results a one-step forcing for producing generic elementary embeddings with critical point ω_1, and developed it into a general method, now known as the **stationary tower**. Using this he showed (by the fall of 1984, *cf.* his [Woo88]) that if there exists a supercompact cardinal (or a strongly compact cardinal), then every set of reals in $\mathbf{L}(\mathbb{R})$ is weakly homogeneously Suslin. (Steel and Woodin would show in 1990 that this conclusion in turn implies $\mathsf{AD}^{\mathbf{L}(\mathbb{R})}$.)

Steel had been working on the problem of finding inner models for supercompact cardinals. Inspired by the results of Foreman, Magidor, Shelah and Woodin, he begin to work on producing models for Woodin cardinals, and had some partial results by the spring of 1985, producing inner models with certain weak variants of Woodin cardinals. These models were generated by sequences of *extenders*, directed systems of ultrafilters which collectively generate elementary embeddings whose images contain more of $\mathbf{V}$ than possible for embeddings generated by a single ultrafilter. Special cases of extenders had appeared in Jensen's proof of the Covering Lemma. The general notion (which first appeared in [Dod82]) is Jensen's simplification of the notion of *hypermeasure*, which was introduced by Mitchell [Mit79]. Steel and Martin saw that the problem of building models with Woodin cardinals was linked to the problem of proving determinacy, and they set their sights on this problem in the late spring of 1985.

One key combinatorial problem related to elementary embeddings is whether infinite iterations of these embeddings produce wellfounded models. Kunen [Kun70] had shown that the answer was positive for iterations derived from a single ultrafilter. With extenders the situation is more complicated, as an iteration does not need to be linear, and can produce a tree of models with no rule for finding a path through the tree leading to a wellfounded model (indeed, this nonlinearity is essential, since otherwise the models would have simply definable wellorderings of their reals). The simplest such tree, a so-called **alternating chain**, is countably infinite and consists of two infinite branches. Martin and Steel saw that the issue of wellfoundedness for the direct limits along the two branches was linked. This observation led to the following theorem, proved in August of 1985.

THEOREM 5.6 (Martin & Steel; [MS89]). Suppose that λ is a Woodin cardinal and A is a λ^+-weakly homogeneously Suslin set of reals. Then for any $\gamma < \lambda$, ${}^{\omega}\omega \backslash A$ is γ-homogeneously Suslin.

It follows from this and the fact that coanalytic sets are homogeneously Suslin in the presence of a measurable cardinal that if there exist n Woodin cardinals below a measurable cardinal, then $\utilde{\mathbf{\Pi}}^1_{n+1}$ sets are determined, and that Projective Determinacy follows from the existence of infinitely many Woodin cardinals.

Combined with Woodin's application of the stationary tower mentioned above, the Martin-Steel theorem implies that $\mathsf{AD}^{\mathbf{L}(\mathbb{R})}$ follows from the existence of a supercompact cardinal. By the end of 1985, Woodin had improved the hypothesis to the existence of infinitely many Woodin cardinals below a measurable cardinal (*cf.* [Lar04]).

THEOREM 5.7 (Woodin). If there exist infinitely many Woodin cardinals below a measurable cardinal, then AD holds in $\mathbf{L}(\mathbb{R})$.

In the spring of 1986, Martin and Steel [MaS94] produced **extender models** (i.e., models of the form $\mathbf{L}[\vec{E}]$, with $\vec{E}$ a sequence of extenders) with n Woodin cardinals and Δ^1_{n+2} wellorderings of the reals. Such a model necessarily has a Σ^1_{n+2} set which is not Lebesgue measurable, and fails to satisfy Π^1_{n+1}-determinacy.

Skipping ahead for a moment, let $(*)_n$ be the statement that for each real x there exists an iterable model M containing x and n Woodin cardinals plus the sharp of $\mathbf{V}^M_\delta$, for δ the largest of these Woodin cardinals. For odd n, the equivalence of $\utilde{\mathbf{\Pi}}^1_{n+1}$-determinacy and $(*)_n$ was proved by Woodin in 1989. That $(*)_n$ implies $\utilde{\mathbf{\Pi}}^1_{n+1}$-determinacy for all n was proved by Neeman [Nee95] in 1994. Roughly, Neeman's methods work by considering a modified game in which one player builds an iteration tree and makes moves in the image of the original game by the embeddings given by the tree. In 1995, Woodin proved that $\utilde{\mathbf{\Pi}}^1_{n+1}$-determinacy implies $(*)_n$ for even $n > 0$.

Woodin followed his Theorem 5.7 by determining the exact consistency strength of AD. The forward direction of Theorem 5.8 below (proved in [KW10]) shows from $\mathsf{ZF}+\mathsf{AD}$ that there exist infinitely many Woodin cardinals in an inner model of a forcing extension (**HOD** of the forcing extension with respect to certain parameters) of $\mathbf{V}$. The proof built on a sequence of results, starting with Solovay's theorem that AD implies that ω_1 is a measurable cardinal, which, as mentioned above, also shows that ω_1 (as defined in $\mathbf{V}$) is measurable in the inner model **HOD**. Becker (*cf.* [BM81]) had shown that, under AD, $\omega_1^{\mathbf{V}}$ is the least measurable in **HOD**. Becker, Martin, Moschovakis and Steel then showed that under $\mathsf{AD}+\mathbf{V}{=}\mathbf{L}(\mathbb{R})$, $\utilde{\delta}^2_1$ is β-strong in **HOD**, where

β is the least measurable cardinal greater than $\underset{\sim}{\delta}_1^2$ in **HOD**.[39] In the 1980s, Woodin showed under the same hypothesis that $\underset{\sim}{\delta}_1^2$ is β-strong in **HOD** for every $\beta < \Theta$ (and that $\underset{\sim}{\delta}_1^2$ is the least ordinal with this property), and that Θ is Woodin in **HOD**.

THEOREM 5.8 (Woodin). The following are equiconsistent.

(i) $\mathsf{ZF} + \mathsf{AD}$.

(ii) There exist infinitely many Woodin cardinals.

The following theorem illustrates the reverse direction of the equiconsistency (*cf.* [Ste09]). It can be seen as a special case of the Derived Model Theorem, discussed in § 6.2. The partial order $\mathrm{Col}(\omega, <\delta)$ consists of all finite partial functions p from $\omega \times \delta$ to δ, with the requirement that $p(n, \alpha) \in \alpha$ for all (n, α) in the domain of p. The order is inclusion. If δ is a regular cardinal, then δ is the ω_1 of any forcing extension by $\mathrm{Col}(\omega, <\delta)$.

THEOREM 5.9 (Woodin). Suppose that λ is a limit of Woodin cardinals, and $G \subseteq \mathrm{Col}(\omega, <\lambda)$ is $\mathbf{V}$-generic filter. Let $\mathbb{R}^* = \bigcup\{\mathbb{R}^{\mathbf{V}[G\restriction\alpha]} : \alpha < \lambda\}$. Then AD holds in $\mathbf{L}(\mathbb{R}^*)$.

The results of § 5.3 illustrate the difficulties in proving the determinacy of Π^1_2 sets. Woodin resolved this problem in 1989. The forward direction of Theorem 5.10 is proved in [KW10]. The proof was inspired in part by a result of Kechris and Solovay [KS85], saying that in models of the form $\mathbf{L}[a]$ for $a \subseteq \omega$, Δ^1_2-determinacy implies the determinacy of all ordinal definable sets of reals. Standard arguments show that if Δ^1_2 determinacy holds, then it holds in $\mathbf{L}[x]$ for some real x. Woodin showed that if $\mathbf{V}$ is $\mathbf{L}[x]$ for some real x, and Δ^1_2-determinacy holds, then $\omega_2^{\mathbf{L}[x]}$ is a Woodin cardinal in **HOD**. Recall (from the end of § 4.2) that Δ^1_2-determinacy and Π^1_2-determinacy are equivalent, by a result of Martin.

THEOREM 5.10 (Woodin). The following are equiconsistent.

(i) $\mathsf{ZFC} + \mathsf{Det}(\Delta^1_2)$.

(ii) ZFC + There exists a Woodin cardinal.

The following theorem illustrates the reverse direction. Its proof can be found in [Nee10, p. 1926]. The partial order $\mathrm{Col}(\omega, \delta)$ is the natural one for making δ countable : it consists of all finite partial functions from ω to δ, ordered by inclusion.

THEOREM 5.11 (Woodin). If δ is a Woodin cardinal and $G \subseteq \mathrm{Col}(\omega, \delta)$ is a $\mathbf{V}$-generic filter, then Δ^1_2-determinacy holds in $\mathbf{V}[G]$.

[39]The cardinal $\underset{\sim}{\delta}_1^2$ is the supremum of the lengths of the $\underset{\sim}{\Delta}_1^2$ prewellorderings of the reals; under $\mathsf{AD} + \mathbf{V}{=}\mathbf{L}(\mathbb{R})$ it is also the largest Suslin cardinal. A cardinal κ is β-**strong** if there is an elementary embedding $j\colon \mathbf{V} \to M$ with critical point κ such that $\mathbf{V}_\beta \subseteq M$, and $<\delta$-**strong** if it is β-strong for all $\beta < \delta$.

§6. Later developments. In this final section we briefly review some of the developments that followed the results of the previous section. As discussed in the introduction, the set of topics presented here is by no means complete. The first subsection briefly introduces a regularity property for sets of reals which is induced by forcing-absoluteness. The second and third discuss forms of determinacy ostensibly stronger than AD, in models larger than $\mathbf{L}(\mathbb{R})$. The next subsection discusses applications of determinacy to the realm of AC, via producing models of AC by forcing over models of determinacy. In the last two we present some results which derive forms of determinacy from their ostensibly weak consequences, or from statements having no obvious relationship to determinacy. Many of the results of the last two subsections are applications of the study of canonical inner models for large cardinals.

6.1. Universally Baire sets. As discussed above in §§ 5.1 & 5.4, homogeneously Suslin and weakly homogeneously Suslin sets of reals played an important role in applications of large cardinals to regularity properties for sets of reals, as early as the 1969 results of Martin and Solovay. Qi Feng, Magidor, and Woodin [FMW92] introduced a related tree representation property for sets of reals. Given a cardinal κ, a set $A \subseteq \omega^\omega$ is κ-**universally Baire** if there exist trees S, T such that $\mathrm{p}[S] = A$ and S and T project to complements in every forcing extension by a partial order of cardinality less than or equal to κ.[40]

Woodin (*cf.* [Kan03, Lar04]) showed that if δ is a Woodin cardinal, then δ-universally Baire sets of reals are $<\delta$-weakly homogeneously Suslin. It follows from the arguments of [MS69] that if $A \subseteq \omega^\omega$ is κ^+-weakly homogeneously Suslin, then it is κ-universally Baire. Combining these facts with Theorem 5.6 gives the following.

THEOREM 6.1. If δ is a limit of Woodin cardinals, then the following are equivalent, for all sets of reals A.

(i) A is $<\delta$-homogeneously Suslin.
(ii) A is $<\delta$-weakly homogeneously Suslin.
(iii) A is $<\delta$-universally Baire.

Feng, Magidor, and Woodin showed that if $\delta_0 < \delta_1$ are Woodin cardinals, then every δ_1-universally Baire set is determined (this follows from Theorem 5.6 and the result of Woodin mentioned before the previous Theorem). Neeman later improved this, showing that if δ is a Woodin cardinal, then all δ-universally Baire sets are determined. In addition to the following theorem, Feng, Magidor and Woodin showed that $\mathsf{Det}(\utilde{\mathbf{\Pi}}^1_1)$ is equivalent to the statement that every $\utilde{\mathbf{\Sigma}}^1_2$ set of reals is universally Baire.

[40]The set A is $<\kappa$-**universally Baire** if it is γ-universally Baire for all $\gamma < \kappa$, and **universally Baire** if it is universally Baire for all κ.

THEOREM 6.2 (Feng, Magidor, and Woodin; [FMW92]). Assume $\mathsf{AD}^{\mathbf{L}(\mathbb{R})}$. Then the following are equivalent.

(i) $\mathsf{AD}^{\mathbf{L}(\mathbb{R})}$ holds in every forcing extension.
(ii) Every set of reals in $\mathbf{L}(\mathbb{R})$ is universally Baire.

Woodin's *Tree Production Lemma* is a powerful means for showing that sets of reals are universally Baire (*cf.* [Lar04]). Woodin's proof of Theorem 5.7 proceeded by applying the lemma to the set $\mathbb{R}^{\#}$. Informally, the lemma can be interpreted as saying that a set of reals A is δ-universally Baire if for every real r generic for a partial order in $\mathbf{V}_\delta$, either r is in the image of A for every $\mathbb{Q}_{<\delta}$-embedding[41] for which r is in the image model, or r is in the image of A for no such embedding.

THEOREM 6.3 (Tree Production Lemma). Suppose that δ is a Woodin cardinal. Let φ and ψ be binary formulas, and let x and y be arbitrary sets, and assume that the empty condition in the stationary tower $\mathbb{Q}_{<\delta}$ forces that for each real r, $M \models \psi(r, j(y)) \Leftrightarrow \mathbf{V}[r] \models \varphi(r, x)$, where $j \colon \mathbf{V} \to M$ is the induced elementary embedding. Then $\{r : \psi(r, y)\}$ is $<\delta$-universally Baire.

6.2. AD^+ **and** $\mathsf{AD}_{\mathbb{R}}$**.** Moschovakis proved that under AD, if λ is less than Θ, A is a set of functions from ω to λ and A is Suslin and co-Suslin, then the game $G_\omega(A)$ is determined, where here the players play elements of λ [Mos81]. Woodin formulated the following axiom, which, assuming AD, holds in every inner model containing the reals whose sets of reals are all Suslin (in $\mathbf{V}$). A set of reals A is said to be ∞-Borel if there exist a set of ordinals S and binary formula φ such that $A = \{x \in \mathbb{R} : \mathbf{L}[x, S] \models \varphi(x, S)\}$. *E.g.*, a Suslin representation for a set of reals witnesses that the set ∞-Borel.

DEFINITION 6.4. AD^+ is the conjunction of the following statements.

1. $\mathsf{DC}({}^{\omega}\omega)$.
2. Every set of reals is ∞-Borel.
3. If $\lambda < \Theta$ and $\pi \colon \lambda^\omega \to \omega^\omega$ is a continuous function, then $\pi^{-1}[A]$ is determined for every $A \subseteq \omega^\omega$.

It is an open question whether AD implies AD^+. It is not known whether $\mathsf{AD}_{\mathbb{R}}$ implies AD^+, although AD^+ does follow from $\mathsf{AD}_{\mathbb{R}} + \mathsf{DC}$.

The following consequences of AD^+ were announced in [Woo99].

THEOREM 6.5 ($\mathsf{ZF} + \mathsf{DC}({}^{\omega}\omega)$). If AD^+ holds and $\mathbf{V} = \mathbf{L}(\wp(\mathbb{R}))$, then

1. the pointclass Σ^2_1 has the scale property,
2. every Σ^2_1 set of reals is the projection of a tree in **HOD**,
3. every true Σ_1-sentence is witnessed by a $\underset{\sim}{\mathbf{\Delta}}^2_1$ set of reals.

---

41 The partial order $\mathbb{Q}_{<\delta}$ is one form of Woodin's stationary tower, mentioned after Definition 5.5.

Woodin's *Derived Model Theorem*, proved around 1986, gives a means of producing models of AD^+. The model $\mathbf{L}(\mathbb{R}^*, \mathrm{Hom}^*)$ in the following theorem is said to be a **derived model** (over the ground model). A tree T is said to be $<\lambda$**-absolutely complemented** if there is a tree S such that $\mathrm{p}[T] = \mathbb{R} \setminus \mathrm{p}[S]$ in all forcing extensions by partial orders of cardinality less than λ.

Given an ordinal λ, $G \subseteq \mathrm{Col}(\omega, <\lambda)$ and $\alpha < \lambda$, we let $G\restriction\alpha$ denote $G \cap \mathrm{Col}(\omega, <\alpha)$. The model $\mathbf{V}(\mathbb{R}^*)$ in the following theorem can be defined as either $\bigcup_{\alpha \in \mathrm{Ord}} L(\mathbf{V}_\alpha, \mathbb{R}^*)$ or $\mathbf{HOD}^{V[G]}_{V \cup \mathbb{R}^*}$. Given a pointclass Γ, M_Γ denotes the collection of transitive sets x such that $\langle x, \in \rangle$ is isomorphic to $\langle \mathbb{R}/E, F/E \rangle$, for some $E, F \in \Gamma$ such that E is an equivalence relation on $\mathbb{R}$ and F is an E-invariant binary relation on $\mathbb{R}$. Models of the form $L(\Gamma, \mathbb{R}^*)$ below are called **derived models**. *Cf.* [Ste09] for an earlier version of the theorem.

THEOREM 6.6 (Derived Model Theorem; Woodin). Let λ be a limit of Woodin cardinals. Let $G \subseteq \mathrm{Col}(\omega, <\lambda)$ be a $\mathbf{V}$-generic filter. Let $\mathbb{R}^*$ be $\bigcup_{\alpha<\lambda} \mathbb{R}^{\mathbf{V}[G\restriction\alpha]}$, Hom^* be the collection of sets of the form $\mathrm{p}[T] \cap \mathbb{R}^*$, for T a $<\lambda$-absolutely complemented tree in $\mathbf{V}[G\restriction\alpha]$ for some $\alpha < \lambda$, and Γ be the collection of sets of reals A in $\mathbf{V}[G]$ such that $\mathbf{L}(A, \mathbb{R}^*) \models \mathsf{AD}^+$. Then

1. $\mathbf{L}(\Gamma, \mathbb{R}^*) \models \mathsf{AD}^+$,
2. Hom^* is the collection of Suslin, co-Suslin sets of reals in $\mathbf{L}(\Gamma, \mathbb{R}^*)$, and
3. $M_\Gamma \prec_{\Sigma_1} \mathbf{L}(\Gamma, \mathbb{R}^*)$.

Woodin also showed that item (3) above is equivalent to AD^+, assuming $\mathsf{AD} + \mathbf{V}{=}\mathbf{L}(\wp(\mathbb{R}))$. The Derived Model Theorem has a converse, also due to Woodin, which says that almany models of AD^+ arise in this fashion.

THEOREM 6.7 (Woodin). Let M be a model of AD^+, and let Γ be the collection of sets of reals which are Suslin, co-Suslin in M. Then in a forcing extension of M there is an inner model N such that $\mathbf{L}(\Gamma, \mathbb{R}^*)$ is a derived model over N.

In unpublished work, Woodin has shown that over AD, $\mathsf{AD}_\mathbb{R}$ is equivalent to some of its ostensibly weak consequences (*cf.* [Woo99]). The implication from (ii) to (i) in the following theorem is due independently to Martin (*cf.* [Mar20]). The implication from (i) to (ii) relies heavily on work of Becker [Bec85]. Recall that Mycielski (*cf.* § 2.3) showed that (i) implies (iii); the implication from (ii) to (iii) is mentioned in § 3.2.

THEOREM 6.8 (Woodin). Assume $\mathsf{ZF}+\mathsf{DC}$. Then the following are equivalent:

(i) $\mathsf{AD}_\mathbb{R}$,
(ii) AD + "every set of reals is Suslin", and
(iii) AD + Uniformization.

Woodin would also produce models of $\mathsf{AD}_\mathbb{R}$ from large cardinals.

THEOREM 6.9 (Woodin). Suppose that there exists a cardinal δ of cofinality ω which is a limit of Woodin cardinals and $<\delta$-strong cardinals. Then there is a forcing extension in which there is an inner model containing the reals and satisfying $\mathsf{AD}_{\mathbb{R}}$.

Steel, using earlier work of Woodin, completed the equiconsistency with the following theorem.

THEOREM 6.10 (Steel). If $\mathsf{AD}_{\mathbb{R}}$ holds, then in a forcing extension there is a proper class model of ZFC in which there exists a cardinal δ of cofinality ω which is a limit of Woodin cardinals and $<\delta$-strong cardinals.

Recall from § 4.1 that Θ is defined to be the least ordinal which is not a surjective image of the reals. Consideration of ordinal definable surjections gives the **Solovay sequence**, $\langle\Theta_\alpha : \alpha \leq \Omega\rangle$. This sequence is defined by letting Θ_0 be the least ordinal which is not the surjective image of an ordinal definable function on the reals, and, for each $\alpha < \Omega$, letting $\Theta_{\alpha+1}$ be the least ordinal which is not a surjective image of $\wp(\Theta_\alpha)$ via an ordinal definable function. Taking limits at limit stages and continuing until $\Theta_\Omega = \Theta$ completes the definition. The consistency strength of $\mathsf{AD}^+ +$ "$\Theta_\alpha = \Theta$" increases with α.

In $\mathbf{L}(\mathbb{R})$, $\Theta_0 = \Theta$. Woodin proved that, assuming AD^+, $\mathsf{AD}_{\mathbb{R}}$ is equivalent to the assertion that the Solovay sequence has limit length. Woodin also showed, under the same assumption, that Θ_α is a Woodin cardinal in **HOD**, for all nonlimit $\alpha \leq \Omega$.

In unpublished work, Woodin showed that if it is consistent that there exists a Woodin limit of Woodin cardinals, then it is consistent that there exist sets of reals A and B such that the models $\mathbf{L}(A, \mathbb{R})$ and $\mathbf{L}(B, \mathbb{R})$ each satisfy AD but $\mathbf{L}(A, B, \mathbb{R})$ does not. Woodin also showed that in this case $\mathbf{L}(\Gamma, \mathbb{R}) \models \mathsf{AD}_{\mathbb{R}} + \mathsf{DC}$, where $\Gamma = \wp(\mathbb{R}) \cap \mathbf{L}(A, \mathbb{R}) \cap \mathbf{L}(B, \mathbb{R})$. Grigor Sargsyan showed that if there exist models $\mathbf{L}(A, \mathbb{R})$ and $\mathbf{L}(B, \mathbb{R})$ as above then there is a proper class model of $\mathsf{AD}_{\mathbb{R}}$ containing the reals in which Θ is regular.

6.3. Long games. As mentioned in Section 5.2, Blass [Bla75] and Mycielski showed that determinacy for games of length ω^2 is equivalent to $\mathsf{AD}_{\mathbb{R}}$. For each $n \in \omega$, determinacy for games of length $\omega + n$ is equivalent to AD (think of the game as being divided in two parts, where in the first part (of length ω) the players try to obtain a position from which they have a winning strategy in the second; the winning strategy in the second part can be uniformly chosen).

Martin and Woodin independently showed that $\mathsf{AD}_{\mathbb{R}}$ is equivalent to determinacy for games of length α for each countable $\alpha \geq \omega^2$. Determinacy for games of length $\omega \cdot 2$ easily gives uniformization, and determinacy for games of length less than ω^2 is easily seen to follow from AD plus uniformization. It follows from this and Theorem 6.8 (assuming DC) that $\mathsf{AD}_{\mathbb{R}}$ is equivalent to determinacy for games of length α for each countable $\alpha \geq \omega \cdot 2$.

While AD does not imply uniformization, the Second Periodicity Theorem (Theorem 3.4) shows that PD implies the uniformization of projective sets. It follows that PD is equivalent to PD for games of length less than ω^2. As noted by Neeman [Nee05], the techniques from the Blass-Mycielski result above can be used to prove the determinacy of games of length ω^2 with analytic payoff from $\mathsf{AD}^{\mathbf{L}(\mathbb{R})}$ plus the existence of $\mathbb{R}^{\#}$.

Steel [Ste88] considered **continuously coded games**, games where each stage of the game is associated with an integer, and the game ends when an associated integer is repeated. Such a game must end after countably many rounds, but runs of the game can have any countable length. Steel proved that $\mathsf{ZF} + \mathsf{AD} + \mathsf{DC}$ + "every set of reals has a scale"+"ω_1 is $\wp(\mathbb{R})$-supercompact" implies the determinacy of all continuously coded games.

None of the results mentioned so far in this section involves proving determinacy directly from large cardinals. Instead they show that some form of determinacy for short games with complicated payoff implies determinacy for longer games with simpler payoff. Proving long game determinacy from large cardinals was pioneered, and extensively developed, by Neeman, who established a number of results on games of variable countable length, and even length ω_1 (*cf.* [Nee04, Nee05, Nee06A]). Neeman's techniques built on the proof of PD from Woodin cardinals by Martin and Steel, using iteration trees. In many cases, his proofs proceed from essentially optimal hypotheses. The proofs of many of these results reduced the determinacy of long games to the iterability of models containing large cardinals.

For example, given $C \subseteq \mathbb{R}^{<\omega_1}$, let $G_{\mathrm{local}}(L, C)$ be the game where players player I and player II alternate playing natural numbers so as to define elements z_ξ of the Baire space. The game ends at the first γ such that γ is uncountable in $\mathbf{L}[z_\xi : \xi < \gamma]$, with player I winning if the sequence $\langle z_\xi : \xi < \gamma\rangle$ is in C. It follows from mild large cardinal assumptions (*e.g.*, the existence of the sharp of every subset of ω_1) that γ must be countable.

Given a pointclass Γ, a set C consisting of countable sequences of reals is said to be Γ **in the codes** if the set of reals coding members of C (under a suitably definable coding) is in Γ.

THEOREM 6.11 (Neeman). *Suppose that there exists a measurable cardinal above a Woodin limit of Woodin cardinals. Then the games $G_{\mathrm{local}}(L, C)$ are determined for all C which are $\Game_\omega({<}\omega^2\text{-}\Pi^1_1)$ in the codes.*

The preceding theorem is obtained by combining the results of [Nee04] and [Nee02A]. The proof proceeds by constructing an iterable class model M with a cardinal ϑ such that ϑ is a Woodin limit of Woodin cardinals in M and countable in $\mathbf{V}$ [Nee02A]. Using inner model theory, Neeman then transformed the iteration strategy of M into a winning strategy in $G_{\mathrm{local}}(L, C)$.

Adapting Kechris and Solovay's proof that Δ^1_2-determinacy implies the existence of a real x such that $\mathbf{L}[x]$ satisfies the determinacy of all ordinal definable sets of reals (discussed before Theorem 5.10), Woodin proved that the amount of determinacy in the conclusion of Theorem 6.11 implies that there exists a set $A \subseteq \omega_1$ such that in $\mathbf{L}[A]$, all games on integers of length ω_1 with payoff definable from reals and ordinals are determined (*cf.* [Nee04, Exercise 7F.15]).

We give one more result of Neeman, proving the determinacy of certain games of length ω_1. In Theorem 6.12 below, $\mathcal{L}^+$ is the language of set theory with one additional unary predicate. Given an integer k and a sequence $\bar{S}$ of stationary sets indexed by $[\omega_1]^{<k}$, $[\bar{S}]$ is the collection of increasing k-tuples $\langle \alpha_0, \ldots, \alpha_{k-1} \rangle$ from ω_1 such that each initial segment of length $j \leq k$ is in $S_{\langle \alpha_0, \ldots, \alpha_{j-l} \rangle}$. The game $\mathrm{G}_{\omega_1,k}(\bar{S}, \varphi)$ is a game of length ω_1 in which the players collaborate to build a function $f : \omega_1 \to \omega_1$. Then player I wins if there is a club C such that $\langle L_{\omega_1}, r \rangle \models \varphi(\alpha_0, \ldots, \alpha_{k-1})$ for all $\langle \alpha_0, \ldots, \alpha_{k-1} \rangle \in [\bar{S}] \cap [C]^k$, and player II wins if there is a club C such that $\langle L_{\omega_1}, r \rangle \models \neg\varphi(\alpha_0, \ldots, \alpha_{k-1})$ for all $\langle \alpha_0, \ldots, \alpha_{k-1} \rangle \in [\bar{S}] \cap [C]^k$. Although there can be runs of the game for which neither player wins, determinacy for this game in the sense of Theorem 6.12 refers to the existence of a strategy for one player or the other that guarantees victory.

The model 0^W is the minimal iterable fine structural inner model M which has a top extender predicate whose critical point is Woodin in M. The existence of such a model is not known to follow from large cardinals.

The last part of the conclusion of Theorem 6.12 extends a result of Martin, who showed that for any recursive enumeration $\langle B_i : i < \omega \rangle$ of the $<\omega^2$-Π^1_1 sets, the set of i such that player I has a winning strategy in $\mathrm{G}_\omega(B_i)$ is recursively isomorphic to $0^\#$.

THEOREM 6.12 (Neeman; [Nee07A]). Suppose that 0^W exists. Let $k < \omega$. Let $\bar{S}$ be a sequence of mutually disjoint stationary sets indexed by $[\omega_1]^{<k}$. Let φ be a $\mathcal{L}^+$ formula with k free variables. Then the game $\mathrm{G}_{\omega_1,k}(\bar{S}, \varphi)$ is determined. Furthermore, the winner of each such game depends only on φ and not on $\bar{S}$, and the set of φ for which the first player has a winning strategy is recursively equivalent to the canonical real coding 0^W.

If one allows the members of $\bar{S}$ all to be ω_1, then there are undetermined games of this type, as observed by Greg Hjorth (*cf.* [Nee07A]). If one allows the members of $\bar{S}$ all to be ω_1 and changes the winning condition for player II to be simply the negation of the winning condition for player I then one can force from a strongly inaccessible limit of measurable cardinals that some game of this type is not determined [Lar05].

Combining Neeman's proof of Theorem 6.12 with his own theory of *hybrid strategy mice*, Woodin proved that if there exist proper class many Woodin limits of Woodin cardinals then AD^+ holds in the **Chang Model**, the smallest inner model of ZF containing the ordinals and closed under countable sequences.

6.4. Forcing over models of determinacy. Steel and Van Wesep showed that by forcing over a model of $\mathsf{AD}_\mathbb{R}$ + "Θ is regular" (the hypothesis they used was actually weaker) one can produce a model of ZFC in which NS_{ω_1} is saturated and $\underset{\sim}{\delta}^1_2 = \omega_2$ [SVW82]. This was the first consistency proof of either of these two statements with ZFC. Martin had conjectured that "for all $n \in \mathbb{N}$, $\underset{\sim}{\delta}^1_n = \aleph_n$" is consistent with ZFC, and this verified the conjecture for the case $n = 2$. Woodin [Woo83] subsequently reduced the hypothesis to AD.

Shelah later showed that it was possible to force the saturation of NS_{ω_1} from a Woodin cardinal [She98]. Woodin proved that the saturation of NS_{ω_1} plus the existence of a measurable cardinal implies that $\underset{\sim}{\delta}^1_2 = \omega_2$ [Woo99]. Woodin then turned his proof into a general method for producing models of ZFC by forcing over models of determinacy. The most general form of this method, a partial order called $\mathbb{P}_{\max}$, consists roughly of a directed system containing all countable models of ZFC with a precipitous ideal on ω_1. In the presence of large cardinals, the resulting extension satisfies all forceable Π_2 sentences in $\mathbf{H}(\omega_2)$, even with predicates for NS_{ω_1} and each set of reals in $\mathbf{L}(\mathbb{R})$. In this model, NS_{ω_1} is saturated and $\underset{\sim}{\delta}^1_2 = \omega_2$. There are many variants of $\mathbb{P}_{\max}$. One of these variants, called $\mathbb{Q}_{\max}$, produces a model in which NS_{ω_1} is $\aleph_1$-dense (*i.e.*, $\wp(\omega_1)/NS_{\omega_1}$ has a dense subset of cardinality $\aleph_1$; this implies saturation), from the assumption that AD holds in $\mathbf{L}(\mathbb{R})$. No other method is known for producing a model of ZFC in which NS_{ω_1} is $\aleph_1$-dense.

Steel showed that under AD, $\mathbf{HOD}^{\mathbf{L}(\mathbb{R})}$ is an extender model below Θ [Ste95A]. Woodin then showed that the entire model $\mathbf{HOD}^{\mathbf{L}(\mathbb{R})}$ is a model of the form $\mathbf{L}[\vec{E}, \Sigma]$, where $\vec{E}$ is a sequence of extenders and Σ is an iteration strategy corresponding to this sequence. Using this approach, Steel showed that for every regular $\kappa < \Theta$, the ω-club filter over κ is an ultrafilter in $\mathbf{L}(\mathbb{R})$. Woodin used this to show that ω_1 is $<\Theta$-supercompact in $\mathbf{L}(\mathbb{R})$. Previously it was known only that ω_1 is λ-supercompact for λ below the supremum of the Suslin cardinals (*cf.* the paragraph after Theorem 4.2).

Woodin also used the inner models approach to show that, in $\mathbf{L}(\mathbb{R})$, ω_1 is huge to κ for each measurable $\kappa < \Theta$, improving results of Becker. Neeman [Nee07B] used this approach to prove, for each $\lambda < \Theta$, the uniqueness of the normal ultrafilter on $\wp_{\aleph_1}(\lambda)$ witnessing the λ-supercompactness of ω_1. Previously this too was known only for $\lambda < \underset{\sim}{\delta}^2_1$ (this is also discussed in the paragraph after Theorem 4.2). Neeman [Nee07B] and Woodin independently used this approach to show that, assuming AD + $\mathbf{V}$=$\mathbf{L}(\mathbb{R})$, one could force without adding reals to obtain ZFC + $\underset{\sim}{\delta}^1_n = \omega_2$, for any $n \geq 3$. It is still unknown whether $\underset{\sim}{\delta}^1_m$ can equal ω_n for any $m \geq n \geq 3$ (under ZFC).

6.5. Determinacy from its consequences. Woodin [Woo82] conjectured that Projective Determinacy follows from the statement that all projective sets are Lebesgue measurable, have the Baire property and can be uniformized by projective functions (all consequences of PD). This conjecture was refuted by Steel in 1997. If one requires the uniformization property for the scaled projective pointclasses, then the conjecture is still open. Woodin did prove the following version of the conjecture in the late 1990s, using work of Steel in inner model theory. Recall that AD implies the first two statements below, and that AD + $\mathbf{V}=\mathbf{L}(\mathbb{R})$ implies the third (*cf.* §§ 2.1 & 3.3).

THEOREM 6.13 (Woodin). Assuming ZF + DC + $\mathbf{V}=\mathbf{L}(\mathbb{R})$, the Axiom of Determinacy follows from the conjunction of the following three statements.

1. Every set of reals is Lebesgue measurable.
2. Every set of reals has the property of Baire.
3. Every Σ^2_1 subset of ${}^2({}^\omega\omega)$ can be uniformized.

Woodin had proved another equivalence in the early 1980s.

THEOREM 6.14 (Woodin). Assume ZF + DC + $\mathbf{V}=\mathbf{L}(\mathbb{R})$. Then the following are equivalent.

(i) AD.
(ii) Turing determinacy.

It is apparently an open question whether AD follows from ZF+DC+$\mathbf{V}=\mathbf{L}(\mathbb{R})$ plus either of (a) for every $\alpha < \Theta$ there is a surjection of ${}^\omega\omega$ onto $\wp(\alpha)$; (b) Θ is inaccessible.

Determinacy would turn out to be necessary for some of its earliest applications. For instance, Steel showed that $\utilde{\Sigma}^1_3$-separation plus the existence of sharps for all reals implies $\utilde{\Delta}^1_2$-determinacy [Ste96]. Applying related work of Steel, Hjorth showed that $\utilde{\Pi}^1_2$-determinacy follows from Wadge determinacy for $\utilde{\Pi}^1_2$ sets [Hjo96A]. Earlier, Harrington had shown that, for each real x, $\utilde{\Pi}^1_1(x)$-Wadge determinacy implies that $x^\#$ exists. It is open whether Wadge determinacy for the projective sets implies PD.

6.6. Determinacy from other statements. Determinacy axioms such as PD and $\mathrm{AD}^{\mathbf{L}(\mathbb{R})}$ imply the consistency of ZFC (plus certain large cardinal statements) and so cannot be proved in ZFC. Empirically, however, these statements appear to follow from every natural statement of sufficient consistency strength. This includes a number of statements ostensibly having little relation to determinacy. In this section we give a few examples of this phenomenon. Most of these arguments use inner model theory, and our presentation relies heavily on [Sch10].

The following theorem shows, among other things, that in the presence of large cardinals, even mere forcing-absoluteness for the theory of $\mathbf{L}(\mathbb{R})$ implies $\mathrm{AD}^{\mathbf{L}(\mathbb{R})}$. The theorem is due to Steel and Woodin independently (*cf.* [Ste02]).

THEOREM 6.15. Suppose that κ is a measurable cardinal. Then the following are equivalent.

(i) For all partial orders $\mathbb{P} \in \mathbf{V}_\kappa$, the theory of $\mathbf{L}(\mathbb{R})$ is not changed by forcing with $\mathbb{P}$.

(ii) For all partial orders $\mathbb{P} \in \mathbf{V}_\kappa$, AD holds in $\mathbf{L}(\mathbb{R})$ after forcing with $\mathbb{P}$.
(iii) For all partial orders $\mathbb{P} \in \mathbf{V}_\kappa$, all sets of reals in $\mathbf{L}(\mathbb{R})$ are Lebesgue measurable after forcing with $\mathbb{P}$.
(iv) For all partial orders $\mathbb{P} \in \mathbf{V}_\kappa$, there is no ω_1-sequence of reals in $\mathbf{L}(\mathbb{R})$ after forcing with $\mathbb{P}$.

A sequence $C = \langle C_\alpha : \alpha < \lambda \rangle$ (for some ordinal λ) is said to be **coherent** if each C_β is a club subset of β, and $C_\alpha = \alpha \cap C_\beta$ whenever α is a limit point of C_β. A **thread** of such a coherent sequence C is a club set $D \subseteq \lambda$ such that $C_\alpha = \alpha \cap D$ for all limit points α of D. The principle $\square(\lambda)$ says that there is a coherent sequence of length λ with no thread. The principle $\square_\kappa$ says that there is a coherent sequence C of length κ^+ such that the ordertype of C_α is at most κ, for each limit $\alpha < \lambda$ (in which case there cannot be a thread). These principles were isolated in the 1960s by Jensen [Jen72], who showed that $\square_\kappa$ holds in $\mathbf{L}$ for all infinite cardinals κ (*cf.* [Dev84, p. 141]).

Todorčević showed that the Proper Forcing Axiom (PFA) implies that $\square(\kappa)$ fails for all cardinals κ of cofinality at least ω_2 [Tod84], from which it follows that $\square_\kappa$ fails for all uncountable cardinals. The failure of these square principles implies the failure of covering theorems for certain inner models, from which one can derive inner models with large cardinals. Using this general approach, Ernest Schimmerling proved that PFA implies $\utilde{\mathbf{\Delta}}^1_2$-determinacy [Sch95]. Woodin extended this proof to show that PFA implies PD.

In 1990, Woodin also showed that PFA plus the existence of a strongly inaccessible cardinal implies $\mathsf{AD}^{\mathbf{L}(\mathbb{R})}$. His proof introduced a technique known as the *core model induction*, an application of descriptive set theory and inner model theory. Roughly, the idea is to inductively work through the Wadge degrees to build canonical inner models which are correct for each Wadge class. The induction works through the gap structure highlighted in [Ste83A]. This general approach had previously been used by Kechris and Woodin [KW83] (*cf.* the end of § 4.2).

Alessandro Andretta, Neeman, and Steel showed that PFA plus the existence of a measurable cardinal implies the existence of a model of $\mathsf{AD}_\mathbb{R}$ containing all the reals and ordinals [ANS01]. Steel showed that if $\square_\kappa$ fails for a singular strong limit cardinal κ, then AD holds in $\mathbf{L}(\mathbb{R})$ [Ste05]. Building on Steel's work, Sargsyan produced a model of $\mathsf{AD}_\mathbb{R}$ + "Θ is regular" from the same hypothesis.

The following theorem is due to Steel. Schimmerling [Sch07] had previously obtained PD from the same assumption.

THEOREM 6.16. If $\kappa \geq \max\{\aleph_2, c\}$ and $\square(\kappa)$ and $\square_\kappa$ fail, then $\mathsf{AD}^{\mathbf{L}(\mathbb{R})}$ holds.

Todorčević (*cf.* [Bek91]) and Boban Veličković showed that PFA implies that $2^{\aleph_0} = 2^{\aleph_1} = \aleph_2$ [Vel92]. This gives another route towards showing that PFA implies that AD holds in $\mathbf{L}(\mathbb{R})$. In May 2011, Andrés Caicedo, Paul Larson,

Grigor Sargsyan, Ralf Schindler, John Steel and Martin Zeman [CLS17] showed that the hypothesis of Theorem 6.16 (with $\kappa = \aleph_2$) can be forced (using $\mathbb{P}_{\max}$) over a model of $\mathsf{AD}_{\mathbb{R}}$ in which Θ and some other member of the Solovay sequence are both regular.

Schimmerling and Zeman used the core model induction to prove the following theorem [SZ01]. They had previously derived Projective Determinacy from the failure of a weaker version of $\square_\kappa$ at a weakly compact cardinal; Woodin had then derived $\mathsf{AD}^{\mathbf{L}(\mathbb{R})}$ from the same hypothesis.

THEOREM 6.17. If κ is a weakly compact cardinal and $\square_\kappa$ fails, then AD holds in $\mathbf{L}(\mathbb{R})$.

As discussed in § 6.4, Woodin showed using a variation of $\mathbb{P}_{\max}$ that over a model of AD one can force to produce a model of ZFC in which the nonstationary ideal on ω_1 is $\aleph_1$-dense. Using the core model induction, he showed that the $\aleph_1$-density of NS_{ω_1} implies $\mathsf{AD}^{\mathbf{L}(\mathbb{R})}$.

Steel had previously shown, using inner models, that Projective Determinacy follows from CH plus the existence of a homogeneous ideal on ω_1 (a weaker assumption that the $\aleph_1$-density of NS_{ω_1}, which is in fact inconsistent with CH, by a theorem of Shelah). He had also shown [Ste96] that if NS_{ω_1} is saturated and there is a measurable cardinal, then $\utilde{\mathbf{\Delta}}^1_2$-determinacy holds. The hypothesis of the measurable cardinal was later removed in collaboration with Jensen.

Using the core model induction, Richard Ketchersid showed that if the restriction of NS_{ω_1} to some stationary set $S \subseteq \omega_1$ is $\aleph_1$-dense, and the restriction of the generic elementary embedding corresponding to forcing with $\wp(S)/\mathrm{NS}_{\omega_1}$ to each ordinal is an element of the ground model, then there is a model of $\mathsf{AD}^+ + \Theta_0{<}\Theta$ containing the reals and the ordinals. Also using this method, Sargsyan would deduce the consistency of $\mathsf{AD}_{\mathbb{R}}$ + "Θ is regular" from the same hypothesis. This gives an equiconsistency, as Woodin has shown how to force the hypothesis over a model of $\mathsf{AD}_{\mathbb{R}}$ + "Θ is regular". In yet another application of the core model induction, Steel and Stuart Zoble [SZ] derived $\mathsf{AD}^{\mathbf{L}(\mathbb{R})}$ from a consequence of Martin's Maximum isolated by Todorčević, known as the *Strong Reflection Principle* at ω_2.

We conclude with three more examples. Silver proved that if κ is a singular cardinal of uncountable cofinality and $2^\alpha = \alpha^+$ for club many $\alpha < \kappa$, then $2^\kappa = \kappa^+$ [Sil75]. Gitik, Schindler and Shelah (*cf.* [GSS06]) showed that if κ is a singular cardinal of uncountable cofinality and the set of $\alpha < \kappa$ for which $2^\alpha = \alpha^+$ is stationary and costationary, then PD holds. They (in the same paper) showed that if $\aleph_\omega$ is a strong limit cardinal and $2^{\aleph_\omega} > \aleph_{\omega_1}$, then PD holds. It is not known whether either of these results can be strengthened to obtain $\mathsf{AD}^{\mathbf{L}(\mathbb{R})}$.

A cardinal κ is said to have the **tree property** if every tree of height κ with all levels of cardinality less than κ has a cofinal branch (*i.e.*, if there are no κ-Aronszajn trees). Foreman, Magidor and Schindler showed that if there exist infinitely many cardinals δ above the continuum such that the tree property holds at δ and at δ^+, then PD holds [FMS01]. The hypothesis of this statement had been shown consistent relative to the existence of infinitely many supercompact cardinals by James Cummings and Foreman [CF98]. It is not known whether the conclusion can be strengthened to $\mathsf{AD}^{\mathbf{L}(\mathbb{R})}$.

Finally, as mentioned in § 2.3, Gitik showed that if there is a proper class of strongly compact cardinals, then there is a model of ZF in which all infinite cardinals have cofinality ω. Using the core model induction, Daniel Busche and Schindler showed that this statement implies that PD holds, and that AD holds in the $\mathbf{L}(\mathbb{R})$ of a forcing extension of **HOD** [BS09].

Acknowledgments. Gunter Fuchs helped with some original sources in German. The author would like to thank Akihiro Kanamori, Alexander Kechris, Richard Ketchersid, Tony Martin, Itay Neeman, Jan Mycielski, Grigor Sargsyan, Robert Solovay and John Steel for making many helpful suggestions.

REFERENCES

JOHN W. ADDISON

[Add58A] *Separation principles in the hierarchies of classical and effective descriptive set theory*, ***Fundamenta Mathematicae***, vol. 46 (1958–9), pp. 123–135.

[Add58B] *Some consequences of the axiom of constructibility*, ***Fundamenta Mathematicae***, vol. 46 (1958–9), pp. 337–357.

JOHN W. ADDISON AND YIANNIS N. MOSCHOVAKIS

[AM68] *Some consequences of the axiom of definable determinateness*, ***Proceedings of the National Academy of Sciences of the United States of America***, vol. 59 (1968), pp. 708–712.

DONALD J. ALBERS AND GERALD L. ALEXANDERSON

[AA85] ***Mathematical People. Profiles and Interviews***, Birkhäuser, Boston, MA, 1985.

ALESSANDRO ANDRETTA, ITAY NEEMAN, AND JOHN R. STEEL

[ANS01] *The domestic levels of* $\mathbf{K}^c$ *are iterable*, ***Israel Journal of Mathematics***, vol. 125 (2001), pp. 157–201.

STEFAN BANACH AND ALFRED TARSKI

[BT24] *Sur la décomposition des ensembles de points en parties respectivement congruentes*, ***Fundamenta Mathematicae***, vol. 6 (1924), pp. 244–277.

JAMES E. BAUMGARTNER, DONALD A. MARTIN, AND SAHARON SHELAH

[BMS84] ***Axiomatic Set Theory. Proceedings of the AMS-IMS-SIAM joint summer research conference held in Boulder, Colo., June 19–25, 1983***, Contemporary Mathematics, vol. 31, American Mathematical Society, 1984.

HOWARD S. BECKER

[Bec78] *Partially playful universes*, in Kechris and Moschovakis [CABAL i], pp. 55–90, reprinted in [CABAL III], pp. 49–85.

[Bec81] AD *and the supercompactness of* $\aleph_1$, ***The Journal of Symbolic Logic***, vol. 46 (1981), pp. 822–841.

[Bec85] *A property equivalent to the existence of scales*, ***Transactions of the American Mathematical Society***, vol. 287 (1985), no. 2, pp. 591–612.

Howard S. Becker and Alexander S. Kechris

[BK84] *Sets of ordinals constructible from trees and the third Victoria Delfino problem*, in Baumgartner et al. [BMS84], pp. 13–29.

Howard S. Becker and Yiannis N. Moschovakis

[BM81] *Measurable cardinals in playful models*, in Kechris et al. [Cabal ii], pp. 203–214, reprinted in [Cabal III], pp. 115–125.

Mohamed Bekkali

[Bek91] ***Topics in Set Theory: Lebesgue measurability, large cardinals, forcing axioms, rho-functions. Notes on lectures by Stevo Todorčević***, Lecture Notes in Mathematics, vol. 1476, Springer-Verlag, Berlin, 1991.

David Blackwell

[Bla67] *Infinite games and analytic sets*, ***Proceedings of the National Academy of Sciences of the United States of America***, vol. 58 (1967), pp. 1836–1837.

[Bla69] *Infinite G_δ-games with imperfect information*, ***Polska Akademia Nauk. Instytut Matematyczny. Zastosowania Matematyki***, vol. 10 (1969), pp. 99–101.

Andreas Blass

[Bla75] *Equivalence of two strong forms of determinacy*, ***Proceedings of the American Mathematical Society***, vol. 52 (1975), pp. 373–376.

L. E. J. Brouwer

[Bro24] *Beweis dass jede volle Funktion gleichmässig stetig ist*, ***Koninklijke Akademie van Wetenschappen te Amsterdam. Proceedings of the Section of Sciences***, vol. 27 (1924), pp. 189–193.

Daniel Busche and Ralf Schindler

[BS09] *The strength of choiceless patterns of singular and weakly compact cardinals*, ***Annals of Pure and Applied Logic***, vol. 159 (2009), no. 1-2, pp. 198–248.

Andrés Eduardo Caicedo, Paul B. Larson, Grigor Sargsyan, Ralf Schindler, John R. Steel, and Martin Zeman

[CLS17] *Square principles in $\mathbb{P}_{\max}$ extensions*, ***Israel Journal of Mathematics***, vol. 217 (2017), no. 1, pp. 231–261.

S. Barry Cooper

[Coo04] ***Computability Theory***, Chapman & Hall/CRC, Boca Raton, FL, 2004.

James Cummings and Matthew Foreman

[CF98] *The tree property*, ***Advances in Mathematics***, vol. 133 (1998), no. 1, pp. 1–32.

Morton Davis

[Dav64] *Infinite games of perfect information*, ***Advances in Game Theory*** (Melvin Dresher, Lloyd S. Shapley, and Alan W. Tucker, editors), Annals of Mathematical Studies, vol. 52, Princeton University Press, 1964, pp. 85–101.

Keith J. Devlin

[Dev84] ***Constructibility***, Perspectives in Mathematical Logic, Springer-Verlag, Berlin, 1984.

Anthony Dodd

[Dod82] ***The Core Model***, London Mathematical Society Lecture Note Series, vol. 61, Cambridge University Press, 1982.

DERRICK ALBERT DUBOSE
[DuB90] *The equivalence of determinacy and iterated sharps*, ***The Journal of Symbolic Logic***, vol. 55 (1990), no. 2, pp. 502–525.

PAUL ERDŐS AND ANDRÁS HAJNAL
[EH58] *On the structure of set mappings*, ***Acta Mathematica Academiae Scientiarum Hungaricae***, vol. 9 (1958), pp. 111–131.
[EH66] *On a problem of B. Jónsson*, ***Bulletin de l'Académie Polonaise des Sciences***, vol. 14 (1966), pp. 19–23.

SOLOMAN FEFERMAN AND AZRIEL LÉVY
[FL63] *Independence results in set theory by Cohen's method II*, ***Notices of the American Mathematical Society***, vol. 10 (1963), p. 593.

QI FENG, MENACHEM MAGIDOR, AND W. HUGH WOODIN
[FMW92] *Universally Baire sets of reals*, in Judah et al. [JJW92], pp. 203–242.

MATTHEW FOREMAN
[For86] *Potent axioms*, ***Transactions of the American Mathematical Society***, vol. 294 (1986), no. 1, pp. 1–28.

MATTHEW FOREMAN, MENACHEM MAGIDOR, AND RALF-DIETER SCHINDLER
[FMS01] *The consistency strength of successive cardinals with the tree property*, ***The Journal of Symbolic Logic***, vol. 66 (2001), no. 4, pp. 1837–1847.

MATTHEW FOREMAN, MENACHEM MAGIDOR, AND SAHARON SHELAH
[FMS88] *Martin's maximum, saturated ideals and nonregular ultrafilters. I*, ***Annals of Mathematics***, vol. 127 (1988), no. 1, pp. 1–47.

HARVEY FRIEDMAN
[Fri71A] *Determinateness in the low projective hierarchy*, ***Fundamenta Mathematicae***, vol. 72 (1971), no. 1, pp. 79–95. (errata insert).
[Fri71B] *Higher set theory and mathematical practice*, ***Annals of Mathematical Logic***, vol. 2 (1971), no. 3, pp. 325–357.

DAVID GALE AND FRANK M. STEWART
[GS53] *Infinite games with perfect information*, ***Contributions to the theory of games, vol. 2***, Annals of Mathematics Studies, no. 28, Princeton University Press, 1953, pp. 245–266.

MOTI GITIK
[Git80] *All uncountable cardinals can be singular*, ***Israel Journal of Mathematics***, vol. 35 (1980), no. 1-2, pp. 61–88.

MOTI GITIK, RALF SCHINDLER, AND SAHARON SHELAH
[GSS06] *PCF theory and Woodin cardinals*, ***Logic Colloquium '02. Joint proceedings of the Annual European Summer Meeting of the Association for Symbolic Logic and the Biannual Meeting of the German Association for Mathematical Logic and the Foundations of Exact Sciences (the Colloquium Logicum) held in Münster, August 3–11, 2002*** (Zoé Chatzidakis, Peter Koepke, and Wolfram Pohlers, editors), Lecture Notes in Logic, vol. 27, Association for Symbolic Logic, 2006, pp. 172–205.

JOHN TOWNSEND GREEN
[Gre78] *Determinacy and the Existence of Large Measurable Cardinals*, ***Ph.D. thesis***, University of California at Berkeley, 1978.

JACQUES HADAMARD

[Had05] *Cinq letters sur la théorie des ensembles*, ***Bulletin de la Societé mathématique de France***, vol. 33 (1905), pp. 261–273.

ANDRÁS HAJNAL

[Haj56] *On a consistency theorem connected with the generalized continuum problem*, ***Zeitschrift für Mathematische Logik und Grundlagen der Mathematik***, vol. 2 (1956), pp. 131–136.

[Haj61] *On a consistency theorem connected with the generalized continuum problem*, ***Acta Mathematica Academiae Scientiarum Hungaricae***, vol. 12 (1961), pp. 321–376.

PAUL R. HALMOS

[Hal50] ***Measure Theory***, D. Van Nostrand Company, Inc., New York, 1950.

LEO A. HARRINGTON

[Har78] *Analytic determinacy and* $0^{\#}$, ***The Journal of Symbolic Logic***, vol. 43 (1978), no. 4, pp. 685–693.

LEO A. HARRINGTON AND ALEXANDER S. KECHRIS

[HK81] *On the determinacy of games on ordinals*, ***Annals of Mathematical Logic***, vol. 20 (1981), pp. 109–154.

FELIX HAUSDORFF

[Hau08] *Grundzüge einer Theorie der geordneten Mengen*, ***Mathematische Annalen***, vol. 65 (1908), pp. 435–505.

[Hau14] *Bemerkung über den Inhalt von Punktmengen*, ***Mathematische Annalen***, vol. 75 (1914), pp. 428–434.

JAMES HENLE, A. R. D. MATHIAS, AND W. HUGH WOODIN

[HMW85] *A barren extension*, ***Methods in mathematical logic. Proceedings of the sixth Latin American symposium on mathematical logic held in Caracas, August 1–6, 1983*** (Carlos A. Di Prisco, editor), Lecture Notes in Mathematics, vol. 1130, Springer-Verlag, Berlin, 1985, pp. 195–207.

GREGORY HJORTH

[Hjo96A] Π^1_2 *Wadge degrees*, ***Annals of Pure and Applied Logic***, vol. 77 (1996), no. 1, pp. 53–74.

PAUL HOWARD AND JEAN E. RUBIN

[HR98] ***Consequences of the Axiom of Choice***, Mathematical Surveys and Monographs, vol. 59, American Mathematical Society, 1998.

STEPHEN JACKSON

[Jac88] AD *and the projective ordinals*, in Kechris et al. [CABAL iv], pp. 117–220, reprinted in [CABAL II], pp. 364–483.

[Jac99] ***A Computation of*** $\underset{\sim}{\delta}^1_5$, vol. 140, Memoirs of the AMS, no. 670, American Mathematical Society, July 1999.

[Jac10] *Structural consequences of* AD, in Kanamori and Foreman [KF10], pp. 1753–1876.

THOMAS JECH

[Jec03] ***Set Theory***, Springer Monographs in Mathematics, Springer-Verlag, Berlin, 2003, the third millennium edition, revised and expanded.

RONALD B. JENSEN

[Jen72] *The fine structure of the constructible hierarchy*, ***Annals of Mathematical Logic***, vol. 4 (1972), pp. 229–308; erratum, p. 443.

H. JUDAH, W. JUST, AND W. HUGH WOODIN

[JJW92] ***Set Theory of the Continuum***, MSRI publications, vol. 26, Springer-Verlag, 1992.

László Kalmár

[Kal28] *Zur Theorie der abstrakten Spiele*, **Acta Scientiarum Mathematicarum (Szeged)**, vol. 4 (1928–29), no. 1–2, pp. 65–85.

Akihiro Kanamori

[Kan95] *The emergence of descriptive set theory*, **From Dedekind to Gödel. Essays on the Development of the Foundations of Mathematics. Proceedings of a conference held at Boston University, Boston, MA, April 5–7, 1992** (Jaakko Hintikka, editor), Synthese Library, vol. 251, Kluwer Academic Publishers, Dordrecht, 1995, pp. 241–262.

[Kan03] **The Higher Infinite. Large Cardinals in Set Theory from Their Beginnings**, second ed., Springer Monographs in Mathematics, Springer-Verlag, Berlin, 2003.

Akihiro Kanamori and Matthew Foreman

[KF10] **Handbook of Set Theory**, Springer-Verlag, 2010.

Alexander S. Kechris

[Kec74] *On projective ordinals*, **The Journal of Symbolic Logic**, vol. 39 (1974), pp. 269–282.

[Kec75] *The theory of countable analytical sets*, **Transactions of the American Mathematical Society**, vol. 202 (1975), pp. 259–297.

[Kec78] AD *and projective ordinals*, in Kechris and Moschovakis [Cabal i], pp. 91–132, reprinted in [Cabal II], pp. 304–345.

[Kec81] *Homogeneous trees and projective scales*, in Kechris et al. [Cabal ii], pp. 33–74, reprinted in [Cabal II], pp. 270–303.

[Kec84] *The axiom of determinacy implies dependent choices in* $\mathbf{L}(\mathbb{R})$, **The Journal of Symbolic Logic**, vol. 49 (1984), no. 1, pp. 161–173.

[Kec95] **Classical Descriptive Set Theory**, Graduate Texts in Mathematics, vol. 156, Springer-Verlag, 1995.

Alexander S. Kechris, Eugene M. Kleinberg, Yiannis N. Moschovakis, and W. H. Woodin

[KKMW81] *The axiom of determinacy, strong partition properties, and nonsingular measures*, in Kechris et al. [Cabal ii], pp. 75–99, reprinted in [Cabal I], pp. 333–354.

Alexander S. Kechris, Benedikt Löwe, and John R. Steel

[Cabal I] *Games, Scales, and Suslin cardinals: the Cabal Seminar, volume I*, Lecture Notes in Logic, vol. 31, Cambridge University Press, 2008.

[Cabal II] *Wadge Degrees and Projective Ordinals: the Cabal Seminar, volume II*, Lecture Notes in Logic, vol. 37, Cambridge University Press, 2012.

[Cabal III] *Ordinal Definability and Recursion Theory: the Cabal Seminar, volume III*, Lecture Notes in Logic, vol. 43, Cambridge University Press, 2016.

Alexander S. Kechris, Donald A. Martin, and Yiannis N. Moschovakis

[KMM81A] *Appendix: Progress report on the Victoria Delfino problems*, in *Cabal Seminar* 77–79 [Cabal ii], pp. 273–274.

[Cabal ii] *Cabal Seminar* 77–79, Lecture Notes in Mathematics, vol. 839, Berlin, Springer-Verlag, 1981.

[KMM83A] *Appendix: Progress report on the Victoria Delfino problems*, in *Cabal Seminar* 79–81 [Cabal iii], pp. 283–284.

[Cabal iii] *Cabal Seminar* 79–81, Lecture Notes in Mathematics, vol. 1019, Berlin, Springer-Verlag, 1983.

Alexander S. Kechris, Donald A. Martin, and John R. Steel

[Cabal iv] *Cabal Seminar* 81–85, Lecture Notes in Mathematics, vol. 1333, Berlin, Springer-Verlag, 1988.

Alexander S. Kechris and Yiannis N. Moschovakis

[KM72] *Two theorems about projective sets*, ***Israel Journal of Mathematics***, vol. 12 (1972), pp. 391–399.

[KM78A] *Appendix. The Victoria Delfino problems*, in *Cabal Seminar* 76–77 [Cabal i], pp. 279–282.

[KM78B] *Notes on the theory of scales*, in *Cabal Seminar* 76–77 [Cabal i], pp. 1–53, reprinted in [Cabal I], pp. 28–74.

[Cabal i] *Cabal Seminar* 76–77, Lecture Notes in Mathematics, vol. 689, Berlin, Springer-Verlag, 1978.

Alexander S. Kechris and Robert M. Solovay

[KS85] *On the relative consistency strength of determinacy hypotheses*, ***Transactions of the American Mathematical Society***, vol. 290 (1985), no. 1, pp. 179–211.

Alexander S. Kechris, Robert M. Solovay, and John R. Steel

[KSS81] *The axiom of determinacy and the prewellordering property*, in Kechris et al. [Cabal ii], pp. 101–125, reprinted in [Cabal II], pp. 118–140.

Alexander S. Kechris and W. Hugh Woodin

[KW83] *Equivalence of partition properties and determinacy*, ***Proceedings of the National Academy of Sciences of the United States of America***, vol. 80 (1983), no. 6 i., pp. 1783–1786.

Stephen C. Kleene

[Kle38] *On notation for ordinal numbers*, ***The Journal of Symbolic Logic***, vol. 3 (1938), pp. 150–155.

[Kle43] *Recursive predicates and quantifiers*, ***Transactions of the American Mathematical Society***, vol. 53 (1943), pp. 41–73.

[Kle55A] *Arithmetical predicates and function quantifiers*, ***Transactions of the American Mathematical Society***, vol. 79 (1955), pp. 312–340.

[Kle55B] *Hierarchies of number-theoretic predicates*, ***Bulletin of the American Mathematical Society***, vol. 61 (1955), pp. 193–213.

[Kle55C] *On the forms of the predicates in the theory of constructive ordinals. II*, ***American Journal of Mathematics***, vol. 77 (1955), pp. 405–428.

Eugene M. Kleinberg

[Kle70] *Strong partition properties for infinite cardinals*, ***The Journal of Symbolic Logic***, vol. 35 (1970), pp. 410–428.

Peter Koellner and W. Hugh Woodin

[KW10] *Large cardinals from determinacy*, in Kanamori and Foreman [KF10], pp. 1951–2119.

[KW] *Foundations of set theory: The search for new axioms*, in preparation.

Motokiti Kondô

[Kon38] *Sur l'uniformization des complementaires analytiques et les ensembles projectifs de la seconde classe*, ***Japanese Journal of Mathematics***, vol. 15 (1938), pp. 197–230.

Dénes Kőnig

[Kőn27] *Über eine Schlussweise aus dem Endlichen ins Unendliche*, ***Acta Scientiarum Mathematicarum (Szeged)***, vol. 3 (1927), no. 2–3, pp. 121–130.

Kenneth Kunen

[Kun70] *Some applications of iterated ultrapowers in set theory*, ***Annals of Mathematical Logic***, vol. 1 (1970), pp. 179–227.

[Kun71A] *Elementary embeddings and infinitary combinatorics*, ***The Journal of Symbolic Logic***, vol. 36 (1971), pp. 407–413.

[Kun71B] *A remark on Moschovakis' uniformization theorem*, circulated note, March 1971.

[Kun71C] *Some singular cardinals*, circulated note, September 1971.

[Kun71D] *Some more singular cardinals*, circulated note, September 1971.

[Kun78] *Saturated ideals*, ***The Journal of Symbolic Logic***, vol. 43 (1978), no. 1, pp. 65–76.

[Kun11] ***Set Theory***, Studies in Logic: Mathematical Logic and Foundations, vol. 34, College Publications, 2011.

KAZIMIERZ KURATOWSKI

[Kur36] *Sur les théorèmes de séparation dans las théorie des ensembles*, ***Fundamenta Mathematicae***, vol. 26 (1936), pp. 183–191.

PAUL B. LARSON

[Lar04] ***The Stationary Tower: Notes on a Course by W. Hugh Woodin***, University Lecture Series, vol. 32, American Mathematical Society, Providence, RI, 2004.

[Lar05] *The canonical function game*, ***Archive for Mathematical Logic***, vol. 44 (2005), no. 7, pp. 817–827.

[Lar12] *A brief history of determinacy*, ***Sets and Extensions in the Twentieth Century*** (Dov M. Gabbay, Akihiro Kanamori, and John Woods, editors), Handbook of the History of Logic, vol. 6, Elsevier, 2012, pp. 457–507.

HENRI LEBESGUE

[Leb05] *Sur les fonctions représentables analytiquement*, ***Journal de Mathématiques Pures et Appliquées***, vol. 1 (1905), pp. 139–216.

[Leb18] *Remarques sur les théories de le mesure et de l'intégration*, ***Annales de l'École Normale supérieure***, vol. 35 (1918), pp. 191–250.

AZRIEL LÉVY

[Lév57] *Indépendance conditionnelle de* **V**=**L** *et d'axiomes qui se rattachent au système de M. Gödel*, ***Comptes rendus hebdomadaires des séances de l'Académie des Sciences***, vol. 245 (1957), pp. 1582–1583.

[Lév60] *A generalization of Gödel's notion of constructibility*, ***The Journal of Symbolic Logic***, vol. 25 (1960), pp. 147–155.

[Lév65A] *Definability in axiomatic set theory. I*, ***Logic, Methodology and Philosophy of Science. Proceedings of the 1964 International Congress*** (Yehoshua Bar-Hillel, editor), Studies in Logic and the Foundations of Mathematics, North-Holland, 1965, pp. 127–151.

[Lév65B] *A hierarchy of formulas in set theory*, ***Memoirs of the American Mathematical Society***, vol. 57 (1965), p. 76.

[Lév79] ***Basic Set Theory***, Springer-Verlag, Berlin, 1979.

ALAIN LOUVEAU AND JEAN SAINT-RAYMOND

[LSR87] *Borel classes and closed games: Wadge-type and Hurewicz-type results*, ***Transactions of the American Mathematical Society***, vol. 304 (1987), no. 2, pp. 431–467.

[LSR88] *The strength of Borel Wadge determinacy*, in Kechris et al. [CABAL iv], pp. 1–30, reprinted in [CABAL II], pp. 74–101.

NIKOLAI LUZIN

[Luz25A] *Les proprietes des ensembles projectifs*, ***Comptes rendus hebdomadaires des séances de l'Académie des Sciences***, vol. 180 (1925), pp. 1817–1819.

[Luz25B] *Sur les ensembles projectifs de M. Henri Lebesgue*, ***Comptes rendus hebdomadaires des séances de l'Académie des Sciences***, vol. 180 (1925), pp. 1318–1320.

[Luz25C] *Sur un problème de M. Emil Borel et les ensembles projectifs de M. Henri Lebesgue: les ensembles analytiques*, ***Comptes rendus hebdomadaires des séances de l'Académie des Sciences***, vol. 164 (1925), pp. 91–94.

[Luz27] *Sur les ensembles analytiques*, ***Fundamenta Mathematicae***, vol. 10 (1927), pp. 1–95.

[Luz30A] *Analogies entre les ensembles mesurables B et les ensembles analytiques*, ***Fundamenta Mathematicae***, vol. 16 (1930), pp. 48–76.

[Luz30B] *Sur le problème de M. J. Hadamard d'uniformisation des ensembles*, ***Comptes rendus hebdomadaires des séances de l'Académie des Sciences***, vol. 190 (1930), pp. 349–351.

NIKOLAI LUZIN AND PETR NOVIKOV

[LN35] *Choix effectif d'un point dans un complemetaire analytique arbitraire, donne par un crible*, ***Fundamenta Mathematicae***, vol. 25 (1935), pp. 559–560.

NIKOLAI LUZIN AND WACŁAW SIERPIŃSKI

[LS18] *Sur quelques propriétés des ensembles* (A), ***Bulletin de l'Académie des Sciences Cracovie, Classe des Sciences Mathématiques, Série A***, (1918), pp. 35–48.

[LS23] *Sur un ensemble non measurable B*, ***Journal de Mathématiques Pures et Appliqueées***, vol. 2 (1923), no. 9, pp. 53–72.

MENACHEM MAGIDOR

[Mag80] *Precipitous ideals and* $\mathbf{\Sigma}^1_4$ *sets*, ***Israel Journal of Mathematics***, vol. 35 (1980), no. 1-2, pp. 109–134.

RICHARD MANSFIELD

[Man70] *Perfect subsets of definable sets of real numbers*, ***Pacific Journal of Mathematics***, vol. 35 (1970), no. 2, pp. 451–457.

[Man71] *A Souslin operation on* Π^1_2, ***Israel Journal of Mathematics***, vol. 9 (1971), no. 3, pp. 367–379.

DONALD A. MARTIN

[Mar68] *The axiom of determinateness and reduction principles in the analytical hierarchy*, ***Bulletin of the American Mathematical Society***, vol. 74 (1968), pp. 687–689.

[Mar70] *Measurable cardinals and analytic games*, ***Fundamenta Mathematicae***, vol. 66 (1970), pp. 287–291.

[Mar75] *Borel determinacy*, ***Annals of Mathematics***, vol. 102 (1975), no. 2, pp. 363–371.

[Mar80] *Infinite games*, ***Proceedings of the International Congress of Mathematicatians, Helsinki 1978*** (Olli Lehto, editor), Academia Scientiarum Fennica, 1980, pp. 269–273.

[Mar83] *The real game quantifier propagates scales*, in Kechris et al. [CABAL iii], pp. 157–171, reprinted in [CABAL I], pp. 209–222.

[Mar85] *A purely inductive proof of Borel determinacy*, ***Recursion Theory***, Proceedings of Symposia in Pure Mathematics, vol. 42, AMS, Providence, RI, 1985, pp. 303–308.

[Mar90] *An extension of Borel determinacy*, ***Annals of Pure and Applied Logic***, vol. 49 (1990), no. 3, pp. 279–293.

[Mar98] *The determinacy of Blackwell games*, ***The Journal of Symbolic Logic***, vol. 63 (1998), no. 4, pp. 1565–1581.

[Mar03] *A simple proof that determinacy implies Lebesgue measurability*, ***Università e Politecnico di Torino. Seminario Matematico. Rendiconti***, vol. 61 (2003), no. 4, pp. 393–397.

[Mar20] *Games of countable length*, 2020, this volume.

DONALD A. MARTIN, YIANNIS N. MOSCHOVAKIS, AND JOHN R. STEEL

[MMS82] *The extent of definable scales*, ***Bulletin of the American Mathematical Society***, vol. 6 (1982), pp. 435–440.

DONALD A. MARTIN, ITAY NEEMAN, AND MARCO VERVOORT

[MNV03] *The strength of Blackwell determinacy*, ***The Journal of Symbolic Logic***, vol. 68 (2003), no. 2, pp. 615–636.

DONALD A. MARTIN AND JEFF B. PARIS

[MP71] $\mathsf{AD} \Rightarrow \exists$ *exactly 2 normal measures on* ω_2, circulated note, March 1971.

DONALD A. MARTIN AND ROBERT M. SOLOVAY

[MS69] *A basis theorem for* Σ^1_3 *sets of reals*, ***Annals of Mathematics***, vol. 89 (1969), pp. 138–160.

DONALD A. MARTIN AND JOHN R. STEEL

[MS83] *The extent of scales in* $\mathbf{L}(\mathbb{R})$, in Kechris et al. [CABAL iii], pp. 86–96, reprinted in [CABAL I], pp. 110–120.

[MS89] *A proof of projective determinacy*, ***Journal of the American Mathematical Society***, vol. 2 (1989), no. 1, pp. 71–125.

[MaS94] *Iteration trees*, ***Journal of the American Mathematical Society***, vol. 7 (1994), no. 1, pp. 1–73.

A. R. D. MATHIAS

[Mat68] *On a generalization of Ramsey's theorem*, ***Notices of the American Mathematical Society***, vol. 15 (1968), p. 931.

[Mat77] *Happy families*, ***Annals of Mathematical Logic***, vol. 12 (1977), no. 1, pp. 59–111.

R. DANIEL MAULDIN

[Mau81] ***The Scottish Book: Mathematics from the Scottish Café***, Birkhäuser, Boston, MA, 1981.

WILLIAM J. MITCHELL

[Mit79] *Hypermeasurable cardinals*, ***Logic Colloquium '78. Proceedings of the Colloquium held in Mons, August 24–September 1, 1978*** (Maurice Boffa, Dirk van Dalen, and Kenneth McAloon, editors), Studies in Logic and the Foundations of Mathematics, vol. 97, North-Holland, Amsterdam, 1979, pp. 303–316.

YIANNIS N. MOSCHOVAKIS

[Mos67] *Hyperanalytic predicates*, ***Transactions of the American Mathematical Society***, vol. 129 (1967), pp. 249–282.

[Mos69A] *Abstract first order computability I*, ***Transactions of the American Mathematical Society***, vol. 138 (1969), pp. 427–463.

[Mos69B] *Abstract first order computability II*, ***Transactions of the American Mathematical Society***, vol. 138 (1969), pp. 464–504.

[Mos70] *Determinacy and prewellorderings of the continuum*, ***Mathematical Logic and Foundations of Set Theory. Proceedings of an international colloquium held under the auspices of the Israel Academy of Sciences and Humanities, Jerusalem, 11–14 November 1968*** (Y. Bar-Hillel, editor), Studies in Logic and the Foundations of Mathematics, North-Holland, Amsterdam-London, 1970, pp. 24–62.

[Mos71] *Uniformization in a playful universe*, ***Bulletin of the American Mathematical Society***, vol. 77 (1971), pp. 731–736.

[Mos73] *Analytical definability in a playful universe*, ***Logic, Methodology, and Philosophy of Science IV. Proceedings of the Fourth International Congress for Logic, Methodology and Philosophy of Science, Bucharest, 29 August–4 September, 1971*** (Patrick Suppes, Leon Henkin, Athanase Joja, and Gr. C. Moisil, editors), North-Holland, 1973, pp. 77–83.

[Mos78] *Inductive scales on inductive sets*, in Kechris and Moschovakis [CABAL i], pp. 185–192, reprinted in [CABAL I], pp. 94–101.

[Mos80] ***Descriptive Set Theory***, Studies in Logic and the Foundations of Mathematics, vol. 100, North-Holland, Amsterdam, 1980.

[Mos81] *Ordinal games and playful models*, in Kechris et al. [CABAL ii], pp. 169–201, reprinted in [CABAL III], pp. 86–114.

[Mos83] *Scales on coinductive sets*, in Kechris et al. [CABAL iii], pp. 77–85, reprinted in [CABAL I], pp. 102–109.

[Mos09] ***Descriptive Set Theory***, second ed., Mathematical Surveys and Monographs, vol. 155, American Mathematical Society, 2009.

JAN MYCIELSKI

[Myc64] *On the axiom of determinateness*, ***Fundamenta Mathematicae***, vol. 53 (1964), pp. 205–224.

[Myc66] *On the axiom of determinateness. II*, ***Fundamenta Mathematicae***, vol. 59 (1966), pp. 203–212.

JAN MYCIELSKI AND HUGO STEINHAUS

[MS62] *A mathematical axiom contradicting the axiom of choice*, ***Bulletin de l'Académie Polonaise des Sciences***, vol. 10 (1962), pp. 1–3.

JAN MYCIELSKI AND STANISŁAW ŚWIERCZKOWSKI

[MŚ64] *On the Lebesgue measurability and the axiom of determinateness*, ***Fundamenta Mathematicae***, vol. 54 (1964), pp. 67–71.

ITAY NEEMAN

[Nee95] *Optimal proofs of determinacy*, ***The Bulletin of Symbolic Logic***, vol. 1 (1995), no. 3, pp. 327–339.

[Nee00] *Unraveling* $\mathbf{\Pi}^1_1$ *sets*, ***Annals of Pure and Applied Logic***, vol. 106 (2000), no. 1-3, pp. 151–205.

[Nee02A] *Inner models in the region of a Woodin limit of Woodin cardinals*, ***Annals of Pure and Applied Logic***, vol. 116 (2002), no. 1-3, pp. 67–155.

[Nee04] ***The Determinacy of Long Games***, de Gruyter Series in Logic and its Applications, vol. 7, Walter de Gruyter, Berlin, 2004.

[Nee05] *An introduction to proofs of determinacy of long games*, ***Logic Colloquium '01. Proceedings of the Annual European Summer Meeting of the Association for Symbolic Logic held in Vienna, August 6–11, 2001*** (Matthias Baaz, Sy-David Friedman, and Jan Krajíček, editors), Lecture Notes in Logic, vol. 20, Association for Symbolic Logic, 2005, pp. 43–86.

[Nee06A] *Determinacy for games ending at the first admissible relative to the play*, ***The Journal of Symbolic Logic***, vol. 71 (2006), no. 2, pp. 425–459.

[Nee06B] *Unraveling* $\mathbf{\Pi}^1_1$ *sets, revisited*, ***Israel Journal of Mathematics***, vol. 152 (2006), pp. 181–203.

[Nee07A] *Games of length* ω_1, ***Journal of Mathematical Logic***, vol. 7 (2007), no. 1, pp. 83–124.

[Nee07B] *Inner models and ultrafilters in* $\mathbf{L}(\mathbb{R})$, ***The Bulletin of Symbolic Logic***, vol. 13 (2007), no. 1, pp. 31–53.

[Nee10] *Determinacy in* $\mathbf{L}(\mathbb{R})$, in Kanamori and Foreman [KF10], pp. 1877–1950.

PETR NOVIKOV

[Nov35] *Sur la séparabilité des ensembles projectifs de seconde class*, ***Fundamenta Mathematicae***, vol. 25 (1935), pp. 459–466.

JOHN C. OXTOBY

[Oxt80] ***Measure and Category***, second ed., Graduate Texts in Mathematics, vol. 2, Springer-Verlag, New York, 1980.

JEFF B. PARIS

[Par72] $\mathsf{ZF} \vdash \Sigma^0_4$ *determinateness*, ***The Journal of Symbolic Logic***, vol. 37 (1972), pp. 661–667.

KAREL PŘÍKRÝ

[Pří76] *Determinateness and partitions*, ***Proceedings of the American Mathematical Society***, vol. 54 (1976), pp. 303–306.

FRANK RAMSEY

[Ram30] *On a problem of formal logic*, ***Proceedings of the London Mathematical Society***, vol. 30 (1930), no. 2, pp. 2–24.

GERALD E. SACKS

[Sac76] *Countable admissible ordinals and hyperdegrees*, ***Advances in Mathematics***, vol. 20 (1976), no. 2, pp. 213–262.

ERNEST SCHIMMERLING

[Sch95] *Combinatorial principles in the core model for one Woodin cardinal*, ***Annals of Pure and Applied Logic***, vol. 74 (1995), no. 2, pp. 153–201.

[Sch07] *Coherent sequences and threads*, ***Advances in Mathematics***, vol. 216 (2007), no. 1, pp. 89–117.

[Sch10] *A core model toolbox and guide*, in Kanamori and Foreman [KF10], pp. 1685–1752.

ERNEST SCHIMMERLING AND MARTIN ZEMAN

[SZ01] *Square in core models*, ***The Bulletin of Symbolic Logic***, vol. 7 (2001), no. 3, pp. 305–314.

ULRICH SCHWALBE AND PAUL WALKER

[SW01] *Zermelo and the early history of game theory*, ***Games and Economic Behavior***, vol. 34 (2001), no. 1, pp. 123–137.

SAHARON SHELAH

[She84] *Can you take Solovay's inaccessible away?*, ***Israel Journal of Mathematics***, vol. 48 (1984), no. 1, pp. 1–47.

[She98] ***Proper and Improper Forcing***, second ed., Perspectives in Mathematical Logic, Springer-Verlag, Berlin, 1998.

SAHARON SHELAH AND W. HUGH WOODIN

[SW90] *Large cardinals imply that every reasonably definable set of reals is Lebesgue measurable*, ***Israel Journal of Mathematics***, vol. 70 (1990), no. 3, pp. 381–394.

JOSEPH R. SHOENFIELD

[Sho61] *The problem of predicativity*, ***Essays on the Foundations of Mathematics. Dedicated to A. A. Fraenkel on his Seventieth Anniversary*** (Y. Bar-Hillel, E. I. J. Poznanski, M. O. Rabin, and A. Robinson, editors), Magnes Press, Jerusalem, 1961, pp. 132–139.

WACŁAW SIERPIŃSKI

[Sie24] *Sur une propriété des ensembles ambigus*, ***Fundamenta Mathematicae***, vol. 6 (1924), pp. 1–5.

[Sie25] *Sur une class d'ensembles*, ***Fundamenta Mathematicae***, vol. 7 (1925), pp. 237–243.

[Sie38] *Fonctions additives non complètement additives et fonctions non mesurables*, ***Fundamenta Mathematicae***, vol. 30 (1938), pp. 96–99.

JACK H. SILVER

[Sil71] *Some applications of model theory in set theory*, ***Annals of Mathematical Logic***, vol. 3 (1971), no. 1, pp. 45–110.

[Sil75] *On the singular cardinals problem*, ***Proceedings of the International Congress of Mathematicians (Vancouver, B.C., 1974), Vol. 1***, Canadian Mathematical Congress, 1975, pp. 265–268.

ROBERT I. SOARE

[Soa87] ***Recursively Enumerable Sets and Degrees***, Perspectives in Mathematical Logic, Springer-Verlag, Berlin, 1987.

ROBERT M. SOLOVAY

[Sol66] *On the cardinality of Σ^1_2 set of reals*, ***Foundations of Mathematics: Symposium papers commemorating the 60th birthday of Kurt Gödel*** (Jack J. Bulloff, Thomas C. Holyoke, and S. W. Hahn, editors), Springer-Verlag, 1966, pp. 58–73.

[Sol67A] *Measurable cardinals and the axiom of determinateness*, lecture notes prepared in connection with the Summer Institute of Axiomatic Set Theory held at UCLA, Summer 1967.

[Sol70] *A model of set-theory in which every set of reals is Lebesgue measurable*, ***Annals of Mathematics***, vol. 92 (1970), pp. 1–56.

[Sol78A] *A Δ^1_3 coding of the subsets of ω_ω*, in Kechris and Moschovakis [CABAL i], pp. 133–150, reprinted in [CABAL II], pp. 346–363.

[Sol78B] *The independence of* DC *from* AD, in Kechris and Moschovakis [CABAL i], pp. 171–184, reprinted in this volume.

JOHN R. STEEL

[Ste81B] *Determinateness and the separation property*, ***The Journal of Symbolic Logic***, vol. 46 (1981), no. 1, pp. 41–44.

[Ste82B] *Determinacy in the Mitchell models*, ***Annals of Mathematical Logic***, vol. 22 (1982), no. 2, pp. 109–125.

[Ste83A] *Scales in* $\mathbf{L}(\mathbb{R})$, in Kechris et al. [CABAL iii], pp. 107–156, reprinted in [CABAL I], pp. 130–175.

[Ste88] *Long games*, in Kechris et al. [CABAL iv], pp. 56–97, reprinted in [CABAL I], pp. 223–259.

[Ste95A] $\mathbf{HOD}^{\mathbf{L}(\mathbb{R})}$ *is a core model below* Θ, ***The Bulletin of Symbolic Logic***, vol. 1 (1995), no. 1, pp. 75–84.

[Ste96] ***The Core Model Iterability Problem***, Lecture Notes in Logic, no. 8, Springer-Verlag, Berlin, 1996.

[Ste02] *Core models with more Woodin cardinals*, ***The Journal of Symbolic Logic***, vol. 67 (2002), no. 3, pp. 1197–1226.

[Ste05] PFA *implies* $\mathsf{AD}^{\mathbf{L}(\mathbb{R})}$, ***The Journal of Symbolic Logic***, vol. 70 (2005), no. 4, pp. 1255–1296.

[Ste08B] *Games and scales. Introduction to Part I*, in Kechris et al. [CABAL I], pp. 3–27.

[Ste08C] *The length-*ω_1 *open game quantifier propagates scales*, in Kechris et al. [CABAL I], pp. 260–269.

[Ste09] *The derived model theorem*, ***Logic Colloquium '06. Proc. of the Annual European Conference on Logic of the Association for Symbolic Logic held at the Radboud University, Nijmegen, July 27–August 2, 2006*** (S. Barry Cooper, Herman Geuvers, Anand Pillay, and Jouko Väänänen, editors), Lecture Notes in Logic, vol. 19, Association for Symbolic Logic, 2009, pp. 280–327.

JOHN R. STEEL AND ROBERT VAN WESEP

[SVW82] *Two consequences of determinacy consistent with choice*, ***Transactions of the American Mathematical Society***, vol. 272 (1982), no. 1, pp. 67–85.

JOHN R. STEEL AND STUART ZOBLE

[SZ] *Determinacy from strong reflection*, in preparation.

MIKHAIL YA. SUSLIN

[Sus17] *Sur une définition des ensembles mesurables* B *sans nombres transfinis*, ***Comptes rendus hebdomadaires des séances de l'Académie des Sciences***, vol. 164 (1917), pp. 88–91.

STEVO TODORČEVIĆ

[Tod84] *A note on the proper forcing axiom*, in Baumgartner et al. [BMS84], pp. 209–218.

STANISŁAW ULAM

[Ula60] ***A Collection of Mathematical Problems***, Interscience Tracts in Pure and Applied Mathematics, vol. 8, Interscience Publishers, New York–London, 1960.

ROBERT VAN WESEP

[Van78A] *Separation principles and the axiom of determinateness*, ***The Journal of Symbolic Logic***, vol. 43 (1978), no. 1, pp. 77–81.

[Van78B] *Wadge degrees and descriptive set theory*, in Kechris and Moschovakis [CABAL i], pp. 151–170, reprinted in [CABAL II], pp. 24–42.

BOBAN VELIČKOVIĆ

[Vel92] *Forcing axioms and stationary sets*, ***Advances in Mathematics***, vol. 94 (1992), no. 2, pp. 256–284.

GIUSEPPE VITALI

[Vit05] *Sul problema della misura dei gruppi di punti di una retta*, ***Tipografia Gamberini e Parmeggiani***, (1905), pp. 231–235.

JOHN VON NEUMANN AND OSKAR MORGENSTERN

[vNM04] ***Theory of Games and Economic Behavior***, Princeton University Press, 2004, Reprint of the 1980 edition.

STAN WAGON

[Wag93] ***The Banach-Tarski paradox***, Cambridge University Press, 1993, corrected reprint of the 1985 original.

PHILIP WOLFE

[Wol55] *The strict determinateness of certain infinite games*, ***Pacific Journal of Mathematics***, vol. 5 (1955), pp. 841–847.

W. HUGH WOODIN

[Woo82] *On the consistency strength of projective uniformization*, ***Proceedings of the Herbrand Symposium. Logic Colloquium '81. Held in Marseille, July 16–24, 1981*** (Jacques Stern, editor), Studies in Logic and the Foundations of Mathematics, vol. 107, North-Holland, Amsterdam, 1982, pp. 365–384.

[Woo83] *Some consistency results in* ZFC *using* AD, in Kechris et al. [CABAL iii], pp. 172–198, reprinted in this volume.

[Woo86] *Aspects of determinacy*, ***Logic, Methodology and Philosophy of Science. VII. Proceedings of the Seventh International Congress held at the University of Salzburg, Salzburg, July 11–16, 1983*** (Ruth Barcan Marcus, Georg J. W. Dorn, and Paul Weingartner, editors), Studies in Logic and the Foundations of Mathematics, vol. 114, North-Holland, Amsterdam, 1986, pp. 171–181.

[Woo88] *Supercompact cardinals, sets of reals, and weakly homogeneous trees*, ***Proceedings of the National Academy of Sciences of the United States of America***, vol. 85 (1988), no. 18, pp. 6587–6591.

[Woo99] ***The Axiom of Determinacy, Forcing Axioms, and the Nonstationary Ideal***, de Gruyter Series in Logic and its Applications, vol. 1, Walter de Gruyter, Berlin, 1999.

ERNST ZERMELO

[Zer04] *Beweis, daß jede Menge wohlgeordnet werden kann*, ***Mathematische Annalen***, vol. 59 (1904), pp. 514–516.

[Zer13] *Über eine Anwendung der Mengenlehre auf die Theorie des Schachspiels*, ***Proceedings of the Fifth International Congress of Mathematicians***, vol. 2, 1913, pp. 501–504.

DEPARTMENT OF MATHEMATICS
MIAMI UNIVERSITY
OXFORD, OHIO 45056
UNITED STATES OF AMERICA
E-mail: larsonpb@muohio.edu

“AD PLUS UNIFORMIZATION” IS EQUIVALENT TO “HALF $\mathsf{AD}_{\mathbb{R}}$”

ALEXANDER S. KECHRIS

Let ω be the set of natural numbers and $\mathbb{R} = {}^{\omega}2$ be the set of “reals”. As usual AD is the assertion that every game on ω is determined and $\mathsf{AD}_{\mathbb{R}}$ is the assertion that every game on $\mathbb{R}$ is determined. Let us also consider an intermediate principle $\mathsf{AD}_{\mathbb{R}}^{1/2}$, which is the assertion that every game in which one of the players plays in ω and the other in $\mathbb{R}$ is determined. Such games appear often in applications instead of full games on $\mathbb{R}$, *cf.*, *e.g.*, [Kec88].

Clearly (in ZF), $\mathsf{AD}_{\mathbb{R}}$ implies $\mathsf{AD}_{\mathbb{R}}^{1/2}$ which in turn implies AD. One other immediate consequence of $\mathsf{AD}_{\mathbb{R}}^{1/2}$ is Unif, which is the assertion: If $S \subseteq \mathbb{R} \times \mathbb{R}$, there is $F : \mathbb{R} \to \mathbb{R}$ such that if there is some y with $S(x, y)$, then $S(x, F(x))$, *i.e.*, every relation on the reals can be uniformized. We prove here that conversely AD+Unif imply $\mathsf{AD}_{\mathbb{R}}^{1/2}$, *i.e.*, we have the following

THEOREM 1.1. Assume ZF+DC. Then

$$\mathsf{AD}+\mathsf{Unif} \iff \mathsf{AD}_{\mathbb{R}}^{1/2}.$$

The relationship between $\mathsf{AD}_{\mathbb{R}}$ and $\mathsf{AD}_{\mathbb{R}}^{1/2}$ has not been fully understood. Woodin (unpublished) has shown (using also results of Becker [Bec85]) that (over ZF+DC) $\mathsf{AD}_{\mathbb{R}}$ and $\mathsf{AD}_{\mathbb{R}}^{1/2}$ are equiconsistent, but it is not known if $\mathsf{AD}_{\mathbb{R}}^{1/2}$ implies $\mathsf{AD}_{\mathbb{R}}$ (over ZF+DC again).

The proof of the above theorem is “local”, so it yields results about various classes of games. For any pointclass Γ, let Γ-AD and Γ-$\mathsf{AD}_{\mathbb{R}}^{1/2}$ denote the determinacy of the corresponding games in Γ and let Γ-Unif be the assertion that any relation in Γ can be uniformized by a function (whose graph is) in Γ. Call a pointclass Γ *nice* if Γ contains all the Borel sets and is closed under Borel substitutions, conjunctions and disjunctions, existential quantification over $\mathbb{R}$, and negation. Then we have

COROLLARY 1.2 (of the proof). Assume ZF+DC, and let Γ be a nice pointclass. Then Γ-AD + Γ-Unif implies Γ-$\mathsf{AD}_{\mathbb{R}}^{1/2}$. (Of course we cannot invert the arrow here unless we assume that games $A \subseteq \mathbb{R} \times {}^{\omega}\mathbb{R}$ in Γ have also strategies in Γ.)

Research partially supported by NSF Grant DMS-8416349.

Large Cardinals, Determinacy, and Other Topics: The Cabal Seminar, Volume IV
Edited by A. S. Kechris, B. Löwe, J. R. Steel
Lecture Notes in Logic, 49

For example, let Γ be the class of projective sets. Then, since by Moschovakis' Theorem (cf. [Mos80]) Γ-AD $\Rightarrow$ Γ-Unif, using the usual notation PD and $\mathsf{PD}_{\mathbb{R}}^{1/2}$ in this case, we have:

$$\mathsf{ZF}{+}\mathsf{DC} \vdash \mathsf{PD} \iff \mathsf{PD}_{\mathbb{R}}^{1/2}.$$

Similarly for $\Gamma = (\underset{\sim}{\mathbf{\Delta}}{}_1^2)^{\mathbf{L}(\mathbb{R})}$ (in which case again Γ-AD $\Rightarrow$ Γ-Unif, by a theorem of Martin and Steel (cf. [MS83])), we have in ZF+DC

$$(\underset{\sim}{\mathbf{\Delta}}{}_1^2)^{\mathbf{L}(\mathbb{R})}\text{-}\mathsf{AD} \iff (\underset{\sim}{\mathbf{\Delta}}{}_1^2)^{\mathbf{L}(\mathbb{R})}\text{-}\mathsf{AD}_{\mathbb{R}}^{1/2}.$$

Recall that $(\underset{\sim}{\mathbf{\Delta}}{}_1^2)^{\mathbf{L}(\mathbb{R})}$-AD $\Leftrightarrow$ $\mathbf{L}(\mathbb{R})$-AD (cf. [KS85]), but of course $\mathbf{L}(\mathbb{R})$-Unif fails.

We now give the proof of the theorem. Assume AD+Unif. Let $A \subseteq \mathbb{R} \times {}^{\omega}\mathbb{R}$ be given and consider the corresponding game G_A:

$$\begin{array}{llllll} \mathrm{I} & \alpha(0) & & \alpha(1) & & \alpha \\ & & & & \cdots & \\ \mathrm{II} & & x_0 & & x_1 & \vec{x} \end{array}$$

$$\alpha(i) \in \omega, x_i \in \mathbb{R}; \text{ Player I wins iff } (a, \vec{x}) \in A.$$

We have to show that this game is determined.

The proof is based on a method of "countable approximations" of games with (some) moves in $\mathbb{R}$, which in different variations (depending on the choice of "approximation") is quite useful in several applications.

For each real y, consider the following "countable approximation" $G_A^{(y)}$ of G_A:

$$\begin{array}{llllll} \mathrm{I} & \alpha(0) & & \alpha(1)\ldots & & \alpha \\ \mathrm{II} & & x_0 & & x_1\ldots & \bar{x} \end{array}$$

$\alpha(i) \in \omega$; $x_i \in \mathbb{R}$, $x_i \leq_{\mathrm{T}} y$ (*i.e.*, x_i is recursive in y); player I wins iff $(\alpha, \bar{x}) \in A$.

This is basically a game on ω, so by AD it is determined for each $y \in \mathbb{R}$. If player II has a winning strategy in $G_A^{(y)}$ for some y, clearly player II has a winning strategy in G_A (and conversely). So assume player I has a winning strategy in $G_A^{(y)}$, for all $y \in \mathbb{R}$. We shall prove that player I has a winning strategy in G_A.

By Unif let F be a function that assigns to each real y a winning strategy $\tau_y \equiv F(y)$ for player I in $G_A^{(y)}$. (Thus $\tau_y \colon {}^{<\omega}\{x : x \leq_{\mathrm{T}} y\} \to \omega$.)

Let us recall here the following result of Martin, (a strengthening of) which will be crucial in our argument.

THEOREM 1.3 (Martin). Assume ZF+DC+AD. Let $A \subseteq \mathbb{R}$. If A is cofinal in the Turing degrees, *i.e.*, $\forall x \in \mathbb{R}\, \exists y \in A (x \leq_{\mathrm{T}} y)$, then A contains a pointed perfect set (*i.e.*, there is a perfect binary tree T such that $x \in [T] \Rightarrow T \leq_{\mathrm{T}} x$ and $[T] \subseteq A$).

PROOF. Consider the game

$$\begin{array}{lcccccc} \mathrm{I} & x(0) & & x(1) & & & \\ & & & & & \cdots & \\ \mathrm{II} & & y(0) & & y(1) & & \end{array}$$

$$x(i), y(i) \in \{0,1\}; \text{ player II wins iff } x \leq_{\mathrm{T}} y \ \&\ y \in A.$$

If player I had a winning strategy τ, player II could play any $y \in A$ with $\tau \leq_{\mathrm{T}} y$ to defeat this strategy. So player II must have a winning strategy σ. For each sequence of moves $u(0), \dots, u(n)$ of player I let $\sigma * u(0), \dots, \sigma * u(n)$ be the corresponding sequence of moves for player II following σ. Fix also a real $\hat{\sigma} \in 2^{\omega}$ with $\sigma \equiv_{\mathrm{T}} \hat{\sigma}$.

Define now the following perfect binary tree T_1:

Let $u_{\varnothing} = \varnothing$. Then let $u_{(0)}, u_{(1)}$ be the first two binary sequences whose even part agrees with $\hat{\sigma}$ (when $u \in {}^{<\omega}2 \cup {}^{\omega}2$ we say that the even part of u agrees with $x \in {}^{\omega}2$ if $u(2i) = x(i)$ for $2i < \mathrm{lh}(u)$), and are such that $\sigma * u_{(0)}, \sigma * u_{(1)}$ are incompatible. Such $u_{(0)}, u_{(1)}$ evidently exist since for $x \in {}^{\omega}2$, $x \leq_{\mathrm{T}} \sigma * x$, thus $x \mapsto \sigma * x$ is not constant on any subset of ${}^{\omega}2$ containing reals of arbitrary high Turing degree. Let then $u_{(00)}, u_{(01)}$ be the first two binary sequences extending $u_{(0)}$ whose even part agrees with $\hat{\sigma}$ and are such that $\sigma * u_{(00)}, \sigma * u_{(11)}$ are incompatible, and similarly define $u_{(10)}, u_{(11)}$, etc. Let T_1 consist of all the subsequences of some u_t, $t \in {}^{<\omega}2$.

Finally let

$$T = \{\sigma * u \ : \ u \in T_1\}.$$

Clearly T is a perfect binary tree, and $[T] \subseteq A$. Also $T \leq_{\mathrm{T}} \sigma$. Let now $y \in [T]$. Then $y = \sigma * x$, for some $x \in [T_1]$. So the even part of x agrees with $\hat{\sigma}$, thus $\sigma \leq_{\mathrm{T}} x$. But also (by the rules of the game) $x \leq_{\mathrm{T}} y$, thus $\sigma \leq_{\mathrm{T}} y$ and so $T \leq_{\mathrm{T}} y$, *i.e.*, T is pointed. ⊣ (Theorem 1.3)

From Martin's theorem we can obtain easily the following

LEMMA 1.4. *Assume* ZF+DC+AD. *Let T be a pointed perfect tree and let $F\colon [T] \to \omega$. Then for each real x there is pointed perfect tree T' with $T' \subseteq T$, $x \leq_{\mathrm{T}} T'$ and $F|[T']$ constant.*

PROOF. First note that for any $G\colon {}^{\omega}2 \to \omega$, there is a pointed perfect S with $G|[S]$ constant. This follows from the fact that by Turing Determinacy some $G^{-1}[\{n\}]$ must be unbounded and the preceding theorem.

Let now $h\colon {}^{\omega}2 \to [T]$ be the canonical homeomorphism and let $G = F \circ h\colon {}^{\omega}2 \to \omega$. Then let T_1 be a pointed perfect tree with $G|[T_1]$ constant. By an easy standard fact, find $T_1' \subseteq T_1$ a pointed perfect tree with $x \leq_{\mathrm{T}} T_1'$. Let now T' be the image of T_1' under h. If $z \in [T']$, then $h^{-1}(z) \in [T_1']$, so $T_1' \leq_{\mathrm{T}} h^{-1}(z)$, thus $T_1' \leq_{\mathrm{T}} z \oplus T$ (note that $h \leq_{\mathrm{T}} T$) and so $T' \leq_{\mathrm{T}} z \oplus T$. But $z \in [T]$, so $T \leq_{\mathrm{T}} z$, thus $T' \leq_{\mathrm{T}} z$, *i.e.*, T' is pointed. Moreover letting $z =$

(leftmost branch of T_1'), we see that $z \leq_T T_1' \leq_T z \oplus T \leq_T z \leq_T T'$. Finally F is clearly constant on $[T']$ and we are done. ⊣ (Lemma 1.4)

We are finally ready to define a strategy for player I in G_A:

Let $F_0 \colon \mathbb{R} \to \omega$ be given by $F_0(y) = \tau_y(\varnothing)$. Choose a pointed perfect tree T_0 such that $F_0 | [T_0]$ is fixed, say with value k_0. Player I plays k_0 as his first move in G_A. Assume now player II plays x_0. Find $T_0^{x_0} \subseteq T_0$ perfect pointed such that $x_0 \leq_T T_0^{x_0}$. Then consider the function $F_1^{x_0} \colon [T_0^{x_0}] \to \omega$ given by $F_1^{x_0}(y) = \tau_y((x_0))$. (Note that for $y \in [T_0^{x_0}]$ we have $x_0 \leq_T y$, thus $\tau_y((x_0))$ is defined.) By the preceding lemma, find $T_1^{x_0} \subseteq T_0^{x_0}$ a pointed perfect tree with $x_0 \leq_T T_1^{x_0}$ and $F_1 | [T_1^{x_0}]$ constant, say with value, k_1. Then player II plays next in G_A this k_1. (Note that by Unif we can choose $T_1^{x_0}$ for each x_0 as asserted.) Let next player II play x_1 in G_A. Find $T_1^{x_0,x_1} \subseteq T_1^{x_0}$ a pointed perfect tree with $x_0 \oplus x_1 \leq_T T_1^{x_0,x_1}$. Then consider the function $F_2^{x_0,x_1} \colon [T_1^{x_0,x_1}] \to \omega$ given by $F_2^{x_0,x_1}(y) = \tau_y((x_0, x_1))$. Again this is well defined, since for $y \in [T_1^{x_0,x_1}]$, $x_0, x_1 \leq_T y$. Then find $T_2^{x_0,x_1}$ a pointed perfect tree with $x_0 \oplus x_1 \leq_T T_2^{x_0,x_1}$ and $F_2 | [T_2^{x_0,x_1}]$ constant, say with value k_2. Player I plays next this k_2, etc.

Thus if player II plays successively $x_0, x_1, \ldots$, we have found pointed perfect sets

$$T_0 \supseteq T_1^{x_0} \supseteq T_2^{x_0,x_1} \supseteq T_3^{x_0,x_1,x_2} \supseteq \ldots,$$

with $x_0 \oplus x_1 \oplus \cdots \oplus x_{n-1} \leq_T T_n^{x_0,x_1,\ldots,x_{n-1}}$, and such that $\tau_y((x_0, \ldots, x_{n-1}))$ is fixed on $T_n^{x_0,x_1,\ldots,x_{n-1}}$, say with value k_n. Player I plays then $k_0, k_1, \ldots$. We claim that this is a winning strategy for player I in G_A, *i.e.*, letting $\vec{k} = (k_0, k_1, \ldots)$, $\vec{x} = (x_0, x_1, \ldots)$, we have $(\vec{k}, \vec{x}) \in A$. To see this note that $\bigcap_n [T_n^{x_0,x_1,\ldots,x_{n-1}}] \neq \varnothing$ and pick $y \in \bigcap_n [T_n^{x_0,x_1,\ldots,x_{n-1}}]$. Then for each n, $x_0 \oplus \cdots \oplus x_{n-1} \leq_T y$, *i.e.*, $x_n \leq_T y$ for each n. Moreover $k_n = \tau_y((x_0, \ldots, x_{n-1}))$, *i.e.*, this is a run of the game $G_A^{(y)}$ in which player I followed τ_y, thus $(\vec{k}, \vec{x}) \in A$ and we are done.

One final remark: One can easily obtain from the preceding proof definability estimates for $\mathsf{AD}_{\mathbb{R}}^{1/2}$ games in terms of definability estimates for uniformization. For example, assuming PD and letting $A \subseteq \mathbb{R} \times {}^{\omega}\mathbb{R}$ be a projective game in which player I has a winning strategy, we have that player I has a projective winning strategy. (A strategy for player I in a map $F \colon {}^{<\omega}\mathbb{R} \to \omega$, *i.e.*, essentially a set of reals.) Of course finer level-by-level estimates and lightface versions can be also easily extracted but we shall not spell them out here.

REFERENCES

HOWARD S. BECKER

[Bec85] *A property equivalent to the existence of scales*, ***Transactions of the American Mathematical Society***, vol. 287 (1985), no. 2, pp. 591–612.

ALEXANDER S. KECHRIS
[Kec88] *A coding theorem for measures*, in Kechris et al. [CABAL iv], pp. 103–109, reprinted in [CABAL I], pp. 398–403.

ALEXANDER S. KECHRIS, BENEDIKT LÖWE, AND JOHN R. STEEL
[CABAL I] *Games, Scales, and Suslin cardinals: the Cabal Seminar, volume I*, Lecture Notes in Logic, vol. 31, Cambridge University Press, 2008.

ALEXANDER S. KECHRIS, DONALD A. MARTIN, AND YIANNIS N. MOSCHOVAKIS
[CABAL iii] *Cabal Seminar* 79–81, Lecture Notes in Mathematics, vol. 1019, Berlin, Springer-Verlag, 1983.

ALEXANDER S. KECHRIS, DONALD A. MARTIN, AND JOHN R. STEEL
[CABAL iv] *Cabal Seminar* 81–85, Lecture Notes in Mathematics, vol. 1333, Berlin, Springer-Verlag, 1988.

ALEXANDER S. KECHRIS AND ROBERT M. SOLOVAY
[KS85] *On the relative consistency strength of determinacy hypotheses*, ***Transactions of the American Mathematical Society***, vol. 290 (1985), no. 1, pp. 179–211.

DONALD A. MARTIN AND JOHN R. STEEL
[MS83] *The extent of scales in* $\mathbf{L}(\mathbb{R})$, in Kechris et al. [CABAL iii], pp. 86–96, reprinted in [CABAL I], pp. 110–120.

YIANNIS N. MOSCHOVAKIS
[Mos80] ***Descriptive Set Theory***, Studies in Logic and the Foundations of Mathematics, vol. 100, North-Holland, Amsterdam, 1980.

DEPARTMENT OF MATHEMATICS
CALIFORNIA INSTITUTE OF TECHNOLOGY
PASADENA, CALIFORNIA 91125
UNITED STATES OF AMERICA
E-mail: kechris@caltech.edu

THE INDEPENDENCE OF DC FROM AD

ROBERT M. SOLOVAY

Introduction. We essentially prove, among other things, that the axiom of dependent choice (DC) does not follow from the axiom of determinacy (AD). The assumption that ZF+$AD_{\mathbb{R}}$ is consistent is required for this theorem, where $AD_{\mathbb{R}}$ is the axiom of determinacy for reals. In fact, the actual theorem proved is "Con(ZF+$AD_{\mathbb{R}}$) implies Con(ZF+$AD_{\mathbb{R}}$+$\neg$DC)". This is obtained by demonstrating that ZF+$AD_{\mathbb{R}}$+DC proves Con(ZF+$AD_{\mathbb{R}}$), and then quoting the second Gödel incompleteness theorem.

The consistency of ZF+$AD_{\mathbb{R}}$ is proved by using $AD_{\mathbb{R}}$ to get indiscernibles for certain models of set theory, and then using $AD_{\mathbb{R}}$+DC to show that some of these models satisfy $AD_{\mathbb{R}}$. Using the indiscernibles we can define the satisfaction relation of the models, hence proving the consistency of $AD_{\mathbb{R}}$. Other theorems along this line are proved similarly.

One of the main facts needed to prove these theorems is the close connection between the amount of choice available and the cofinality of Θ, where Θ is the least ordinal which is not the surjective image of $\mathbb{R}$. Indeed, to obtain inner models of $AD_{\mathbb{R}}$ thin enough to have a definable satisfaction relation we shall need that the cofinality of Θ is $> \omega$. The necessary results in this direction are proved in §3.

This line of research was suggested by work of Ramez Sami directed towards proving DC from AD + $\mathbf{V}$=$\mathbf{L}(\mathbb{R})$. In particular, the proof of Lemma 2.8 is modeled after arguments of Sami.

§1. Preliminaries. $\mathbb{R}$ is the set of reals. For the purposes of this paper, "reals" are elements of ${}^{\omega}\omega$. We fix some reasonable encoding of pairs of reals as reals. If x and y are reals, we frequently write $\langle x, y\rangle$ to refer to the real coding the ordered pair of x and y rather than to the ordered pair itself.

An **inner model** of ZF is a transitive class model of ZF which contains all ordinals. By $\mathbf{L}(\mathbb{R})$, we denote the smallest inner model of ZF that contains

---

The results in this paper were sketched in two lectures to the Caltech-UCLA logic seminar, and then presented in more detail to Greg Ennis, whose notes formed the basis for the final form of the presentation.

Large Cardinals, Determinacy, and Other Topics: The Cabal Seminar, Volume IV
Edited by A. S. Kechris, B. Löwe, J. R. Steel
Lecture Notes in Logic, 49

$\mathbb{R}$ as an element. We write $\wp(\mathbb{R})$ for the power set of $\mathbb{R}$; then $\mathbf{L}(\wp(\mathbb{R}))$ is the smallest inner model of ZF that contains $\mathbb{R}$ and $\wp(\mathbb{R})$ as elements.

For $A \subseteq \mathbb{R}^\omega$, the game G_A is defined as follows: players I and II alternate turns picking reals, producing a sequence $\langle x_0, x_1, x_2, \ldots\rangle$. Player I wins if $\langle x_n : n \in \omega\rangle \in A$, player II wins otherwise. We say that G_A is **determined** if either player I or player II has a winning strategy. We write $\mathsf{AD}_\mathbb{R}$ for the axiom asserting that $\forall A \subseteq \mathbb{R}^\omega$ (G_A is determined).

Clearly $\mathsf{AD}_\mathbb{R}$ implies AD. It is of course inconsistent with full choice, but it does imply certain forms of restricted choice. In particular let $\mathsf{AC}_\mathbb{R}$ be the following:

for all sequences $\langle S_x : x \in \mathbb{R}\rangle$ of non-empty sets of reals there is an $F\colon \mathbb{R} \to \mathbb{R}$ such that for all x, we have $F(x) \in S_x$. $\quad (\mathsf{AC}_\mathbb{R})$

Then $\mathsf{AD}_\mathbb{R}$ implies $\mathsf{AC}_\mathbb{R}$. (Proof: Consider the two-step game in which player II wins if and only if $x_1 \in S_{x_0}$. Clearly player I has no winning strategy, so player II's winning strategy yields the desired choice function.)

Let X be any nonempty set. The axiom of dependent choice on X (DC_X) is as follows:

Let $\mathcal{S}$ be a collection of finite sequences from X with $\varnothing \in \mathcal{S}$ and without $\subseteq$-maximal element. Then there is an $f : \omega \to X$ such that for all n, $f{\restriction}n \in \mathcal{S}$. $\quad (\mathsf{DC}_X)$

We have $\mathsf{AC}_\mathbb{R} \implies \mathsf{DC}_\mathbb{R}$ and hence $\mathsf{AD}_\mathbb{R} \implies \mathsf{DC}_\mathbb{R}$.

The full axiom of dependent choice (DC) is the statement "DC_X holds for every non-empty set X".

For $A, B \subseteq \mathbb{R}$ we say A is **Wadge reducible** to B ($A \leq_\mathrm{W} B$) if there is a continuous $f\colon \mathbb{R} \to \mathbb{R}$ such that for all x, $x \in A \iff f(x) \in B$. We write $A \equiv_\mathrm{W} B$ if $A \leq_\mathrm{W} B$ and $B \leq_\mathrm{W} A$, and the Wadge degree of A is its Wadge equivalence class. The Wadge hierarchy is the collection of Wadge degrees with the induced partial order, and the **modified** Wadge hierarchy is obtained from the Wadge hierarchy by identifying the degree of A with the degree of $\mathbb{R} - A$. Wadge showed that AD+$\mathsf{DC}_\mathbb{R}$ implies the modified Wadge hierarchy is a linear ordering, and Martin showed that, in fact, it's well ordered. For these and other facts see [Van78B].

We use the symbol $\twoheadrightarrow$ to indicate an *onto* map.

Definition 1.1. We define $\Theta := \sup\{\xi \in \mathrm{Ord} : \exists f\colon \mathbb{R} \twoheadrightarrow \xi\}$. Under AC+GCH, $\Theta = \omega_2$. Under AD it is much larger. A major theme of this paper is that the cofinality of Θ is related to the amount of choice available. A simple but useful fact is the following:

Lemma 1.2 (AD+$\mathsf{DC}_\mathbb{R}$). *The ordinal Θ is the order type of the modified Wadge hierarchy.*

Proof. Let γ be the order type of the modified Wadge hierarchy. If A has Wadge degree $\eta < \gamma$, then we can define $h\colon \mathbb{R} \twoheadrightarrow \eta + 1$ by letting $h(x)$ be the

Wadge ordinal of the set Wadge reducible to A via the continuous function coded by x. Hence $\gamma \leq \Theta$.

To show $\Theta \leq \gamma$, we use the following fact: there is a uniform, canonical procedure which takes a set $A \subseteq \mathbb{R}$ to a set $A' \subseteq \mathbb{R}$ with A' of higher Wadge degree than A, $\mathbb{R} - A$. (Proof: let f_x be the continuous function coded by x and put $\langle 0, x\rangle \in A' \iff f_x(\langle 0, x\rangle) \notin A$, $\langle 1, x\rangle \in A' \iff f_x(\langle 1, x\rangle) \in A$.) Now if $0 < \lambda < \Theta$, let $\varphi\colon \mathbb{R} \twoheadrightarrow \lambda$. Define $\{A_\eta : \eta \leq \lambda\} \subseteq \wp(\mathbb{R})$ by induction on η: $A_\eta = \{\langle y, z\rangle : \varphi(y) < \eta \wedge z \in A_{\varphi(y)}\}'$. Then $\eta < \delta \leq \lambda \Longrightarrow A_\eta <_{\mathrm{W}} A_\delta$, hence $\lambda < \gamma$. ⊣

§2. Choice and the cofinality of Θ. Given $A \subseteq \mathbb{R}$, let $A_{x,y} = \{z \in \mathbb{R} : \langle x, y, z\rangle \in A\}$. **Collection** is the following choice-like axiom:

$$\forall x \in \mathbb{R}\, \exists A \subseteq \mathbb{R}\, (\langle x, A\rangle \in U) \Longrightarrow \exists B\, \forall x \in \mathbb{R}\, \exists y \in \mathbb{R}\, (\langle x, B_{x,y}\rangle \in U).$$

LEMMA 2.1. Collection implies Θ is regular.

PROOF. Assume, toward a contradiction, that Collection holds and that Θ is singular. Since Θ is singular, there is a cofinal function $G\colon \mathbb{R} \to \Theta$. Thus for every x there is an A such that A is a prewellordering of $\mathbb{R}$ of length $G(x)$. By collection, there is a $B \subseteq \mathbb{R}$ such that

$$\forall x\, \exists y \in \mathbb{R}\, (B_{x,y} \text{ is a prewellordering of } \mathbb{R} \text{ of length } G(x)).$$

Define $F : \mathbb{R}^3 \to \Theta$ by

$$F(x, y, z) = \begin{cases} \text{rank of } z \text{ in } B_{x,y} & \text{if } B_{x,y} \text{ is a pwo of } \mathbb{R} \\ 0 & \text{otherwise.} \end{cases}$$

Then F is onto, a contradiction. ⊣

THEOREM 2.2 (AD+DC$_\mathbb{R}$). Collection is equivalent to the regularity of Θ.

PROOF. Assume that Θ is regular. We show Collection holds.

Suppose $\forall x \in \mathbb{R}\, \exists A \subseteq \mathbb{R}\, (\langle x, A\rangle \in U)$ holds. Let $f(x)$ be the least ξ such that there is an $A \subseteq \mathbb{R}$ of Wadge degree ξ with $\langle x, A\rangle \in U$. Since Θ is regular the range of f is bounded by some $\gamma < \Theta$ (using, of course, Lemma 1.2). Let B be any set of Wadge degree γ. Define $G\colon \mathbb{R} \to \wp(\mathbb{R})$ by letting $G(x)$ be the set Wadge reducible to B via the continuous function coded by x. Let

$$\langle x, y, z\rangle \in B^* \iff z \in G(y).$$

Then $\forall x\, \exists A \in \operatorname{ran}(G)\, (\langle x, A\rangle \in U)$ implies $\forall x\, \exists y\, (\langle x, B^*_{x,y}\rangle \in U)$. ⊣

Thus under AD$_\mathbb{R}$, the regularity of Θ is equivalent to a choice-like axiom. We next want to show that under AD$_\mathbb{R}$ + **V**=**L**($\wp(\mathbb{R})$), DC is equivalent to $\operatorname{cf}(\Theta) > \omega$. One direction is easy and requires no strong hypotheses:

LEMMA 2.3. DC implies $\operatorname{cf}(\Theta) > \omega$.

PROOF. Assume DC and that $\mathrm{cf}(\Theta) = \omega$. We shall derive the absurd conclusion that there is a map of $\mathbb{R}$ onto Θ.

Suppose $F\colon \omega \to \Theta$ cofinally. Then for all n there is an S such that S is a prewellordering of $\mathbb{R}$ of length $F(n)$. By DC (in fact, by countable choice), let h be a function with domain ω such that n, $h(n)$ is a prewellordering of $\mathbb{R}$ of length $F(n)$). Now define $G\colon \mathbb{R} \twoheadrightarrow \Theta$ by letting $G(x)$ be the rank of $\lambda m.x(m+1)$ in the pwo $h(x(0))$ (using Church's lambda notation and thinking of $\mathbb{R}$ as ${}^{\omega}\omega$). ⊣

THEOREM 2.4. $\mathsf{AD} + \mathsf{DC}_{\mathbb{R}} + \mathbf{V}{=}\mathbf{L}(\wp(\mathbb{R})) + \mathrm{cf}(\Theta) > \omega \vdash \mathsf{DC}$.

PROOF. Throughout the rest of § 2, we assume AD, $\mathsf{DC}_{\mathbb{R}}$, $\mathbf{V}{=}\mathbf{L}(\wp(\mathbb{R}))$, $\mathrm{cf}(\Theta) > \omega$ and $\neg\mathsf{DC}$, heading for a contradiction.

LEMMA 2.5. $\neg\mathsf{DC}_{\wp(\mathbb{R})}$.

PROOF. This follows from the following facts whose proofs will be suppressed:

1. If there is some $h\colon X \twoheadrightarrow Y$, then DC_X implies DC_Y.
2. For all ordinals λ, DC_X implies $\mathsf{DC}_{X\times\lambda}$.
3. $\mathbf{V}{=}\mathbf{L}(\wp(\mathbb{R}))$ implies that for all nonempty X there is an ordinal λ and some $h\colon \wp(\mathbb{R}) \times \lambda \twoheadrightarrow X$. ⊣

LEMMA 2.6. Θ is singular.

PROOF. Fix a collection of tuples from $\wp(\mathbb{R})$ $\mathcal{S}_0$ which violates $\mathsf{DC}_{\wp(\mathbb{R})}$. Define $\mathcal{S} \subseteq \mathcal{S}_0$ as follows: $\varnothing \in \mathcal{S}$, and if $\langle X_1, \dots, X_n\rangle \in \mathcal{S}$, then

$$\langle X_1, \dots, X_n, Y\rangle \in \mathcal{S} \iff \text{a) } \langle X_1, \dots, X_n, Y\rangle \in \mathcal{S}_0 \text{ and}$$
$$\text{b) } |Y|_{\mathrm{W}} \text{ is minimal among things satisfying a).}$$

Clearly $\mathcal{S}$ is a counterexample to $\mathsf{DC}_{\wp(\mathbb{R})}$.

Now for $\langle X_1, \dots, X_n\rangle \in \mathcal{S}$ define $\gamma_0(X_1, \dots, X_n) = |X_1, \dots, X_n|_{\mathrm{W}}$, and for $m > 0$ define

$$\gamma_m(X_1, \dots, X_n) = \sup\{|X_1, \dots, X_n, Y_1, \dots, Y_m|_{\mathrm{W}} : \langle X_1, \dots, X_n, Y_1, \dots, Y_m\rangle \in \mathcal{S}\}.$$

CLAIM 2.7. There is an m such that $\gamma_m(\varnothing) = \Theta$.

PROOF OF CLAIM. Suppose that for all m, $\gamma_m(\varnothing) < \Theta$. Since $\mathrm{cf}(\Theta) > \omega$, $\exists\eta < \Theta$ such that $\sup\{\gamma_m(\varnothing) : m \in \omega\} < \eta$. Thus $\mathcal{S}$ is a collection of tuples from $\wp_\eta(\mathbb{R})$ where

$$\wp_\eta(\mathbb{R}) = \{A \subseteq \mathbb{R} : |A|_{\mathrm{W}} < \eta\}.$$

But since there is a map $\mathbb{R} \twoheadrightarrow \wp_\eta(\mathbb{R})$, we have that $\mathsf{DC}_{\mathbb{R}}$ implies $\mathsf{DC}_{\wp_\eta(\mathbb{R})}$, hence $\mathcal{S}$ cannot be a counterexample for $\mathsf{DC}_{\wp(\mathbb{R})}$. ⊣ (Claim)

Now let m be least such that there is a sequence $\langle X_1, \ldots, X_n \rangle \in \mathcal{S}$ with $\gamma_m(X_1, \ldots, X_n) = \Theta$. Fix such a $\langle X_1, \ldots, X_n \rangle$. Clearly $m \geq 1$. Let $Q = \{Y : \langle X_1, \ldots, X_n, Y \rangle \in \mathcal{S}\}$. For $Y \in Q$ define $\eta(Y) = \gamma_{m-1}(X_1, \ldots, X_n, Y)$. By minimality of m, $\eta(Y) < \Theta$ for all $Y \in Q$. But then $\sup\{\eta(Y) : Y \in Q\} = \gamma_m(X_1, \ldots, X_n) = \Theta$. Hence η is cofinal in Θ. But Q is a fixed subset of a fixed Wadge degree, hence there is a map $\mathbb{R} \twoheadrightarrow Q$. Hence Θ is singular. ⊣

LEMMA 2.8. *Let $\lambda = \mathrm{cf}(\Theta)$. There is a countably additive measure ν on λ such that for some ordinal β, the ultrapower β^λ/ν is not well founded.*

PROOF. By Lemma 2.6, there is a $g \colon \mathbb{R} \twoheadrightarrow \lambda$. Let $\mathcal{D}$ be the Turing degrees and μ the Martin measure on $\mathcal{D}$. Define $h \colon \mathcal{D} \to \lambda$ by

$$h(d) := \sup\{g(x) : x \leq_{\mathrm{T}} d\}.$$

Note that, by our standing assumptions, λ has cofinality $> \omega$. So the displayed supremum is indeed $< \lambda$.

Let ν be the measure on λ induced by h from μ. By $\mathsf{DC}_{\mathbb{R}}$, ν is countably additive. Let $\mathcal{S}$ be a collection of tuples from $\wp(\mathbb{R})$ violating $\mathsf{DC}_{\wp(\mathbb{R})}$. For $\xi < \Theta$, let $\mathcal{S}_\xi = \{\langle X_1, \ldots, X_n \rangle \in \mathcal{S} : \forall i \leq n\,(|X_i|_{\mathrm{W}} < \xi)\}$. Using $\mathsf{DC}_{\wp_\xi(\mathbb{R})}$ we obtain a rank function on $\mathcal{S}_\xi$ for $\xi < \Theta$ as follows: define an inductive operator Φ on $\mathcal{S}_\xi$ by

$$\langle X_1, \ldots, X_n \rangle \in \Phi(U) \iff \forall Y\,(\langle X_1, \ldots, X_n, Y \rangle \in \mathcal{S}_\xi \implies \langle X_1, \ldots, X_n, Y \rangle \in U).$$

Let Φ^∞ be the least fixed point of the operator Φ. Suppose $\mathcal{S}_\xi - \Phi^\infty \neq \varnothing$. Then

$$\forall \langle X_1, \ldots, X_n \rangle \in \mathcal{S}_\xi - \Phi^\infty\, \exists Y\,(\langle X_1, \ldots, X_n, Y \rangle \in \mathcal{S}_\xi - \Phi^\infty).$$

By $\mathsf{DC}_{\wp_\xi(\mathbb{R})}$, there is an f such that for all n, $f{\restriction}n \in \mathcal{S}_\xi - \Phi^\infty$. But then $\mathcal{S}$ does not violate $\mathsf{DC}_{\wp(\mathbb{R})}$.

Hence $\mathcal{S}_\xi \subseteq \Phi^\infty$ and we can define for $\langle X_1, \ldots, X_n \rangle \in \mathcal{S}_\xi$: $|X_1, \ldots, X_n|_\xi$ is the least η such that $\langle X_1, \ldots, X_n \rangle$ occurs in the ηth iterate of Φ. (This is of course the standard rank function on $\mathcal{S}_\xi$).

Let $\beta = \sup\{|X_1, \ldots, X_n|_\xi : \xi < \Theta, \langle X_1, \ldots, X_n \rangle \in \mathcal{S}_\xi\}$. Let $k \colon \lambda \to \Theta$ be cofinal and order preserving, and for $\langle X_1, \ldots, X_n \rangle \in \mathcal{S}$ define

$$|X_1, \ldots, X_n|_\nu = [\lambda\xi.|X_1, \ldots, X_n|_{k(\xi)}]_\nu \in \beta^\lambda/\nu.$$

Note that $|X_1, \ldots, X_n|_{k(\xi)}$ is actually only defined for ξ such that $k(\xi) > \sup\{|X_i|_{\mathrm{W}} : i \leq n\}$, but these ξ's form a set of ν-measure 1.

Clearly if $\langle X_1, \ldots, X_n, Y \rangle$, $\langle X_1, \ldots, X_n \rangle \in \mathcal{S}$ then $|X_1, \ldots, X_n, Y|_\nu < |X_1, \ldots, X_n|_\nu$. Hence $\{|X_1, \ldots, X_n|_\nu : \langle X_1, \ldots, X_n \rangle \in \mathcal{S}\}$ is a subset of β^λ/ν with no least element. ⊣

We can now complete the proof of Theorem 2.4. Fix $S \subseteq \beta^\lambda/\nu$, $S \neq \varnothing$ such that S has no least element. For $\xi < \Theta$, let S_ξ be those members of S which have representatives ordinal definable from a set of Wadge degree ξ and a real. By slightly modifying the definition of S, we may suppose that $S_0 \neq \varnothing$. (Add the equivalence class of a large constant function to S.)

CLAIM 2.9. For all $\xi < \Theta$, every nonempty $A \subseteq S_\xi$ has a least element.

PROOF OF CLAIM. Fix ξ, $A \subseteq S_\xi$. Let

$$B = \{h \in \beta^\lambda : h \text{ is ordinal definable from } C \text{ and a real, and } [h]_\nu \in A\},$$

where C is a set of Wadge degree ξ. There is a fixed function F such that every element of B is $F(\eta, C, x)$ for some ordinal η and some real x. As B is a set we need only a set of such η's, hence B is the surjective image of $\mathbb{R} \times \alpha$ for some ordinal α. Hence DC_B. If A has no least element, then for all $h \in B$ there is a $g \in B$ such that $[h]_\nu > [g]_\nu$, so by DC_B there is a function G with domain ω such that for all n, we have $[G(n)]_\nu > [G(n+1)]_\nu$ which violates the countable additivity of ν. ⊣ (Claim)

Let b_ξ be the least element of S_ξ. Suppose there is a $\xi_0 < \Theta$ such that for all $\eta < \Theta$, $b_{\xi_0} \leq b_\eta$. Since $\mathbf{V}=\mathbf{L}(\wp(\mathbb{R}))$, $S = \bigcup_{\eta<\Theta} S_\eta$. Hence b_{ξ_0} is the least element of S, contrary to our assumption on S.

Hence $\forall\xi < \Theta\, \exists\eta < \Theta\, (\eta > \xi$ and $b_\xi > b_\eta)$. Let $\xi_0 = 0$ and ξ_{n+1} be the least ordinal $> \xi_n$ such that $b_{\xi_{n+1}} < b_{\xi_n}$. Since $\mathrm{cf}(\Theta) > \omega$, $\gamma = \sup\{\xi_n : n < \omega\} < \Theta$. But then $\{b_{\xi_n} : n < \omega\}$ is a subset of S_γ with no least element. This contradiction completes the proof of Theorem 2.4. ⊣

§3. Inner models for $\mathsf{AD}_\mathbb{R}$. For $\xi < \Theta$, let $\wp_\xi(\mathbb{R}) = \{A \subseteq \mathbb{R} : |A|_W < \xi\}$. The main result of this section is the construction of a function $G : \Theta \to \Theta$ such that if $\xi < \Theta$ is a limit ordinal closed under G then $\mathbf{L}(\wp_\xi(\mathbb{R})) \models \mathsf{AD}_\mathbb{R}$ (assuming $\mathsf{AD}_\mathbb{R}$).

Recall that an inner model of ZF is a transitive class model of ZF containing all ordinals. $\mathbf{L}(\wp_\xi(\mathbb{R}))$ is the smallest inner model of ZF containing $\mathbb{R}$ and $\wp_\xi(\mathbb{R})$ as elements.

LEMMA 3.1 ($\mathsf{AD}_\mathbb{R}$). Let $A \subseteq \mathbb{R}$. Then there is a $B \subseteq \mathbb{R}$ such that for every game Wadge reducible to A there is a winning strategy for some player which is Wadge reducible to B.

PROOF. Consider the following game: Player I plays x_0 which codes a game G_{x_0} Wadge reducible to A. (*I.e.*, x_0 codes the continuous function which reduces the payoff set of G_{x_0} to A). Player II then chooses to be player I or player II in G_{x_0}. Then, player I, player II play G_{x_0}. Clearly player I has no winning strategy, since if he did, some G_{x_0} would have winning strategies for both players. So player II has a winning strategy, which can be coded as a $B \subseteq \mathbb{R}$. Clearly this B works. ⊣

LEMMA 3.2 ($\mathsf{AD}_{\mathbb{R}}$). For all $A \subseteq \mathbb{R}$ there is a $B \subseteq \mathbb{R}$ such that B is not ordinal definable from A and a real.

PROOF. Suppose A is such that for all $B \subseteq \mathbb{R}$, B is ordinal definable from A and a real. For each $x \in \mathbb{R}$, let $S_x = \{y \in \mathbb{R} : y \text{ is ordinal definable from } x, A\}$. By Myhill-Scott, S_x is well-orderable for each x. So by AD, each S_x is countable. By $\mathsf{AC}_{\mathbb{R}}$, there is a function F such that $\forall x \in \mathbb{R}\,(F(x) \in \mathbb{R} - S_x)$. By assumption, F is ordinal definable from A and a real x_0, hence $F(x_0)$ is ordinal definable from A and x_0, hence $F(x_0) \in S_{x_0}$, contradiction. ⊣

The following definition is non-standard, but useful for our proofs.

DEFINITION 3.3. Let A and B be subsets of $\mathbb{R}$. Then $\mathbf{L}(A, B)$ is the smallest inner model, M, of ZF such that:

1. $A \cap M \in M$;
2. $B \in M$.

We think of the structure $\mathbf{L}(A, B)$ as having the following similarity type: In addition to the two-place predicates $=$ and $\in$, there is a one-place predicate, A, (whose extension is the set A) as well as a distinguished constant, B, (which, of course, denotes the set B).

COROLLARY 3.4 ($\mathsf{AD}_{\mathbb{R}}$). For all $A \subseteq \mathbb{R}$, $\mathbf{V} \neq \mathbf{L}(A, \mathbb{R})$.

PROOF. The statement $\mathbf{V}{=}\mathbf{L}(A, \mathbb{R})$ implies that everything is ordinal definable from A and a real. ⊣

We shall see later that in fact $\mathsf{AD}_{\mathbb{R}}$ implies "$A^{\#}$ exists" for every $A \subseteq \mathbb{R}$.

THEOREM 3.5 ($\mathsf{AD}_{\mathbb{R}}$). There is a monotone $G : \Theta \to \Theta$, such that if $\eta < \Theta$ is a limit ordinal closed under G, then $\mathbf{L}(\wp_\eta(\mathbb{R})) \models \mathsf{AD}_{\mathbb{R}}$ and $\wp(\mathbb{R}) \cap \mathbf{L}(\wp_\eta(\mathbb{R})) = \wp_\eta(\mathbb{R})$.

PROOF. Let $G_0(\eta)$ be the least ξ such that no set of reals of Wadge degree ξ is ordinal definable from a real and a set of Wadge degree η.

Let $G_1(\eta)$ be the least ξ such that every game of Wadge complexity $\leq \eta$ has a winning strategy of complexity $\leq \xi$.

Let $G(\eta) = \max(G_0(\eta), G_1(\eta))$. Suppose $\eta < \Theta$ is a limit ordinal closed under G. If $B \subseteq \mathbb{R}$ is in $\mathbf{L}(\wp_\eta(\mathbb{R}))$, then there is an A with $|A|_{\mathrm{W}} = \delta < \eta$ such that in $\mathbf{L}(\wp_\eta(\mathbb{R}))$, B is ordinal definable from A and a real. Hence $|B|_{\mathrm{W}} \leq G_0(\delta) < \eta$, *i.e.*, $B \in \wp_\eta(\mathbb{R})$.

Now given any game in $\mathbf{L}(\wp_\eta(\mathbb{R}))$, by the above its Wadge complexity is $< \eta$. Since η is closed under G_1, there is a winning strategy for the game in $\wp_\eta(\mathbb{R})$. Hence the game is determined in $\mathbf{L}(\wp_\eta(\mathbb{R}))$, *i.e.*, $\mathbf{L}(\wp_\eta(\mathbb{R})) \models \mathsf{AD}_{\mathbb{R}}$. ⊣

§4. A measure on the set of countable subsets of $\mathbb{R}$. A subset of $\mathbb{R}$ is *countable* if and only if it is finite or countably infinite. Let $\Omega = \{S \subseteq \mathbb{R} : S \text{ is countable}\}$.

For $A \subseteq \Omega$, consider the following game G_A: Player I and player II alternately play (possibly empty) finite subsets of $\mathbb{R}$, and player II wins if and only if the countable set produced by both players is in A. We shall refer to this countable set of reals produced by the two players as the *outcome* of the corresponding play of the game. Since finite subsets of $\mathbb{R}$ can be coded as individual reals, $\mathsf{AD}_{\mathbb{R}}$ implies that all G_A are determined, for $A \subseteq \Omega$.

Define for $A \subseteq \Omega$,

$$A \in U \iff \text{player II has a winning strategy in } G_A.$$

LEMMA 4.1 ($\mathsf{AD}_{\mathbb{R}}$). The set U is a countably complete ultrafilter on Ω.

PROOF. It will suffice to establish the following four claims:

1. $\varnothing \notin U$.
2. Suppose that $A \in U$ and $A \subseteq B \subseteq \Omega$. Then $B \in U$.
3. Suppose that $A \notin U$. Then $\Omega - A \in U$.
4. Let $\langle A_n : n \in \omega \rangle$ be a sequence of members of U. Let $B = \bigcap_n A_n$. Then $B \in U$.

The first two claims are evident.

Towards claim 3, suppose that $A \notin U$. Let τ be a w.s. for player I in G_A. Player II wins $G_{\Omega - A}$ by employing the following strategy, σ: Player I plays s_0. Player II ignores this play and plays s_1 according to τ. Player I plays s_2. Player II plays $s_3 = \tau(s_0 \cup s_2)$. From this point on player II plays τ against player I's moves. The outcome of a play in which player II employs σ is the outcome of a play in which player I employs τ. Hence this outcome lies in $\Omega - A$.

It follows that σ is a winning strategy for Player II in $G_{\Omega - A}$. Thus $A \notin U$ implies $\Omega - A \in U$.

We turn to claim 4: suppose $A_n \in U$ for $n \in \omega$. We show first that there is a sequence of strategies $\langle \tau_n : n \in \omega \rangle$ such that $\forall n$ (τ_n is a w.s. for player II in G_{A_n}). (This would be evident if we were assuming DC.)

This can be seen by considering the following auxiliary game: player I plays an integer $n \in \omega$; player II then passes. Then the two players commence a play of G_{A_n}. By $\mathsf{AD}_{\mathbb{R}}$, this game is determined. It is evident that player I cannot win. But from a winning strategy for player II in this auxiliary game the desired sequence of strategies $\langle \tau_n : n \in \omega \rangle$ can easily be extracted.

We proceed to show how player II can win G_B. Let $\langle \cdot, \cdot \rangle : \omega \times \omega \leftrightarrow \omega$ be a recursive pairing function such that if $m < k$, then $\langle n, m \rangle < \langle n, k \rangle$. For all $n, m < \omega$, on the $\langle n, m \rangle$th move of player II, player II plays according to τ_n, treating all reals played by either player since player II's $\langle n, m - 1 \rangle$th move as a single play of player I. (We leave the slight adjustments needed if $m = 0$ to the reader.) Thus on player II's plays at steps $\langle n, 0 \rangle, \langle n, 1 \rangle, \ldots$ player II is ensuring that the outcome of this play of G_B is in A_n. ⊣

In the following three definitions, it is understood that V is a countably complete ultrafilter on Ω.

DEFINITION 4.2. A filter V is **fine** if and only if $\forall x \in \mathbb{R}\,(\{S : x \in S\} \in V)$.

DEFINITION 4.3. A filter V is **normal** if and only if whenever $T : \Omega \to \Omega$ satisfies $\forall S \in \Omega$, $T(S) \subseteq S$, and $\{S : T(S) \neq \varnothing\} \in V$ then $\exists x \in \mathbb{R}$ such that $\{S : x \in T(S)\} \in V$.

DEFINITION 4.4. A filter V has the **diagonal intersection property** if and only if whenever $\langle M_x : x \in \mathbb{R}\rangle$ is a collection of elements of V, then $\{S \in \Omega : \forall x \in S\,(S \in M_x)\} \in V$.

That U is fine is quite easy and we shall leave the proof of this to the reader. Before establishing that U is normal, it will be useful to prove the following lemma.

LEMMA 4.5. Let V be a countably complete ultrafilter on Ω. Then the following are equivalent;

1. V is normal.
2. V has the diagonal intersection property.

PROOF. First suppose that V is normal and that $\langle M_x : x \in \mathbb{R}\rangle$ is a collection of elements of V.

Define $T : \Omega \to \Omega$ as follows: $T(S) = \{x \in S : S \notin M_x\}$. To prove that 1 implies 2, we have to show that $T(S) = \varnothing$ for V-almost every $S \in \Omega$.

Suppose not. Then by the normality of V, there is an $x \in \mathbb{R}$ such that $x \in T(S)$ for V-almost every $S \in \Omega$. *I.e.*, $S \notin M_x$ for V-almost every $S \in \Omega$. But this contradicts our assumption that $M_x \in V$.

Now suppose that V is not normal, but that the diagonal intersection property holds of V. We shall reach a contradiction.

Since V is not normal, there is $T : \Omega \to \Omega$ such that

1. $T(S) \subseteq S$ for all $S \in \Omega$;
2. $T(S) \neq \varnothing$ for V-almost every $S \in \Omega$; and
3. for $x \in \mathbb{R}$, if $M_x = \{S \in \Omega : x \notin T(S)\}$, then $M_x \in V$ for all $x \in \mathbb{R}$.

Hence by the diagonal intersection property, for V-almost every $S \in \Omega$ we have $S \in M_x$ for all $x \in S$. In other words, for V-almost all S the following obtain:

1. $S \cap T(S) = \varnothing$;
2. $T(S) \subseteq S$;
3. $T(S) = \varnothing$.

But this contradicts our assumption that $T(S) \neq \varnothing$ for V-almost every $S \in \Omega$. ⊣

LEMMA 4.6. The filter U is normal.

PROOF. By Lemma 4.5, it suffices to show that U has the diagonal intersection property. Suppose then that $\langle M_x : x \in \mathbb{R}\rangle$ is a collection of elements of U.

We must show that the set

$$B = \{S \in \Omega : \forall x \in S\,(S \in M_x)\} \in U.$$

We first establish that there is an indexed family of strategies, $\{\tau_x : x \in \mathbb{R}\}$, such that for all $x \in \mathbb{R}$, τ_x is a w.s. for player II in G_{M_x}.

To see this consider the following auxiliary game: player I plays $x \in \mathbb{R}$, player II passes, and then player I and player II play G_{M_x}. By $\mathsf{AD}_{\mathbb{R}}$, this game is determined. It is evident that player I cannot win this game. But from a winning strategy for player II, the desired indexed family of strategies is easily extracted.

The proof of the current lemma is now similar to the proof of countable completeness. Again, player II must ensure that the set produced is in each of countably many sets, *i.e.*, S must be in M_x for each $x \in S$. The difference between this and the countable completeness situation is that the countably many sets player II is trying to get into are not given before the game is played but are determined as play progresses. But this is no real problem: each time an x appears in a play of either player, player II resolves to play via τ_x infinitely often (*i.e.*, a dovetailing procedure). When player II plays τ_x for some x, all moves of both players since the last time player II used τ_x are treated as a single move of player I. By playing in this way player II ensures that $S \in M_x$ for all $x \in S$. ⊣

Lemma 4.7. For U-almost all S, S is closed under ordered pairs.

Proof. Suppose not. For $S \in \Omega$, let $T(S) = \{x \in S : (\exists y \in S)(\langle x, y\rangle \notin S)\}$. Then for U-almost every S, $T(S) \neq \varnothing$. Hence, by normality of U there is an $x \in \mathbb{R}$ such that $x \in T(S)$ for U-almost every S.

For $S \in \Omega$ let $T_1(S) = \{y \in S : \langle x, y\rangle \notin S\}$. By our choice of x, it follows that $T_1(S) \neq \varnothing$ for U-almost every S.

But then, by normality, we can find $y \in \mathbb{R}$ such that $y \in T_1(S)$ for U-almost every S.

In other words, $\langle x, y\rangle \notin S$ for U-almost every S. But this contradicts the fact that U is fine. ⊣

Lemma 4.8 ($\mathsf{AD}_{\mathbb{R}}$). For any $A \subseteq \mathbb{R}$, $\{S \in \Omega : \mathbb{R} \cap \mathbf{L}(A, S) = S\} \in U$.

Proof. Suppose not. Then for almost all S there is a $y \in \mathbb{R} \cap \mathbf{L}(A, S) - S$. By Lemma 4.7, S is closed under ordered pairs for U-almost every S. For such S we have a canonical map τ (which is a class of $\mathbf{L}(A, S)$) enumerating $\mathbf{L}(A, S)$, *i.e.*,

$$\mathbf{L}(A, S) = \{\tau(\xi, x) : \xi \in \mathrm{Ord}, x \in S\}.$$

Thus for almost all S there are $x \in S$ and ξ such that $\tau(\xi, x) \notin S$. Let $T(S) = \{x \in S : \exists\xi\,(\tau(\xi, x) \notin S)\}$. By normality, there is an $x_0 \in \mathbb{R}$ such that $\{S : x_0 \in T(S)\} \in U$. For S such that $x_0 \in T(S)$, let ξ_S be the least ξ such that $\tau(\xi, x_0) \notin S$. Let $y_S = \tau(\xi_S, x_0)$.

Using the countable additivity of U, it is easy to see that there is a fixed $y \in \mathbb{R}$ which equals y_S for U-almost every S. So $y \notin S$ for U-almost every S. But this contradicts the fineness of U. ⊣

§5. Sharps for sets of reals.

5.1. Introduction. In the next few sections, our goal is to prove, in $\mathsf{ZF}+\mathsf{AD}_{\mathbb{R}}$, the proposition: "for every $A \subseteq \mathbb{R}$, $A^\#$ exists". But it will take us a while to explain precisely the meaning of the sentence "$A^\#$ exists".

Here are the two consequences of "$A^\#$ exists" which we shall need for our applications:

1. There is a definable satisfaction relation (satisfying the usual inductive clauses) for the structure $\mathbf{L}(A, \mathbb{R})$.
2. There is an ordinal λ such that $\mathbf{L}_\lambda(A, \mathbb{R})$ is a ZF model and an elementary submodel of $\mathbf{L}(A, \mathbb{R})$.

We fix one particular $A \subseteq \mathbb{R}$ and work toward the goal of establishing "$A^\#$ exists". We begin with some preliminary remarks.

5.2. We shall assume that the reader is familiar with the theory of $0^\#$ (and its trivial relativization to the notion of $x^\#$ for $x \in \mathbb{R}$.) A treatment of this material can be found in many standard texts, for example, [Jec03, Chapter 18], [Kan94, § 9], or [Mos80, § 8H].

We shall also need the fact that ZF+AD proves that for all $x \in \mathbb{R}$, $x^\#$ exists. This proof can be found in [Kan94, Corollary 28.3].

5.3. Our construction of $A^\#$ will employ the measure U of § 4. For U-almost every $S \in \Omega$, we shall define $\langle A \cap S, S \rangle^\#$ by the following stratagem: We shall code the pair $\langle A \cap S, S \rangle$ by a real x. Then we shall use the real $x^\#$ to define the sharp of $\langle A \cap S, S \rangle$.

Finally, we shall define $A^\#$ by the following prescription:

$$A^\# = \{x \in \mathbb{R} : x \in \langle A \cap S, S \rangle^\# \text{ for } U \text{ almost every } S \in \Omega\}$$

5.4. Henceforth, we shall concentrate on those S such that $\mathbb{R} \cap \mathbf{L}(A, S) = S$. Let us call such S **excellent**. The important examples are:

1. $S = \mathbb{R}$.
2. U-almost all $S \in \Omega$ (*cf.* Lemma 4.8).

5.5. We begin with some preliminary material on $\mathbf{L}(A, S)$. For the moment we work in the following theory ZF_A:

1. The predicates of ZF_A are $=$, $\in$ and a one-place predicate A. We shall refer to the language of ZF_A as $\mathcal{L}_\mathsf{A}$. (It is our intention that $\mathsf{A}(x)$ express "$x \in A$".)

2. The axioms of $\mathsf{ZF_A}$ fall into two groups. First, there is an axiom that asserts $\forall x(\mathsf{A}(x) \Longrightarrow (x \in \mathbb{R}))$.
3. Second, there are all the axioms of ZF. (The symbol A is allowed to appear in the instances of the comprehension and replacement schemas).

We first define, for λ an ordinal, the set $\mathbf{L}_\lambda(A, \mathbb{R})$:

1. ($\lambda = 0$.) For x a set, let $\mathrm{tcl}(x)$ be the smallest transitive set y such that $x \subseteq y$. Let $A_0 = \{x \in \mathbb{R} : \mathsf{A}(x)\}$. Then $\mathbf{L}_0(A, \mathbb{R}) = \mathrm{tcl}(\mathbb{R}) \cup \{A_0\}$.
2. ($\lambda = \beta + 1$.) Then $\mathbf{L}_\lambda(A, \mathbb{R})$ consists of all the first order definable (allowing parameters) subsets of the structure $\mathbf{L}_\beta(A, \mathbb{R})$ in the usual language of set-theory (with just the predicates $=$ and $\in$).
3. (λ is a limit ordinal.) Then $\mathbf{L}_\lambda(A, \mathbb{R}) = \bigcup_{\beta<\lambda} \mathbf{L}_\beta(A, \mathbb{R})$.

5.6. We formulate a proposition which expresses "$\mathbf{V}{=}\mathbf{L}(A, \mathbb{R})$" in $\mathcal{L}_\mathsf{A}$.

It is just: For every set, x, there is an ordinal λ such that $x \in \mathbf{L}_\lambda(A, \mathbb{R})$.

Of course, we shall refer to this sentence as:

$$\mathbf{V}{=}\mathbf{L}(A, \mathbb{R})$$

5.7. The following proposition expresses, more or less, that $\mathbf{L}(A, \mathbb{R})$ is generated by the reals and the ordinals.

Proposition 5.1 ($\mathsf{ZF_A}$). There is a class sized function

$$\tau\colon \mathrm{Ord} \times \mathbb{R} \twoheadrightarrow \mathbf{L}(A, \mathbb{R})$$

which is definable in $\mathcal{L}_\mathsf{A}$. (It uses the predicate A.)

Proof. Most of the details of this proof will be left to the reader. The essential point is that each set in $\mathbf{L}_\lambda(A, \mathbb{R})$ is denoted (in many possible ways) by an "abstraction term" which can be coded by a finite sequence of ordinals and reals (and hence, by an ordinal and a real). ⊣

5.8. Let S be an excellent set of reals (*cf.* § 5.4.) Our next task is to give a precise explication of the phrase "$\langle A \cap S, S\rangle^\#$ exists". We shall define a predicate $\Phi(S, B)$ which is intended to express "$B = \langle A \cap S, S\rangle^\#$. We shall prove in $\mathsf{ZF_A}$ that if $\Phi(S, B)$ and $\Phi(S, B')$ then $B = B'$. Then "$\langle A \cap S, S\rangle^\#$ exists" will be shorthand for $(\exists B)\Phi(S, B)$.

We shall proceed as follows. We shall define a certain first-order language $\mathcal{L}_S$. It will be straightforward to Gödel code the objects of the various syntactical categories of $\mathcal{L}_S$ (terms and formulas) as reals. (We shall usually blur the distinction between, for example, a sentence of $\mathcal{L}_S$ and its Gödel code.) Then $\langle A \cap S, S\rangle^\#$ will consist of the Gödel codes of a certain complete theory in the language $\mathcal{L}_S$.

§6. The syntax of $\mathcal{L}_S$.

6.1. Before plunging into the details of the syntax of $\mathcal{L}_S$ we make some preliminary remarks.

1. The terms and formulas of $\mathcal{L}_S$ will "under the hood" be finite sequences from a certain alphabet Σ. We shall give simultaneous inductive definitions of the notions of "term" and "formula" of $\mathcal{L}_S$ and spell out the precise description of Σ at the same time. (We could, if we wanted, specify Σ in advance, but it makes for a more concise presentation not to do so.)
2. If $\sigma \in \Sigma$, we write $\langle \sigma \rangle$ for the sequence of length one whose sole constituent is σ. We shall frequently blur the distinction between the symbol $\sigma \in \Sigma$ and the corresponding one element sequence $\langle \sigma \rangle$.
3. Although we shall not make this explicit, we are thinking of the precise rendering of the terms and formulas as finite sequences as using the device of "Polish notation" (*cf.* the treatment of the syntax of first order theories in [Sho67]).
4. We shall be using a variant of the "definite description operator" as a device for forming terms. (So if $(\exists!\, x)\varphi(x)$ then $\iota\, x\; \varphi(x)$ will denote the unique y such that $\varphi(y)$.) This has the effect that we must define the notions of "term" and "formula" for $\mathcal{L}_S$ by a simultaneous induction. (We used a similar device in [Sol67B].)

6.2. We begin the inductive simultaneous definition of terms and formulas of $\mathcal{L}_S$ and of the alphabet Σ by listing those ingredients which are also part of the syntax of $\mathcal{L}_{\mathsf{A}}$:

1. For each $i \in \omega$ there is a symbol $v_i \in \Sigma$. (These are the "variables".) If v_i is a variable, $\langle v_i \rangle$ is a term.
2. There are two quantifier symbols in Σ, "$\exists$" and "$\forall$". If Q is a quantifier, v is a variable and ψ is a formula, then $Q\, v\, \psi$ is a formula.
3. There are four binary propositional connectives in Σ: $\wedge$ (and), $\vee$ (or), $\Longrightarrow$ (implies) and $\Longleftrightarrow$ (if and only if). If C is a binary propositional connective and ψ and χ are formulas, then $\psi\; C\; \chi$ is a formula. (Notice that in this informal description of the syntax of $\mathcal{L}_S$ we are *not* rendering things in "Polish notation". The translation of our definitions from "mathematical English" to "Polish notation" is left to the reader.)
4. There is one unary propositional connective, $\neg$, in Σ. If ψ is a formula, so is $\neg\, \psi$.
5. There are two binary predicate symbols, $=$, and $\in$ in Σ. If t_1 and t_2 are terms and P is a binary predicate symbol, then $t_1\; P\; t_2$ is a formula.
6. There is a symbol, A, in Σ. If t is a term, $\mathsf{A}\, t$ is a formula. The intended semantics is that "$\mathsf{A}\, x$" expresses "$x \in A$".

6.3. The following are the new ingredients of the syntax of $\mathcal{L}_S$ that do not appear in the language $\mathcal{L}_{\mathsf{A}}$.

1. For each $i \in \omega$ there is a symbol c_i in Σ. If $i \in \omega$, $\langle c_i \rangle$ is a term. The intended semantics is that the c_i's denote an increasing ω-sequence from a closed cofinal class of ordinal indiscernibles for $\mathbf{L}(A, S)$.
2. For each $x \in S$, there is a symbol $\mathbf{x}$ in Σ. If $x \in S$, $\langle \mathbf{x} \rangle$ is a term. The intended semantics is that $\mathbf{x}$ denotes the real x.
3. There is a symbol $\imath$ in Σ. $\imath$ is a binder (like the quantifiers). If $\varphi(x)$ is a formula, having at most the one free variable x, then $\imath x\, \varphi(x)$ is a term (having *no* free variables). The intended semantics for $\imath x\, \varphi(x)$ is as follows. If there is exactly one x such that $\varphi(x)$, then $\imath x\, \varphi(x)$ is the unique y such that $\varphi(y)$. Otherwise, $\imath x\, \varphi(x)$ is 0 (the empty set).

§7. The syntactical requirements of Φ.

7.1. We now start the description of Φ (the list of requirements that $\langle A \cap S, S \rangle^\#$ must satisfy).

Strictly speaking, $\langle A \cap S, S \rangle^\#$ is a set of reals, the Gödel codes for a collection of sentences. But we shall mostly ignore this distinction in what follows and speak as if $\langle A \cap S, S \rangle^\#$ *is* a set of sentences.

So the following is the beginning of the description of $\Phi(S, B)$. It will be convenient to have a name for this partial description. We shall refer to it as $\Phi_1(S, B)$.

1. If $x \in B$, x is the Gödel code for some sentence of $\mathcal{L}_S$.
2. If φ is a sentence of $\mathcal{L}_S$ precisely one of $\varphi, \neg\varphi$ belongs to B.
3. B is closed under modus ponens: If $\varphi, \varphi \implies \chi$ belong to B, then so does χ.
4. If φ is a logically valid sentence of $\mathcal{L}_S$, then φ in B. (The definition of "logically valid" in the current context is perhaps a little delicate. The following alternative formulation will suffice: Suppose that φ_1 is a logically valid formula of the language of $\mathcal{L}_\mathsf{A}$. Suppose that $w_1, \dots w_n$ are the variables appearing free in φ_1 and that no variable appears both free and bound in φ_1. Let $t_1, \dots, t_n$ be closed terms of $\mathcal{L}_S$. Let φ be the sentence obtained by replacing (throughout φ_1) each w_i by t_i. Then φ is in B.)
5. Each axiom of ZF appears in B. (In the replacement and comprehension schemas all the instances that can be formulated in $\mathcal{L}_S$ should appear in B.)
6. The sentence $\mathbf{V}{=}\mathbf{L}(A, \mathbb{R})$ is in B (*cf.* § 5.6).

7.2. It follows from the requirements just listed that if φ is a sentence of $\mathcal{L}_S$ and $\mathsf{ZF} + \mathbf{V}{=}\mathbf{L}(A, \mathbb{R}) \vdash \varphi$, then $\varphi \in B$. Also if φ_1, φ_2 are sentences of $\mathcal{L}_S$ and $\mathsf{ZF} + \mathbf{V}{=}\mathbf{L}(A, \mathbb{R}) \vdash (\varphi_1 \iff \varphi_2)$ then $\varphi_1 \in B$ if and only if $\varphi_2 \in B$.

We shall use this fact in the following way. We shall describe a sentence φ in "mathematical English" and assert that it is in B. The remark of the proceeding paragraph entails that for any reasonable translation of φ into the formal language of $\mathcal{L}_S$, the interpretation of this as saying that this formal translation lies in B will have the same effect.

7.3. The following requirement assures that our description operator works correctly. The description operator plays an important role in ensuring that $\mathcal{L}_S$ "has enough terms".

Let $\varphi(x)$ be a formula of $\mathcal{L}_S$ containing free at most the variable x.

1. If "$(\exists!x)(\varphi(x))$" (suitably translated into the language $\mathcal{L}_S$) appears in B then so does "$\varphi(\iota x \varphi(x))$".
2. If "$(\exists!x)(\varphi(x))$" does not appear in B, then "$\iota x\, \varphi(x) = 0$" appears in B.

We shall use this to define, for each $n \in \omega$, a term $\mathbf{n}$ which "denotes n".

7.4. The following condition will be useful in ensuring that the "term-models" we shall consider presently compute the truth of existential statements correctly:

(The witness condition) Suppose that the sentence "$(\exists x)\varphi(x)$" is in B. Then there is a closed term t such that:

1. The sentence "$\varphi(t)$" is in B.
2. If the indiscernible c_i appears in t (viewed as a sequence of symbols), then it appears in $\varphi(x)$ as well.

7.5.

1. The sentence "$(\forall x)(\mathsf{A}x \Longrightarrow x \in \mathbb{R})$" is in B.
2. Let $x \in S$. Then the sentence "$\mathsf{A}\,\mathbf{x}$" is in B if and only if $x \in A$.
3. Let $x \in S$. Suppose that $x(n) = m$. (Here, $n, m \in \omega$.) Then the sentence "$\mathbf{x}(\mathbf{n}) = \mathbf{m}$" is in B.
4. Let t be a closed term. Suppose that the sentence "$t \in \omega$" is in B. Then for some $n \in \omega$, the sentence "$t = \mathbf{n}$" is in B.
5. Let t be a closed term. Suppose that the sentence "$t \in \mathbb{R}$" is in B. Then for some $x \in S$, the sentence "$t = \mathbf{x}$" is in B.

7.6. We continue to describe Φ_1. The following are the basic properties of the indiscernibles c_i:

1. The sentence "c_1 is an ordinal" appears in B.
2. The c_i's are indiscernibles. Suppose φ is a sentence appearing in B and that $c_{i_1}, \dots, c_{i_k}$ are the symbols for indiscernibles appearing in φ (considered as a finite sequence of symbols). We assume $i_1 < i_2 \cdots < i_k$. Let $j_1, \dots, j_k$ be elements of ω with $j_1 < j_2 < \cdots < j_k$. Let φ' be obtained from φ by replacing each c_{i_r} that appears in φ by c_{j_r}. Then φ' appears in B.

   (The next two conditions are known as the "remarkability properties" of the c_i's.)

3. (R1) Let t be a closed term and suppose that the sentence "t is an ordinal" appears in B. Suppose that the symbols for indiscernibles appearing in t are among $c_0, \ldots, c_k$. Then the sentence "$t < c_{k+1}$" appears in B.
4. (R2) Let t be a closed term. Suppose:
   (a) The sentence "t is an ordinal" appears in B.
   (b) The indiscernible symbols appearing in t are among $c_0, \ldots c_{m+n-1}$.
   (c) The sentence "$t < c_m$" appears in B.

   Let t' be the term obtained from t by replacing c_{m+j} (wherever it appears in t) by c_{m+n+j} (for each j with $0 \leq j < n$).

   Then the sentence $t = t'$ appears in B.

This completes our description of Φ_1 (the syntactic conditions in our axiomatization Φ of $\langle A, S\rangle^\#$).

§8. **The term model construction.** Recall the construction (for ξ an ordinal) of the models $\Gamma(0^\#, \xi)$: $0^\#$ is a set of sentences in a language with constants $\langle c_n : n < \omega\rangle$. To get the underlying set of $\Gamma(0^\#, \xi)$ we form a new language with constants $\langle d_\eta : \eta < \xi\rangle$, and then we define an equivalence relation on terms of the form

$$\tau_\varphi(d_{\eta_1}, \ldots, d_{\eta_k})$$

(where τ_φ is a Skolem function) by referring to $0^\#$ on similar terms involving the c_n's. The underlying set of $\Gamma(0^\#, \xi)$ consists of the equivalence classes of this equivalence relation. (It is then necessary to define an $\in$ relation on this model. We shall not stop to recall how this is done.)

If $\Phi_1(S, B)$ we shall do a similar construction getting a term model $\Gamma(B, \xi)$ for each limit ordinal ξ. Our final requirement (in Φ) will then be that, for every limit ordinal ξ, the model $\Gamma(B, \xi)$ is well-founded.

We turn to the details.

8.1. Let ξ be a limit ordinal. We define a variant of the language $\mathcal{L}_S$, which we shall denote $\mathcal{L}_S[\xi]$. The only difference in this new language is that we shall have indiscernibility constants c_α for all $\alpha < \xi$.

We define the notion of "term" and "formula" of this new language in total analogy with the definitions of the analogous concepts for $\mathcal{L}_S$.

We define a set of sentences of $\mathcal{L}_S[\xi]$ (which we denote by $B[\xi]$) as follows:

Let φ be a sentence of $\mathcal{L}_S[\xi]$. Let $c_{\alpha_1}, \ldots, c_{\alpha_n}$ be the indiscernibles appearing in (the sequence) φ. (We assume that $\alpha_1 < \cdots < \alpha_n$.) Let $j_1 < \cdots < j_n$ be similarly ordered ordinals less than ω. Let φ' be the sentence of $\mathcal{L}_S$ obtained by replacing c_{α_i} by c_{j_i} throughout φ.. Then φ shall be in $B[\xi]$ if and only if φ' is in B.

Of course, it is important that this definition does not depend on the choice of the increasing n-tuple of integers $j_1 < \cdots < j_n$. This follows from our indiscernibility assumption about B (*cf.* §7.6).

Define an equivalence relation $\approx$ on the closed terms of $\mathcal{L}_S[\xi]$ as follows: we say that $t_1 \approx t_2$ if and only if the sentence "$t_1 = t_2$" lies in $B[\xi]$. (Of course, the reader should check that this *is* an equivalence relation.) We denote the equivalence class of the term t by $[t]$. The underlying set of $\Gamma(B, \xi)$ will consist of all the equivalence classes of $\approx$.

Define a binary relation $\in$ on $\Gamma(B, \xi)$ as follows: we say that $[t_1] \in [t_2]$ if and only if the sentence "$t_1 \in t_2$" lies in $B[\xi]$. (Again the reader should verify that this does not depend on the choices of representatives in the equivalence classes $[t_1]$ and $[t_2]$.)

Finally, we define a unary predicate A on $\Gamma(B, \xi)$ as follows: we say that $\mathsf{A}([t])$ if and only if the sentence "$\mathsf{A}(t)$" lies in $B[\xi]$. (Again the reader should verify this does not depend on the choice of representative from $[t]$.)

We have now completed our description of the structure $\Gamma(B, \xi)$. (It is a structure of the appropriate similarity type for the language $\mathcal{L}_{\mathsf{A}}$.)

8.2. Our next task is to formulate and prove a proposition to the effect that the set of sentences $B[\xi]$ correctly describes the model $\Gamma(B, \xi)$.

Suppose that φ is a formula of $\mathcal{L}_{\mathsf{A}}$ having free at most the variables $x_1, \dots, x_n$. (We shall suppose that none of these variables occur *bound* in φ.) Let $t_1, \dots, t_n$ be closed terms of $\mathcal{L}_S$. Then by $\varphi(t_1, \dots, t_n)$ we mean the sentence of $\mathcal{L}_S$ that results when each x_i is replaced throughout φ by t_i.

By the expression $\Gamma(B, \xi) \models \varphi(t_1, \dots, t_n)$ we mean that the formula φ is satisfied in $\Gamma(B, \xi)$ when the variable x_i is interpreted by the element $[t_i]$ of $\Gamma(B, \xi)$.

PROPOSITION 8.1. Let $\varphi, t_1, \dots, t_n$ be as just stated. Then the following are equivalent:

1. $\varphi(t_1, \dots, t_n) \in B[\xi]$ and
2. $\Gamma(B, \xi) \models \varphi([t_1], \dots, [t_n])$.

PROOF. We first consider the special case when the symbol $\forall$ does not appear in φ. In that case our proof proceeds by induction on the length of φ.

If φ is an atomic formula, our claim is immediate from the way the structure $\Gamma(B, \xi)$ was defined. The case when φ is a Boolean combination of shorter formulas is easy. As a sample we consider the case when φ is $\psi \wedge \chi$.

On the one hand, we have that $\varphi(t_1, \dots, t_n) \in B[\xi]$ if and only if $\psi(t_1, \dots, t_n) \in B[\xi]$ and $\chi(t_1, \dots, t_n) \in B[\xi]$. On the other hand, we have $\Gamma(B, \xi) \models \varphi([t_1], \dots, [t_n])$ if and only if $\Gamma(B, \xi) \models \psi([t_1], \dots, [t_n])$ and $\Gamma(B, \xi) \models \chi([t_1], \dots, [t_n])$. So our claim for φ follows immediately from the corresponding claims for the shorter ψ and χ.

Now consider the case when φ has the form $(\exists x)\psi(x)$. Suppose first that $\Gamma(B, \xi) \models \varphi$. We show that $\varphi \in B[\xi]$. Indeed for some closed term t of $\mathcal{L}_S$, we must have $\Gamma(B, \xi) \models \psi([t])$. By our inductive hypothesis, $\psi(t) \in B[\xi]$. But $B[\xi]$ is closed under logical consequence (since B is; *cf.* § 7.2). It follows that φ (which is $\exists x \psi(x)$) is in $B[\xi]$.

The other direction is slightly more difficult. Suppose that $\varphi \in B[\xi]$. We must show that $\Gamma(B,\xi) \models \varphi$. Let the indiscernibles appearing in φ be $c_{\alpha_1}, \dots c_{\alpha_n}$ with $\alpha_1 < \dots \alpha_n$. We define a map π whose domain is the set of finite sequences s of symbols (from Σ) such that the only indiscernibles appearing in s are from $\{c_{\alpha_1}, \dots, c_{\alpha_n}\}$. π simply replaces c_{α_i} wherever it appears in s, by c_i and leaves the other symbols in s unchanged. It is clear that π is an injection. The domain of π^{-1} consists of those sequences s' such that the only indiscernibles appearing in s' are from $\{c_1, \dots, c_n\}$.

Let $\varphi' = \pi(\varphi)$. (So φ' has the form $(\exists x)\psi'(x)$.) Then $\varphi' \in B$. By the witness condition (*cf.* § 7.4), there is a term t' such that

1. "$\psi'(t')$" is in B and
2. the only indiscernibles appearing in t' are among those appearing in φ.

Let $t = \pi^{-1}(t')$. (This makes sense because the only indiscernibles appearing in t' are among $c_1, \dots, c_n$.) Then $\psi(t)$ appears in $B[\xi]$. Hence by our inductive hypothesis, $\Gamma(B,\xi) \models \psi([t])$. It follows that $\Gamma(B,\xi) \models \varphi$. Our treatment of this case is complete.

We now know that if φ is a formula in which $\forall$ does not appear, then the claim of the proposition is true. The general case is easily handled. Let φ be a formula of $\mathcal{L}_{\mathsf{A}}$. Let φ' be a formula of $\mathcal{L}_{\mathsf{A}}$ which is logically equivalent to φ and in which $\forall$ does not appear. Let $x_1, \dots, x_n$ and $t_1, \dots, t_n$ be as in the statement of the proposition. Then because $B[\xi]$ is deductively closed, $\varphi(t_1, \dots, t_n)$ is in $B[\xi]$ if and only if $\varphi'(t_1, \dots, t_n)$ is in $B[\xi]$. By the special case of the proposition we have already proved, $\varphi'(t_1, \dots, t_n)$ is in $B[\xi]$ if and only if $\Gamma(B,\xi) \models \varphi'([t_1], \dots, [t_n])$. And clearly $\Gamma(B,\xi) \models \varphi'([t_1], \dots, [t_n])$ if and only if $\Gamma(B,\xi) \models \varphi([t_1], \dots, [t_n])$. Our proof of the proposition is complete. ⊣

8.3. The final requirement on Φ. Let ξ be a limit ordinal. It follows from Proposition 8.1 and the requirements imposed on B in § 7.1 that $\Gamma(B,\xi)$ is a model of ZF + $\mathbf{V}{=}\mathbf{L}(A, \mathbb{R})$.

Our final requirement is that this model is well-founded (or equivalently, that the ordinals of the model are well-ordered.)

§9. Uniqueness of sharps. The arguments of the present section can be carried out in ZF_{A}.

9.1. Let S be an excellent set of reals (*cf.* § 5.4). Suppose that B_1 and B_2 are subsets of $\mathbb{R}$ such that $\Phi(S, B_1)$ and $\Phi(S, B_2)$. We shall show that $B_1 = B_2$.

For the moment, let B be such that $\Phi(S, B)$. We shall develop some results about such a B and then apply it to the B_1 and B_2 of the preceding paragraph.

Let ξ be a limit ordinal. Our assumptions imply that the term model $\Gamma(B,\xi)$ is isomorphic to a transitive set of the form $\mathbf{L}_\lambda(A, S)$. We shall identify the term model with this transitive set.

Suppose now that ξ_1 and ξ_2 are limit ordinals with $\xi_1 < \xi_2$. Let λ_i be the ordinals of $\Gamma(B, \xi_i)$ (viewed as a transitive set).

Define an elementary embedding $j\colon \mathbf{L}_{\lambda_1}(A, S) \mapsto \mathbf{L}_{\lambda_2}(A, S)$ by $j([t]) := [t]$ for any closed term t of $\mathcal{L}_S[\xi_1]$. Let me spell out the definition of j just given in more detail. Let $x_1 \in \mathbf{L}_{\lambda_1}(A, S)$. Then x_1 corresponds to an element $[t]$ in the term model $\Gamma(B, \xi_1)$. We pick a representative $t \in [t]$. (Of course, t is not uniquely determined.) t determines an element (again noted $[t]$) in the term model $\Gamma(B, \xi_2)$. This element corresponds (via transitivization) to an element x_2 in $\mathbf{L}_{\lambda_2}(A, S)$. It is this element x_2 which we take as $j(x_1)$.

That j is well-defined (*i.e.*, does not depend on the choice of representative t) follows immediately from Proposition 8.1. Similarly, it also follows from Proposition 8.1 that j is elementary.

PROPOSITION 9.1.
1. In $\Gamma(B, \xi_2)$, c_{ξ_1} denotes λ_1.
2. Every ordinal less than λ_1 is in the range of j.
3. The function j is the identity on the ordinals less than λ_1.
4. The function j is the identity map.
5. In $\Gamma(B, \xi_2)$, $[c_{\xi_1}] = \sup\{[c_\alpha] : \alpha < \xi_1\}$.

PROOF. These claims will follow from the "remarkability" properties imposed on B in § 7.6.

It follows from (R1) that the ordinal (say γ) denoted by c_{ξ_1} is greater than any ordinal in the range of j. Then from (R2) and the fact that ξ_1 is a limit ordinal, it follows that every ordinal less than γ is in the range of j. So j maps the ordinals of $\mathbf{L}_{\lambda_1}(A, S)$ (*i.e.*, λ_1) one-to-one and onto the ordinals less than γ. It follows that $\gamma = \lambda_1$. Our first two claims are now evident.

The third claim follows since j is an order preserving map on ordinals. The fourth claim follows since the models which are the domain and range of j satisfy "$\mathbf{V}=\mathbf{L}(A, \mathbb{R})$". And claim 5 is an immediate consequence of claims 1 and 2. ⊣

9.2. Satisfaction for $\mathbf{L}(A, S)$. Of course, defining in ZF the satisfaction relation for a proper class sized structure can be problematic. (Tarski's theorem on the undefinability of truth shows this is impossible for the structure $\langle \mathbf{V}, \in \rangle$.) However, if there is a B such that $\Phi(S, B)$, we shall see that it *is* possible to define a satisfaction relation for $\mathbf{L}(A, S)$. The key tool is Proposition 9.1.

9.3. Though we shall blur this point in the discussion that follows, what we need to do is provide a *single* formula that "computes" from the Gödel number of a formula φ and a sequence of elements which give the "meanings" of the free variables of φ the truth value of the resulting sentence.

First consider the language $\mathcal{L}_\mathsf{A}$. Let $\varphi(x_1, \ldots, x_n)$ be a formula of $\mathcal{L}_\mathsf{A}$ having the indicated free variables, and let $a_1, \ldots, a_n$ be elements of the $\mathbf{L}(A, S)$.

Our prescription for determining if $\varphi(a_1, \ldots, a_n)$ holds in $\mathbf{L}(A, S)$ is as follows: Find a limit ordinal ξ so large that $a_1, \ldots, a_n$ are elements of (the transitivization of) $\Gamma(B, \xi)$. Then the truth value assigned to $\varphi(a_1, \dots, a_n)$ in $\mathbf{L}(A, S)$ is its truth value in $\Gamma(B, \xi)$.

Note that it follows from Proposition 9.1 that this prescription does not depend on the choice of ξ. Note also that this prescription insures that the inclusion of $\Gamma(B, \xi)$ in $\mathbf{L}(A, S)$ is an elementary embedding.

9.4. We now consider the definition of satisfaction in $\mathbf{L}(A, S)$ for a suitable variant of $\mathcal{L}_S$. (The reader should compare the definitions that follow to the definitions at the beginning of § 8.1.)

Let Ord be the class of all ordinals. We introduce a variant of $\mathcal{L}_S$ which we note $\mathcal{L}_S[\text{Ord}]$. (The language $\mathcal{L}_S[\text{Ord}]$ will have a proper class of symbols.) The only difference in this new language is that we shall have indiscernability constants c_α for all $\alpha \in \text{Ord}$.

We are going to assign to each closed term t of $\mathcal{L}_S[\text{Ord}]$ an element of $\mathbf{L}(A, S)$ as follows: Let ξ be a limit ordinal which is greater than any α such that c_α appears in t. Then t determines an element of the term model $\Gamma(B, \xi)$. Via the identification of $\Gamma(B, \xi)$ with a transitive set, this element of $\Gamma(B, \xi)$ determines an element of $\mathbf{L}(A, S)$.

9.5. We define $B[\text{Ord}]$ in total analogy with the definition of $B[\xi]$, for limit ordinals ξ, in § 8.1. The following lemma is the analogue of Proposition 8.1.

LEMMA 9.2. Let φ be a formula of $\mathcal{L}_\mathsf{A}$ containing the free variables $x_1, \dots, x_n$. (We assume no variable appears both free and bound in φ.) Let $t_1, \dots, t_n$ be closed terms of $\mathcal{L}_S[\text{Ord}]$ and denote their denotations in $\mathbf{L}(A, S)$ by $[t_1], \dots, [t_n]$. Then the following are equivalent:

1. $\mathbf{L}(A, S) \models \varphi([t_1], \dots, [t_n])$.
2. $\varphi(t_1, \dots, t_n) \in B[\text{Ord}]$.

PROOF. Let ξ be a limit ordinal larger than any α such that c_α appears in $\varphi(t_1, \dots, t_n)$. Then the following are equivalent:

1. $\varphi(t_1, \dots, t_n) \in B[\text{Ord}]$.
2. $\varphi(t_1, \dots, t_n) \in B[\xi]$. (Because of the way $B[\xi]$ and $B[\text{Ord}]$ are derived from B.)
3. $\Gamma(B, \xi) \models \varphi([t_1], \dots, [t_n])$. (By Proposition 8.1.)
4. $\mathbf{L}(A, S) \models \varphi([t_1], \dots, [t_n])$.

   There are two points here:
   (a) By the way we have defined the denotation of terms of $\mathcal{L}_S[\text{Ord}]$ in $\mathbf{L}(A, S)$, $[t_i]$ denotes the same thing in the transitivization of $\Gamma(B, \xi)$ as in $\mathbf{L}(A, S)$.
   (b) By Proposition 9.1, the transitivization of $\Gamma(B, \xi)$ is an elementary submodel of $\mathbf{L}(A, S)$ (with respect to the language $\mathcal{L}_\mathsf{A}$). ⊣

9.6. The club of Silver indiscernibles. Let $C = \{[c_\alpha] : \alpha \in \text{Ord}\}$. It follows easily from Proposition 9.1 that C is a closed unbounded class of ordinals. We shall refer to C as the club of Silver indiscernibles appropriate to B. Let F

be the class sized function with domain Ord that enumerates C in increasing order. (So $F(\alpha) = [c_\alpha]$.) Let $D = \{\alpha : F(\alpha) = \alpha\}$. Then D is a club class of ordinals.

9.7. We now return to the situation where we have *two* sets, B_1 and B_2 such that $\Phi(S, B_1)$ and $\Phi(S, B_2)$. Of course, our eventual goal is to show that $B_1 = B_2$.

Let C_i be the club of Silver indiscernibles appropriate to B_i. Let F_i be the class sized function that enumerates C_i in increasing order, and let D_i be the club class of ordinals that are fixed by F_i. Let $D = D_1 \cap D_2$. Of course, D is also a club class of ordinals.

Let $\langle \gamma_i : i \in \omega \rangle$ be the first ω elements of D. Let γ be their sup. Then for *either* $i = 1$ or $i = 2$, we have $\mathbf{L}_\gamma(A, S)$ is the transitivization of $\Gamma(B_i, \gamma)$. Moreover, the element of $\mathbf{L}_\gamma(A, S)$ that is the denotation of c_{γ_i} (with respect to either B_1 or B_2) is γ_i.

9.8. We first consider the interpretation of the subset of the language $\mathcal{L}_S$ without description terms. Consider then a sentence φ of $\mathcal{L}_S$ which contains no description terms. Say that the only c_j's appearing in φ are among $c_0, \dots, c_n$. Let φ' be the sentence obtained from φ by replacing c_j by c_{γ_j} (for $j = 0, \dots, n$). throughout φ. We have the following chain of equivalences (for either $i = 1$ or $i = 2$):

1. $\varphi \in B_i$;
2. $\varphi' \in B_i[\gamma]$ (by the definition of $B_i[\gamma]$; *cf.* § 8.1);
3. $\Gamma(B_i, \gamma) \models \varphi'(c_{\gamma_0}, \dots, c_{\gamma_n})$ (by Proposition 8.1);
4. $\mathbf{L}_\gamma(A, S) \models \varphi'(\gamma_0, \dots, \gamma_n)$ (by our remarks at the end of § 9.7).

The upshot is that both "$\varphi \in B_1$" and "$\varphi \in B_2$" are equivalent to the same proposition *which does not depend on i*. Hence these two propositions are equivalent.

9.9. We now consider the case when φ is allowed to contain description terms. We shall soon establish the following lemma:

LEMMA 9.3. Let φ' be a sentence of $\mathcal{L}_{\mathsf{A}}$. Then there is a sentence φ' of $\mathcal{L}_{\mathsf{A}}$ containing no description terms and not depending on i such that the equivalence $\varphi \iff \varphi'$ lies in both B_1 and B_2.

It should be clear that this lemma, together with the special case proved in § 9.8, suffice to establish that $B_1 = B_2$.

Our proof of Lemma 9.3 proceeds by eliminating, one by one, each description term in favor of a suitable definition. To carry out the formal details, we first need a measure of how deeply nested the description terms in a term or formula are.

So let φ be a term or formula of $\mathcal{L}_S$. We define, by induction on the length of φ, the **rank** of φ:

1. If φ is a variable, one of the c_i's, or one of the **x**'s, then the rank of φ is 0;

2. if φ is a description term $\iota x \psi(x)$, then the rank of φ is $1 + \text{rank}(\psi(x))$;
3. if φ is an atomic sentence "$t_1 R t_2$" where R is one of the binary predicates $\in$ or $=$, then the rank of φ is $\max(\text{rank}(t_1), \text{rank}(t_2))$;
4. if φ is an atomic sentence, "$\mathsf{A}(t)$", then the rank of φ is the rank of t;
5. if φ has the form "$Qx\psi$" where Q is one of the quantifiers $\exists$ or $\forall$, then the rank of φ is the rank of ψ;
6. if φ has the form "$\psi C \chi$" where C is a binary proposition connective, then the rank of φ is the max of rank(ψ) and rank(χ);
7. if φ is "$\neg\psi$", then rank(φ) is rank(ψ).

9.10. We shall also need another invariant which will serve to measure our progress in eliminating description terms. We call it the "$\star$-rank". The well-ordered set in which the $\star$-rank will take its values will be $\omega \times \omega$ equipped with the lexicographic order. Let then φ be a sentence of $\mathcal{L}_S$. The $\star$-rank of φ is (n, m) where

1. n is the rank of φ (as defined in § 9.9);
2. m is the number of description terms in φ of rank n (if the same description term appears several times in φ we only count it once).

If φ has rank 0 then it contains no description terms and has $\star$-rank $(0, 0)$. The following lemma says, in effect, that if φ is a sentence that *does* contain description terms then we can find an equivalent sentence of lesser $\star$-rank.

Lemma 9.4. Let φ be a sentence of $\mathcal{L}_S$ that *does* contain description terms. Then there is a sentence φ' of $\mathcal{L}_S$ whose $\star$-rank is less than that of φ and such that the equivalence $\varphi \iff \varphi'$ lies in both B_1 and B_2.

It is clear that Lemma 9.3 follows easily from Lemma 9.4. We simply apply the latter lemma to φ again and again until we reach a formula equivalent to φ of $\star$-rank $(0, 0)$.

We turn to the proof of Lemma 9.4.

Let then φ be a sentence of $\mathcal{L}_S$ containing description terms. Let τ be a description term appearing in φ whose rank is as large as possible. Say τ is $\iota x \psi(x)$.

Let y be a fresh variable, not appearing in φ. Define a formula $\vartheta(y)$ which expresses "$y = \tau$" as follows:

1. If $(\exists! x)\psi(x)$, then $\psi(y)$.
2. If not, then $y = \varnothing$.

We can now describe φ' (of the lemma we are trying to prove): Let φ_1 be obtained by replacing τ, wherever it appears in φ by y. Then φ' is the sentence $(\exists y)(\vartheta(y) \wedge \varphi_1)$. This completes our proof of Lemma 9.4 and with that our proof of the "uniqueness of sharps".

§10. Existence of sharps: the countable case. We work for the moment in $\mathsf{ZFA}+\mathsf{AD}_{\mathbb{R}}$ (though we shall drop shortly to a much weaker theory). Let S be a countable excellent set of reals (*cf.* § 5.4). We shall establish that there exists a set of reals, B such that $\Phi(S, B)$.

10.1. Recall that we are construing reals, in this paper, as elements of ${}^{\omega}\omega$. For the moment, let $(\cdot,\cdot)$ be some fixed coding of pairs of integers as integers.

We view each real x as coding a countable sequence of reals as follows: $(x)_i$ shall be the real $\lambda n\, x((i, n))$.

We can code the pair $\langle A \cap S, S\rangle$ by a real x by requiring

1. $S = \{(x)_{i+1} : i \in \omega\}$ and
2. $A \cap S = \{(x)_{i+1} : x(0, i+1) > 0\}$.

We fix such a real, x, coding $\langle A \cap S, S\rangle$ as just described for the remainder of this section.

As explained in § 5.2, it follows from $\mathsf{AD}_{\mathbb{R}}$ that "$x^{\#}$ exists". We shall drop to the inner model $\mathbf{L}(x^{\#})$ for the remainder of this section.

1. This is harmless. The proposition "$\langle A, S\rangle^{\#}$ exists is absolute under the transition from our outer model where $\mathsf{AD}_{\mathbb{R}}$ obtains to the inner model $\mathbf{L}(x^{\#})$.
2. This transition is useful. The axiom of choice holds in $\mathbf{L}(x^{\#})$.

Notice that x knows enough to define $\mathbf{L}(A, S)$. Indeed, all we need to define this model is $A \cap S$ which is encoded in x.

10.2. Let γ be $\aleph_1$ (as computed in $\mathbf{L}(x^{\#})$). We are going to interpret $\mathcal{L}_S$ in the structure $\mathbf{L}_{\gamma}(A, S)$.

Precisely what we mean by "interpret" here needs some spelling out. For each syntactic object (term or formula) t of $\mathcal{L}_S$ we shall define a "local satisfaction relation". The definition of this local satisfaction relation will proceed by induction on the length of t.

The key point is that we view the "indiscernibility constants" (the c_i's) appearing in t as another flavor of variable and we require (before determining the meaning of t) to know what ordinals are assigned to these c_i's.

10.3. Let t be a term or formula of $\mathcal{L}_S$. An *assignment* for t will be a function f such that:

1. The domain of f is the set of x such that either
   (a) x is a variable appearing free in t or
   (b) x is an "indiscernible symbol" (a c_i) that occurs in t.
2. f takes values in (the underlying set of) $\mathbf{L}_{\gamma}(A, S)$.

10.4. If t is a term our "local satisfaction relation" will assign to each assignment a value for t in $\mathbf{L}_{\gamma}(A, S)$. If t is a formula, the "local satisfaction relation" will assign to each assignment a "truth value" for t in {true, false}.

The definition of these local satisfaction relations is done by induction on the length of t. The clauses in this definition are quite easy to surmise and I shall leave the detailed spelling out of this definition to the reader.

10.5. We now define the set of sentences, B, of $\mathcal{L}_S$ which will turn out to be $\langle A \cap S, S\rangle^{\#}$. Of course, one must verify the various clauses of $\Phi(S, B)$ before we can identify B with $\langle A \cap S, S\rangle^{\#}$.

It is evident that our definition of the satisfaction relation for $\mathcal{L}_S$ in $\mathbf{L}_\gamma(A, S)$ can be given in $\mathbf{L}(x)$ (using the parameters x and γ). (x allows us to lay our hands on S and $A \cap S$. Since S is excellent, this is all we need to define $\mathbf{L}(A, S)$.)

We let I be the set of Silver indiscernibles for x which are less than γ. Let φ be a sentence of $\mathcal{L}_S$, and let $c_{i_1}, \dots, c_{i_k}$ be the indiscernibility constants appearing in φ. Because the elements of I are indiscernibles for $\mathbf{L}(x)$, the local satisfaction relation for φ gives the same truth value for any assignment of indiscernibles $\eta_1, \dots, \eta_k$ from I to the c_i's such that $\eta_1 < \dots < \eta_k$. We put $\varphi \in B$ just in case this common truth value is "true".

10.6. For the most part the conditions in Φ follow easily from analogous properties of $x^{\#}$. We consider just those clauses which, perhaps, are not evident.

First, we consider the "witness condition" (*cf.* § 7.4). Suppose then that "$(\exists x)\varphi(x)$" is in B. We describe a t that shows the witness condition holds.

We shall use the class sized function τ described in § 5.7. We fix a suitable assignment of indiscernibles from I to the indiscernible constants appearing in $\varphi(x)$.

According to Proposition 5.1, there is a real y and an ordinal $\xi < \gamma$ such that $\mathbf{L}_\gamma(A, S) \models \varphi(\tau(\xi, y))$. (We must interpret the indiscernible constants using the assignment mentioned in the preceding paragraph.)

So the following formula $\chi(x)$ can be satisfied in exactly one way: "ξ is the least ordinal such that $\mathbf{L}_\gamma(A, S) \models \varphi(\tau(\xi, y))$ and $x = \tau(\xi, y)$". But then $t = \iota x \chi(x)$ satisfies all the demands of the witness condition.

10.7. The final condition that we shall consider is that of § 8.3: If ξ is a limit ordinal, then the term model $\Gamma(B, \xi)$ is well-founded.

Exactly as in the theory of $0^{\#}$, it suffices to verify this for *countable* ξ. So fix a countable limit ordinal ξ. We shall use the following global assignment of meanings for the c_i's for $i < \xi$. Assign to c_i the ith member of the set I of Silver indiscernibles for $x^{\#}$. In this way, our satisfaction relation determines a map of the term model $\Gamma(B, \xi)$ into the well-founded model $\mathbf{L}_\gamma(A, S)$. Hence this term model is well-founded.

This completes our discussion of the proof of existence of $\langle A, S\rangle$ in the case of countable excellent sets of reals.

§11. The existence of $A^{\#}$. By $A^{\#}$ we mean the set of reals that we have been calling, in the preceding sections, $\langle A, \mathbb{R}\rangle^{\#}$. It is the appropriate "sharp" for the class model $\mathbf{L}(A, \mathbb{R})$.

We are working once again in the theory $\mathsf{ZF_A}+\mathsf{AD}_{\mathbb{R}}$. We first indicate the prescription we shall use to define $A^{\#}$. We then discuss why this prescription works.

We shall employ the measure U of § 4. For U-almost all $S \in \Omega$, S is excellent, so $\langle A \cap S, S\rangle^{\#}$ has been defined in § 10.

Define a set of Gödel codes for sentences, B, by $\varphi \in B$ if and only if $\varphi \in \langle A \cap S, S\rangle^{\#}$ for U-almost every $S \in \Omega$. The set B will turn out to be $A^{\#}$. We must verify the various clauses of $\Phi(\mathbb{R}, B)$.

11.1. For most of these clauses, the verification follows easily from the fact that the corresponding clause holds for $\langle A \cap S, S\rangle^{\#}$. We discuss the few clauses which are, perhaps, not evident.

Consider first the "witness condition" (*cf.* § 7.4). Suppose that the sentence "$(\exists x)\varphi(x)$" is in B. Then for U-almost every S in Ω, "$(\exists x)\varphi(x)$" is in $\langle A \cap S, S\rangle^{\#}$.

But the witness condition holds for $\langle A \cap S, S\rangle^{\#}$. Hence for almost every $S \in \Omega$, we can define a non-empty subset $T(S) \subseteq S$ such that

1. if $x \in T(S)$, x is the Gödel code for a term t such that $\varphi(t) \in \langle A, S\rangle^{\#}$, and
2. any c_i that appears in t appears also in $\varphi(x)$.

We now invoke the fact that U is normal (*cf.* Definition 4.3 & Lemma 4.6.) It follows that there is a single x which is in $T(S)$ for U-almost every $S \in \Omega$. Let t be the term of which x is the Gödel code. Then t instantiates this case of the witness condition for B.

11.2. The final condition of Φ that we shall discuss is that of § 8.3: If ξ is a limit ordinal, then the term model $\Gamma(B, \xi)$ is well-founded.

Let λ be the least cardinal, in $\mathbf{L}$, which is greater than ξ. (We just need an ordinal greater than ξ that is "suitably closed". This λ will do the trick.)

Each of the terms of $\mathcal{L}_S[\xi]$ (*cf.* § 8.1) can be coded by a finite sequence each of whose members is either in $\mathbb{R}$ or λ. Hence it is easy to define a surjective map of $\mathbb{R} \times \lambda$ onto the set of terms of $\mathcal{L}_S[\xi]$.

Although it is not clear that we have DC (and we shall show presently that there are models of $\mathsf{AD}_{\mathbb{R}}$ in which DC fails) we do have $\mathsf{DC}_{\mathbb{R}}$ (*cf.* the discussion in § 1). Hence we have $\mathsf{DC}_{\mathbb{R}\times\lambda}$ (*cf.* the discussion during the proof of Lemma 2.5, especially the second itemized fact in that proof).

So fix a limit ordinal ξ such that $\Gamma(B, \xi)$ is not well-founded. We shall derive a contradiction.

Using $\mathsf{DC}_{\mathbb{R}\times\lambda}$, we can get a sequence of terms $\langle t_i : i \in \omega\rangle$ from $\mathcal{L}_S[\xi]$ such that

1. the sentence "t_i is an ordinal" lies in B for all $i \in \omega$ and
2. the sentence "$t_i > t_{i+1}$" lies in B for all $i \in \omega$.

But then the same sequence of terms serves to demonstrate that for U-almost all $S \in \Omega$, the term model $\Gamma(\langle A \cap S, S\rangle^{\#}, \xi)$ is not well-founded. We have reached our desired contradiction.

This completes our discussion of the proof, in $\mathsf{ZF_A}$+$\mathsf{AD}_{\mathbb{R}}$, of the fact that for every set of reals A, $A^{\#}$ exists.

§12. The main theorems.

12.1. By this point, we have already implicitly established the two goals mentioned in § 5.1. It is worth our while to pause a moment and make this explicit.

In § 9.3, we defined a satisfaction relation for $\mathbf{L}(A, S)$ from B provided the following conditions hold:

1. S is excellent for A (*cf.* § 5.4) and
2. $\Phi(S, B)$.

We take $\mathbb{R}$ for S. (That $\mathbb{R}$ is excellent for A is noted in § 5.4.)

The whole point of the last few sections has been to construct $A^{\#}$ so that $\Phi(\mathbb{R}, A^{\#})$. (The set of reals A has been suppressed from the notation (including that for Φ) of the last few sections.)

So we have indeed defined a satisfaction relation for $\mathbf{L}(A, \mathbb{R})$ using $A^{\#}$.

The second point we want to make is that, for ξ a limit ordinal, the transitivization of $\Gamma(A^{\#}, \xi)$ is a rank-initial elementary submodel (which is a *set*) of $\mathbf{L}(A, \mathbb{R})$. This should be clear from the way that we defined the satisfaction relation for $\mathbf{L}(A, \mathbb{R})$ in § 9.3 (*cf.* especially Proposition 9.1).

12.2. Each of our main results has the form $T_1 \vdash \mathrm{Con}(T_2)$. Here T_1 will be some theory at least as strong as ZF+$\mathsf{AD}_{\mathbb{R}}$ and T_2 will be a theory at least as strong as ZF.

In each case, we shall proceed by giving a transitive proper class model, M, of T_2. To proceed from this to the conclusion "$\mathrm{Con}(T_2)$" certainly takes some care. (Gödel's proper class model of ZFC+GCH, established in ZF, certainly did *not* establish ZF $\vdash$ Con(ZFC+GCH).)

However, in the cases we consider M will be a definable subclass of a model of the form $\mathbf{L}(A, \mathbb{R})$ for some set of reals A. We can conclude $\mathrm{Con}(T_2)$ by either of the following routes:

1. As discussed in the preceding subsection, there is an ordinal λ such that $\mathbf{L}_\lambda(A, \mathbb{R})$ is an elementary submodel of $\mathbf{L}(A, \mathbb{R})$. ($\mathbf{L}_\lambda(A, \mathbb{R})$ is the transitivization of $\Gamma(A^{\#}, \omega)$.) The same definition that yields M inside $\mathbf{L}(A, \mathbb{R})$ will yield a set-sized model, M', of T_2 when carried out inside $\mathbf{L}_\lambda(A, \mathbb{R})$. But from the set-sized model M' of T_2 we easily get $\mathrm{Con}(T_2)$ by standard theorems.
2. Alternatively, one can adapt the usual proof that "models yield consistency" to class-sized models *provided* we have a satisfaction relation for the class-sized model.

12.3.

THEOREM 12.1. ZF+$\mathsf{AD}_{\mathbb{R}}$ $\vdash$ Con(ZF+AD).

PROOF. We work in ZF+$\mathsf{AD}_{\mathbb{R}}$. The model we shall use to establish this result is $\mathbf{L}(\mathbb{R})$ (which, of course, is just $\mathbf{L}(\mathbb{R},\mathbb{R})$.) In view of the discussion of the proceeding section it suffices to see that $\mathbf{L}(\mathbb{R}) \models \mathsf{AD}$.

Of course, $\mathsf{AD}_{\mathbb{R}}$ implies AD. Let S be a set of reals lying in $\mathbf{L}(\mathbb{R})$. Because $\mathbf{L}(\mathbb{R})$ contains all reals, from the fact that the integer game corresponding to S is determined in our outer model of $\mathsf{AD}_{\mathbb{R}}$, it follows that it is determined in $\mathbf{L}(\mathbb{R})$. Hence AD holds in $\mathbf{L}(\mathbb{R})$. ⊣

This result can be slightly sharpened to the following:

THEOREM 12.2. $\mathsf{ZF}+\mathsf{AD}_{\mathbb{R}} \vdash \mathrm{Con}(\mathsf{ZF}+\mathsf{AD}+\mathsf{DC})$

PROOF. Work in ZF+$\mathsf{AD}_{\mathbb{R}}$. It suffices to show that $\mathbf{L}(\mathbb{R}) \models \mathsf{DC}$. But $\mathsf{AD}_{\mathbb{R}} \vdash \mathsf{DC}_{\mathbb{R}}$ and because $\mathbf{L}(\mathbb{R})$ contains all the reals, it inherits DC_R from our outer model of ZF+$\mathsf{AD}_{\mathbb{R}}$. But DC_R implies full DC if $\mathbf{V}=\mathbf{L}(\mathbb{R})$ (*cf.* Fact 2 of Lemma 2.5). ⊣

12.4. Let us refer, from now on, to our outer model of $\mathsf{AD}_{\mathbb{R}}$ as V.

LEMMA 12.3 ($\mathsf{AD}_{\mathbb{R}}$). Let $A \subseteq \mathbb{R}$ have Wadge degree η. Then the model $\mathbf{L}(\wp_\eta(\mathbb{R}))$ is definable (from the parameter A) in $\mathbf{L}(A,\mathbb{R})$.

PROOF. The model $\mathbf{L}(A,\mathbb{R})$ contains every real. It follows first that if B, C are subsets of $\mathbb{R}$ with $B \in \mathbf{L}(A,\mathbb{R})$ and C Wadge reducible to B, then C is in $\mathbf{L}(A,\mathbb{R})$. Moreover, the notion of Wadge reducibility is absolute from V to $\mathbf{L}(A,\mathbb{R})$.

It follows that the Wadge rank of a set of reals is also absolute from V to $\mathbf{L}(A,\mathbb{R})$. Hence $\wp_\eta(\mathbb{R})]$ is a set of $\mathbf{L}(A,\mathbb{R})$. But then, of course, $\mathbf{L}(\wp_\eta(\mathbb{R}))$ is a definable class of $\mathbf{L}(A,\mathbb{R})$. ⊣

LEMMA 12.4 ($\mathsf{AD}_{\mathbb{R}}$). Let $\eta < \Theta$. Then the class model $\mathbf{L}(\wp_\eta(\mathbb{R}))$ has a definable satisfaction relation.

PROOF. Let A be a set of reals of Wadge rank η. By the preceding lemma, $\mathbf{L}(\wp_\eta(\mathbb{R}))$ is definable from the parameter A in $\mathbf{L}(A,\mathbb{R})$. We have a satisfaction relation for $\mathbf{L}(A,\mathbb{R})$ obtained from $A^{\#}$ as discussed above. A satisfaction relation for $\mathbf{L}(\wp_\eta(\mathbb{R}))$ is easily obtained from this satisfaction relation for $\mathbf{L}(A,\mathbb{R})$. ⊣

THEOREM 12.5. $\mathsf{ZF} + \mathsf{AD}_{\mathbb{R}} + \mathrm{cf}(\Theta) > \omega \vdash \mathrm{Con}(\mathsf{AD}_{\mathbb{R}})$.

PROOF. Work in $\mathbf{L}(\wp(\mathbb{R}))$. Let G be the function given by Theorem 3.5. Then $\mathrm{cf}(\Theta) > \omega$ implies $\exists \eta < \Theta$ (η is closed under G). Hence, by Theorem 3.5, $\mathbf{L}(\wp_\eta(\mathbb{R})) \models \mathsf{AD}_{\mathbb{R}}$. Since the satisfaction relation of $\mathbf{L}(\wp_\eta(\mathbb{R}))$ is definable, the theorem follows. ⊣

COROLLARY 12.6. $\mathsf{ZF}+\mathsf{AD}_{\mathbb{R}}+\mathsf{DC} \vdash \mathrm{Con}(\mathsf{ZF}+\mathsf{AD}_{\mathbb{R}})$.

PROOF. DC implies $\mathrm{cf}(\Theta) > \omega$ (*cf.* Lemma 2.3). ⊣

COROLLARY 12.7. $\mathrm{Con}(\mathsf{ZF}+\mathsf{AD}_{\mathbb{R}}) \Longrightarrow \mathrm{Con}(\mathsf{ZF}+\mathsf{AD}_{\mathbb{R}}+\neg\mathsf{DC})$.

PROOF. The following proof can easily be formalized in second order number theory.

Assume that $\mathsf{ZF}+\mathsf{AD}_{\mathbb{R}}$ is consistent. Then, by Gödel's second incompleteness theorem, $\mathsf{ZF}+\mathsf{AD}_{\mathbb{R}}$ does not prove $\mathrm{Con}(\mathsf{ZF}+\mathsf{AD}_{\mathbb{R}})$. Hence there is a model M of $\mathsf{ZF}+\mathsf{AD}_{\mathbb{R}}+\neg\,\mathrm{Con}(\mathsf{ZF}+\mathsf{AD}_{\mathbb{R}})$.

I say that $M \models \neg\mathsf{DC}$. Suppose not. Then $M \models \mathsf{ZF}+\mathsf{AD}_{\mathbb{R}}+\mathsf{DC}$. Hence, by the immediately preceding corollary, $M \models \mathrm{Con}(\mathsf{ZF}+\mathsf{AD}_{\mathbb{R}})$. But this contradicts our assumptions about M.

So $M \models \mathsf{ZF}+\mathsf{AD}_{\mathbb{R}}+\neg\mathsf{DC}$. It follows that the theory $\mathsf{ZF}+\mathsf{AD}_{\mathbb{R}}+\neg\mathsf{DC}$ is consistent. ⊣

THEOREM 12.8. $\mathsf{ZF}+\mathsf{AD}_{\mathbb{R}}+$"$\Theta$ regular" $\vdash \mathrm{Con}(\mathsf{ZF}+\mathsf{AD}_{\mathbb{R}}+\mathsf{DC})$.

PROOF. Work in $\mathbf{L}(\wp(\mathbb{R}))$, and let G be as before. As Θ is regular, there is an $\eta < \Theta$ of cofinality ω_1 which is closed under G. Then $\mathbf{L}(\wp_\eta(\mathbb{R})) \models \mathsf{AD}_{\mathbb{R}}$. But by Theorem 3.5, $\mathbf{L}(\wp_\eta(\mathbb{R})) \models \eta = \Theta$, hence $\mathbf{L}(\wp_\eta(\mathbb{R})) \models \mathrm{cf}(\Theta) > \omega$. So by Theorem 2.4, $\mathbf{L}(\wp_\eta(\mathbb{R})) \models \mathsf{DC}$. By Lemma 12.4, the satisfaction relation of $\mathbf{L}(\wp_\eta(\mathbb{R}))$ is definable, hence the theorem follows. ⊣

12.5. By Gödel's second incompleteness theorem, we thus get the following:

THEOREM 12.9.

1. If $\mathsf{ZF}+\mathsf{AD}$ is consistent, then $\mathsf{ZF}+\mathsf{AD}$ does not prove $\mathsf{AD}_{\mathbb{R}}$.
2. If $\mathsf{ZF}+\mathsf{AD}_{\mathbb{R}}$ is consistent, then $\mathsf{ZF}+\mathsf{AD}_{\mathbb{R}}$ does not prove $\mathrm{cf}(\Theta) > \omega$, hence $\mathsf{ZF}+\mathsf{AD}_{\mathbb{R}}$ does not prove DC.
3. If $\mathsf{ZF}+\mathsf{AD}_{\mathbb{R}}+\mathsf{DC}$ is consistent, then $\mathsf{ZF}+\mathsf{AD}_{\mathbb{R}}+\mathsf{DC}$ does not prove Θ is regular.

§13. Open questions; recent results.

13.1. The current paper is a revision of a paper I wrote in the mid 1970's. (As I write these words, it is early November of 2014.) Not surprisingly there has been some significant progress since that date. I have chosen to list the open questions precisely as they appeared in the earlier version of the paper and then comment on more recent results (many of which, alas, are still unpublished.)

13.2. Open questions.

1. Does AD imply $\mathsf{DC}_{\mathbb{R}}$? We conjecture that it does not.
2. Recall the functions G_0 and G_1 used to define the function G of § 5. Is $G_0(\eta) < G_1(\eta)$?

It is known (Mycielski-Blass) that $\mathsf{AD}_{\mathbb{R}}$ is equivalent to $\mathsf{AD}[\omega^2]$, which asserts the determinacy of all games in which players play integers in a sequence of ω^2 moves [Bla75, Myc64, Myc66]. It is known that the analogous axiom $\mathsf{AD}[\omega_1]$ is false.

3. Conjecture: $\mathsf{ZF}+\mathsf{AD}[\omega^3] \vdash \mathrm{Con}(\mathsf{ZF}+\mathsf{AD}[\omega^2]+\Theta$ regular).
4. $\mathsf{AD}[\omega^3]$ should give stronger measures.
5. Does $\mathsf{AD}_{\mathbb{R}}$+"Θ is regular" imply every set of reals has a scale? (For the notion of a scale, *cf.* [Kec78].) We know that AD does not imply every set has a scale. In fact, if $\mathbf{V}=\mathbf{L}(\mathbb{R})$, there is a set without a scale.

13.3. Subsequent results.

13.3.1. *Kechris on* DC. Kechris has shown that the theory $\mathsf{ZF}+\mathsf{AD}+\mathbf{V}=\mathbf{L}(\mathbb{R})$ proves DC. The result has been published in [Kec84].

13.3.2. *Woodin on* DC. In the present paper, I have proved, among other things that $\mathrm{Con}(\mathsf{ZF}+\mathsf{AD}_{\mathbb{R}})$ implies $\mathrm{Con}(\mathsf{ZF}+\mathsf{AD}+\neg\mathsf{DC})$ (*cf.* Corollary 12.7). Woodin has proved the following sharper result in unpublished work: $\mathrm{Con}(\mathsf{ZF}+\mathsf{AD})$ implies $\mathrm{Con}(\mathsf{ZF}+\mathsf{AD}+\neg\mathsf{DC})$.

The following sketch of his proof is extracted from a letter of Woodin to the author:

> The proof was by forcing over **HOD** of $\mathbf{L}(\mathbb{R})$ well above Θ to obtain a symmetric extension in which DC or even countable choice fails and not add any subsets of Θ. Call this model N. Then $N(\mathbb{R}) \models \mathsf{AD}$ since $N[\mathbb{R}]$ and $\mathbf{HOD}(\mathbb{R}) = \mathbf{L}(\mathbb{R})$ have the same sets of reals.

13.3.3. *Woodin and Martin on Scales and* $\mathsf{AD}_{\mathbb{R}}$. The following is extracted from a letter of Woodin to the author:

THEOREM 13.1. The following are equivalent in ZF+DC:

1. AD + "Every set has a scale",
2. $\mathsf{AD}_{\mathbb{R}}$, and
3. AD + "Uniformization".

Woodin comments that the implication "(1)⇒(2)" is a theorem of Martin's and the implication "(3)⇒(1)" needs DC. According to Martin, his proof required AC_ω. Here is an extract from a letter from Martin to the author:

> Probably what Hugh was talking about was my proof that AD + "Scales" + AC_ω implies that all games on ω of countable length are determined.

Martin's proof appears as [Mar20, Theorem 9] in this volume. According to that paper, Martin's result was independently proved by Woodin at about the same time as Martin's original proof and Woodin has subsequently found a new proof that does not use AC_ω.

REFERENCES

ANDREAS BLASS

[Bla75] *Equivalence of two strong forms of determinacy*, ***Proceedings of the American Mathematical Society***, vol. 52 (1975), pp. 373–376.

THOMAS JECH

[Jec03] ***Set Theory***, Springer Monographs in Mathematics, Springer-Verlag, Berlin, 2003, the third millennium edition, revised and expanded.

AKIHIRO KANAMORI

[Kan94] ***The Higher Infinite. Large Cardinals in Set Theory from Their Beginnings***, Perspectives in Mathematical Logic, Springer-Verlag, Berlin, 1994.

ALEXANDER S. KECHRIS

[Kec78] AD *and projective ordinals*, in Kechris and Moschovakis [CABAL i], pp. 91–132, reprinted in [CABAL II], pp. 304–345.

[Kec84] *The axiom of determinacy implies dependent choices in* $\mathbf{L}(\mathbb{R})$, ***The Journal of Symbolic Logic***, vol. 49 (1984), no. 1, pp. 161–173.

ALEXANDER S. KECHRIS, BENEDIKT LÖWE, AND JOHN R. STEEL

[CABAL II] *Wadge Degrees and Projective Ordinals: the Cabal Seminar, volume II*, Lecture Notes in Logic, vol. 37, Cambridge University Press, 2012.

ALEXANDER S. KECHRIS AND YIANNIS N. MOSCHOVAKIS

[CABAL i] *Cabal Seminar* 76–77, Lecture Notes in Mathematics, vol. 689, Berlin, Springer-Verlag, 1978.

DONALD A. MARTIN

[Mar20] *Games of countable length*, 2020, this volume.

YIANNIS N. MOSCHOVAKIS

[Mos80] ***Descriptive Set Theory***, Studies in Logic and the Foundations of Mathematics, vol. 100, North-Holland, Amsterdam, 1980.

JAN MYCIELSKI

[Myc64] *On the axiom of determinateness*, ***Fundamenta Mathematicae***, vol. 53 (1964), pp. 205–224.

[Myc66] *On the axiom of determinateness. II*, ***Fundamenta Mathematicae***, vol. 59 (1966), pp. 203–212.

JOSEPH R. SHOENFIELD

[Sho67] ***Mathematical Logic***, Addison-Wesley, 1967.

ROBERT M. SOLOVAY

[Sol67B] *A nonconstructible* Δ^1_3 *set of integers*, ***Transactions of the American Mathematical Society***, vol. 127 (1967), no. 1, pp. 50–75.

ROBERT VAN WESEP

[Van78B] *Wadge degrees and descriptive set theory*, in Kechris and Moschovakis [CABAL i], pp. 151–170, reprinted in [CABAL II], pp. 24–42.

DEPARTMENT OF MATHEMATICS
UNIVERSITY OF CALIFORNIA
BERKELEY, CALIFORNIA 94720-3840
UNITED STATES OF AMERICA
E-mail: solovay@gmail.com

GAMES OF COUNTABLE LENGTH

DONALD A. MARTIN

We work throughout the paper in Zermelo-Fraenkel set theory without the Axiom of Choice. For our main theorem and its consequences, we shall need to assume as a hypothesis the Countable Axiom of Choice (AC_ω), which is the restriction of the Axiom of Choice to countable sets of sets.

Background information about determinacy and the Axiom of Determinacy can be found in each of [Mos80, Kan94, Kec95, CABAL I].

Let α be an ordinal number. We mean by an **α-game** a game in which player I and player II take turns playing natural numbers, with player I playing at 0 and limit ordinals, until a sequence of length α—*i.e.*, an element of ${}^\alpha\omega$—has been produced. Let $\mathsf{AD}(\alpha)$ be the assertion that all α-games are determined. Thus the Axiom of Determinacy (AD) is $\mathsf{AD}(\omega)$. By a result independently due to Andreas Blass and Jan Mycielski, the Axiom of Determinacy for Reals ($\mathsf{AD}_\mathbb{R}$) is equivalent to $\mathsf{AD}(\omega^2)$.

The notion of a **Suslin** subset of ${}^\omega\omega$ is defined below. By Scales we mean the assertion that every subset of ${}^\omega\omega$ is Suslin. Being Suslin is equivalent to *admitting a scale* in the sense of Moschovakis, and this explains the name "Scales".

We shall show (Theorem 10) that

$$\mathsf{AD}+\mathsf{AC}_\omega+\text{Scales implies that for all } \alpha < \omega_1,\ \mathsf{AD}(\alpha) \text{ holds.}$$

This result was proved independently by Hugh Woodin, by a proof that is different from ours. Both proofs date from the 1980s. After writing the present paper, we learned from Woodin that he had much more recently found a still different proof that does not need AC_ω. Given a payoff set A, this proof uses its hypotheses to get a model M of ZFC with many Woodin cardinals that is Σ^2_1 correct for A and in which A is homogeneously Suslin. A theorem of [Nee04, Exercise 2E.14] implies that countable length games with payoff set A are determined in M. Despite its weaker result, our short and more direct proof may be of some interest.

In our main theorem (Theorem 9), the hypothesis Scales will be replaced by an apparently stronger hypothesis, Homogeneous Scales. But it follows from, for example, [MS08, Theorem 1.1] (stated as Theorem 2 below) that the two hypotheses are equivalent in the presence of AD.

Large Cardinals, Determinacy, and Other Topics: The Cabal Seminar, Volume IV
Edited by A. S. Kechris, B. Löwe, J. R. Steel
Lecture Notes in Logic, 49

The property of being Suslin and that of being **homogeneously Suslin** are defined as follows. Let T be a tree on $\omega \times \delta$ for some ordinal δ. Define

$$[T] = \{(x, f) : (\forall n \in \omega)(x\restriction n, f\restriction n) \in T\};$$
$$\mathrm{p}[T] = \{x \in {}^{\omega}\omega : (\exists f)(x, f) \in [T]\};$$
$$T_s = \{t \in {}^{\mathrm{lh}(s)}\delta : (s, t) \in T\} \text{ (for } s \in {}^{<\omega}\omega).$$

A **homogeneity system** for T is a system $\langle \mu_s : s \in {}^{\omega}\omega\rangle$ such that

(i) for each s, μ_s is a countably additive $\{0, 1\}$-valued measure (*i.e.*, a countably complete ultrafilter) on T_s;
(ii) if $s_1 \subseteq s_2$ and $\mu_{s_1}(X) = 1$ then $\mu_{s_2}(\{t : t\restriction \text{length}(s_1) \in X\}) = 1$;
(iii) for all $x \in \mathrm{p}[T]$ and all sequences $\langle X_n : n \in \omega\rangle$, if $\mu_{x\restriction n}(X_n) = 1$ for every n, then there is an $f \in {}^{\omega}\delta$ such that $f\restriction n \in X_n$ for all n.

A subset of ${}^{\omega}\omega$ is **Suslin** if it is $\mathrm{p}[T]$ for some tree T. A set is **homogeneously Suslin** if it is $\mathrm{p}[T]$ for a tree T that has a homogeneity system. Homogeneous Scales is the assertion that every subset of ${}^{\omega}\omega$ is homogeneously Suslin. As is mentioned above, being Suslin is equivalent to admitting a scale. See any one of [KM78B, Ste08B, Mos80, Kan94, Kec81] for the definition of a *scale*.

The hypothesis Scales implies certain choice principles. Uniformization is the assertion that every subset of ${}^{\omega}\omega \times {}^{\omega}\omega$ can be uniformized, *i.e.*, that for all $A \subseteq {}^{\omega}\omega \times {}^{\omega}\omega$ there is a $B \subseteq A$ such that $\{x : (\exists y)(x, y) \in B\} = \{x : (\exists y)(x, y) \in A\}$ and B is the graph of a function. The Axiom of Dependent Choice for Reals ($\mathsf{DC}_{\mathbb{R}}$) asserts, for any relation R, that if $(\forall x \in {}^{\omega}\omega)(\exists y \in {}^{\omega}\omega)R(x, y)$ then there is a sequence $\langle x_i : i \in \omega\rangle$ such that $(\forall i \in \omega)R(x_i, x_{i+1})$.

LEMMA 1. Scales implies Uniformization, and Uniformization implies $\mathsf{DC}_{\mathbb{R}}$.

PROOF. The first assertion is proved, or essentially proved, in each of [KM78B, Mos80, Kan94, Kec95]. The proof of the second assertion is fairly obvious. ⊣

THEOREM 2 (Martin & Steel; [MS08]). $\mathsf{AD}+\mathsf{DC}_{\mathbb{R}}$ implies that every Suslin, co-Suslin subset of ${}^{\omega}\omega$ is homogeneously Suslin.

COROLLARY 3. Scales implies Homogeneous Scales.

Corollary 3 will let us weaken the hypotheses of Theorem 9 below to those of Theorem 10. The following major result of Woodin, proved a few years after the results mentioned so far, provides even more weakening.

THEOREM 4 (Woodin). AD+Uniformization implies Scales.

COROLLARY 5. $\mathsf{AD}_{\mathbb{R}}$ implies Scales. Indeed $\mathsf{AD}(\omega \cdot 2)$ implies Scales.

PROOF. Uniformization is easily seen to be equivalent to the determinacy of games of length 2 in which the players play reals (or elements of ${}^{\omega}\omega$).

Such determinacy follows trivially from $\mathsf{AD}_{\mathbb{R}}$ and also follows trivially from $\mathsf{AD}(\omega \cdot 2)$. ⊣

The cardinal number Θ is the least non-zero ordinal number not the surjective image of ${}^{\omega}\omega$. The following well-known fact will be used in the proof of Theorem 9.

Lemma 6. AD+$\mathsf{Uniformization}$ implies that every wellorderable set of subsets of ${}^{\omega}\omega$ has cardinal number $< \Theta$.

Proof. Let $\mathcal{A}$ be a wellordered set of subsets of ${}^{\omega}\omega$. Let $\kappa = \text{Card}(\mathcal{A})$ and let $\alpha \mapsto A_\alpha$ be a bijection from κ to $\mathcal{A}$. Let

$$R = \{(x, y) : x \in {}^{\omega}\omega \wedge y \in {}^{\omega}\omega \wedge y \notin \mathbf{OD}[x, \{(\alpha, z) : \alpha < \kappa \wedge z \in A_\alpha\}]\}.$$

(Here, $\mathbf{OD}[z_1, z_2]$ is the class of all sets ordinal definable from z_1 and z_2.) By $\mathsf{Uniformization}$, let f uniformize R. Let B be a subset of ${}^{\omega}\omega$ coding f.

We show that $A \leq_W B$ for every $A \in \mathcal{A}$. If this is false, then by Wadge's Lemma there is an $A \in \mathcal{A}$ such that $B \leq_W \neg A$. For such an A, let x be an element of ${}^{\omega}\omega$ coding a function witnessing $B \leq_W \neg A$. Then B is definable from x and A. Since A is ordinal definable from $\{(\alpha, z) : \alpha < \kappa \wedge z \in A_\alpha\}$, it follows that B—and so f—is ordinal definable from x and $\{(\alpha, z) : \alpha < \kappa \wedge z \in A_\alpha\}$. This gives us the contradiction that $f(x) \in \mathbf{OD}[x, \{(\alpha, z) : \alpha < \kappa \wedge z \in A_\alpha\}]$.

For each continuous $g : {}^{\omega}\omega \to {}^{\omega}\omega$, let $\Phi(g)$ be the g-preimage of B. By what we just proved, $\mathcal{A} \subseteq \text{ran}(\Phi)$. Hence $\text{Card}(\mathcal{A}) < \Theta$. ⊣

A key step in the proof of Theorem 9 will begin with a set $A^* \subseteq {}^{\omega}\omega$ and a tree and homogeneity system witnessing that A^* is homogeneously Suslin. From these ingredients, we shall get a tree T on $\omega \times \Theta$ such that:

(a) $\text{p}[T] = \neg A^*$;
(b) for every cardinal number $\kappa < \Theta$, T has a homogeneity system whose measures are all κ-complete.

We now list some facts which together will yield this step.

The main ingredient in getting T is what Jackson calls the **Martin-Solovay tree** construction. For this construction, *cf.* [Jac08, p. 302]. Here is the version of this construction we shall need.

Let $i \mapsto s_i$ be an enumeration of ${}^{<\omega}\omega$ such that $s_i \subseteq s_j \to i \leq j$. Let S be a tree on $\omega \times \delta$ for some ordinal δ. Let $\vec{\mu} = \langle \mu_s : s \in {}^{<\omega}\omega \rangle$ be a homogeneity system for S. If $\gamma \geq \delta^+$, we define a tree $T(S, \vec{\mu}, \gamma)$ on $\omega \times \gamma$. Let $(s, w) \in T(S, \vec{\mu}, \gamma)$ just in case there is a function f from $\{(s_i, t) : i \leq \text{lh}(s) \wedge (s_i, t) \in S\}$ to γ such that

(i) f is order-preserving with respect to the Brouwer-Kleene ordering on $\text{dom}(f)$;
(ii) for each $i \leq \text{lh}(s)$, if $\mu_{s_i}(\{t : (s_i, t) \in S\}) = 1$, then the function f_{s_i} given by $f_{s_i}(t) = f(s_i, t)$ represents $w(i)$ in the ultrapower of V by μ_{s_i}.

Theorem 7. Let δ, S, and $\vec{\mu}$ be as above. Let γ be such that $\delta^+ \leq \gamma$.

(1) $p[T(S, \vec{\mu}, \gamma)] = \neg p[S]$.
(2) If there is a cardinal κ such that $\delta^+ \leq \kappa \leq \Theta$ and $\kappa \to (\kappa)^\kappa$, then $p[T(S, \vec{\mu}, \Theta)]$ has a homogeneity system with κ-complete measures.

PROOF. For the proof of something a little more general than part (1), *cf.* [Jac08, pp. 302–303]. For a proof of a variant of a more general version of part (2), *cf.* [Jac08, pp. 304–305]. ⊣

To get the cardinal we need for our application of (2), we shall use the the following theorem of [KKMW81]:

THEOREM 8. For every $\alpha < \Theta$, there is a κ with $\kappa \leq \alpha < \Theta$ such that $\kappa \to (\kappa)^\kappa$.

We now turn to our main result (which is, as we have already said, independently due to Woodin).

THEOREM 9. Assume $\mathsf{AD}+\mathsf{AC}_\omega+$Homogeneous Scales. Then $\mathsf{AD}(\alpha)$ holds for every countable ordinal α.

PROOF. We prove the theorem by induction on α. Fix a countable α, and assume that $\mathsf{AD}(\beta)$ holds for every $\beta < \alpha$. If α is 0 or a successor, then $\mathsf{AD}(\alpha)$ follows easily. We may thus assume that α is a limit ordinal.

Let

$$0 = \alpha_0 < \alpha_1 < \cdots$$

with $\lim_i \alpha_i = \alpha$. Let $k \mapsto \beta_k$ be a one-one correspondence between ω and α. Choose this correspondence so that $\beta_k < \alpha_{k+1}$ for all k. For each element x of ${}^\alpha\omega$, let $x^* \in {}^\omega\omega$ be given by

$$x^*(k) = x(\beta_k).$$

For $n \in \omega$ and $q \in {}^{\alpha_n}\omega$, let q^* be the common value of $x^* \restriction n$ for all $x \in {}^\alpha\omega$ such that $q \subseteq x$. (Note that, while x^* determines x, q^* determines only finitely much of q.) For any subset A of ${}^\alpha\omega$, let $A^* \subseteq {}^\omega\omega$ be given by

$$A^* = \{x^* : x \in A\}.$$

Note that $\neg A^* = (\neg A)^*$.

Now fix $A \subseteq {}^\alpha\omega$. Let G be the α-game in which player I's winning set is A. Let S and $\langle \mu_s : s \in {}^{<\omega}\omega\rangle$ witness that A^* is homogeneously Suslin. Let $T = T(S, \vec{\mu}, \Theta)$.

Consider a game $\tilde{G}$ in which the players produce an $x \in {}^\alpha\omega$ as in G and player II also produces an $f : \omega \to \Theta$. Let player II's move $f(n)$ be made immediately before the move $x(\alpha_{n+1})$ is made. Player II wins just in case $(x^*, f) \in [T]$.

For positions p in $\tilde{G}$, let q_p be the x-part of p and let t_p be the f-part of p. If q and t are, respectively, the x-part and the f-part of a position in $\tilde{G}$, call that position $p(q, t)$. Say that a position p in $\tilde{G}$ is **significant** if $p = \varnothing$ or if there is an $n > 0$ such that the last move in p is $f(n-1)$. Note that, in either case, $q_p \in {}^{\alpha_n}\omega$ for some $n \geq 0$.

We assign ordinal numbers $\|p\|$ to some of the significant positions p in the game $\tilde{G}$. Set $\|p\| = 0$ just in case p (is significant and) $(q_p^*, t_p) \notin T$. For $\gamma > 0$ inductively define $\|p\| \leq \gamma$ to hold just in case player II does not have a winning strategy for the following game $H(p, \gamma)$. If $q_p \in {}^{\alpha_n}\omega$ then the players start $H(p, \gamma)$ at p and play the next $(\alpha_{n+1} - \alpha_n)$ moves in $\tilde{G}$. Player I wins if and only if, for every move ξ by player II in the resulting position p' in $\tilde{G}$, $\|p'{}^\frown\langle\xi\rangle\| < \gamma$ (*i.e.*, $\|p'{}^\frown\langle\xi\rangle\|$ is defined and there is a $\delta < \gamma$ $\|p'{}^\frown\langle\xi\rangle\| \leq \delta$).

Since each $H(p, \gamma)$ is an $(\alpha_{n+1} - \alpha_n)$-game, our induction hypothesis implies that each is determined.

Say that a significant p in $\tilde{G}$ is **ranked** if $\|p\|$ is defined. Let λ be greater than $\|p\|$ for every significant ranked p.

Suppose that p' is the position in $\tilde{G}$ resulting from a winning play for player II in some $H(p, \lambda)$. By the winning condition for $H(p, \lambda)$ and the definition of λ, there is an ordinal ξ such that $p'{}^\frown\langle\xi\rangle$ is an unranked significant position in $\tilde{G}$. Call such a ξ a **good** move for player II at p'. Call an unranked significant position p **minimal** if each of player II's ordinal moves has been good and has been the least good move.

Fix $n \in \omega$. There is a function that assigns a winning strategy τ_p for player II in $H(p, \lambda)$ to each minimal unranked significant position p in $\tilde{G}$ with $q_p \in {}^{\alpha_n}\omega$. To see this, consider the following game. Player I must first choose a minimal unranked significant p with $q_p \in {}^{\alpha_n}\omega$. Then the players play $H(p, \lambda)$. This game is essentially an α_{n+1}-game, and so our induction hypothesis guarantees that it is determined. Obviously player I cannot have a winning strategy. A winning strategy for player II gives us our $p \mapsto \tau_p$.

By AC_ω, we can get $p \mapsto \tau_p$ for all n simultaneously.

Assume that the initial position in $\tilde{G}$ is unranked. We get a winning strategy $\tilde{\tau}$ for player II in $\tilde{G}$ as follows. Whenever a minimal unranked significant p is reached, player II follows the strategy τ_p and then plays the least good move in the resulting position p' in $\tilde{G}$. No significant p consistent with $\tilde{\tau}$ is ranked, so none satisfies $\|p\| = 0$. Hence if x and p are the components of a play consistent with $\tilde{\tau}$ then $(x^*, f) \in [T]$ and so $x^* \in \neg A^*$ and $x \in \neg A$. The obvious strategy for G derived from $\tilde{\tau}$ is thus a winning strategy in G for player II.

For each $t \in {}^{<\omega}\Theta$, let $Q(t)$ be the set of all pairs (q, q') such that

(i) there is a ranked significant position p in $\tilde{G}$ with $t_p = t$ and $q_p = q$, *i.e.*, with $p = p(q, t)$;
(ii) q' is a play of $H(p, \|p\|)$ that is a win for player I.

Let $\mathcal{Q} = \{Q(t) : t \in {}^{<\omega}\Theta\}$. The set $\mathcal{Q}$ is a wellorderable set of subsets of ${}^{<\alpha}\omega$. Using a bijection between α and ω, we can get a wellorderable collection $\mathcal{Q}'$ of subsets of ${}^{\omega}\omega$ with $\text{Card}(\mathcal{Q}') = \text{Card}(\mathcal{Q})$. By Lemma 6, $\text{Card}(\mathcal{Q}') < \Theta$.

By Theorem 8, let $\kappa \to (\kappa)^\kappa$ with $\mathrm{Card}(\mathcal{Q}) < \kappa < \Theta$. By Part (2) of Theorem 7, let $\langle \nu_r \mid r \in {}^{<\omega}\omega\rangle$ be a homogeneity system for T such that each ν_r is κ-complete.

For each $n \in \omega$, consider the following game $\hat{G}_n$. On pain of losing, player II must first choose a $q \in {}^{\alpha_n}\omega$ such that

$$\nu_{q^*}(\{t \in {}^n\Theta : p(q,t) \text{ is ranked}\}) = 1.$$

It is not hard to see that $\nu_{q^*}(\{t \in {}^n\Theta : p(q,t) \text{ has rank } 0\})$ is never 0, and and so $H(p(q,t), \|p(q,t)\|)$ is defined for any non-losing (not immediately losing) first move q by player II. For such a q, the remainder of the play of $\hat{G}$ consists of a play of $H(p(q,t), \|p(q,t)\|)$, and we define the winner of the one game to be the winner of the other.

$\hat{G}_n$ is essentially an α_{n+1}-game, and so it is determined. For any non-losing first move q, player I has a winning set for $H(p(q,t), \|p(q,t)\|)$ that—by the κ-completeness of ν_{q^*}—is constant on a set X with $\nu_{q^*}(X) = 1$. Thus player II cannot have a winning strategy for $\hat{G}_n$, and so player I has a winning strategy. By AC_ω, there is a function that assigns to each n a winning strategy $\hat{\sigma}_n$ for player I for $\hat{G}_n$.

Now assume that the initial position in $\tilde{G}$ is ranked. We define a strategy σ for player I for $\hat{G}_n$ as follows. Assume inductively that $\sigma(q)$ has been defined for every q of length $< \alpha_n$ and assume that every position of length α_n consistent with σ is a not immediately losing first move for player II in $\hat{G}_n$. The latter assumption holds trivially for $n = 0$. Let q_n have length α_n and be consistent with σ. Starting at q_n, the strategy σ follows $\hat{\sigma}_n$, taking player I's first move in $\hat{G}_n$ to be q_n. Let q_{n+1} be any position of length α_{n+1} reached. Since $\hat{\sigma}_n$ is a winning strategy, there is a set X with $\nu_{q_n^*}(X) = 1$ and such that, for every move ξ in the position $p(q_{n+1}, t)$ in $\tilde{G}$,

$$\|p(t^\frown\langle\xi\rangle, q_{n+1})\| < \|p(t, q_n)\|.$$

This not only verifies our induction hypothesis for $n+1$, it also shows that for any play x consistent with σ there are sets X_n, $n \in \omega$, such that:

(i) $(\forall n)\nu_{x^*\restriction n}(X_n) = 1$;

(ii) $(\forall n)(\forall t \in X_{n+1})\|p(t, x\restriction\alpha_{n+1})\| < \|p(t\restriction n, x\restriction n)\|$.

The existence of such sets, together with clause (iii) of the definition of a homogeneity system—implies that no play consistent with σ can be a win for player II in G. ⊣

THEOREM 10. $\mathsf{AD}+\mathsf{AC}_\omega+\mathsf{Scales}$ implies that $\mathsf{AD}(\alpha)$ holds for all countable α.

PROOF. This follows from the theorem, Theorem 2, and Lemma 1. ⊣

COROLLARY 11. Each of the following implies that $\mathsf{AD}(\alpha)$ holds for every countable ordinal α:

(a) $\mathsf{AD}+\mathsf{AC}_\omega+\mathsf{Uniformization}$;

(b) $\mathsf{AD}_{\mathbb{R}}+\mathsf{AC}_\omega$;
(c) $\mathsf{AD}(\omega\cdot 2)+\mathsf{AC}_\omega$.

PROOF. (a) follows from Theorem 10 and Theorem 4. (b) and (c) follow from (a) and Corollary 5. ⊣

Of course, the proof of Woodin mentioned at the beginning of the paper yields Theorem 10 and the corollary with "+ AC_ω" removed.

REFERENCES

STEPHEN JACKSON

[Jac08] *Suslin cardinals, partition properties, homogeneity. Introduction to Part II*, in Kechris et al. [CABAL I], pp. 273–313.

AKIHIRO KANAMORI

[Kan94] ***The Higher Infinite. Large Cardinals in Set Theory from Their Beginnings***, Perspectives in Mathematical Logic, Springer-Verlag, Berlin, 1994.

ALEXANDER S. KECHRIS

[Kec81] *Homogeneous trees and projective scales*, in Kechris et al. [CABAL ii], pp. 33–74, reprinted in [CABAL II], pp. 270–303.

[Kec95] ***Classical Descriptive Set Theory***, Graduate Texts in Mathematics, vol. 156, Springer-Verlag, 1995.

ALEXANDER S. KECHRIS, EUGENE M. KLEINBERG, YIANNIS N. MOSCHOVAKIS, AND W. H. WOODIN

[KKMW81] *The axiom of determinacy, strong partition properties, and nonsingular measures*, in Kechris et al. [CABAL ii], pp. 75–99, reprinted in [CABAL I], pp. 333–354.

ALEXANDER S. KECHRIS, BENEDIKT LÖWE, AND JOHN R. STEEL

[CABAL I] *Games, Scales, and Suslin cardinals: the Cabal Seminar, volume I*, Lecture Notes in Logic, vol. 31, Cambridge University Press, 2008.

[CABAL II] *Wadge Degrees and Projective Ordinals: the Cabal Seminar, volume II*, Lecture Notes in Logic, vol. 37, Cambridge University Press, 2012.

ALEXANDER S. KECHRIS, DONALD A. MARTIN, AND YIANNIS N. MOSCHOVAKIS

[CABAL ii] *Cabal Seminar* 77–79, Lecture Notes in Mathematics, vol. 839, Berlin, Springer-Verlag, 1981.

ALEXANDER S. KECHRIS AND YIANNIS N. MOSCHOVAKIS

[KM78B] *Notes on the theory of scales*, in *Cabal Seminar* 76–77 [CABAL i], pp. 1–53, reprinted in [CABAL I], pp. 28–74.

[CABAL i] *Cabal Seminar* 76–77, Lecture Notes in Mathematics, vol. 689, Berlin, Springer-Verlag, 1978.

DONALD A. MARTIN AND JOHN R. STEEL

[MS08] *The tree of a Moschovakis scale is homogeneous*, in Kechris et al. [CABAL I], pp. 404–420.

YIANNIS N. MOSCHOVAKIS

[Mos80] ***Descriptive Set Theory***, Studies in Logic and the Foundations of Mathematics, vol. 100, North-Holland, Amsterdam, 1980.

ITAY NEEMAN

[Nee04] ***The Determinacy of Long Games***, de Gruyter Series in Logic and its Applications, vol. 7, Walter de Gruyter, Berlin, 2004.

John R. Steel
[Ste08B] *Games and scales. Introduction to Part I*, in Kechris et al. [Cabal I], pp. 3–27.

DEPARTMENT OF MATHEMATICS
UNIVERSITY OF CALIFORNIA
LOS ANGELES, CALIFORNIA 90095
UNITED STATES OF AMERICA
E-mail: dam@math.ucla.edu

SOME CONSISTENCY RESULTS IN ZFC USING AD

W. HUGH WOODIN

For the most part the uses of the axiom of determinacy (AD) have been to settle natural questions that arise about sets under its influence, *i.e.*, (certain) sets of reals. This combined with the fact that to assume AD requires restricting ones attention to a fragment of the universe in which the axiom of choice fails, would seem to indicate that AD has little to offer in the way of solutions to problems in more conventional set theory. Set theorists as a rule ignore constraints of definability in choosing objects for their amusement nor do they wish to abandon the axiom of choice.

Recently, however, there have been several applications of AD to obtain consistency results in ZFC (cf. [SVW82, Woo]). These methods revolve around starting with a model of ZF + AD and constructing a forcing extension in which ZFC holds, the hope being that enough of the influence of AD will extend to produce some desired property in the generic extension.

We shall be concerned with the results obtained by Steel and Van Wesep [SVW82], they show the consistency of ZFC together with ω_2 is the second uniform indiscernible and the nonstationary ideal on ω_1 is ω_2-saturated. The latter solves a well known problem within the theory of saturated ideals. For their result, Steel and Van Wesep needed to assume the consistency of ZF + AD + $\mathbb{R}$-AC, where $\mathbb{R}$-AC is the axiom of choice for families indexed by the reals. This theory is substantially stronger than ZF + AD. We reduce the assumption needed to just the consistency of ZF + AD. As we have suggested their method is to start with a model of ZF + AD + $\mathbb{R}$-AC and then construct a forcing extension in which ZFC holds. The forcing is mild enough so that in the generic extension ω_2 is the second uniform indiscernible and the nonstationary ideal on ω_1 is ω_2-saturated. Basically we simply show that ZF + AD + $\mathbf{V}{=}\mathbf{L}(\mathbb{R})$ suffices to carry out their forcing arguments. We also isolate a single combinatorial principle which in ZFC implies both that ω_2 is the second uniform indiscernible and that the nonstationary ideal on ω_1 is ω_2-saturated. This principle we show holds in the generic extension, The point of this is twofold. First it offers a means to those uninterested in AD for mining

Research partially supported by NSF Grant MCS 80-21468.

Large Cardinals, Determinacy, and Other Topics: The Cabal Seminar, Volume IV
Edited by A. S. Kechris, B. Löwe, J. R. Steel
Lecture Notes in Logic, 49

the combinatorial riches of this model and second it suggests that a theory weaker than AD may suffice for these consistency results.

Finally we extend the results of [SVW82] to show that if ZF + AD is consistent then so is ZFC + MA + ¬CH + "the nonstationary ideal on ω_1 is ω_2-saturated".

We shall for the most part be working in ZF + DC + AD throughout this paper. Notation for the most part will be as in [SVW82] and we assume familiarity with the elementary aspects of set theory in the context of AD, as presented in [SVW82].

§1. As usual we define the reals as elements of ω^ω. Giving ω the discrete topology naturally induces a product topology on ω^ω, it is with respect to this topology that we define the notion of category. Of course ω^ω with this topology is homeomorphic to the space of irrationals so in addition we have naturally a notion of Lebesgue measure on ω^ω.

Suppose $\alpha < \beta$ are countable ordinals. Let T_α denote the space $(\omega + \alpha)^\omega$ and let $T_{\alpha,\beta} = \prod_{\alpha \le \delta < \beta} T_\delta$. The spaces T_α, $T_{\alpha,\beta}$ with their natural topologies are each homeomorphic to ω^ω (for T_α choose the topology induced by the discrete topology on $\omega + \alpha$, for $T_{\alpha,\beta}$ choose the product topology). Suppose $f \in T_{\alpha,\beta}$. Thus f is a function with domain $[\alpha, \beta) = \{\delta : \alpha \le \delta < \beta\}$ such that for each $\delta \in [\alpha, \beta)$, $f(\delta) \in T_\delta$. We denote $f(\delta)$ by f_δ. Note that f_δ is also a function, $f_\delta \colon \omega \to \omega + \delta$. We are using notation from [SVW82].

We define the basic notion of forcing introduced in [SVW82].

Definition 1.1. Let P denote the set of conditions defined as follows. A condition is a pair $(f, \vec{X})$ such that $f \in T_{0,\alpha}$ for some $\alpha < \omega_1$, and $\vec{X} = \langle X_\beta : \beta < \omega_1 \rangle$, where:

(1) For each $\delta < \alpha$, $f_\delta \colon \omega \to \omega + \delta$ is onto.
(2) For each $\beta < \omega_1$, $X_\beta \subseteq T_{0,\beta}$ such that if $\beta > \alpha$ then $\{h \in T_{\alpha,\beta} : f^\frown h \in X_\beta\}$ is comeager in $T_{\alpha,\beta}$.
(3) (coherence) For each $\delta, \beta, \alpha \le \delta < \beta$;
    (a) for $g \in X_\beta$, if $f \subseteq g$ then $g_\delta \colon \omega \to \omega + \delta$ is onto and $g \restriction \delta \in X_\delta$.
    (b) for $g \in X_\delta$, if $f \subseteq g$ then $\{h \in T_{\delta,\beta} : g^\frown h \in X_\beta\}$ is comeager in $T_{\delta,\beta}$.

Define the order on P by

$$(g, \vec{Y}) \le (f, \vec{X}) \text{ iff } f \subseteq g,\ g \in \bigcup X_\beta \text{ and for each } \beta < \omega_1,\ Y_\beta \subseteq X_\beta.$$

Let $\mathbb{P}$ denote the partial order $(P, \le)$.

Let $P' = \{(f, \vec{X}) : (f, \vec{X}) \text{ satisfies (1) and (2) in the definition of } P\}$. Let $\mathbb{P}'$ denote P' together with the obvious order extending the order on P.

If we assume that every set of reals has the property of Baire then $\mathbb{P}$ is dense in $\mathbb{P}'$. Specifically if $(f, \vec{X}) \in P'$ then there is a refinement $\vec{Y}$ of $\vec{X}$ such that $(f, \vec{Y}) \in P$, $(f, \vec{Y}) \le (f, \vec{X})$ in $\mathbb{P}'$. The construction of $\vec{Y}$ is straightforward using the following observations:

(1) (Kuratowski, Ulam) Suppose $A \subseteq \mathbb{R} \times \mathbb{R}$ is comeager, then $\{x : A_x$ is comeager$\}$ is comeager. $A_x = \{y : (x, y) \in A\}$. This is simply the Fubini theorem for category.
(2) Suppose $\langle A_\zeta : \zeta < \lambda \rangle$ is a wellordered sequence of comeager sets of reals. Then $\bigcap A_\zeta$ is comeager.

Note that (1) is true in ZF + DC, (2) is true in ZF + DC if in addition one assume that every set of reals has the property of Baire. Under this additional assumption the converse of (1) is true as well.

Thus assuming that every set of reals has the property of Baire, forcing with $\mathbb{P}$ is equivalent to forcing with $\mathbb{P}'$. All of this is noted in [SVW82] to which we refer the reader for more details.

We now define a set of conditions Q_{ω_1}; $t \in Q_{\omega_1}$ iff t is a function with domain a finite subset of ω_1, such that for $\delta \in \operatorname{dom} t$, $t(\delta) \in (\omega + \delta)^{<\omega}$. Order the elements of Q_{ω_1} by $t_1 \leq t_2$ iff $t_2 \subseteq t_1$. Let $\mathbb{Q}_{\omega_1}$ denote the corresponding partial order $(Q_{\omega_1}, \leq)$.

Suppose α, β are countable ordinals with $\alpha < \beta$. Each element, t, of Q_{ω_1} naturally defines a basic open set in $T_{\alpha,\beta}$. We denote this by $[t]_{\alpha,\beta}$. Thus $[t]_{\alpha,\beta} = \{f \in T_{\alpha,\beta} : t(\delta) \subseteq f_\delta$ for each $\delta \in \operatorname{dom} t \cap [\alpha, \beta)\}$. Frequently when the subscripts are implicit from context we denote $[t]_{\alpha,\beta}$ by $[t]$. Of course $[t]_{\alpha,\beta}$ depends only on $t \restriction [\alpha, \beta)$. Let $Q_{\alpha,\beta} = \{t \in Q_{\omega_1} : \operatorname{dom} t \subseteq [\alpha, \beta)\}$ and let $\mathbb{Q}_{\alpha,\beta}$ denote the associated partial order. Note that the completion of $\mathbb{Q}_{\alpha,\beta}$ is isomorphic to the algebra of regular open subsets of $T_{\alpha,\beta}$.

Though most of our considerations are within the theory of ZF + AD + DC, some of the theorems we prove require less. Much in the spirit of [SVW82] we indicate weaker hypotheses when it is easy to do so. We abbreviate 'all sets of reals have the property of Baire' by C.

The following lemmas are proved in [SVW82] and require only ZF + DC + C. We include them with proofs for the sake of completeness.

Lemma 1.2 (ZF + DC + C). *Assume $D \subseteq P$ is open and dense in $\mathbb{P}$. Suppose $(f, \vec{X}) \in P$ with $f \in T_{0,\alpha}$ and for $\beta > \alpha$ let $S_\beta = \{h \in T_{\alpha,\beta} : f^\frown h \in X_\beta$ and $(f^\frown h, \vec{Y}) \in D$ for some $\vec{Y}\}$. Then for sufficiently large β, S_β is comeager in $T_{\alpha,\beta}$.*

Proof. We are given that every set of reals has the property of Baire, hence if S_β is not comeager in $T_{\alpha,\beta}$ there must exist $t \in Q_{\alpha,\beta}$ such that $S_\beta \cap [t]$ is meager. We claim that for each $t \in Q_{\omega_1}$ there exist β_t such that for $\beta > \beta_t$, $S_\beta \cap [t]$ is not meager in $T_{\alpha,\beta}$. If not then for each β such that $S_\beta \cap [t]$ is meager let $Z_\beta = \{f^\frown h : h \notin S_\beta \cap [t]\}$. Let $Z_\beta = X_\beta$ otherwise. $(f, \vec{Z}) \in P'$ so choose $(f, \vec{Y}) \in P$ with $Y_\beta \subseteq Z_\beta$ for each β. Choose $g \in Y_\beta$ such that $t \in Q_{\alpha,\beta}$ and $g \in [t]_{\alpha,\beta}$. Then $(f^\frown g, \vec{Y}) \in P$ and has no extension in D.

Thus for each $t \in Q_{\omega_1}$ there exists β_t such that for $\beta > \beta_t$, $S_\beta \cap [t]$ is meager in $T_{\alpha,\beta}$. Choose a limit ordinal γ, $\alpha < \gamma < \omega_1$, such that for $\beta < \gamma$ and

$t \in Q_{\alpha,\beta}$, $\beta_t < \gamma$. Hence $S_\gamma \cap [t]$ is not meager for each $t \in Q_{\alpha,\gamma}$ and therefore S_γ is comeager in $T_{\alpha,\gamma}$. Finally this implies that S_β is comeager in $T_{\alpha,\beta}$ for $\beta > \gamma$. ⊣

LEMMA 1.3 (ZF + DC + C). The partial order $\mathbb{P}$ is (ω, ∞)-distributive.

PROOF. Let $\langle D_n : n < \omega \rangle$ be a sequence of open dense sets in $\mathbb{P}$. Suppose $(f, \vec{X}) \in \mathbb{P}$ with $f \in T_{0,\alpha}$. It suffices to find $(g, \vec{Y}) \leq (f, \vec{X})$ such that $(g, \vec{Y}) \in D_n$ for each n. For each n and β, $\alpha < \beta < \omega_1$, define S^n_β relative to D_n and $(f, \vec{X})$ as S_β is defined in Lemma 1.2. By Lemma 1.2 there exists $\overline{\beta} < \omega_1$ such that for each n, $S^n_{\overline{\beta}}$ is comeager in $T_{\alpha,\overline{\beta}}$. Choose $h \in \bigcap_n S^n_{\overline{\beta}}$. Using DC choose for each n, $\vec{Y}_n$ such that $(f^\frown h, \vec{Y}_n) \leq (f, \vec{X})$ and $(f^\frown h, \vec{Y}_n) \in D_n$. For each $\beta < \omega_1$ let $Y_\beta = \bigcap_n (\vec{Y}_n)_\beta$. $(f^\frown h, \vec{Y})$ is as required. ⊣

We shall now use AD to prove two additional theorems regarding $\mathbb{P}$. First we establish some mild coding, we need to code both elements of $T_{\alpha,\beta}$ and simple subsets of $T_{\alpha,\beta}$ by reals in a reasonably effective fashion.

Let $\mathbf{H}_{\omega_1}$ denote all sets with countable transitive closure, *i.e.*, $\mathbf{H}_{\omega_1}$ is the set of all hereditarily countable sets. Suppose $a \in \mathbf{H}_{\omega_1}$ and let $\mathrm{TC}(a)$ denote the transitive closure of a. The set a can be easily coded by coding the countable structure $\langle \mathrm{TC}(a), a, \in \rangle$. Thus we choose as codes for a, structures $\langle \omega, A, E \rangle$ (where $A \subseteq \omega$, $E \subseteq \omega \times \omega$) that are isomorphic to $\langle \mathrm{TC}(a), a, \in \rangle$. Codes of a we regard in an obvious fashion as reals, this simply amounts to identifying elements of $\wp(\omega) \times \wp(\omega \times \omega)$ with reals.

This method of encoding defines a partial map π of $\mathbb{R}$ onto $\mathbf{H}_{\omega_1}$ with domain a Π^1_1 set of reals. Further π chosen in this manner is easily seen to be Δ_1 over $\mathbf{H}_{\omega_1}$. Recall that a subset of $\mathbf{H}_{\omega_1}$ is Σ_1 over $\mathbf{H}_{\omega_1}$ if it is definable in the structure $\langle \mathbf{H}_{\omega_1}, \in \rangle$ by a Σ_1-formula without parameters, similarly a subset is Π_1 over $\mathbf{H}_{\omega_1}$ if it can be defined by a Π_1-formula without parameters. A subset of $\mathbf{H}_{\omega_1}$ is Δ_1 if it is both Σ_1 and Π_1 over $\mathbf{H}_{\omega_1}$. A subset is $\utilde{\Sigma}_1$ if it is definable by a Σ_1-formula with parameters from $\mathbf{H}_{\omega_1}$, similarly for $\utilde{\Pi}_1$, $\utilde{\Delta}_1$. In light of the fact that the coding map π is Δ_1, it follows that a set is $\utilde{\Sigma}_1$ iff it can be defined by a Σ_1-formula with real parameters.

Fix $B \subseteq \mathbb{R} \times \mathbf{H}_{\omega_1}$ that is Σ_1 over $\mathbf{H}_{\omega_1}$ and universal for $\utilde{\Sigma}_1$ subsets of $\mathbf{H}_{\omega_1}$, *i.e.*, such that for each $A \subseteq \mathbf{H}_{\omega_1}$, $\utilde{\Sigma}_1$ over $\mathbf{H}_{\omega_1}$, there is a real, x, with $A = B_x$ where $B_x = \{a \in \mathbf{H}_{\omega_1} : \langle x, a \rangle \in B\}$.

Note that $T_{\alpha,\beta} \subseteq \mathbf{H}_{\omega_1}$ and therefore π defines a coding of the spaces $T_{\alpha,\beta}$. Also it is easily verified that Borel subsets of $T_{\alpha,\beta}$ are $\utilde{\Delta}_1$ over $\mathbf{H}_{\omega_1}$. Hence Borel subsets of $T_{\alpha,\beta}$ are coded by reals via the universal set B, *i.e.*, $x \in \mathbb{R}$ codes $A \subseteq T_{\alpha,\beta}$ provided that $A = B_x$. We can assume by judicious choice of B that for $x \in \operatorname{dom} \pi$, $B_x = \{\pi(x)\}$, therefore we view the coding induced by B as extending π.

We are perhaps being overly technical, we are simply coding $\underset{\sim}{\Sigma}_1$ subsets of $\mathbf{H}_{\omega_1}$ by reals via a universal set with the additional requirement that the coding of ordinals be a standard one, the latter for boundedness considerations.

Finally conditions $(f, \vec{X}) \in P$ we regard as subsets of $\mathbf{H}_{\omega_1}$ by identifying $(f, \vec{X})$ with the corresponding set, $\{\langle f, g\rangle : g \in \bigcup X_\alpha\}$. Let $\underset{\sim}{\Sigma}^1_1(\mathbf{H}_{\omega_1}) = \{A \subseteq \mathbf{H}_{\omega_1} : A \text{ is } \underset{\sim}{\Sigma}_1 \text{ over } \mathbf{H}_{\omega_1}\}$.

THEOREM 1.4 (ZF + AD + DC). *The set $P \cap \mathbf{L}(\mathbb{R})$ is dense in $\mathbb{P}$. In fact $P \cap \underset{\sim}{\Sigma}_1(\mathbf{H}_{\omega_1})$ is dense in $\mathbb{P}$.*

PROOF. Suppose $(f, \vec{X})$ is a condition in P with $f \in T_\alpha$. For each $\beta < \omega_1$, $\beta > \alpha$, let $X_{\alpha,\beta} = \{h \in T_{\alpha,\beta} : f^\frown h \in X_\beta\}$. Thus $X_{\alpha,\beta}$ is comeager in $T_{\alpha,\beta}$. We construct a sequence $\langle Z_{\alpha,\beta} : \beta < \omega_1, \beta > \alpha\rangle$ in $\mathbf{L}(\mathbb{R})$ such that for each $\beta > \alpha$, $Z_{\alpha,\beta}$ is comeager in $T_{\alpha,\beta}$ and $Z_{\alpha,\beta} \subseteq X_{\alpha,\beta}$. To find $\langle Z_{\alpha,\beta} : \beta < \omega_1, \beta > \alpha\rangle$ consider the following Solovay-type game:

I	II
n_0	m_0
n_1	m_1
$\vdots$	$\vdots$
x	y

Player II wins iff whenever x codes a countable ordinal $\gamma > \alpha$ then y codes a sequence $\langle Z_{\alpha,\beta} : \beta < \beta_0, \beta > \alpha\rangle$ where $\beta_0 \geq \gamma$ and each $Z_{\alpha,\beta}$ is comeager in $T_{\alpha,\beta}$, $Z_{\alpha,\beta} \subseteq X_{\alpha,\beta}$. View y as coding $\bigcup Z_{\alpha,\beta}$. Thus player II wins iff whenever x codes a countable ordinal $\gamma > \alpha$, *i.e.*, $B_x = \{\gamma\}$, $\gamma > \alpha$, then for all $\beta \leq \gamma$, $\beta > \alpha$, $B_y \cap T_{\alpha,\beta}$ is comeager and contained in $X_{\alpha,\beta}$.

By standard arguments player I cannot have a winning strategy since any strategy for player I can be forced to produce a real that fails to code an ordinal or must always produce codes for ordinals bounded by some fixed countable ordinal γ_0. In the latter case player II can easily defeat the strategy by playing a code for a sequence $\langle Z_{\alpha,\beta}\rangle$ of length longer than γ_0. Implicit in this is the fact that each $X_{\alpha,\beta}$ contains a comeager Borel set.

Thus since by AD the game is determined, player II must have a winning strategy. Let s be such a strategy (which we regard as a real) and for $x \in \mathbb{R}$ let $s(x)$ denote the response by s to player I playing x.

Let $Z_{\alpha,\beta} = \{h \in T_{\alpha,\beta} : h \in B_{s(x)}$ for some real x, a code of $\gamma > \beta\}$. Thus for each $\beta < \omega_1$, $\beta > \alpha$, $Z_{\alpha,\beta}$ is comeager in $T_{\alpha,\beta}$ and is a subset of $X_{\alpha,\beta}$. Let $Z_\beta = \{f^\frown h : h \in Z_{\alpha,\beta}\}$. $(f, \vec{Z}) \in P'$ and clearly $(f, \vec{Z}) \in \mathbf{L}(\mathbb{R})$. Thus $(f, \vec{Z})$ can be refined in $\mathbf{L}(\mathbb{R})$ to $(f, \vec{Z}')$ with $(f, \vec{Z}') \in P$. Finally it follows that $(f, \vec{Z}') \leq (f, \vec{X})$ in $\mathbb{P}$.

It is apparent from the definition of $(f, \vec{Z})$ that $(f, \vec{Z})$ is $\underset{\sim}{\Sigma}_1$ over $\mathbf{H}_{\omega_1}$. We claim that the refinement $(f, \vec{Z}')$ can be chosen so that $(f, \vec{Z}')$ is $\underset{\sim}{\Sigma}_1$ over $\mathbf{H}_{\omega_1}$.

To see this simply let $Z'_{\alpha,\beta} = \{h \in T_{\alpha,\beta} : h$ is generic over $\mathbf{L}[s, f]$ for the partial order $\mathbb{Q}_{\alpha,\beta}\}$ for $\beta < \omega_1$, $\beta > \alpha$. Let $Z'_\beta = \{f^\frown h : h \in Z'_{\alpha,\beta}\}$. It is routine to verify that $(f, \vec{Z}') \in P$. Further it follows that $(f, \vec{Z}')$ is $\underset{\sim}{\mathbf{\Sigma}}_1$ over $\mathbf{H}_{\omega_1}$, the key parameter is $\langle s, f\rangle^\#$. ⊣

Theorem 1.4 is in effect a coding lemma for certain elements of P. We now want to consider terms for subsets of ω_1 in the forcing language for $\mathbb{P}$.

Suppose $A \subseteq Q_{\omega_1} \times \omega_1$. Associated to A is a term for a subset of ω_1. Toward defining this term suppose G is generic over $\mathbf{V}$ for $\mathbb{P}$. The generic object G defines in a natural fashion an ω_1 sequence of reals which may be regarded as an element of T_{0,ω_1}, generalizing our notion slightly. This sequence is defined from G by $\bigcup\{f : (f, \vec{X}) \in G$ for some $\vec{X}\}$. The entire generic object is easily recovered from this sequence hence we identify G with this sequence. Using the set A we can define a subset of ω_1 in $\mathbf{V}[G]$ as follows. $\alpha \in S$ iff $\langle t, \alpha\rangle \in A$ for some $t \in Q_{\omega_1}$ with $G \in [t]$. Denote the corresponding term in $\mathbf{V}^{\mathbb{P}}$ by τ_A.

Lemma 1.5 (ZF + DC + C). *Suppose $(f, \vec{X}) \in P$ and τ is a term for a subset of ω_1, $\tau \in \mathbf{V}^{\mathbb{P}}$. Then for some $A \subseteq Q_{\omega_1} \times \omega_1$ and condition $(f, \vec{X}) \in P$, $(f, \vec{Y}) \le (f, \vec{X})$ and $(f, \vec{Y}) \Vdash \tau = \tau_A$.*

Proof. Fix α such that $f \in T_{0,\alpha}$.

Define A as follows. $\langle t, \gamma\rangle \in A$ iff for some $\beta > \alpha$, $t \in Q_{\alpha,\beta}$ and the set, $\{h \in T_{\alpha,\beta} : h \in [t]$ and for some $\vec{Y}$, $(f^\frown h, \vec{Y}) \Vdash \gamma \in \tau\}$, is comeager in $[t]$.

For $\beta > \alpha$ let $Z_{\alpha,\beta} = \{h \in T_{\alpha,\beta} : f^\frown h \in X_\beta$ and for all $\vec{Y}, \gamma < \omega_1$, if $(f^\frown h, \vec{Y}) \Vdash$ "$\gamma \in \tau$" then $h \in [t]$ for some $t \in Q_{\alpha,\beta}$ with $\langle t, \gamma\rangle \in A\}$. $Z_{\alpha,\beta}$ is comeager in $T_{\alpha,\beta}$ for each $\beta > \alpha$, therefore let $Z_\beta = \{f^\frown h : h \in Z_{\alpha,\beta}\}$. Choose a refinement $\vec{Y}$ of $\vec{Z}$ such that $(f, \vec{Y}) \in P$. Hence $(f, \vec{Y}) \le (f, \vec{X})$ and $(f, \vec{Y}) \Vdash \tau = \tau_A$. ⊣

The generic object for $\mathbb{P}$ can also be viewed as a filter over $\mathbb{Q}_{\omega_1}$, *i.e.*, if G is generic for $\mathbb{P}$ we can identify G with $\{t \in Q_{\omega_1} : G \in [t]\}$ which is a filter over $\mathbb{Q}_{\omega_1}$. Lemma 1.5 simply says that terms for subsets of ω_1 in $\mathbf{V}^{\mathbb{P}}$ correspond to terms for subsets of ω_1 in $\mathbf{V}^{\mathbb{Q}_{\omega_1}}$.

Subsets of $Q_{\omega_1} \times \omega_1$ are in essence subsets of ω_1. By an early theorem of Solovay, assuming AD every subset of ω_1 is constructible from a real. This in the presence of sharps is equivalent to saying that every subset of ω_1 is $\underset{\sim}{\mathbf{\Sigma}}_1$ over $\mathbf{H}_{\omega_1}$ and so assuming AD, every $A \subseteq Q_{\omega_1} \times \omega_1$ is $\underset{\sim}{\mathbf{\Sigma}}_1$ over $\mathbf{H}_{\omega_1}$ and constructible from a real.

Theorem 1.6 (ZF + AD + DC). *Suppose $(f, X) \in P$ and $\tau \in \mathbf{V}^{\mathbb{P}}$ is a term for a subset of ω_1. Then there is a condition $(f, Y) \le (f, X)$, $(f, Y) \in \underset{\sim}{\mathbf{\Sigma}}_1(\mathbf{H}_{\omega_1})$ and a set $A \subseteq Q_{\omega_1} \times \omega_1$, $A \in \underset{\sim}{\mathbf{\Sigma}}_1(\mathbf{H}_{\omega_1})$, such that $(f, Y) \Vdash \tau = \tau_A$.*

PROOF. Immediate by the preceding remarks, Theorem 1.4 and Lemma 1.5. ⊣

Steel and Van Wesep prove in [SVW82] that if $\mathbf{V} \models \mathsf{ZF} + \mathsf{AD} + \mathbb{R}\text{-}\mathsf{AC}$ and G is generic over $\mathbf{V}$ for $\mathbb{P}$ then in the generic extension $\mathbf{V}[G]$, ω_1-DC holds. As usual we view G as an ω_1 sequence of reals. By Theorem 1.4 G is generic over $\mathbf{L}(\mathbb{R})$ for $\mathbb{P}$ defined in $\mathbf{L}(\mathbb{R})$ (*i.e.*, $\mathbb{P} \cap \mathbf{L}(\mathbb{R})$) and by Theorem 1.6, $\mathbf{V}[G]$ and $\mathbf{L}(\mathbb{R}, G)$ have the same subsets of ω_1 so that in fact $\mathbf{L}(\mathbb{R}, G)$ is just $\mathbf{L}(\wp(\omega_1))$ defined in $\mathbf{V}[G]$. Hence $\mathbf{L}(\mathbb{R}, G) \models \omega_1$-$\mathsf{DC}$. But this is a statement in $\mathbf{L}(\mathbb{R})$ about forcing with $\mathbb{P}$ and one would hope that this is provable about $\mathbf{L}(\mathbb{R})$ just assuming AD, *i.e.*, this should be a theorem of $\mathsf{ZF} + \mathsf{AD} + \mathbf{V}{=}\mathbf{L}(\mathbb{R})$.

We now define two choice principles which we shall show holds in $\mathbf{L}(\mathbb{R})$ assuming AD.

For each infinite countable ordinal α let $S_\alpha \subseteq \mathbb{R}$ be the set of those reals coding α. The set S_α naturally defines a Baire space homeomorphic to a comeager subset of α^ω, *i.e.*, S_α with its natural topology is homeomorphic to $\{f \in \alpha^\omega : f : \omega \to \alpha \text{ is onto}\}$. Note that for $\alpha \neq \beta$, $S_\alpha \cap S_\beta = \varnothing$.

DEFINITION 1.7. (1) Let $**\mathbb{R}$-AC denote the axiom: For each $f : \mathbb{R} \to \mathbf{V}$ there is a choice function $g : \mathbb{R} \to \mathbf{V}$ such that $\{x \in \mathbb{R} : g(x) \in f(x) \text{ or } f(x) = \varnothing\}$ is comeager.

(2) Let $**\omega_1$-AC denote the axiom: For each $f : \mathbb{R} \to \mathbf{V}$ there is a choice function $g : \mathbb{R} \to \mathbf{V}$ such that $\{x \in \mathbb{R} : g(x) \in f(x) \text{ or } f(x) = \varnothing\} \cap S_\alpha$ is comeager in S_α for every infinite countable ordinal α.

LEMMA 1.8 ($\mathsf{ZF} + \mathsf{AD} + \mathsf{DC}$). Assume $\mathbf{V}{=}\mathbf{L}(\mathbb{R})$. Then $**\mathbb{R}$-AC and $**\omega_1$-AC hold.

PROOF. Let $f : \mathbb{R} \to \mathbf{V}$ be given, we seek the appropriate choice function.

Since $\mathbf{V}{=}\mathbf{L}(\mathbb{R})$ every set is ordinal definable from a real and therefore we can assume $f : \mathbb{R} \to \wp(\mathbb{R})$. But then f is in essence a subset of $\mathbb{R} \times \mathbb{R}$ and in this case $**\mathbb{R}$-AC is simply uniformization on a comeager set, a well known consequence of AD.

We now find a choice function as required by $**\omega_1$-AC. Fix $f : \mathbb{R} \to \wp(\mathbb{R})$ as given, assume $f(x) \neq \varnothing$ for every $x \in \mathbb{R}$ and let $A \subseteq \mathbb{R} \times \mathbb{R}$ be the corresponding subset of $\mathbb{R} \times \mathbb{R}$. We seek a function $g : \mathbb{R} \to \mathbb{R}$ such that $\{x \in \mathbb{R} : g(x) \in f(x)\} \cap S_\alpha$ is comeager in S_α for all infinite ordinals $\alpha < \omega_1$. To find g it suffices to find $H \subseteq A$, H a $\utilde{\Sigma}^1_2$ set, *i.e.*, simple, such that $\{x \in \mathbb{R} : H_x \neq \varnothing\} \cap S_\alpha$ is comeager in S_α for every $\alpha > \omega$, $\alpha < \omega_1$, where for $x \in \mathbb{R}$, $H_x = \{y \in \mathbb{R} : \langle x, y \rangle \in H\}$. Since H is a $\utilde{\Sigma}^1_2$ set we can find (and this is a theorem of ZF) a function $g : \mathbb{R} \to \mathbb{R}$ such that for all $x \in \mathbb{R}$, $H_x = \varnothing$ or $g(x) \in H_x$. Clearly g is as required.

To construct H consider the following integer game which is a Solovay game:

$$\begin{array}{cc} \mathrm{I} & \mathrm{II} \\ n_0 & m_0 \\ n_1 & m_1 \\ \vdots & \vdots \\ z & w \end{array}$$

Player II wins if whenever z codes an ordinal α, w codes a $\utilde{\Sigma}^1_2$ set $H \subseteq A$ such that $\{x \in \mathbb{R} : H_x \neq \varnothing\} \cap S_\gamma$ is comeager in S_γ for all infinite $\gamma \leq \alpha$. More precisely, using the $\utilde{\Sigma}_1$-universal set $B \subseteq \mathbb{R} \times \mathbf{H}_{\omega_1}$, player II wins if whenever z codes an ordinal α, *i.e.*, $B_z = \{\alpha\}$, $B_w \subseteq A$ and $\{x \in \mathbb{R} : (B_w)_x \neq \varnothing\} \cap S_\gamma$ is comeager in S_γ for all infinite $\gamma \leq \alpha$. Note that $\utilde{\Sigma}^1_2$ subsets of $\mathbb{R} \times \mathbb{R}$ are $\utilde{\Sigma}_1$ over $\mathbf{H}_{\omega_1}$ and conversely. Equivalently we could have used a universal $\utilde{\Sigma}^1_2$ subset of $\mathbb{R} \times (\mathbb{R} \times \mathbb{R})$ for the decoding of w.

By $**\mathbb{R}$-AC and standard arguments player I cannot have a winning strategy in this game, the situation is similar to that in the proof of Theorem 1.4. Hence by AD player II has a winning strategy. Let s be a winning strategy for player II and for $z \in \mathbb{R}$ let $s(z)$ denote the response by s to player I playing z. Define $H \subseteq \mathbb{R} \times \mathbb{R}$ by $\{\langle x, y\rangle \in \mathbb{R} \times \mathbb{R} : \langle x, y\rangle \in B_{s(z)}$ for some $z \in \mathbb{R}\}$. Clearly $H \subseteq A$ and by the definition of H, H is $\utilde{\Sigma}_1$ over $\mathbf{H}_{\omega_1}$ in particular $H \subseteq \mathbb{R} \times \mathbb{R}$ is $\utilde{\Sigma}^1_2$. Finally $\{x \in \mathbb{R} : H_x \neq \varnothing\} \cap S_\gamma$ is comeager in S_γ for all infinite $\gamma < \omega_1$. ⊣

We remark that assuming ZF + AD + DC, Lemma 1.8 is true in a more general context than $\mathbf{V}=\mathbf{L}(\mathbb{R})$, more precisely Lemma 1.8 holds whenever $\mathbf{V}=\mathbf{L}(\wp(\mathbb{R}))$ and Θ is regular. Recall $\Theta = \sup\{\zeta \in \mathrm{Ord}$: there is an onto map $h: \mathbb{R} \to \zeta\}$.

For all countable ordinals $\alpha < \beta$ let $T^*_{\alpha,\beta}$ denote the comeager subset of $T_{\alpha,\beta}$ defined by $\{h \in T_{\alpha,\beta}$: for all $\gamma \in \operatorname{dom} h$, $h(\gamma): \omega \to \omega + \gamma$ is onto$\}$. Note that for all $\beta > \omega_1$, $T^*_{\alpha,\beta}$ is definably homeomorphic to S_β.

Lemma 1.9 (ZF + DC + C + $**\mathbb{R}$-AC). Suppose $(f, \vec{X}) \in P$ and $(f, \vec{X}) \Vdash \exists Z\varphi(Z)$ where $\varphi(Z)$ is a formula in the forcing language for $\mathbb{P}$. Then there is a refinement $\vec{Y}$ of $\vec{X}$ and a term $\tau \in \mathbf{V}^{\mathbb{P}}$ such that $(f, \vec{Y}) \Vdash \varphi(\tau)$.

Proof. Fix α such that $f \in T_{0,\alpha}$.

Clearly the set $\{(g, \vec{Y}) \in P : (g, \vec{Y}) \Vdash \varphi(\tau)$ for some term $\tau \in \mathbf{V}^{\mathbb{P}}\}$ is open and dense below $(f, \vec{X})$. Choose by Lemma 1.2 an infinite countable ordinal $\beta > \alpha$ such that the set $\{h \in T_{\alpha,\beta+1}$: there is a term $\tau \in \mathbf{V}^{\mathbb{P}}$ and a refinement $\vec{Y}$ of $\vec{X}$ such that $(f^\frown h, \vec{Y}) \Vdash \varphi(\tau)\}$ is comeager in $T_{\alpha,\beta+1}$. The Baire space $T^*_{\alpha,\beta+1}$ is homeomorphic to S_β which in turn is homeomorphic to a comeager subset of $\mathbb{R}$. Hence by $**\mathbb{R}$-AC there is a choice function g such that $\operatorname{dom} g \subseteq T_{\alpha,\beta+1}$ is comeager in $T_{\alpha,\beta+1}$ and for $h \in \operatorname{dom} g$, $g(h)$ is a pair $\langle \tau_h, \vec{Y}_h \rangle$ such that $(f^\frown h, \vec{Y}_h) \in P$ and $(f^\frown h, \vec{Y}_h) \Vdash \varphi(\tau_h)$. Using the function g it is routine to define τ, $\vec{Y}$ as desired. ⊣

Lemma 1.9 asserts that the forcing language for $\mathbb{P}$ is in a weak sense full. This has as an immediate corollary that forcing with $\mathbb{P}$ preserves DC. To prove that the forcing language is full requires $\mathbb{R}$-AC, this is the method in [SVW82].

LEMMA 1.10 ($\mathsf{ZF} + \mathsf{DC} + \mathsf{C} + **\mathbb{R}\text{-}\mathsf{AC}$). Suppose $\mathbf{V}[G]$ is a generic extension of $\mathbf{V}$ obtained by forcing with $\mathbb{P}$. Then $\mathbf{V}[G] \models \mathsf{DC}$.

PROOF. Immediate by Lemma 1.8. ⊣

Using $**\omega_1$-AC we can improve Lemma 1.10 and show that in addition $\mathbf{V}[G] \models \omega_1$-AC.

LEMMA 1.11 ($\mathsf{ZF} + \mathsf{DC} + \mathsf{C} + **\omega_1\text{-}\mathsf{AC}$). Suppose $\mathbf{V}[G]$ is a generic extension of $\mathbf{V}$ obtained by forcing with $\mathbb{P}$. Then $\mathbf{V}[G] \models \omega_1$-AC.

PROOF. Let τ_F be a term in $\mathbf{V}^{\mathbb{P}}$ for a function F with domain ω_1. For each $\gamma < \omega_1$ let $\tau_F(\gamma)$ denote the corresponding term for $F(\gamma)$. We assume that for each $\gamma < \omega_1$, $\tau_F(\gamma)$ is a term for a nonempty set.

Fix $(f, \vec{X}) \in P$ with $f \in T_{0,\alpha}$. For infinite $\beta < \omega_1$, $\beta > \alpha$, the Baire space $T^*_{\alpha,\beta+1}$ is definably homeomorphic to S_β. Hence by Lemma 1.9 and $**\omega_1$-AC there is a choice function g such that for infinite $\beta > \omega_1$ with $\beta > \alpha$, $\operatorname{dom} g \cap T_{\alpha,\beta+1}$ is comeager in $T_{\alpha,\beta+1}$ and for $h \in \operatorname{dom} g \cap T_{\alpha,\beta+1}$, $g(h)$ is a pair $\langle \tau_h, \vec{Y}_h \rangle$ such that $(f^\frown h, \vec{Y}_h) \leq (f, \vec{X})$ and $(f^\frown h, \vec{Y}_h) \Vdash \tau_h \in \tau_F(\beta)$.

By a routine argument using the 'Fubini' theorem for category *etc.*, there is a refinement $\vec{Y}$ of $\vec{X}$ such that $\bigcup_\beta Y_{\beta+1} \subseteq \operatorname{dom} g$ and further for $h \in \bigcup_\beta Y_{\beta+1}$, $(f^\frown h, \vec{Y}_h) \geq (f, \vec{Y})$. Finally g defines a term τ_H for a function H that is forced by $(f, \vec{Y})$ to be a choice function for F, *i.e.*, $(f, \vec{Y}) \Vdash \tau_H$ is a choice function for τ_F.

Hence $\mathbf{V}[G] \Vdash \omega_1$-AC. Of course we have used implicitly that ω_1 is not collapsed by forcing with $\mathbb{P}$. ⊣

Since AD implies that $**\omega_1$-AC holds in $\mathbf{L}(\mathbb{R})$ we have shown, assuming AD, that forcing with $\mathbb{P}$ over $\mathbf{L}(\mathbb{R})$ recovers ω_1-AC. We prove that forcing with $\mathbb{P}$ over $\mathbf{L}(\mathbb{R})$ yields ω_1-DC in the generic extension. To do this we need deeper consequences of AD about $\mathbf{L}(\mathbb{R})$.

A subset $A \subseteq \mathbb{R} \times \mathbb{R}$ can be uniformized if there is a choice function $g \colon \mathbb{R} \to \mathbb{R}$ such that for all $x \in \mathbb{R}$, $g(x) \in A_x$ or $A_x = \varnothing$.

We work now in $\mathbf{L}(\mathbb{R})$. Recall that a set of reals is $\underset{\sim}{\boldsymbol{\Sigma}}^2_1$ if it can be defined by a Σ^2_1-formula with real parameters (equivalently if it is Σ_1 over $\mathbf{L}_\Theta(\mathbb{R})$ allowing real parameters).

The theorem of AD about $\mathbf{L}(\mathbb{R})$ that we need is due to D. Martin and J. Steel (cf. [MMS82]) and implies that assuming AD, sets of reals $\underset{\sim}{\boldsymbol{\Sigma}}^2_1$ in $\mathbf{L}(\mathbb{R})$ can be uniformized in $\mathbf{L}(\mathbb{R})$. Our use of this will be by exploiting a trick due to R. Solovay (cf. [KSS81, Appendix B]), the key idea of which is the following. Suppose λ is an ordinal. Then there is an ordinal $\alpha < \Theta$ such that $\mathbf{L}_\alpha(\mathbb{R})$ is elementarily equivalent to $\mathbf{L}_\lambda(\mathbb{R})$ and such that every set of reals in $\mathbf{L}_\alpha(\mathbb{R})$ is $\utilde{\Sigma}^2_1$ in $\mathbf{L}(\mathbb{R})$, in fact we can assume something stronger, that the structure $\langle \mathbf{L}_\alpha(\mathbb{R}), \in \rangle$ is isomorphic to a structure $\langle \mathbb{R}, E \rangle$ with $E \subseteq \mathbb{R} \times \mathbb{R}$, E a $\utilde{\Sigma}^2_1$ set. This is a theorem of ZF + DC and is proved by a reflection argument using the facts that Θ is a regular cardinal and that every set is ordinal definable from a real. It is a variant of the basis theorem for $\utilde{\Sigma}^2_1$ subsets of $\wp(\mathbb{R})$, *i.e.*, every $\utilde{\Sigma}^2_1$ subset of $\wp(\mathbb{R})$ contains a $\utilde{\Delta}^2_1$ set of reals, this is also a theorem of ZF + DC (+$\mathbf{V}$=$\mathbf{L}(\mathbb{R})$) due to R. Solovay.

THEOREM 1.12 (ZF + AD + DC + $\mathbf{V}$=$\mathbf{L}(\mathbb{R})$). *Suppose* $\mathbf{V}[G]$ *is a generic extension of* $\mathbf{V}$ *obtained by forcing with* $\mathbb{P}$. *Then* $\mathbf{V}[G] \models \omega_1$-DC.

PROOF. As usual we identify G with the corresponding ω_1-sequence of reals. Suppose $(f, \vec{X}) \in P$ and $(f, \vec{X})$ belongs to the generic filter defined by G. We abuse notation slightly and indicate this by $(f, \vec{X}) \in G$.

By Lemma 1.10, $\mathbf{L}(\mathbb{R}, G) \models$ DC. Hence to show that $\mathbf{L}(\mathbb{R}, G) \models \omega_1$-DC it suffices to show that in $\mathbf{L}(\mathbb{R}, G)$ any ω-closed subtree of $\mathbb{R}^{<\omega_1}$ has an unbounded branch, which for nontrivial subtrees is simply to require that there exists an ω_1-branch. Fix $\tau \in \mathbf{L}(\mathbb{R})^{\mathbb{P}}$ a term for such a tree $T \in \mathbf{L}(\mathbb{R}, G)$. We find in $\mathbf{L}(\mathbb{R})$ a set $D \subseteq \mathbb{R} \times \mathbb{R}$ such that any uniformization function of D in $\mathbf{L}(\mathbb{R})$ yields an ω_1-branch through T in $\mathbf{L}(\mathbb{R}, G)$.

Let $A = \{\langle f, g \rangle : f \in T^*_{0,\beta+1}$ for some β, $g \in \mathbb{R}^{<\omega_1}$, $\operatorname{dom} g \subseteq \beta$, and for some $x \in \mathbb{R}$ and $\vec{Y}$, $(f, \vec{Y}) \Vdash g^\frown x \in \tau\}$. For $\langle f, g \rangle \in A$ let $F(\langle f, g \rangle) = \{\langle x, \vec{Y} \rangle : x \in \mathbb{R}, (f, \vec{Y}) \in P$ and $(f, \vec{Y}) \Vdash g^\frown x \in \tau\}$. This defines a function $F : A \to \mathbf{L}(\mathbb{R})$.

Suppose H is a choice function for F, *i.e.*, $H : A \to \mathbf{L}(\mathbb{R})$ with $H(\langle f, g \rangle) \in F(\langle f, g \rangle)$ for all $\langle f, g \rangle \in A$. Regard H as a pair of function H_1, H_2 thus $H(\langle f, g \rangle) = \langle H_1(\langle f, g \rangle), H_2(\langle f, g \rangle) \rangle$. From H one can define in $\mathbf{L}(\mathbb{R}, G)$ a function $H^* : T \to \mathbb{R}$ such that for all $g \in T$, $g^\frown H^*(g) \in T$. To find H^* work in $\mathbf{L}(\mathbb{R}, G)$. For each $g \in T$ choose the least ordinal γ such that if $f = G \restriction \gamma$ then $\langle f, g \rangle \in A$ and $(f, H_2(\langle f, g \rangle)) \in G$. Define $H^*(g) = H_1(\langle f, g \rangle)$. Since $H \in \mathbf{L}(\mathbb{R})$, H^* is defined for all $g \in T$.

Using the function H^* it is easy to construct an ω_1-branch through T in $\mathbf{L}(\mathbb{R}, G)$.

Note, this is a key point, that A is isomorphic to a subset of $\mathbb{R}$ (*i.e.*, the set A is in 1 to 1 correspondence with a set of reals) hence finding H reduces in a canonical fashion to uniformizing a certain subset of $\mathbb{R} \times \mathbb{R}$ which we take as D.

Suppose $\mathbf{L}(\mathbb{R}, G) \models$ "ω_1-DC fails". By the obvious homogeneity if $\mathbb{P}$, $[\![\omega_1\text{-DC fails}]\!]_{\mathbb{P}} = 1$. This is a statement of $\mathbf{L}(\mathbb{R})$ hence by the remarks preceding the statement of this theorem there is an ordinal $\xi < \Theta$ such that $\mathbf{L}_\xi(\mathbb{R}) \models \mathsf{ZF}^- + \mathsf{AD} + \mathsf{DC}$+"$[\![\omega_1\text{-}\mathsf{DC}\text{ fails}]\!]_{\mathbb{P}} = 1$" and with the additional property that every set of reals in $\mathbf{L}_\xi(\mathbb{R})$ is $\underset{\sim}{\mathbf{\Sigma}}^2_1$ in $\mathbf{L}(\mathbb{R})$. ZF^- refers to ZF-replacement, *i.e.*, a large enough fragment of ZF. Fix ξ with these properties.

It is an immediate consequence of Theorem 1.4 that $\mathbb{P} \cap \mathbf{L}_\xi(\mathbb{R})$ is dense in $\mathbb{P}$. Therefore $\mathbf{L}_\xi(\mathbb{R}, G)$ is a generic extension of $\mathbf{L}_\xi(\mathbb{R})$ via forcing with $\mathbb{P}$ and so $\mathbf{L}_\xi(\mathbb{R}, G) \models$ "ω_1-DC fails". Choose $T \in \mathbf{L}_\xi(\mathbb{R}, G)$ such that in $\mathbf{L}_\xi(\mathbb{R}, G)$, T is an ω-closed subtree of $\mathbb{R}^{<\omega_1}$ with no unbounded branch. Fix $\tau \in \mathbf{L}_\xi(\mathbb{R})^{\mathbb{P}}$ a term for T and proceeding as above define in $\mathbf{L}_\xi(\mathbb{R})$, A, F and D. But then D is $\underset{\sim}{\mathbf{\Sigma}}^2_1$ in $\mathbf{L}(\mathbb{R})$ hence in $\mathbf{L}(\mathbb{R})$ the set D can be uniformized. This yields in $\mathbf{L}(\mathbb{R})$ a choice function for F and this in turn produces a branch through T in $\mathbf{L}(\mathbb{R}, G)$. Finally by Theorem 1.6 this branch, being in essence a subset of ω_1, must lie in $\mathbf{L}_\xi(\mathbb{R}, G)$, a contradiction. ⊣

Having established that in the presence of AD forcing with $\mathbb{P}$ over $\mathbf{L}(\mathbb{R})$ yields ω_1-DC, the consistency results of [SVW82] easily shown. Before indicating how we offer a different perspective of forcing with $\mathbb{P}$.

Lemma 1.5 clearly suggests an intimate relationship between forcing with $\mathbb{P}$ and forcing with $\mathbb{Q}_{\omega_1}$, we shall prove a theorem clarifying this but first we state a theorem due independently to H. Friedman.

THEOREM 1.13 ($\mathsf{ZF} + \mathsf{DC} + \mathsf{C} + {**}\mathbb{R}\text{-}\mathsf{AC}$). *Suppose c is a Cohen real over $\mathbf{V}$. Then there is a generic elementary embedding:*

$$j : \mathbf{V} \to M \subseteq \mathbf{V}[c] \text{ where } c \in M.$$

PROOF. Since every set of reals has the property of Baire the complete Boolean algebra generated by the Cohen conditions is isomorphic to $\wp(\mathbb{R})/\vartheta$ where ϑ denotes the ideal of meager sets. A standard generic ultrapower argument using $**\mathbb{R}$-AC instead of the full axiom of choice yields j. ⊣

In the special case of $\mathbf{V}=\mathbf{L}(\mathbb{R})$ the generic elementary embedding of Theorem 1.13 is easily (and uniquely) defined. Suppose c is a Cohen real over $\mathbf{L}(\mathbb{R})$ and let $\mathbb{R}_c$ denote the set of reals as defined in $\mathbf{V}[c]$. Each set in $\mathbf{L}(\mathbb{R})$ is ordinal definable from a real, this plus the observation that $j\restriction\mathrm{Ord}$ must be the identity uniquely defines j. If φ is a formula that defines in $\mathbf{L}(\mathbb{R})$ a set A, φ with parameters $x \in \mathbb{R}$, $\gamma \in \mathrm{Ord}$, then φ defines $j(A)$ in $\mathbf{L}(\mathbb{R}_c)$.

Forcing with $\mathbb{Q}_{\omega_1}$ is equivalent as far as adding reals to forcing with $\mathbb{C}_{\omega_1}$, the partial order of Cohen conditions for adding ω_1 reals (if one assumes the axiom of choice $\mathbb{C}_{\omega_1}$ and $\mathbb{Q}_{\omega_1}$ are in fact isomorphic). Suppose $G \subseteq Q_{\omega_1}$ is generic over $\mathbf{L}(\mathbb{R})$. Then Theorem 1.13 easily extends to yield a generic elementary embedding, $j : \mathbf{L}(\mathbb{R}) \to \mathbf{L}(\mathbb{R}_G)$ where $\mathbb{R}_G$ denotes the set of reals in $\mathbf{V}[G]$.

THEOREM 1.14 ($\mathsf{ZF} + \mathsf{DC} + \mathsf{C} + {**}\mathbb{R}\text{-}\mathsf{AC}$). *Assume $\mathbf{V}=\mathbf{L}(\mathbb{R})$ and suppose $G \subseteq Q_{\omega_1}$ is generic over $\mathbf{L}(\mathbb{R})$ for the partial order $\mathbb{Q}_{\omega_1}$. Let $\mathbb{R}_G$ denote the set* of reals in $\mathbf{V}[G]$. Then G defines a generic filter over $\mathbf{L}(\mathbb{R}_G)$ for the partial order $\mathbb{P}$ defined in $\mathbf{L}(\mathbb{R}_G)$.

PROOF. For sets of ordinals $s \in \mathbf{L}(\mathbb{R}_G)$ let $\mathbb{R}_s$ denote the set of reals in $\mathbf{L}(\mathbb{R}, s)$. The embedding $j: \mathbf{L}(\mathbb{R}) \to \mathbf{L}(\mathbb{R}_G)$ induces unique elementary embeddings $j_1: \mathbf{L}(\mathbb{R}) \to \mathbf{L}(\mathbb{R}_s)$, $j_2: \mathbf{L}(\mathbb{R}_s) \to \mathbf{L}(\mathbb{R}_G)$ with $j = j_2 \circ j_1$. We denote j_2 by j_s.

Let P_G denote P defined in $\mathbf{L}(\mathbb{R}_G)$ and let $\mathbb{P}_G$ denote the corresponding partial order. Similarly define P_s, $\mathbb{P}_s$ for sets of ordinals $s \in \mathbf{L}(\mathbb{R}_G)$, *i.e.*, P_s denotes P defined in $\mathbf{L}(\mathbb{R}_s)$, *etc.*

In $\mathbf{L}(\mathbb{R}, G)$, G clearly defines a filter over $\mathbb{P}_G$. Suppose $D \subseteq P_G$ is dense and open in $\mathbb{P}_G$, $D \in \mathbf{L}(\mathbb{R}_G)$. Thus D is ordinal definable from a real x_0, so fix a formula φ with parameters only x_0 and possibly some ordinal, such that φ defines D. Choose a countable ordinal α large enough so that $x_0 \in \mathbf{L}(\mathbb{R}, G\restriction\alpha)$ where $G\restriction\alpha = G \cap Q_{0,\alpha}$. The restriction $G\restriction\alpha$ canonically defines $f \in T_{0,\alpha}$ which we interpret as a set of ordinals. Hence φ defines in $\mathbf{L}(\mathbb{R}_f)$ a dense, open set $D_f \subseteq P_f$. Further $j_f(D_f) = D$.

Work in $\mathbf{L}(\mathbb{R}_f)$ and choose by Lemma 1.2 $\beta > \alpha$ such that the set $A = \{h \in T_{\alpha,\beta} : (f^\frown h, \vec{Y}) \in D_f$ for some sequence $\vec{Y}\}$ is comeager in $T_{\alpha,\beta}$.

Let h be the element of $T_{\alpha,\beta}$ defined by $G \cap Q_{\alpha,\beta}$. Let $g = f^\frown h$ which as usual we can regard as a set of ordinals. The embedding $j_f: \mathbf{L}(\mathbb{R}_f) \to \mathbf{L}(\mathbb{R}_G)$ induces unique elementary embeddings $j_1: \mathbf{L}(\mathbb{R}_f) \to \mathbf{L}(\mathbb{R}_g)$, $j_g: \mathbf{L}(\mathbb{R}_g) \to \mathbf{L}(\mathbb{R}_G)$ with $j_f = j_g \circ j_1$. Further $h \in j_1(A)$ hence in $\mathbf{L}(\mathbb{R}_g)$ there exists a sequence $\vec{Y}_g$ such that $(g, \vec{Y}_g) \in j_1(D_f) = D_g$.

For each countable ordinal $\gamma > \beta$ let h_γ denote that element of $T_{\beta,\gamma}$ determined by $G \cap Q_{\beta,\gamma}$. It follows that for each γ, $g^\frown h_\gamma \in j_g((\vec{Y}_g)_\gamma)$ hence $j_g((g, \vec{Y}_g))$ belongs to the filter over $\mathbb{P}_G$ defined by G. But $j_g((g, \vec{Y}_g)) \in j_g(D_g) = D$. ⊣

The assumption $\mathbf{V}{=}\mathbf{L}(\mathbb{R})$ is not necessary in Theorem 1.14 though the proof is perhaps more straightforward under this additional assumption.

An immediate corollary to Theorem 1.14 is that forcing with $\mathbb{P}$ does not collapse any cardinals. This is also proved in [SVW82], though by a different argument. One can also use Theorem 1.14 to reprove the (ω, ∞)-distributivity of $\mathbb{P}$.

As noted in [SVW82] if one replaces the spaces $T_{\alpha,\beta}$ by $\prod_{\alpha \leq \delta < \beta} \omega^\omega$ then there is a partial order $\mathbb{P}^*$ analogous to $\mathbb{P}$. Forcing with $\mathbb{P}^*$ over a ground model of ZF + DC + C + $**\mathbb{R}$-AC adds an ω_1-sequence of distinct reals without adding any new reals or collapsing any cardinals. Further there is an obvious variant of Theorem 1.14 for $\mathbb{P}^*$ using C_{ω_1} instead of Q_{ω_1}. The main point of this is that for arbitrary cardinals λ it indicates a procedure for generalizing the definition of $\mathbb{P}^*$ in order to define $\mathbb{P}^*_\lambda$, an (ω, ∞)-distributive partial order with the property that forcing with $\mathbb{P}^*_\lambda$ adds a λ-sequence of distinct reals without collapsing any cardinals. We are of course assuming ZF + DC + C + $**\mathbb{R}$-AC, for the sake of simplicity assume as well that $\mathbf{V}{=}\mathbf{L}(\mathbb{R})$. Define $\mathbb{P}^*_\lambda$ so that if G

is a generic λ-sequence of Cohen reals over $\mathbf{L}(\mathbb{R})$ then G defines a generic filter over $\mathbf{L}(\mathbb{R}_G)$ for $j(\mathbb{P}^*_\lambda)$ *i.e.*, $\mathbb{P}^*_\lambda$ defined in $\mathbf{L}(\mathbb{R}_G)$. The desired properties of $\mathbb{P}^*_\lambda$ will then follow. We leave the details to the reader.

Thus assuming $\mathsf{ZF}+\mathsf{DC}+\mathsf{C}+{**}\mathbb{R}\text{-}\mathsf{AC}$ it is possible to create in a generic fashion *arbitrarily long* wellordered sequences of distinct reals without introducing new reals or collapsing cardinals.

§2. Throughout this section we work in $\mathsf{ZF}+\mathsf{AD}+\mathsf{DC}$.

Assume G is generic over $\mathbf{L}(\mathbb{R})$ for $\mathbb{P}$. Then by Theorem 1.12, $\mathbf{L}(\mathbb{R}, G) \models \omega_1\text{-}\mathsf{DC}$. We use this model to prove two consistency results in $\mathsf{ZF}+\omega_1\text{-}\mathsf{DC}$. The point is that some of the combinatorial consequences of AD carry over to $\mathbf{L}(\mathbb{R}, G)$, in fact we shall show that a form of determinacy holds in $\mathbf{L}(\mathbb{R}, G)$.

The first result concerns infinitary partition properties. Suppose κ is an infinite cardinal and that $S \subseteq \kappa$. For $\alpha \leq \kappa$ let $[S]^\alpha$ denote the set of all subsets of S with ordertype α. We use the standard notation $\kappa \to (\kappa)^\alpha_\beta$ to indicate that for any map $e\colon [\kappa]^\alpha \to \beta$ there is a subset S of κ, S of ordertype κ, such that $e\restriction[S]^\alpha$ is constant.

Assuming $\mathsf{AD}+\mathsf{DC}$ there exist many cardinals κ such that $\kappa \to (\kappa)^\kappa_2$ (cf. [KKMW81]). Thus granting the consistency of $\mathsf{ZF}+\mathsf{AD}+\mathsf{DC}$ it is impossible to refute in $\mathsf{ZF}+\mathsf{DC}$ the existence of uncountable cardinals κ for which $\kappa \to (\kappa)^\kappa_2$. If however one assumes ω_1-DC, many partition properties must fail for all uncountable cardinals as is implicit in the following theorem.

THEOREM 2.1 ($\mathsf{ZF}+\mathsf{DC}$). *Assume there is an uncountable sequence of distinct reals. Then for all uncountable cardinals* κ, $\kappa \not\to (\kappa)^{\omega_1}_2$.

PROOF. A slight refinement of the results in [KKMW81], using similar methods, shows that if for some uncountable cardinal κ, $\kappa \to (\kappa)^{\omega_1}_2$ then every ω_1-Suslin set of reals is determined. Recall that a set $A \subseteq \omega^\omega$ is ω_1-Suslin if there is a tree T on $\omega \times \omega_1$ such that $A = \{x \in \omega^\omega : \text{for some } f \in \omega_1^\omega, \langle x, f\rangle \text{ defines an infinite branch through } T\}$.

Using an ω_1-sequence of distinct reals it is straightforward to construct an ω_1-Suslin set of reals that is not determined. ⊣

In contrast to Theorem 2.1 we show that it is possible in the presence of ω_1-DC for there to exist uncountable cardinals κ for which $\kappa \to (\kappa)^\alpha_{<\kappa}$ for all $\alpha < \omega_1$.

THEOREM 2.2 ($\mathsf{ZF}+\mathsf{AD}+\mathsf{DC}+\mathbf{V}=\mathbf{L}(\mathbb{R})$). *Suppose* $\mathbf{V}[G]$ *is a generic extension of* $\mathbf{V}$ *obtained by forcing with* $\mathbb{P}$. *Let* κ *be any regular cardinal of* $\mathbf{V}$, $\kappa \geq \omega_2$, *such that in* $\mathbf{V}$, $\kappa \to (\kappa)^\alpha_{<\kappa}$ *for all* $\alpha < \omega_1$. *Then in* $\mathbf{V}[G]$, $\kappa \to (\kappa)^\alpha_{<\kappa}$ *for all* $\alpha < \omega_1$.

PROOF. We work for the moment in $\mathbf{L}(\mathbb{R})$ and isolate the relevant combinatorial occurrence responsible for the theorem. We shall proceed by successive approximation.

Since $\kappa \to (\kappa)^{\xi}_{<\kappa}$ for all $\xi < \omega_1$, for each $\alpha < \omega_1$ there is a canonical κ-complete measure defined over $[\kappa]^\alpha$, *i.e.*, $S \subseteq [\kappa]^\alpha$ has measure 1 if for some $C \subseteq \kappa$, C closed and unbounded in κ, $[C]^\alpha \subseteq S$.

Claim 2.3. Suppose $F\colon [\kappa]^\alpha \to \wp(\mathbb{R})$ is a function such that the set $\{s \in [\kappa]^\alpha : F(s) \text{ is comeager}\}$ is of measure 1. Then there is a closed set $C \subseteq \kappa$, unbounded in κ, and a comeager set $A \subseteq \mathbb{R}$, such that for $x \in [C]^\alpha$, $A \subseteq F(s)$.

Proof of Claim 2.3. Using $**\mathbb{R}$-AC define a partial function $H\colon \mathbb{R} \to \wp(\kappa)$ with comeager domain, such that for $x \in \operatorname{dom} H$, $H(x)$ is closed and unbounded in κ, $[H(x)]^\alpha \subseteq \{s \in [\kappa]^\alpha : x \in F(s)\}$ or $[H(x)]^\alpha \subseteq \{s \in [\kappa]^\alpha : x \notin F(s)\}$. Let $C = \big\{\gamma < \kappa : \text{the set } \{x \in \mathbb{R} : \gamma \in H(x)\} \text{ is comeager}\big\}$ and let $A = \{x \in \mathbb{R} : C \subseteq H(x)\}$. Since the wellordered intersection of comeager sets is comeager it follows that C is closed and unbounded in κ and that A is comeager. ⊣ (Claim 2.3)

Claim 2.4. Suppose $F\colon [\kappa]^\alpha \to \mathbf{L}(\mathbb{R})$ is a function such that on a set of measure 1, $F(s) \cap T_{0,\zeta+1}$ is comeager in $T_{0,\zeta+1}$ for each $\zeta < \omega_1$. Then there is a closed, unbounded set $C \subseteq \kappa$ and a set A, $A \cap T_{0,\zeta+1}$ comeager in $T_{0,\zeta+1}$ for each $\zeta < \omega_1$, such that for $s \in [C]^\alpha$, $A \subseteq F(s)$.

Proof of Claim 2.4. Using $**\omega_1$-AC define a function $H\colon \operatorname{dom} H \to \wp(\kappa)$ such that $\operatorname{dom} H \cap T_{0,\zeta+1}$ is comeager in $T_{0,\zeta+1}$ for each $\zeta < \omega_1$ and such that for $f \in \operatorname{dom} H$, $H(f)$ is closed, unbounded in κ, $[H(f)]^\alpha \subseteq \{s \in [\kappa]^\alpha : f \in F(s)\}$ or $[H(f)]^\alpha \subseteq \{s \in [\kappa]^\alpha : f \notin F(s)\}$. For each $\zeta < \omega_1$ let $C_\zeta = \{\gamma < \kappa : \text{the set } \{f \in T_{0,\zeta+1} : \gamma \in H(f)\} \text{ is comeager in } T_{0,\zeta+1}\}$ and let $A_\zeta = \{f \in T_{0,\zeta+1} : C_\zeta \subseteq H(f)\}$. Thus for each $\zeta < \omega_1$, C_ζ is closed, unbounded in κ, and $A_\zeta \cap T_{0,\zeta+1}$ is comeager in $T_{0,\zeta+1}$. Define $C = \bigcap_\zeta C_\zeta$, $A = \bigcup_\zeta A_\zeta$. Since $\kappa > \omega_2$, C is closed and unbounded in κ. ⊣ (Claim 2.4)

We shall actually need the following variant of Claim 2.4 which is an immediate corollary of Claim 2.4.

Claim 2.5. Suppose $F\colon [\kappa]^\alpha \to P$ is a function such that on a set of measure 1, $F(s)$ is a condition of the form $(\varnothing, \vec{X})$. Then there is a closed, unbounded set $C \subseteq \kappa$ and a condition $(\varnothing, \vec{Y}) \in P$ such that for $s \in [C]^\alpha$, $(\varnothing, \vec{Y}) \leq F(s)$.

Proof of Claim 2.5. Immediate using Claim 2.4. ⊣ (Claim 2.5)

We use Claim 2.5 to prove Theorem 2.2.

Since $\mathbb{P}$ is (ω, ∞)-distributive, for each $\alpha < \omega_1$, $[\kappa]^\alpha$ is the same computed in $\mathbf{L}(\mathbb{R})$ or $\mathbf{L}(\mathbb{R}, G)$. Suppose that for some $\alpha < \omega_1$, $\kappa \not\to (\kappa)^{\alpha}_{<\kappa}$ in $\mathbf{L}(\mathbb{R}, G)$. Choose in $\mathbf{L}(\mathbb{R}, G)$ $\lambda < \kappa$ and a map $e\colon [\kappa]^\alpha \to \lambda$ with no homogeneous set. Choose a term $\tau \in \mathbf{L}(\mathbb{R})^{\mathbb{P}}$ for e, by Lemma 1.9 we can assume that for some sequence $\vec{X}$, $(\varnothing, \vec{X}) \Vdash$"$\tau$ defines a map $e\colon [\kappa]^\alpha \to \lambda$ with no homogeneous set". For $s \in [\kappa]^\alpha$ let $\tau(s)$ denote the term corresponding to $e(s)$.

We again work in $\mathbf{L}(\mathbb{R})$. Define a map $e^*\colon [\kappa]^\alpha \to \lambda \times Q_{\omega_1} \times \omega_1$ by $e^*(s) = \langle \gamma, t, \beta \rangle$ for the 'first' triple $\langle \gamma, t, \beta \rangle$ such that the set $\{f \in T_{0,\beta} :$ for some $\vec{Y}$, $(f, \vec{Y}) \Vdash \tau(s) = \gamma\}$ is comeager in $[t]$.

Since $\kappa \to (\kappa)^\zeta_{<\kappa}$ for all $\zeta < \omega_1$ there is a closed, unbounded subset C of κ and a triple $\langle \gamma_0, t_0, \beta_0 \rangle \in \lambda \times Q_{\omega_1} \times \omega_1$ such that for $s \in [C]^\alpha$, $e^*(s) = \langle \gamma_0, t_0, \beta_0 \rangle$.

By Claim 2.5 we can find $C^* \subseteq C$, closed and unbounded, $f^* \in T_{0,\beta_0}$, and a sequence $\vec{Y}^*$, such that $(f^*, \vec{Y}^*) \in P$ and for $s \in [C^*]^\alpha$, $(f^*, \vec{Y}^*) \Vdash \tau(s) = \gamma$. Hence $(f^*, \vec{Y}^*) \Vdash$ "C^* is homogeneous for e", a contradiction. $\dashv$ (Theorem 2.2)

Suppose S is a set of ordinals. Define an equivalence relation on reals by $x \sim_S y$ if $\mathbf{L}[x, S] = \mathbf{L}[y, S]$. For $x \in \mathbb{R}$ let $[x]_S$ denote the equivalence class of x. We call the equivalence classes S-degrees. This is a standard generalization of the familiar notion of Turing degrees. A cone of S-degrees is a set of S-degrees defined by $\{[x]_S : x_0 \in \mathbf{L}[S, x]\}$ for some $x_0 \in \mathbb{R}$. We say that S-degree determinacy holds if for any set, A, of S-degrees either A contains a cone or the complement of A contains a cone. In addition, to avoid (some) trivialities, we require that there must be no maximal S-degree. It is a fundamental theorem of AD, due to D. Martin, that Turing (degree) determinacy holds. Thus assuming AD, S-degree determinacy holds for any S.

Assuming ω_1-DC, S-degree determinacy must fail for any S such that the S-degrees are countable, *i.e.*, such that $[x]_S$ is countable for each $x \in \mathbb{R}$. This however does not rule out the possibility of S-degree determinacy if the corresponding degrees are uncountable.

THEOREM 2.6 (ZF + AD + DC + $\mathbf{V}$=$\mathbf{L}(\mathbb{R})$). *Assume $G \subseteq Q_{\omega_1}$ defines a generic filter over $\mathbf{V}$ for $\mathbb{P}$. Then in $\mathbf{V}[G]$, G-degree determinacy holds.*

PROOF. For reals x, y define $x \leq_{\mathbf{L}} y$ if $x \in \mathbf{L}[y]$ and let $[x]_{\mathbf{L}} = \{z \in \mathbb{R} : \mathbf{L}[z] = \mathbf{L}[x]\}$. Thus $[x]_{\mathbf{L}}$ is the constructible degree of x which we call the $\mathbf{L}$-degree of x. Note that the corresponding form of (degree) determinacy, $\mathbf{L}$-degree determinacy, is an immediate consequence of Turing determinacy hence of AD.

As usual we identify elements of $T^*_{0,\alpha+1}$ with reals. This is done in a natural fashion. Further suppose x, y are reals, we let $x * y$ denote the real coding the pair $\langle x, y \rangle$ in some canonical fashion.

We isolate in a series of claims the main technical lemma that we shall need.

CLAIM 2.7. *Suppose $F : \mathbb{R} \to \wp(\mathbb{R})$ is a function such that for $x \in \mathbb{R}$, $F(x)$ is comeager. Then for every $x_0 \in \mathbb{R}$ there us a real $x_1 \geq_{\mathbf{L}} x_0$ and a comeager set $A \subseteq \mathbb{R}$ such that for any real $x \geq_{\mathbf{L}} x_1$ and any $c \in A$, $\mathbf{L}[x, c] = \mathbf{L}[y, c]$ for some $y \in \mathbb{R}$, $y \geq_{\mathbf{L}} x_0$, and $c \in F(y)$.*

PROOF OF CLAIM 2.7. Suppose not and assume the claim fails for some real, x_0. Then (in the worst case) by $**\mathbb{R}$-AC and $\mathbf{L}$-degree determinacy there is a partial function $H: \mathbb{R} \to \mathbb{R}$ with comeager domain such that for $c \in \operatorname{dom} H$ and any real $x \geq_{\mathbf{L}} H(c)$ there is no real $y \in \mathbf{L}[x, c]$ for which $\mathbf{L}[y, c] = \mathbf{L}[x, c]$, $y \geq_{\mathbf{L}} x_0$, and $c \in F(y)$.

However since every set of reals has the property of Baire, the function H must agree with a Borel function, h, on some comeager Borel set D. Let y be a real coding x_0, h, and D. Choose $c \in F(y)$ such that c is a Cohen real over $\mathbf{L}[y]$. Thus $c \in D$ and so $h(c) = H(c)$. Further $\mathbf{L}[y, c] = \mathbf{L}[y, h(c), c]$. Let $x = y * h(c)$. Therefore $H(c) \leq_{\mathbf{L}} x$ but $\mathbf{L}[x, c] = \mathbf{L}[y, c]$ and $c \in F(y)$, a contradiction since $x_0 \leq_{\mathbf{L}} y$. ⊣ (Claim 2.7)

We shall need an ω_1-form of Claim 2.7.

Claim 2.8. Suppose $F: \mathbb{R} \to \wp(\mathbb{R})$ is a function such that for $x \in \mathbb{R}$, $F(x) \cap T_{0,\alpha+1}$ is comeager in $T_{0,\alpha+1}$ for each $\alpha < \omega_1$. Then for every $x_0 \in \mathbb{R}$ there is a real $x_1 \geq_{\mathbf{L}} x_0$ and a set A, $A \cap T_{0,\alpha+1}$ comeager in $T_{0,\alpha+1}$ for each $\alpha < \omega_1$, such that for any real $x \geq_{\mathbf{L}} x_1$ and any $f \in A$, $\mathbf{L}[x, f] = \mathbf{L}[y, f]$ for some real y, $y \geq_{\mathbf{L}} x_0$ and $f \in F(y)$.

Proof of Claim 2.8. Fix x_0. Let $A_1 = \{\langle f, z\rangle : f \in T^*_{0,\alpha+1}$ for some $\alpha < \omega_1$, $z \in \mathbb{R}$ and for any $x \geq_{\mathbf{L}} z$, $\mathbf{L}[x, f] = \mathbf{L}[y, f]$ for some $y \in \mathbb{R}$, $y \geq_{\mathbf{L}} x_0$ and $f \in F(y)\}$.

Since the spaces $T^*_{0,\alpha+1}$ are each homeomorphic to a comeager set of reals, by Claim 2.7, $\{f : \langle f, z\rangle \in A_1$ for some $z\} \cap T_{0,\alpha+1}$ is comeager in $T_{0,\alpha+1}$ for each $\alpha < \omega_1$.

The set A_1 defines a partial function $J: \mathbf{H}_{\omega_1} \to \wp(\mathbb{R})$ (in the codes, a partial function $J^*: \mathbb{R} \to \wp(\mathbb{R})$). By the proof of Lemma 1.8 there is a choice (partial) function H, $\operatorname{dom} H \cap T_{0,\alpha+1}$ comeager in $T_{0,\alpha+1}$ for each $\alpha < \omega_1$, such that H is $\mathbf{\Sigma}_1$ over $\mathbf{H}_{\omega_1}$ (in the codes, a $\utilde{\mathbf{\Sigma}}^1_2$ partial function $H^*: \mathbb{R} \to \mathbb{R}$). Let x_1 be a real coding x_0 and a parameter sufficient to define H in a Σ_1 fashion. Let $A = \operatorname{dom} H$. Suppose $f \in A$, $f \in T^*_{0,\alpha+1}$, and $x \geq_{\mathbf{L}} x_1$. Then $\mathbf{L}[x, f] = \mathbf{L}[x, H(f), f] = \mathbf{L}[z, f]$ where $z = x * H(f)$. The point being that by absoluteness considerations $H(f) \in \mathbf{L}[x, f]$. Thus $z \geq_{\mathbf{L}} H(f)$ and so $\mathbf{L}[z, f] = \mathbf{L}[y, f]$ for some $y \in \mathbb{R}$, $y \geq_{\mathbf{L}} x_0$, and $f \in F(y)$. But then $\mathbf{L}[x, f] = \mathbf{L}[y, f]$ and therefore A and x_1 have the required properties. ⊣ (Claim 2.8)

It is a variant of Claim 2.8 that is the main lemma that we shall need.

Claim 2.9. Suppose $F: \mathbb{R} \to P$ is a function such that for $x \in \mathbb{R}$, $F(x)$ is a condition of the form $(\varnothing, \vec{X})$. Then for every real x_0, there is a real $x_1 \geq_{\mathbf{L}} x_0$ and a condition $(\varnothing, \vec{Y}) \in P$ such that for any real $x \geq_{\mathbf{L}} x_1$ and any condition $(f, \vec{Z}) \leq (\varnothing, \vec{Y})$, if $\operatorname{dom} f = \alpha + 1$ (*i.e.*, $f \in T^*_{0,\alpha+1}$) for some $\alpha < \omega_1$ then there is a real $y \in \mathbf{L}[x, f]$ for which $\mathbf{L}[x, f] = \mathbf{L}[y, f]$, $x_0 \leq_{\mathbf{L}} y$ and $(f, \vec{X}) \leq F(y)$ for some sequence $\vec{X}$.

Proof of Claim 2.9. Immediate using Claim 2.8. ⊣ (Claim 2.9)

In addition to Claim 2.9 we shall also need:

CLAIM 2.10. Fix $\alpha < \omega_1$. Suppose $F : \mathbb{R} \to \wp(T_{0,\alpha+1})$ is a function such that for $x \in \mathbb{R}$, $F(x)$ is comeager in $T_{0,\alpha+1}$. Then for every $x \in \mathbb{R}$ the set $\{c \in \mathbb{R} :$ for some $f, g \in T^*_{0,\alpha+1}$, $\mathbf{L}[x, c, f] = \mathbf{L}[x, g, f]$ and $f \in F(x * c) \cap F(x * g)\}$ is comeager in $\mathbb{R}$.

PROOF OF CLAIM 2.10. Routine. ⊣ (Claim 2.10)

We continue with the proof of Theorem 2.6. Fix $\tau \in \mathbf{L}(\mathbb{R})^{\mathbb{P}}$ a term for a set of G-degrees. Assume toward a contradiction, using Lemma 1.9, that for some sequence $\vec{X}$, $(\varnothing, \vec{X}) \Vdash$"$\tau$ defines a partition of the G-degrees neither piece of which contains a cone".

We work in $\mathbf{L}(\mathbb{R})$ and fix some additional notation. Let RO Q_{ω_1} denote the set of elements of the complete Boolean algebra generated by Q_{ω_1}. Similarly for $\alpha < \beta < \omega_1$ let RO $Q_{\alpha,\beta}$ denote the set of elements in the completion of $\mathbb{Q}_{\alpha,\beta}$. Thus in a natural fashion RO $Q_{\omega_1} = \bigcup_{\alpha<\omega_1}$ RO $Q_{0,\alpha}$. We depart slightly from previous conventions and for $b \in$ RO Q_{ω_1} let $[b] = \{f :$ for some $\alpha < \beta < \omega_1$ and $t \in Q_{\alpha,\beta}$, $t \leq b$, and $f \in [t]_{\alpha,\beta}\}$. The difference is that for $t \in Q_{\omega_1}$, if dom $t \not\subseteq$ dom f then $f \notin [t]$.

For each $x \in \mathbb{R}$ define $b_x \in$ RO Q_{ω_1} such that for some sequence $\vec{Y}$ and all conditions $(f, \vec{Z}) \leq (\varnothing, \vec{Y})$, $(f, \vec{Z}) \Vdash$ "$[x]_G \in \tau$" iff $f \in [b_x]$. Clearly b_x depends only on $[x]_{\mathbf{L}}$, hence assume with no loss of generality that for some real, y_0, and all $x \geq_{\mathbf{L}} y_0$, $b_x \neq 0$. We claim the map $[x]_{\mathbf{L}} \mapsto b_x$ is constant on a cone (of $\mathbf{L}$-degrees). To show this we first note that on a cone, $b_x \in$ RO $Q_{0,\alpha}$ for some $\alpha < \omega_1^x$, *i.e.*, for some α countable in $\mathbf{L}[x]$. To see this suppose not. Then this must fail on a cone, hence for some real, z_0, and any $x \geq_{\mathbf{L}} z_0$, $b \notin$ RO $Q_{0,\alpha}$ for any $\alpha < \omega_1^x$. For each $c \in \mathbb{R}$ let $e(c)$ be the least α such that $b_{c*z_0} \in$ RO $Q_{0,\alpha}$. Thus on a comeager set A, the range of e is bounded by some ordinal $\alpha_0 < \omega_1$. Define for each $c \in \mathbb{R}$, $F(c) \subseteq T_{0,\alpha_0+1}$ by $F(c) = \{h \in T^*_{0,\alpha_0+1} : h \in [b_{z_0*c}]$ iff for some sequence $\vec{Z}$, $(h, \vec{Z}) \Vdash$ "$[z_0 * c]_G \in \tau$"$\}$. For $t \in Q_{0,\alpha_0+1}$ and $f \in T_{0,\alpha_0+1}$ let f_t denote the element of $T_{0,\alpha+1}$ obtained by perturbing f by t, hence $f_t \in [t]$. For every $c \in \mathbb{R}$, $F(c)$ is comeager in T_{0,α_0+1}. By shrinking if necessary we can assume in addition that for $f \in F(c)$ and $t \in Q_{0,\alpha_0+1}$, $f_t \in F(c)$. Finally by Claim 2.10, for some $f, g \in T^*_{0,\alpha_0+1}$ and some $c \in A$, $\mathbf{L}[z_0, c, f] = \mathbf{L}[z_0, g, f]$ and $f \in F(z_0 * c) \cap F(z_0 * g)$. Thus it follows that $b_{z_0*g} = b_{z_0*c}$ and so in particular $b_{z_0*g} \in$ RO Q_{0,α_0+1} but $\omega_1^{z_0*g} > \alpha_0 + 1$ a contradiction since $z_0 * g \geq_{\mathbf{L}} z_0$.

Hence on a cone, $b_x \in$ RO $Q_{0,\alpha}$ for some $\alpha < \omega_1^x$. But then on a cone, $e(x) < \omega_1^x$. By standard arguments it follows that $e(x)$ is constant on a cone and so if $b_0 = \bigvee \{t \in Q_{\omega_1} :$ on a cone $t \leq b_x\}$ then $b_x = b_0$ on a cone. Fix $x_0 \in \mathbb{R}$ such that for $x \geq_{\mathbf{L}} x_0$, $b_x = b_0$.

For each $x \in \mathbb{R}$ let $F_x = \{f : f \in T_{0,\alpha}$ for some $\alpha < \omega_1$ and $f \in [b_x]$ iff for some sequence $\vec{Z}$, $(f, \vec{Z}) \Vdash [x]_G \in T\}$.

This defines in a natural fashion a map $F: \mathbb{R} \to P$ such that for $x \in \mathbb{R}$, $F(x)$ is a condition of the form $(\varnothing, \vec{X})$. By Claim 2.9 there is a real, x_1, and a condition $(\varnothing, \vec{Y}) \in P$ such that for any real $x \geq_{\mathbf{L}} x_1$ and any condition $(f, \vec{Z}) \leq (\varnothing, \vec{Y})$, if $\operatorname{dom} f = \alpha + 1$ for some $\alpha < \omega_1$ then there is a real $y \in \mathbf{L}[x, f]$ for which $\mathbf{L}[x, f] = \mathbf{L}[y, f]$, $x_0 \leq_{\mathbf{L}} y$ and $(f, \vec{X}) \leq F(y)$ for some sequence $\vec{X}$. Assume by refining if necessary that $(\varnothing, \vec{Y}) \Vdash$"$\tau$ defines a partition of the G-degrees neither piece of which contains a cone". Choose $g \in \bigcup_\beta Y_\beta$ such that $g \in [b_0]$. We claim that $(g, \vec{Y}) \Vdash$"τ contains the cone of G-degrees generated by x_1". Suppose not and choose $x \geq_{\mathbf{L}} x_1$ and a condition $(f, \vec{Z}) \leq (g, \vec{Y})$ with $\operatorname{dom} f = \alpha + 1$ for some $\alpha < \omega_1$, such that $(f, \vec{Z}) \Vdash$"$[x]_G \notin \tau$". Therefore for some $y \in \mathbb{R}$, $\mathbf{L}[x, f] = \mathbf{L}[y, f]$, $y \geq_{\mathbf{L}} x_0$ and $(f, \vec{X}) \leq F(y)$ for some sequence $\vec{X}$. Hence, since $f \in [b_0]$ and $y \geq_{\mathbf{L}} x_0$, for some condition $(f, \vec{X}) \leq (f, \vec{Z})$, $(f, \vec{X}) \Vdash$"$[y]_G \in \tau$". But $\mathbf{L}[f, x] = \mathbf{L}[f, y]$ so $(f, \vec{X}) \Vdash$"$[x]_G = [y]_G$" and hence $(f, \vec{X}) \Vdash$"$[x]_G \in \tau$", a contradiction.

Thus $(g, \vec{Y}) \Vdash$"τ contains the cone generated by x_1" but $(g, \vec{Y}) \leq (\varnothing, \vec{Y})$ contradicting that $(\varnothing, \vec{Y}) \Vdash$"$\tau$ does not contain a cone". ⊣ (Theorem 2.6)

§3. Following the basic approach of [SVW82] we show the consistency of ZFC+"ω_2 is the second uniform indiscernible" + "the nonstationary ideal on ω_1 is ω_2-saturated" assuming the consistency of ZF + AD. In fact were this our only goal we could easily finish by using Theorem 1.12 and the relevant proofs of [SVW82] (the use of $\mathbb{R}$-AC in [SVW82] is really only in establishing the appropriate version of Theorem 1.12). Our approach is slightly different than that of [SVW82], we work through a combinatorial intermediary:

> For all $x \in \mathbb{R}$, $x^\#$ exists, and for some (filter) $G \subseteq Q_{\omega_1}$ and all $A \subseteq \omega_1$, $A \in \mathbf{L}[x][G]$ for some $x \in \mathbb{R}$ with G generic over $\mathbf{L}[x]$ for $\mathbb{Q}_{\omega_1}$. $(*)$

Assuming AD, ω_2 is the second uniform indiscernible and the nonstationary ideal on ω_1 is ω_2-saturated (trivially since AD implies that the filter of closed, unbounded, subsets of ω_1 is an ultrafilter). In fact assuming AD something even stronger is true: For every $x \in \mathbb{R}$, $x^\#$ exists, and for all $A \subseteq \omega_1$, $A \in \mathbf{L}[x]$ for some $x \in \mathbb{R}$. This of course must fail in ZFC, $(*)$ is an attempt to find a version more palatable with the axiom of choice.

THEOREM 3.1 (ZF+AD+DC+$\mathbf{V}$=$\mathbf{L}(\mathbb{R})$). *Suppose* $\mathbf{V}[G]$ *is a generic extension of* $\mathbf{V}$ *obtained by forcing with* $\mathbb{P}$. *Then* $\mathbf{V}[G] \models (*)$.

PROOF. This theorem can be proved in a variety of ways. We use Theorem 1.13 and 1.14 and use the relevant notation.

Suppose $G_1 \subseteq Q_{\omega_1}$ defines a generic filter over $\mathbf{V}$ for $\mathbb{Q}_{\omega_1}$. Then by Theorem 1.14, $\mathbf{L}(\mathbb{R}_{G_1})[G_1]$ is a generic extension of $\mathbf{L}(\mathbb{R}_{G_1})$ for forcing with $\mathbb{P}_{G_1}$ ($\mathbb{P}$ defined in $\mathbf{L}(\mathbb{R}_{G_1})$).

We show that $(*)$ holds in $\mathbf{L}(\mathbb{R}_{G_1})[G_1]$ and in fact that G_1 is the appropriate witness. It is easily verified that in $\mathbf{L}(\mathbb{R}_{G_1})[G_1] \subseteq \mathbf{V}[G_1]$, for every real, x, $x^{\#}$ exists. Hence to verify $(*)$ it suffices to show that for all $A \subseteq \omega_1$, $A \in \mathbf{V}[G_1]$, $A \in \mathbf{L}[x][G_1]$ for some real, x, with G_1 generic over $\mathbf{L}[x]$ for $\mathbb{Q}_{\omega_1}$. Fix $A \subseteq \omega_1$, $A \in \mathbf{V}[G_1]$. Choose a term $\tau_A \in \mathbf{V}^{\mathbb{Q}_{\omega_1}}$ for A. Working in $\mathbf{V}$, let $S_A = \{\langle p, \alpha\rangle : p \in Q_{\omega_1}, \alpha < \omega_1$, and $p \Vdash \alpha \in \tau_A\}$. Hence $S_A \in \mathbf{L}[x_0]$ for some real, $x_0 \in \mathbf{V}$ and therefore $A \in \mathbf{L}[x_0][G_1]$. But G_1 is generic over $\mathbf{V}$ for $\mathbb{Q}_{\omega_1}$ so G_1 is generic over $\mathbf{L}[x_0]$ for $\mathbb{Q}_{\omega_1}$.

Thus $(*)$ holds in $\mathbf{L}(\mathbb{R}_{G_1})[G_1]$. Therefore by Theorem 1.13, forcing over $\mathbf{L}(\mathbb{R})$ with $\mathbb{P}$ must yield $(*)$ in the generic extension. ⊣

By a recent theorem of A. Kechris (cf. [Kec84]) ZF + AD implies DC in $\mathbf{L}(\mathbb{R})$. Thus we obtain as a corollary to the previous theorem:

THEOREM 3.2. Assume ZF + AD is consistent. Then so is ZFC + $(*)$.

PROOF. By Theorem 3.1, if ZF + AD is consistent then so is ZF + ω_1-DC + $(*)$. Suppose $\mathbf{V} \models$ ZF + ω_1-DC + $(*)$. Then $\mathbf{L}(\wp(\omega_1)) \models$ ZF + ω_1-DC + $(*)$. Forcing over $\mathbf{L}(\wp(\omega_1))$ it is possible to recover the axiom of choice without adding new subsets of ω_1 (thereby preserving $(*)$), *i.e.*, in $\mathbf{L}(\wp(\omega_1))$ let $T = \{f : \delta \to \wp(\omega_1) : \delta < \omega_2\}$. Define for $f, g \in T$, $f \leq g$ iff $g \subseteq f$. Suppose $G \subseteq T$ is generic over $\mathbf{L}(\wp(\omega_1))$ for the partial order $\langle T, \leq\rangle$. Then since ω_1-DC holds in $\mathbf{L}(\wp(\omega_1))$, $\mathbf{L}(\wp(\omega_1))$ and $\mathbf{L}(\wp(\omega_1))[G]$ have the same subsets of ω_1 and so $\mathbf{L}(\wp(\omega_1))[G] \models$ ZFC + $(*)$. ⊣

Before proceeding we generalize some notation. Suppose α, β are ordinals with $\alpha < \beta$. Define $Q_{\alpha,\beta}$ in the obvious fashion extending the definition in the case of α, β countable, *i.e.*, $Q_{\alpha,\beta} = \{f : f$ is a function with $\operatorname{dom} f \subseteq [\alpha, \beta)$, $\operatorname{dom} f$ finite, and for $\delta \in \operatorname{dom} f$, $f(\delta) \in (\omega+\delta)^{<\omega}\}$. Let $\mathbb{Q}_{\alpha,\beta}$ denote the corresponding partial order. Let $\mathrm{RO}\, Q_{\alpha,\beta}$ denote the elements of the completion of $\mathbb{Q}_{\alpha,\beta}$, *etc.* We shall on occasion denote $Q_{0,\alpha}$ by Q_α.

THEOREM 3.3 (ZFC). Assume $(*)$. Then ω_2 is the second uniform indiscernible and the nonstationary ideal on ω_1 is ω_2-saturated. In fact $\wp(\omega_1)/\mathrm{NS} \cong \mathrm{RO}\, Q_{\omega_2}$.

PROOF. To show that ω_2 is the second uniform indiscernible it suffices to show that for each ordinal, α, with $\omega_1 < \alpha < \omega_2$, α is collapsed to ω_1 inside some $\mathbf{L}[x]$, *i.e.*, that there is an onto function $h\colon \omega_1 \to \alpha$ with $h \in \mathbf{L}[x]$ for some $x \in \mathbb{R}$. This is an immediate consequence of $(*)$ using the fact that Q_{ω_1} is c.c.c.

The proof that $\wp(\omega_1)/\mathrm{NS} \cong \mathrm{RO}\, Q_{\omega_2}$ is based upon the corresponding proof in [SVW82].

For the remainder of this proof ω_1 and ω_2 refer to the ω_1 and the ω_2 of $\mathbf{V}$.

Define a map $I : \wp(\omega_1)/\mathrm{NS} \cong \mathrm{RO}\, Q_{\omega_1,\omega_2}$ as follows. Fix $G_0 \subset Q_{\omega_1}$ as given by $(*)$. Suppose $A \subseteq \omega_1$. Choose $x \in \mathbb{R}$, $A \in \mathbf{L}[x][G_0]$ with G_0 generic over

$\mathbf{L}[x]$ for $\mathbb{Q}_{\omega_1}$. Choose a term $\tau \in \mathbf{L}[x]^{\mathbb{Q}_{\omega_1}}$ for A. Regard τ as a subset of ω_1 (*i.e.*, as a subset of $Q_{\omega_1} \times \omega_1$). Let $j\colon \mathbf{L}[x] \to \mathbf{L}[x]$ be an elementary embedding with $j(\alpha) = \alpha$ for $\alpha < \omega_1$ and $j(\omega_1) = \omega_2$. Thus $j(\mathbb{Q}_{\omega_1}) = \mathbb{Q}_{\omega_2}$ and so $j(\tau)$ is a term in $\mathbf{L}[x]^{\mathbb{Q}_{\omega_2}}$ for a subset of ω_2. Note that $\mathbb{Q}_{\omega_2} \cong \mathbb{Q}_{\omega_1} \times \mathbb{Q}_{\omega_1,\omega_2}$ and therefore in $\mathbf{L}[x][G_0]$, $j(\tau)$ defines naturally a term $\tau^* \in \mathbf{L}[x][G_0]^{\mathbb{Q}_{\omega_1,\omega_2}}$ for a subset of ω_2. Define $I(\tau) = [\![\omega_1 \in \tau^*]\!]_{\mathbb{Q}_{\omega_1,\omega_2}}$ (*i.e.*, $[\![\omega_1 \in \tau^*]\!]$ computed in $\mathbf{L}[x][G_0]^{\mathbb{Q}_{\omega_1,\omega_2}}$).

It is routine to verify that the map I is well defined and further that I defines a Boolean isomorphism of $\wp(\omega_1)/\mathrm{NS}$ into $\mathrm{RO}\, Q_{\omega_1,\omega_2}$. This suffices for showing that the nonstationary ideal on ω_1 is ω_2-saturated, with a little more work one can show the map I is onto, the basic observation is that terms in $\mathbf{V}^{\mathbb{Q}_\alpha}$ for elements of $\mathrm{RO}\, Q_{\alpha,\beta}$ correspond canonically to elements of $\mathrm{RO}\, Q_\beta$. We briefly indicate the argument.

Fix $b \in Q_{\omega_1,\omega_2}$. Hence $b \in \mathrm{RO}\, Q_{\omega_1,\lambda}$ for $\lambda < \omega_2$. Code b by a subset $A_b \subseteq \omega_1$, view A_b as coding a collapse of λ to ω_1 *etc.* (for instance pick $A_b \subseteq \omega_1$ coding a structure $\langle \omega_1, E, S\rangle \cong \langle \mathbf{L}_\lambda, \in, S\rangle$ where $S = \{t \in Q_{\omega_1,\lambda} : t \le b\}$). Choose $x \in \mathbb{R}$ with $A_b \in \mathbf{L}[x][G_0]$ and for which G_0 is generic over $\mathbf{L}[x]$ for $\mathbb{Q}_{\omega_1}$. Hence $b \in \mathbf{L}[x][G_0]$ and so pick a term $\tau \in \mathbf{L}[x]^{\mathbb{Q}_{\omega_1}}$ for b. Working in $\mathbf{L}[x]$ define $w = \{p \cup q : p \in Q_{\omega_1},\ q \in Q_{\omega_1,\lambda}, \text{ and } p \Vdash q \le \tau\}$. Let $c \in \mathrm{RO}\, Q_\lambda$ be that element of $\mathrm{RO}\, Q_\lambda$ defined by w. Choose in $\mathbf{L}[x]$ a code $A_c \subseteq \omega_1$ of c. Thus it follows that if $\delta < \omega_1$ is an indiscernible for $\mathbf{L}[x]$ then $A_c \cap \delta$ codes an element c_δ of $\mathrm{RO}\, Q_{\omega_1}$. Define $A = \{\delta : c_\delta \in G_0^*$, the filter over $\mathrm{RO}\, Q_{\omega_1}$ generated by $G_0\}$. Finally it is straightforward to verify that $I(A) = b$.

The partial orders $\mathbb{Q}_{\omega_2}$, $\mathbb{Q}_{\omega_1,\omega_2}$ are isomorphic hence $\wp(\omega_1)/\mathrm{NS} \cong \mathrm{RO}\, Q_{\omega_2}$. ⊣

Using Theorem 3.3 it is possible to deduce from $(*)$ some useful variants of $(*)$.

Assume $\models \mathsf{ZFC} + (*)$. Let $G_0 \subseteq Q_{\omega_1}$ be as given by $(*)$. Suppose $G \subseteq Q_{\omega_1,\omega_2}$ defines a filter over $\mathbb{Q}_{\omega_1,\omega_2}$, generic over $\mathbf{V}$. Via the isomorphism $I\colon \wp(\omega_1)/\mathrm{NS} \cong \mathrm{RO}\, Q_{\omega_1,\omega_2}$ as constructed in the proof of Theorem 3.3 there is a generic elementary embedding:

$$j\colon \mathbf{V} \to M \subseteq \mathbf{V}[G] \text{ with } M^\omega \subseteq M \text{ in } \mathbf{V}[G].$$

This is simply the embedding corresponding to the appropriate generic ultrapower of $\mathbf{V}$.

$G_0 \times G$ defines in a natural fashion a filter, $G_0 \otimes G$, over $\mathbb{Q}_{\omega_2}$. It is routine to verify that $j(G_0) = G_0 \otimes G$.

For $\alpha < \omega_1$ let $(G_0)_\alpha = G_0 \cap Q_{\alpha,\omega_1}$. We claim that for some $\alpha < \omega_1$ and all $A \subseteq \omega_1$, $A \in \mathbf{L}[x][(G_0)_\alpha]$ for some $x \in \mathbb{R}$ with $(G_0)_\alpha$ generic over $\mathbf{L}[x^\#]$ for $\mathbb{Q}_{\alpha,\omega_1}$ (or even with $(G_0)_\alpha$ generic over $\mathbf{HOD}_x^{\mathbf{L}(\mathbb{R})}$ for $\mathbb{Q}_{\alpha,\omega_1}$, where $\mathbf{HOD}_x^{\mathbf{L}(\mathbb{R})}$ denotes $\mathbf{HOD}_x$ computed in $\mathbf{L}(\mathbb{R})$).

To see this note that $j(G_0)_{\alpha+1} = G \cap Q_{\omega_1+1,\omega_2}$ if $\alpha = \omega_{\mathbf{V}}^1$ and therefore has this property in M.

The partial orders $\mathbb{Q}_{\omega_1}$, $\mathbb{Q}_{\alpha,\omega_1}$ are isomorphic (in $\mathbf{L}$) hence $(G_0)_\alpha$ may be viewed as a filter on $\mathbb{Q}_{\omega_1}$. Thus $(*)$ is equivalent to the variant: For all $x \in \mathbb{R}$, $x^\#$ exists, and for some (filter) $G \subseteq Q_{\omega_1}$ and all $A \subseteq \omega_1$, $A \in \mathbf{L}[x][G]$ for some $x \in \mathbb{R}$ with G generic over $\mathbf{L}[x^\#]$ for $\mathbb{Q}_{\omega_1}$.

It is straightforward to show that ZFC + MA (where MA stands for Martin's Axiom) refutes $(*)$. However a slight weakening of $(*)$ seems sufficient to avoid this difficulty.

> For all $x \in \mathbb{R}$, $x^\#$ exists, and for all $A \subseteq \omega_1$, there exists a filter $G \subseteq Q_{\omega_1}$ and a real, x, such that $A \in \mathbf{L}[x][G]$ with G generic over $\mathbf{L}[x]$ for $\mathbb{Q}_{\omega_1}$. $(**)$

THEOREM 3.4. Assume ZFC + $(*)$ is consistent. Then so is ZFC + MA + $\neg$CH + $(**)$.

PROOF. Assume $\mathbf{V} \models$ ZFC + $(*)$. Suppose P is a c.c.c. partial order of size $\aleph_1$. View P as given by an order $<_P$ on ω_1, *i.e.*, $P = \langle \omega_1, <_P \rangle$. It suffices to show that for any P if $G_P \subseteq P$ is a filter, generic over $\mathbf{V}$ then $\mathbf{V}[G_P] \models (**)$. (Actually one can show $\mathbf{V}[G_P] \models (*)$.)

Suppose $S \subseteq \omega_1$, $S \in \mathbf{V}[G_P]$. Fix a term $\tau_S \in \mathbf{V}^P$ for S. We regard τ_S as a subset of $P \times \omega_1$, *i.e.*, of $\omega_1 \times \omega_1$.

Invoking $(*)$ (more precisely its 'useful' variant) fix a filter $G_0 \subseteq Q_{\omega_1}$ such that for all $C \subseteq \omega_1$, $C \in \mathbf{L}[x][G_0]$ for some $x \in \mathbb{R}$ with G_0 generic over $\mathbf{L}[x^\#]$ for $\mathbb{Q}_{\omega_1}$.

Choose $x \in \mathbb{R}$ such that $P, \tau_S \in \mathbf{L}[x][G_0]$ and for which G_0 is generic over $\mathbf{L}[x^\#]$ for $\mathbb{Q}_{\omega_1}$. Select in $\mathbf{L}[x]$ a term τ for P and let $A = \mathbb{Q}_{\omega_1} * P$ the iteration defined in $\mathbf{L}[x]$ using τ. The pair $\langle G_0, G_P \rangle$ defines a filter, $G_0 \otimes G_P$, on A that is generic over $\mathbf{L}[x^\#]$. Further $S \in \mathbf{L}[x^\#][G_0 \otimes G_P]$. Note that A is $\omega_1^{\mathbf{V}}$-c.c. in $\mathbf{L}[x^\#]$ hence in $\mathbf{L}[x^\#]$, $\mathrm{RO}(A) \cong \mathrm{RO}\, Q_{\omega_1^{\mathbf{V}}}$ (since $A \in \mathbf{L}[x]$). Therefore $\mathbf{L}[x^\#][G_0 \otimes G_P] = \mathbf{L}[x^\#][G_1]$ for some $G_1 \subseteq Q_{\omega_1}$, G_1 generic over $\mathbf{L}[x^\#]$ for $\mathbb{Q}_{\omega_1}$. $S \in \mathbf{L}[x^\#][G_1]$ and so we are done. $\dashv$

We now consider the problem of Martin's Axiom and the saturation of the nonstationary ideal on ω_1.

THEOREM 3.5. Assume ZFC + $(*)$ is consistent. Then so is ZFC + MA + $\neg$CH + "The nonstationary ideal on ω_1 is ω_2-saturated".

PROOF. We fix some notation. Suppose P, Q are (separative) partial orders with $P \subseteq Q$ and P relatively complete in Q, *i.e.*, $\mathrm{RO}(P)$ is a complete subalgebra of $\mathrm{RO}(Q)$. Suppose $G \subseteq P$ is a filter, generic over $\mathbf{V}$. Then we denote by G/P the quotient partial order Q/G computed in $\mathbf{V}[G]$.

By a chain $\langle P_\alpha : \alpha < \lambda\rangle$ of partial orders we shall always mean a chain for which P_α is relatively complete in P_β for all $\alpha < \beta$, *i.e.*, chains will correspond to iterations of forcing.

Assume $\mathbf{V}_0 \models \mathsf{ZFC} + (*)$. We work in $\mathbf{V}_0$ and assume $2^{\aleph_0} = 2^{\aleph_1} = \aleph_2$. Construct a chain of partial orders $\langle P_\alpha : \alpha < \omega_2\rangle$ such that each partial order P_α is c.c.c. and of size $\aleph_1$, $P_\beta = \bigcup_{\alpha<\beta} P_\alpha$ for limit $\beta < \omega_2$, and such that if $P = \bigcup_\alpha P_\alpha$ then $\mathbf{V}_0^P \models \mathsf{MA}$. Assume each P_α is given as an order $<_\alpha$ on $\omega_1 \cdot \alpha$, *i.e.*, $P_\alpha = \langle \omega_1 \cdot \alpha, <_\alpha\rangle$. Thus implicit in P_α is the chain $\langle P_\beta : \beta < \alpha\rangle$.

We claim $\mathbf{V}_0^P \models$ "The nonstationary ideal on ω_1 is ω_2-saturated". Suppose $G \subset \mathbb{P}$ is $\mathbf{V}_0$-generic and for each $\alpha < \omega_2$, let $G_\alpha = G \cap \mathbb{P}_\alpha$. Arguing as in the proof of Theorem 3.4, since $(*)$ holds in $\mathbf{V}_0$ and since each partial order $\mathbb{P}_\alpha$ is c.c.c. in $\mathbf{V}_0$ and of cardinality ω_1 in $\mathbf{V}_0$, for each $\alpha < \omega_2$, $\mathbf{V}_0[G_\alpha] \models (*)$.

Let $\mathbb{B} = \wp(\omega_1)/\mathrm{NS}$ as defined in $\mathbf{V}_0[G]$ and for each $\alpha < \omega_2$, let $\mathbb{B}_\alpha = \wp(\omega_1)/\mathrm{NS}$ as defined in $\mathbf{V}_0[G_\alpha]$.

Stationary subsets of ω_1 are preserved in passing from $\mathbf{V}_0[G_\alpha]$ to $\mathbf{V}_0[G]$ for each $\alpha < \omega_2$. Thus in $\mathbf{V}[G]$, each $\mathbb{B}_\alpha$ is a Boolean subalgebra of $\mathbb{B}$ and $\mathbb{B} = \bigcup_\alpha \mathbb{B}_\alpha$. The key point is that if $\alpha < \omega_2$ and α has countable cofinality then $\bigcup_{\beta<\alpha} \mathbb{B}_\beta$ is dense in $\mathbb{B}_\alpha$. This follows from the fact that if $S \subset \omega_1$ is a set in $\mathbf{V}[G_\alpha]$ then S is a countable union of sets S_i for $i < \omega$ such that each $i < \omega$, $S_i \in \mathbf{V}_0[G_\beta]$ for some $\beta < \alpha$.

For each $\alpha < \omega_2$, $\mathbb{B}_\alpha$ is ω_2-c.c. in $\mathbf{V}_0[G_\alpha]$ since $(*)$ holds in $\mathbf{V}_0[G_\alpha]$. Therefore for each $\alpha < \omega_2$, $\mathbb{B}_\alpha$ is necessarily ω_2-c.c. in $\mathbf{V}_0[G]$. Further the set of $\alpha < \omega_2$ such that α has countable cofinality is trivially stationary in ω_2 and so now it follows that $\mathbb{B}$ is ω_2-c.c. in $\mathbf{V}[G]$. ⊣

Much in the spirit of [SVW82] we summarize some of these independence results in the following theorem.

Theorem 3.6. The following are equiconsistent:

1) $\mathsf{ZF} + \mathsf{AD}$.
2) $\mathsf{ZFC} + \mathsf{AD}^{\mathbf{L}(\mathbb{R})} + \omega_2$ is the second uniform indiscernible + the nonstationary ideal on ω_1 is ω_2-saturated.
3) $\mathsf{ZFC} + \mathsf{AD}^{\mathbf{L}(\mathbb{R})} + \mathsf{MA} + \neg\mathsf{CH} + \omega_2$ is the second uniform indiscernible + the nonstationary ideal on ω_1 is ω_2-saturated.

As we have indicated $(*)$ is a ZFC version of:

> For every $x \in \mathbb{R}$, $x^\#$ exists, and for all $A \subseteq \omega_1$, $A \in \mathbf{L}[x]$ for some $x \in \mathbb{R}$. (‡)

We shall show that assuming $\mathsf{ZFC} + (*)$, $\mathbf{L}(\mathbb{R}) \models (‡)$. In fact, we shall prove something stronger but first we isolate what is necessary in $\mathbf{L}(\mathbb{R})$ in order for it to be possible to force over $\mathbf{L}(\mathbb{R})$ to obtain a model of $\mathsf{ZFC} + (*)$. Clearly the following in addition to (‡) will suffice:

(1) $\mathsf{C} + {**}\mathbb{R}\text{-}\mathsf{AC}$.

(2) Forcing with $\mathbb{P}$ yields ω_1-DC.

Let $\mathfrak{F}$ denote the filter over the reals generated by wellordered intersections of comeager subsets of $\mathbb{R}$. Assume $\mathfrak{F}$ is nontrivial ($\varnothing \notin \mathfrak{F}$) and let $\mathbb{P}^*$ denote the partial order defined as $\mathbb{P}$ is defined, using $\mathfrak{F}$ in place of the comeager filter. Upon examination of the relevant proofs it becomes apparent that (1) and (2) can be replaced by:

(1′) $\mathfrak{F}$ is nontrivial and for every set $A \subseteq \mathbb{R}$ there is a Borel set B and a set $D \in \mathfrak{F}$ such that $A \cap D = B \cap D$. For any function $f : \mathbb{R} \to \mathbf{V}$ there is a (partial) choice function $g : \mathbb{R} \to \mathbf{V}$, $\operatorname{dom} g \in \mathfrak{F}$, such that for $x \in \operatorname{dom} g$, $g(x) \in f(x)$ or $f(x) = \varnothing$.

(2′) Forcing with $\mathbb{P}^*$ yields ω_1-DC.

Note that (1′) simply asserts that every set of reals has the property of Baire relative to a set in $\mathfrak{F}$ (this is the $\mathfrak{F}$-version of C) and that the appropriate version of $**\mathbb{R}$-AC holds. Also it is possible to isolate choice principles in the spirit of $**\omega_1$-AC that are equivalent to (2′) (in ZF + DC + (1′)), we leave the details of this to the curious reader.

THEOREM 3.7. Assume ZFC + $(*)$. Then $\mathbf{L}(\mathbb{R}) \models (\ddagger)$. In fact $\mathbf{L}(\mathbb{R}) \models$ (1′) + (2′) + $(\ddagger)$ and $\mathbf{L}(\wp(\omega_1))$ is a generic extension of $\mathbf{L}(\mathbb{R})$ via forcing with $\mathbb{P}^*$ as defined in $\mathbf{L}(\mathbb{R})$.

PROOF. Assume $\mathbf{V} \models$ ZFC + $(*)$.

Fix $G_0 \subseteq \mathbb{Q}_{\omega_1}$ as given by $(*)$. Suppose $G \subseteq \mathbb{Q}_{\omega_1,\omega_2}$ defines a filter, generic over $\mathbf{V}$ for $\mathbb{Q}_{\omega_1,\omega_2}$. Let $I : \wp(\omega_1)/\mathrm{NS} \cong \mathrm{RO}\, Q_{\omega_1,\omega_2}$ be the isomorphism as constructed in the proof of Theorem 3.3. Hence there is a generic elementary embedding:

$$j : \mathbf{V} \to M \subseteq \mathbf{V}[G] \text{ with } M^\omega \subseteq M \text{ in } \mathbf{V}[G].$$

Further $j(G_0) = G_0 \otimes G$ the filter over $\mathbb{Q}_{\omega_2}$ determined by $G_0 \times G$.

Let $\mathbf{L}(\mathbb{R})$ denote the $\mathbf{L}(\mathbb{R})$ of $\mathbf{V}[G]$. To show that $\mathbf{L}(\mathbb{R}) \models$ (1′) it suffices to show that $\mathbf{L}(\mathbb{R}_G) \models$ (1′) since $\mathbf{L}(\mathbb{R}_G)$ is also the $\mathbf{L}(\mathbb{R})$ of M. That $\mathbf{L}(\mathbb{R}_G) \models (1)'$ is immediate via the following claim:

CLAIM 3.8. Assume $\mathbf{V}_0 \models$ ZFC and that $\mathbf{V}_0[g]$ is a generic extension of $\mathbf{V}_0$ obtained by adding a generic ω_1-sequence of Cohen reals. Let $\mathbf{L}(\mathbb{R}_g)$ denote the $\mathbf{L}(\mathbb{R})$ of $\mathbf{V}_0[g]$. Then $\mathbf{L}(\mathbb{R}_g) \models$ (1′).

PROOF OF CLAIM 3.8. Routine. ⊣ (Claim 3.8)

Note that $\mathbf{V}[G] = \mathbf{V}[G_1][G_2]$ where $G_1 = G \cap Q_{\omega_1+1}$, $G_2 = G \cap Q_{\omega_1+1,\omega_2}$. Hence it follows by the claim that $\mathbf{L}(\mathbb{R}_G) \models$ (1′) and so $\mathbf{L}(\mathbb{R}) \models$ (1′).

Let $\mathbb{P}^*_0$ denote the partial order $\mathbb{P}^*$ as defined in $\mathbf{L}(\mathbb{R})$. by an argument analogous to the proof of Theorem 1.14 it follows that $G_0 \otimes G$ defines a filter over $j(\mathbb{P}^*_0)$, generic over $\mathbf{L}(\mathbb{R}_G)$.

Hence G_0 defines a filter over $\mathbb{P}_0^*$ that is generic over $\mathbf{L}(\mathbb{R})$. By $(*)$ the generic extension $\mathbf{L}(\mathbb{R})[G_0] = \mathbf{L}(\wp(\omega_1))$. Thus $\mathbf{L}(\mathbb{R})[G_0] \models \omega_1$-DC, this proves $\mathbf{L}(\mathbb{R}) \models (2')$.

Finally to show that $\mathbf{L}(\mathbb{R}) \models (\ddagger)$, observe that since $\mathbf{L}(\mathbb{R})[G_0] \models (*)$ an analysis in $\mathbf{L}(\mathbb{R})$ of the forcing language for $\mathbb{P}_0^*$ will yield $(\ddagger)$. ⊣ (Theorem 3.7)

Thus by Theorem 3.7, ZFC $+ (*)$ is equiconsistent with ZF $+$ DC $+ (1') + (2') + (\ddagger)$. Observe that ZF $+$ DC $+ (1') + (2')$ is equiconsistent with ZF (add ω_1 Cohen reals to $\mathbf{L}$, the new $\mathbf{L}(\mathbb{R})$ satisfies $(1') + (2')$). Thus ignoring possibly significant interference effects between $(1') + (2')$ and $(\ddagger)$, the consistent strength of ZFC $+ (*)$ could well approximate that of ZF $+$ DC $+ (\ddagger)$. We conjecture that ZF $+$ AD $\vdash$ Con(ZF $+$ DC $+ (\ddagger)$), in fact we shall be foolish enough to suggest a scenario for a proof. Assume ZF $+$ AD $+$ $\mathbf{V}{=}\mathbf{L}(\mathbb{R})$. The model that we are interested in is **HOD**. Fix $\lambda = \Theta^{\mathbf{L}(\mathbb{R})}$. Suppose $G \subseteq Q_\lambda$ is generic over **HOD** for $\mathbb{Q}_\lambda$. Let $\mathbf{L}(\mathbb{R}_G)$ denote the $\mathbf{L}(\mathbb{R})$ of $\mathbf{HOD}[G]$. We conjecture that $\mathbf{L}[\mathbb{R}_G] \models (\ddagger)$.

There is actually some evidence that ZF $+$ PD $\vdash$ Con(ZF $+$ DC $+ (\ddagger)$), where PD denotes the axiom of projective determinacy.

We use ZFC $+ (*)$ to produce one final independence result. For each $n < \omega$ let $\Sigma_n(\omega_1)$ denote the class of those subsets of ω_1 that are definable over $\langle \wp(\omega_1), \in \rangle$ by a Σ_n-formula without parameters. Similarly define the classes $\Pi_n(\omega_1)$. These classes are ω_1-versions of the more conventional classes $\Sigma^1_n(\omega)$, $\Pi^1_n(\omega)$. We work within ZFC and consider the question of which of the classes $\Sigma_n(\omega_1)$, $\Pi_n(\omega_1)$ $n < \omega$ have the prewellordering property.

Recall that assuming PD, for each $n < \omega$, PWO($\Sigma^1_{2n}(\omega)$) and PWO ($\Pi^1_{2n+1}(\omega)$), cf. [Mos80] for details.

THEOREM 3.9. Assume ZFC $+ (*) +$ PD. Then for each n, PWO($\Sigma_{2n}(\omega_1)$) and PWO($\Pi_{2n+1}(\omega_1)$).

PROOF (SKETCH). Note that as a consequence of $(*)$, a set of $S \subseteq \omega_1$ that is $\Sigma_k(\omega_1)$ defines in the codes a set $S^* \subseteq \mathbb{R}$ that is Σ^1_{k+2} and conversely. By PD for each integer k, PWO($\Sigma^1_{2k}(\omega^\omega)$) and PWO($\Pi^1_{2k+1}(\omega^\omega)$), cf. [Mos80] for the relevant details. Using 'generic' codes of countable ordinals it follows from this that for each $n < \omega$, PWO($\Sigma_{2n}(\omega_1)$) and PWO($\Pi_{2n+1}(\omega_1)$). ⊣

Let $\Pi_1(\wp(\omega_1))$ denote the class of those subsets of $\wp(\omega_1)$, Π_1 over $\langle \wp(\omega_1), \in \rangle$. Assume ZFC $+ (*) +$ PD. Then it can be shown that PWO($\Pi_1(\wp(\omega_1))$) fails as does PWO($\Sigma_1(\wp(\omega_1))$).

Many of the questions posed in [SVW82] remain unanswered. We add one to the list. Does ZFC $+$ DC $+$ AD $+$ $\mathbf{V}{=}\mathbf{L}(A, \mathbb{R})$ $(A \subset \mathbb{R})$ suffice for Theorem 1.12? This may seem (and be) a rather technical question however it tests the power of AD. The point being that assuming $\mathsf{AD}_{\mathbb{R}}$, the conclusion of Theorem 1.12 holds in $\mathbf{L}(A, \mathbb{R})$ for each set of reals $A \subseteq \mathbb{R}$. It is therefore natural to ask if AD suffices to show this in $\mathbf{L}(A, \mathbb{R})$.

We note that prior to the results here the consistency of Martin's Axiom with the existence of an ω_2-saturated ideal on ω_1 was open. This problem will surely fall to more conventional assumptions. The situation for the nonstationary ideal is less clear.[1]

The problem of Martin's Axiom and the existence of ω_2-saturated ideals is an instance of a more general question, can there exist a c.c.c. indestructible ω_2-saturated ideal? Note that assuming $\mathsf{ZFC} + (*)$, "$\wp(\omega_1)/\mathrm{NS} \cong \mathrm{RO}\, Q_{\omega_2}$" is true in any c.c.c. extension via a partial order of size $\aleph_1$ (this because $(*)$ holds in any such forcing extension, cf. Theorem 3.4)

The consistency of $\mathsf{ZFC} + (*)$ can also be obtained from the consistency of $\mathsf{ZF} + \mathsf{DC}+$ Every set of reals is Suslin $+\ \aleph_1$ is measurable. This follows via a recent result of A. Kechris which states that if $\aleph_1$ is measurable then a subset $A \subseteq \omega_1$ is constructible from a real if and only if the set of reals defined by A (in the codes) is Suslin and CoSuslin.

Corrections from the original version of the paper and subsequent results. Theorem 3.5 as originally formulated is false. That formulation claimed that if $\mathsf{ZFC} + (*)$ is consistent then so is $\mathsf{ZFC} + \mathsf{MA} + \neg\mathsf{CH} + \wp(\omega_1)/\mathrm{NS} \cong \mathrm{RO}\, Q_{\omega_2}$.

In fact, if MA holds (and CH fails) then for any normal ideal ω_2-saturated ideal I on ω_1, the complete Boolean algebra $\mathbb{B}$ given by the quotient, $\wp(\omega_1)/I$ has the property that if $\mathbb{B}^*$ is a complete countably generated subalgebra of $\mathbb{B}$ without atoms then $\mathbb{B}^* = \mathbb{B}$.

The results in this paper have been substantially generalized in the development of $\mathbb{P}_{\max}$-theory, [Woo99] and [Woo10]. This theory began with theorem that if the nonstationary ideal on ω_1 is ω_2-saturated and $(\wp(\omega_1))^{\#}$ exists then ω_2 is necessarily the second uniform indiscernible.

REFERENCES

Matthew Foreman, Menachem Magidor, and Saharon Shelah

[FMS88] *Martin's maximum, saturated ideals and nonregular ultrafilters. I*, ***Annals of Mathematics***, vol. 127 (1988), no. 1, pp. 1–47.

Alexander S. Kechris

[Kec84] *The axiom of determinacy implies dependent choices in* $\mathbf{L}(\mathbb{R})$, ***The Journal of Symbolic Logic***, vol. 49 (1984), no. 1, pp. 161–173.

Alexander S. Kechris, Eugene M. Kleinberg, Yiannis N. Moschovakis, and W. H. Woodin

[KKMW81] *The axiom of determinacy, strong partition properties, and nonsingular measures*, in Kechris et al. [Cabal ii], pp. 75–99, reprinted in [Cabal I], pp. 333–354.

[1] By the results of Foreman, Magidor, and Shelah on Martin's Maximum, Martin's Maximum implies that the nonstationary ideal on ω_1 is ω_2-saturated [FMS88]. Further assuming there is a supercompact cardinal, Martin's Maximum holds in a semi-proper forcing extension of $\mathbf{V}$.

Alexander S. Kechris, Benedikt Löwe, and John R. Steel

[Cabal I] *Games, Scales, and Suslin cardinals: the Cabal Seminar, volume I*, Lecture Notes in Logic, vol. 31, Cambridge University Press, 2008.

[Cabal II] *Wadge Degrees and Projective Ordinals: the Cabal Seminar, volume II*, Lecture Notes in Logic, vol. 37, Cambridge University Press, 2012.

Alexander S. Kechris, Donald A. Martin, and Yiannis N. Moschovakis

[Cabal ii] *Cabal Seminar* 77–79, Lecture Notes in Mathematics, vol. 839, Berlin, Springer-Verlag, 1981.

Alexander S. Kechris, Robert M. Solovay, and John R. Steel

[KSS81] *The axiom of determinacy and the prewellordering property*, in Kechris et al. [Cabal ii], pp. 101–125, reprinted in [Cabal II], pp. 118–140.

Donald A. Martin, Yiannis N. Moschovakis, and John R. Steel

[MMS82] *The extent of definable scales*, ***Bulletin of the American Mathematical Society***, vol. 6 (1982), pp. 435–440.

Yiannis N. Moschovakis

[Mos80] ***Descriptive Set Theory***, Studies in Logic and the Foundations of Mathematics, vol. 100, North-Holland, Amsterdam, 1980.

John R. Steel and Robert Van Wesep

[SVW82] *Two consequences of determinacy consistent with choice*, ***Transactions of the American Mathematical Society***, vol. 272 (1982), no. 1, pp. 67–85.

W. Hugh Woodin

[Woo99] ***The Axiom of Determinacy, Forcing Axioms, and the Nonstationary Ideal***, de Gruyter Series in Logic and its Applications, vol. 1, Walter de Gruyter, Berlin, 1999.

[Woo10] ***The Axiom of Determinacy, Forcing Axioms, and the Nonstationary Ideal***, revised ed., de Gruyter Series in Logic and its Applications, vol. 1, Walter de Gruyter, Berlin, 2010.

[Woo] *An* $\aleph_1$ *dense ideal on* $\aleph_1$, in preparation.

DEPARTMENT OF MATHEMATICS
UNIVERSITY OF CALIFORNIA
BERKELEY, CALIFORNIA 94720-3840
UNITED STATES OF AMERICA
E-mail: woodin@math.berkeley.edu

SUBSETS OF $\aleph_1$ CONSTRUCTIBLE FROM A REAL

ALEXANDER S. KECHRIS

The purpose of this paper is to give a necessary and sufficient condition for a subset A of $\aleph_1$ to be constructible from a real in terms of structural properties of the code set of A, valid under the assumption that an appropriate measurable cardinal exists. This can be combined with recent results of Woodin to provide upper bounds for the consistency strength of theories of the form $\mathsf{ZFC} + \forall x \in {}^{\omega}\omega(x^{\#}$ exists$)$ + "every subset of $\aleph_1$ with code set in Γ is constructible from a real," for various pointclasses Γ.

For each $A \subseteq \aleph_1$ let

$$A^* = \{w \in \mathrm{WO} : |w| \in A\},$$

where WO is the set of reals coding wellordering of ω, and for $w \in \mathrm{WO}$ we let $|w|$ be the ordinal of the wellordering coded by w. The **code set** of A is then

$$\langle A \rangle = \{0^\frown w : w \in A^*\} \cup \{1^\frown w : w \in (\aleph_1 \backslash A)^*\},$$

i.e., the disjoint union of A^* and $(\aleph_1 \backslash A)^*$.

A set of reals $P \subseteq {}^{\omega}\omega$ is called **Suslin** if there is a tree T on $\omega \times \lambda$, λ an ordinal, with $P = \mathrm{p}[T] = \{x \in {}^{\omega}\omega : \exists f \in {}^{\omega}\lambda(x, f) \in [T]\}$. If the tree T can be taken to be *homogeneous*, we call P **homogeneously Suslin**. For the definition of homogeneous trees, *cf.* [Kec81]. If finally $P = \mathrm{p}[R] = \{x : \exists y(x, y) \in R\}$, where $R \subseteq {}^{\omega}\omega \times {}^{\omega}\omega$ and R is homogeneously Suslin, we call P **weakly homogeneously Suslin**.

We have now the following characterization.

THEOREM 1.1 (ZF+DC).

i) If $\aleph_1$ is measurable, then $A \subseteq \aleph_1$ is constructible from a real iff $\langle A \rangle$ is Suslin.
ii) If there exists a measurable cardinal, then $A \subseteq \aleph_1$ is constructible from a real iff $\langle A \rangle$ is weakly homogeneously Suslin.

Research partially supported by NSF Grants MCS-8117804 and DMS-8416349.

Large Cardinals, Determinacy, and Other Topics: The Cabal Seminar, Volume IV
Edited by A. S. Kechris, B. Löwe, J. R. Steel
Lecture Notes in Logic, 49

Let us mention now some consequences. Woodin (unpublished) has shown that

$$\text{Con}(\mathsf{ZFC} + \text{there are infinitely many strong cardinals}) \text{ implies } \text{Con}(\mathsf{ZFC} + \text{“every projective set is Suslin”}),$$

so that we have Con(ZFC + “there are infinitely many strong cardinals”) implies Con(ZFC + $\forall x(x^{\#}$ exists)+ “every projective subset of $\aleph_1$ is constructible for a real”).

Recall that κ is **strong** if for all λ, there is a $j : \mathbf{V} \to M$ such that $\text{crit}(j) = \kappa$ and $\mathbf{V}_\lambda \subseteq M$. On the other hand, it is known that ZFC + $\forall x(x^{\#}$ exists) + “every projective subset of $\aleph_1$ is constructible for a real” is (consistency-wise) quite strong, stronger at least than the large cardinals for which core models have been constructed (which includes at least κ with $o(\kappa) = \kappa^+$).

Another consequence is that from the consistency of ZF + DC + “every set of reals is Suslin” + “$\aleph_1$ is measurable” one obtains the consistency of ZFC + $*$—*cf.* [Woo83] for this principle and its consequences.

We proceed now to prove the theorem. For technical convenience and without loss of generality, we shall assume that $A \subseteq \aleph_1 \setminus \omega$.

First we need some preliminaries. Let $\tau_0, \tau_1, \tau_2, \ldots$ be a 1-1 enumeration of ${}^{<\omega}\omega$, with $\tau_0 = \varnothing$, $\text{lh}(\tau_i) \leq i$, $\tau_i \supsetneqq \tau_j \Rightarrow i > j$. For $\sigma \in {}^{<\omega}\omega$, $\text{lh}(\sigma) = n$ let

$$T_\sigma = \{\tau_i \in {}^{<\omega}\omega : i < n \wedge \forall j(\varnothing \neq \tau_j \subseteq \tau_i \Rightarrow \sigma(j) = 0\} \cup \{\varnothing\}.$$

Then T_σ is a finite tree on ω and $\sigma \subseteq \sigma' \Rightarrow T_\sigma \subseteq T_{\sigma'}$. For $\sigma \neq \varnothing$, define the following ordering $<_\sigma$ on n:

$$i <_\sigma j \Leftrightarrow (i, j < n \wedge \tau_i, \tau_j \notin T_\sigma \wedge i < j) \vee (\tau_i \notin \tau_\sigma \wedge \tau_j \in T_\sigma) \vee (\tau_i, \tau_j \in T_\sigma \wedge \tau_i <_{\text{BK}} \tau_j),$$

where $<_{\text{BK}}$ is the Brouwer-Kleene ordering on ${}^{<\omega}\omega$. Clearly 0 is the top element of $<_\sigma$ and $\sigma \subseteq \sigma' \Rightarrow <_\sigma \subseteq <_{\sigma'}$ (since for $i < n = \text{lh}(\sigma)$, $\tau_i \in T_\sigma \Leftrightarrow \tau_i \in T_{\sigma'}$ in view of $\text{lh}(\tau_i) \leq \tau_i$). For $\alpha \in {}^\omega\omega$, let $<_\alpha = \bigcup_n <_{\alpha\restriction n}$. Then $<_\alpha$ is the Brouwer-Kleene ordering of $T_\alpha = \{\tau_i \in {}^{<\omega}\omega : \forall j(\varnothing \neq \tau_j \subseteq \tau_i \Rightarrow \alpha(j) = 0)\} \cup \{\varnothing\}$, transferred to ω via the coding, with the rest of ω thrown in at the bottom with its usual ordering. Let $\text{WO} := \{\alpha : T_\alpha \text{ well founded}\}$ and $|\alpha|$ be the rank of 0 in $<_\alpha$. Thus, $\{|\alpha| : \alpha \in \text{WO}\} = \omega_1 \setminus \omega$. Let

$$\text{SH} = \{(\sigma, u) \in {}^{<n}\omega \times {}^{<n}\text{Ord} : u\colon n \to \text{Ord} \wedge u \text{ is order preserving for } <_\sigma\}.$$

Then clearly

i) $\text{WO} = \text{p[SH]} = \text{p[SH}\restriction\kappa]$, for any $\kappa \geq \aleph_1$.
ii) $(\sigma, u) \in \text{SH} \wedge \text{lh}(\sigma) = n \Rightarrow u(0) > u(1) \wedge u(0) > u(2) \wedge \cdots \wedge u(0) > u(n-1)$.
iii) $(w, f) \in [\text{SH}] \Rightarrow |w| \leq f(0)$.

iv) If $\pi_\sigma : n \to n$ is the permutation defined by

$$i <_\sigma j \Leftrightarrow \pi_\sigma(i) < \pi_\sigma(j),$$

then

$$(\sigma, u) \in \mathrm{SH} \Leftrightarrow \exists v \in {}^n[\mathrm{Ord}] \ (u = v \circ \pi_\sigma).$$

So if κ is a measurable cardinal and μ_n is the n-fold cartesian product of a normal measure μ on κ, then $\mathrm{SH}\restriction\kappa$ is homogeneous with homogeneity measures $\mu_\sigma = (\pi_\sigma)_*(\mu_n)$.

Finally if $P \subseteq {}^\omega\omega$ is any Π^1_1 set and $F : {}^\omega\omega \to {}^\omega\omega$ is a Lipschitz function such that $F^{-1}[\mathrm{WO}] = P$, consider the tree induced by F, *i.e.*,

$$(\sigma, u) \in \mathrm{SH}_F \Leftrightarrow (F(\sigma), u) \in \mathrm{SH}.$$

Then $P = \mathrm{p}[\mathrm{SH}_F]$ and SH_F inherits all the above properties of SH.

Given $\varnothing \neq A \subseteq \aleph_1 \setminus \omega$ consider its Solovay game S_A:

$$\begin{array}{lll} \mathrm{I} & \mathrm{II} & \mathrm{II} \text{ wins iff} \\ x & y, \alpha & x \in \mathrm{WO} \Rightarrow y \in \mathrm{WO} \wedge |x| \leq |y| \wedge \forall n[(\alpha)_n \in \mathrm{WO}] \wedge \\ & & A \cap (|y| + 1) \subseteq \{|(\alpha)_n| : n \in \omega\} \subseteq A. \end{array}$$

By the usual boundedness, player I cannot have a winning strategy. Also if player II has a winning strategy, then A is constructible from a real.

Proof of i). Assume $\aleph_1$ is measurable. If $A \subseteq \aleph_1 \setminus \omega$ is constructible from a real, then using $\forall x \in {}^\omega\omega$ ($x^\#$ exits) it is easy to check that $\langle A \rangle$ is $\underset{\sim}{\Sigma}^1_2$, so it is Suslin.

Conversely assume that $\langle A \rangle$ is Suslin. Let

$$Q(v, \alpha) \Leftrightarrow \forall n[(\alpha)_n \in \mathrm{WO}] \wedge v \in \mathrm{WO} \wedge \\ \forall z \leq_T v[z \in A^* \wedge |z| \leq |v| \Rightarrow \exists k(|z| = |(\alpha_k)|] \wedge \forall n[(\alpha)_n \in A^*],$$

thus

$$Q(v, \alpha) \Leftrightarrow \forall n[(\alpha)_n \in \mathrm{WO}] \wedge v \in \mathrm{WO} \wedge \\ A \cap (|v| + 1) \subseteq \{|(\alpha)_n| : n \in \omega\} \subseteq A.$$

Moreover since $\langle A \rangle$ is Suslin, so is Q, say $Q = \mathrm{p}[T]$, T a tree on $\omega \times \lambda$.

Let also

$$R(w, v) \Leftrightarrow w, v \in \mathrm{WO} \wedge |w| \leq |v|.$$

Then R is Π^1_1, so let F be Lipschitz with $F^{-1}[\mathrm{WO}] = R$ and let $S' = \mathrm{SH}_F \restriction \aleph_1$. Then $R = \mathrm{p}[S']$ and S' is homogeneous with homogeneity measures $\mu'_{\sigma,\tau}$; $\sigma, \tau \in \bigcup_n ({}^n\omega \times {}^n\omega)$. Let also $S = \mathrm{SH}\restriction\aleph_1$, so that $\mathrm{p}[S] = \mathrm{WO}$ and S has homogeneity measures μ_σ, $\sigma \in {}^{<\omega}\omega$.

Now consider the following closed game, motivated by a game of Martin (*cf.* [Kec81, § 7]):

$$\begin{array}{ll} \text{I} & \text{II} \\ w, f & v, g, \alpha, h \end{array} \qquad \begin{array}{l} w, v, \alpha \in {}^{\omega}\omega;\, f, g \in {}^{\omega}\aleph_1, h \in {}^{\omega}\lambda; \\ \\ \text{II wins iff} \\ \forall n \geq 1[(w\restriction n, f\restriction n) \in S \Rightarrow \\ (w\restriction n, v\restriction n, g\restriction n) \in S' \wedge (v\restriction n, \alpha\restriction n, h\restriction n) \in T]. \end{array}$$

This game is of course determined. We want to show that player I cannot have a winning strategy.

Assume that he did, towards a contradiction, and call it G.

His first move by G is α_0, ξ_0 with $(\alpha_0, \xi_0) \in S$. Choose then $v_0 \in \text{WO}$ with $v_0 \in \text{WO}$ and $|v_0| \geq \xi_0$ and then choose α_0 with $Q(v_0, \alpha_0)$ and h_0 with $(v_0, \alpha_0, h_0) \in [T]$. We can then use the homogeneity of S' (and the fact that, after his first move, player I following G has to play ordinals $< \xi_0$, as long as player II plays in the appropriate trees, thus the countable additivity of the homogeneity measures on S' suffices) to find a g_0 such that v_0, g_0, α_0, h_0 defeats G. The details of this kind of argument are spelled out in a similar situation in [Kec81] (*cf.* proof of direction ($\Rightarrow$) in 7.1).

So player II has a winning strategy. Then consider the game

$$\begin{array}{ll} \text{I} & \text{II} \\ w, f & v, \alpha \end{array} \qquad \begin{array}{l} w, v, \alpha \in {}^{\omega}\omega;\, f \in {}^{\omega}\aleph_1; \\ \\ \text{II wins iff} \\ (w, f) \in [S] \Rightarrow v \in \text{WO} \wedge |w| \leq |v| \wedge Q(v, \alpha). \end{array}$$

Clearly player II has a winning strategy in that game too, call it F. (F is obtained by a winning strategy of player II in the preceding closed game by forgetting about the g, h, which are witnesses that $|w| \leq |v|$ and $Q(v, \alpha)$.)

Using F define finally the following strategy Σ for player II in the Solovay game S_A:

$$\Sigma(w\restriction n+1) = (v\restriction n+1, \alpha\restriction n+1),$$

where on some set $X(w\restriction n+1)$ of $\mu_{w\restriction n+1}$-measure 1 and for all $f\restriction n+1 \in X(w\restriction n+1)$, $F(w\restriction n+1, f\restriction n+1) = (v\restriction n+1, \alpha\restriction n+1)$. In other words Σ is obtained by integrating F over the homogeneity measures for S.

We claim that this is winning for player II, thereby completing the proof. Indeed, assume w, v, α have been played with player I playing w and v, α determined by Σ. If $w \notin \text{WO}$, there is nothing to prove. Else $w \in \text{WO}$, so by homogeneity find f with $f\restriction n+1 \in X(w\restriction n+1)$, for all n. Then clearly w, f, v, α is a run of the preceding game in which player II followed F, thus since $(w, f) \in [S]$ we have $v \in \text{WO} \wedge |w| \leq |v| \wedge Q(v, \alpha)$, *i.e.*, player II won this run of the Solovay game and we are done. $\dashv$

Proof of ii). Let κ be the least measurable cardinal. Assume now $\langle A\rangle$ is weakly homogeneously Suslin. Since weakly homogeneously Suslin sets are closed under conjunctions, disjunctions and countable intersections it follows that the relation Q we defined before is weakly homogeneously Suslin, say

$$(v,\alpha)\in Q\Leftrightarrow \exists\beta\exists h(v,\alpha,\beta,h)\in[T],$$

where T is a tree on $\omega\times\lambda$, T homogeneous with homogeneity measures $\mu''_{\sigma,\tau,\rho}$, $\sigma,\tau,\rho\in\bigcup_n({}^n\omega\times{}^n\omega\times{}^n\omega)$. Let now $S=\mathrm{SH}{\restriction}\kappa$, $S'=\mathrm{SH}_F{\restriction}\kappa$ with homogeneity measures μ_σ, $\mu'_{\sigma,r}$.

Consider again the game

$$\begin{array}{lll} \text{I} & \text{II} & \text{II wins iff} \\ w,f & v,g,\alpha,\beta,h & \forall n\geq 1[(w{\restriction}n,f{\restriction}n)\in S\Rightarrow \\ & & (w{\restriction}n,v{\restriction}n,g{\restriction}n)\in S'\wedge(v{\restriction}n,\alpha{\restriction}n,\beta{\restriction}n,h{\restriction}n)\in T]. \end{array}$$

It will be of course enough to show that player I has no winning strategy. Say he had one and call it G, towards a contradiction.

Define now a strategy

$$\sigma(v{\restriction}n,\alpha{\restriction}n,\beta{\restriction}n)=(w{\restriction}n+1,f{\restriction}n+1)$$

so that for each n, $(w{\restriction}n+1,f{\restriction}n+1)\in S$, as follows:

$$\sigma(\varnothing,\varnothing,\varnothing)=(w(0),f(0))\stackrel{\text{def}}{=}G(\varnothing,\varnothing,\varnothing,\varnothing,\varnothing)\in S'.$$

Given $(v{\restriction}1,\alpha{\restriction}1,\beta{\restriction}1)$ consider $S'(w{\restriction}1,f{\restriction}1)$ and $T(v{\restriction}1,\alpha{\restriction}1,\beta{\restriction}1)$. Then for each $h{\restriction}1\in T(v{\restriction}1,\alpha{\restriction}1,\beta{\restriction}1)$ and $g{\restriction}1\in S'(w{\restriction}1,v{\restriction}1)$ consider $G(v{\restriction}1,g{\restriction}1,\alpha{\restriction}1,\beta{\restriction}1,h{\restriction}1)$. We claim that on a set of measure 1 of $(h{\restriction}1,g{\restriction}1)$ in the product measure $\mu''_{v{\restriction}1,\alpha{\restriction}1,\beta{\restriction}1}\times\mu'_{w{\restriction}1,v{\restriction}1}$ we have that $G(v{\restriction}1,g{\restriction}1,\alpha{\restriction}1,\beta{\restriction}1,h{\restriction}1)$ is fixed, say with value $(w{\restriction}2,f{\restriction}2)\in S$, which we define to be our $\sigma(v{\restriction}1,\alpha{\restriction}1,\beta{\restriction}1)$. This is because for $(w{\restriction}2,f{\restriction}2)\in S$, $f(1)<f(0)<\kappa$ and the product measure is countably complete thus κ-complete since κ is the least measurable.

Proceeding similarly we define $\sigma(v{\restriction}n,\alpha{\restriction}n,\beta{\restriction}n)$. Put now

$$\begin{aligned}\sigma'(v{\restriction}n,\alpha{\restriction}n,\beta{\restriction}n)&=w{\restriction}n+1\\ &:=\text{the first coordinate of }\sigma(v{\restriction}n,\alpha{\restriction}n,\beta{\restriction}n),\end{aligned}$$

and let $\sigma'(v,\alpha,\beta)=\bigcup_n\sigma'(v{\restriction}n,\alpha{\restriction}n,\beta{\restriction}n)$. Then clearly $\sigma'(v,\alpha,\beta)\in\mathrm{WO}$ for all v, α, β, thus by boundedness find $\xi_0<\aleph_1$ with

$$\sup\{|\sigma'(v,\alpha,\beta)|:v,\alpha,\beta\in{}^\omega\omega\}<\xi_0.$$

Pick now $v_0\in\mathrm{WO}$ with $\xi_0<|v_0|$ and α_0 with $Q(v_0,\alpha_0)$. Then pick β_0 with $(v_0,\alpha_0,\beta_0)\in\mathrm{p}[T]$. Let X_{n+1} be sets of $\mu''_{v_0{\restriction}n+1,\alpha_0{\restriction}n+1,\beta_0{\restriction}n+1}$-measure 1 so that if $h{\restriction}n+1\in X_{n+1}$ then for a set of $\mu'_{w_0{\restriction}n+1,v_0{\restriction}n+1}$-measure 1 worth of $g{\restriction}n+1$

we have

$$G(v_0\restriction n+1, g\restriction n+1, \alpha_0\restriction n+1, \beta_0\restriction n+1, h\restriction n+1)$$
$$= \sigma(v_0\restriction n+1, \alpha_0\restriction n+1, \beta_0\restriction n+1) = (w_0\restriction n+2, f_0\restriction n+2).$$

By the homogeneity of T find h_0 with $h_0\restriction n+1 \in X_{n+1}$, all n. Then

$$G(v_0\restriction n+1, g\restriction n+1, \alpha_0\restriction n+1, \beta_0\restriction n+1, h_0\restriction n+1)$$
$$= \sigma(v_0\restriction n+1, \alpha_0\restriction n+1, \beta_0\restriction n+1)$$

for $g\restriction n+1 \in Y_{n+1}$, where Y_{n+1} has measure 1 in $\mu'_{w_0\restriction n+1, v_0\restriction n+1}$. Since $w_0 = \sigma'(v_0, \alpha_0, \beta_0)$, we have $|w_0| \leq \xi \leq |v_0|$, and using the homogeneity of S' we can find $g_0\restriction n+1 \in Y_{n+1}$ for all n, *i.e.*,

$$G(v_0\restriction n+1, g_0\restriction n+1, \alpha_0\restriction n+1, \beta_0\restriction n+1, h_0\restriction n+1)$$
$$= \sigma(v_0\restriction n+1, \alpha_0\restriction n+1, \beta_0\restriction n+1) = (w_0\restriction n+2, f_0\restriction n+2)$$

for all n. Then if player II plays v_0, g_0, α_0, β_0, h_0 he defeats G and we are done. ⊣

REFERENCES

Alexander S. Kechris

[Kec81] *Homogeneous trees and projective scales*, in Kechris et al. [Cabal ii], pp. 33–74, reprinted in [Cabal II], pp. 270–303.

Alexander S. Kechris, Benedikt Löwe, and John R. Steel

[Cabal II] *Wadge Degrees and Projective Ordinals: the Cabal Seminar, volume II*, Lecture Notes in Logic, vol. 37, Cambridge University Press, 2012.

Alexander S. Kechris, Donald A. Martin, and Yiannis N. Moschovakis

[Cabal ii] *Cabal Seminar* 77–79, Lecture Notes in Mathematics, vol. 839, Berlin, Springer-Verlag, 1981.

[Cabal iii] *Cabal Seminar* 79–81, Lecture Notes in Mathematics, vol. 1019, Berlin, Springer-Verlag, 1983.

W. Hugh Woodin

[Woo83] *Some consistency results in* ZFC *using* AD, in Kechris et al. [Cabal iii], pp. 172–198, reprinted in this volume.

DEPARTMENT OF MATHEMATICS
CALIFORNIA INSTITUTE OF TECHNOLOGY
PASADENA, CALIFORNIA 91125
UNITED STATES OF AMERICA
E-mail: kechris@caltech.edu

AD AND THE UNIQUENESS OF THE SUPERCOMPACT MEASURES ON $\wp_{\omega_1}(\lambda)$

W. HUGH WOODIN

An early consequence of determinacy was the supercompactness of ω_1. Specifically, Solovay [Sol78B] showed that assuming the determinacy of real games ($\mathsf{AD}_{\mathbb{R}}$) then for each $\lambda < \Theta$ there is a supercompact measure on $\wp_{\omega_1}(\lambda)$ where $\Theta = \sup\{\zeta : \text{there is an onto map } f : \mathbb{R} \to \zeta\}$.

Recall that $\wp_{\omega_1}(\lambda) = \{\sigma \subseteq \lambda : \sigma \text{ is countable}\}$ and that a measure μ on $\wp_{\omega_1}(\lambda)$ is supercompact measure if it is normal and fine which is to say that μ satisfies the following two conditions:

(1) (normality) Suppose $f : \wp_{\omega_1}(\lambda) \to \lambda$ is a function such that $\{\sigma : f(\sigma) \in \sigma\}$ is of (μ) measure one. Then for some $\alpha < \lambda$, $\{\sigma : f(\sigma) = \alpha\}$ is of measure one.

(2) (fineness) For each $\alpha < \lambda$, $\{\sigma : \alpha \in \sigma\}$ is of measure one.

Harrington and Kechris [HK81] improved Solovay's result by weakening the assumption to the determinacy of integer games (AD). However the price paid is a weaker result. They proved that if λ lies below a Suslin cardinal then ω_1 is λ supercompact. Becker [Bec81] then showed (assuming AD) that if λ is a Suslin cardinal then the supercompact measure on $\wp_{\omega_1}(\lambda)$ is in fact unique.

This left open the general question of uniqueness of the supercompact measure on $\wp_{\omega_1}(\lambda)$ for arbitrary λ. A typical case is $\lambda = \omega_2$. Assuming AD, ω_2 admits exactly two normal measures. Each of these may be used to construct a supercompact measure on $\wp_{\omega_1}(\omega_2)$ using the normal measure that AD provides on ω_1. It is not immediately clear that these two supercompact measures should be equal.

We show that the supercompact measure on $\wp_{\omega_1}(\omega_2)$ is unique assuming AD. Our arguments are general enough to show uniqueness for the case of an arbitrary λ below a Suslin cardinal assuming AD or for any $\lambda < \Theta$ assuming $\mathsf{AD}_{\mathbb{R}}$.

We assume familiarity with the fundamentals of set theory in the context of determinacy, for a reference, cf. [Mos80]. We work in $\mathsf{ZF} + \mathsf{DC}$ throughout this paper.

Research partially supported by NSF Grant MCS80-21468.

Large Cardinals, Determinacy, and Other Topics: The Cabal Seminar, Volume IV
Edited by A. S. Kechris, B. Löwe, J. R. Steel
Lecture Notes in Logic, 49

§1. Suppose $\lambda < \Theta$. The use of determinacy to construct a supercompact measure on $\wp_{\omega_1}(\lambda)$ is through a simulation of the following ordinal game. Fix $A \subseteq \wp_{\omega_1}(\lambda)$. Associated to A is the game G_A:

$$\begin{array}{lll} \text{I} & \text{II} & \\ \alpha_0 & \beta_0 & \alpha_i < \lambda,\ \beta_i < \lambda \\ \alpha_1 & \beta_1 & \text{I wins} \leftrightarrow \{\alpha_0, \beta_0, \dots\} \in A \\ \vdots & \vdots & \end{array}$$

The players alternately play ordinals less than λ. Player I wins provided $\{\alpha_0, \beta_0, \dots\} \in A$.

Fix a map $\pi\colon \mathbb{R} \to \lambda$ that is onto.

We say that a real r codes the ordinal α if $\pi(r) = \alpha$. We can now simulate the ordinal game G_A with the obvious real game. Let G_A^r denote this real game, where there is an obvious implicit dependence on π.

We define the notion of a quasi-strategy for Player I in G_A^r.

Definition 1.1. A quasi-strategy (for Player I) is a set, τ, of finite sequences of reals such that if $p \in \tau$ is of even length then $p^\frown x \in \tau$ for each $x \in \mathbb{R}$, otherwise $p^\frown x \in \tau$ for some $x \in \mathbb{R}$.

Given a quasi-strategy τ let

$$p_\pi(\tau) = \big\{\sigma \in \wp_{\omega_1}(\lambda) : \text{for some } f \in \mathbb{R}^\omega,\ f\restriction n \in \tau \text{ for each } n \text{ and } \sigma = \{\pi(f(n)) : n \in \omega\}\big\}.$$

Thus $p_\pi(\tau)$ denotes the set of elements of $\wp_{\omega_1}(\lambda)$ that can be enumerated by a play against τ. Note that $p_\pi(\tau)$ depends upon the notion of coding, *i.e.*, on the map $\pi\colon \mathbb{R} \to \lambda$.

For $A \subseteq \wp_{\omega_1}(\lambda)$ we say that G_A^r is determined provided there is a quasi-strategy τ such that $p_\pi(\tau) \subseteq A$ or $p_\pi(\tau) \subseteq \wp_{\omega_1}(\lambda)\backslash A$.

Let

$$\mathfrak{F} = \{p_\pi(\tau) : \tau \text{ is a quasi-strategy and } \pi\colon \mathbb{R} \to \lambda \text{ is onto}\}.$$

$\mathfrak{F}$ is easily seen to be a fine filter over $\wp_{\omega_1}(\lambda)$.

If one assumes the determinacy of real games ($\mathsf{AD}_{\mathbb{R}}$) then for each $A \subseteq \wp_{\omega_1}(\lambda)$, G_A^r is determined. The point is that if Player II wins the real game G_A^r then Player I must win G_B^r, where $B = \wp_{\omega_1}(\lambda)\backslash A$. Hence $\mathfrak{F}$ defines an ultrafilter over $\wp_{\omega_1}(\lambda)$. A standard dovetailing of strategies show that the ultrafilter is normal (cf. [Sol78B]).

It follows from the results of Harrington and Kechris [HK81] that assuming AD, for each $A \subseteq \wp_{\omega_1}(\lambda)$, G_A^r is determined given that λ lies below a Suslin cardinal. Hence again $\mathfrak{F}$ defines a normal measure.

We show that any normal fine measure on $\wp_{\omega_1}(\lambda)$ must extend $\mathfrak{F}$. The desired uniqueness results will then be immediate.

THEOREM 1.2 (ZF + DC). Suppose λ is an ordinal less than Θ and that τ is a (real) quasi-strategy on λ. Let μ be a supercompact measure on $\wp_{\omega_1}(\lambda)$. Then $\mu(p_\pi(\tau)) = 1$.

PROOF. Fix τ, μ and λ. Let $\pi\colon \mathbb{R} \to \lambda$ be the underlying notion of coding.

Assume toward a contradiction that $\mu(p(\tau)) = 0$. Let $B = \wp_{\omega_1}(\lambda)\backslash p_\pi(\tau)$. Hence $\mu(B) = 1$.

For each $\sigma \in B$ consider the following game g_σ:

I	II	
α_0	s_0	$\alpha_i \in \sigma,\ s_i \in \tau,\ s_{i+1}$ extends
α_1	s_1	s_i and $\alpha_i \in \{\pi(r) : r \in s_i\} \subseteq \sigma$.
$\vdots$	$\vdots$	

Player I plays ordinals in σ, Player II plays elements of τ. Player II must play such that s_{i+1} extends s_i and $\alpha_i \in \{\pi(r) : r \in s_i\} \subseteq \sigma$.

Player I wins if at some stage Player II cannot play.

The game g_σ is open for Player I, therefore it is determined. More precisely, either Player I or Player II has a winning "many valued" strategy. If such is the case for Player II then using DC one can easily show that $\sigma \in p_\pi(\tau)$. Hence Player I has the winning "many valued" strategy. Since Player I plays ordinals he must therefore have a winning strategy. Further this strategy may be chosen canonically in terms of decreasing rank. Let ρ_σ denote this strategy. View ρ_σ as a set of (legal) positions in g_σ.

Suppose $q = \langle \alpha_0 s_0 \dots \alpha_i s_i \rangle$ is such that for μ-almost all σ, $q \in \rho_\sigma$. Hence for each such σ there is an $\alpha \in \sigma$ such that $q^\frown\alpha \in \rho_\sigma$. This defines a function $f\colon \wp_{\omega_1}(\lambda) \to \lambda$ with $f(\sigma) \in \sigma$ on a set of μ-measure one. Thus by the normality of μ there is an $\alpha^* < \lambda$ so that for μ-almost all σ, $q^\frown\alpha^* \in \rho_\sigma$.

Using DC and the fineness of μ construct a sequence $\langle \alpha_0 s_0 \dots \alpha_i s_i \dots \rangle$ such that for all i and μ-almost all σ, $\langle \alpha_0 s_0 \dots \alpha_i s_i \rangle$ is a position in g_σ with $\langle \alpha_0 s_0 \dots \alpha_i s_i \rangle \in \rho_\sigma$.

Thus by the countable completeness of μ there is a fixed σ^* such that for all i, $\langle \alpha_0 s_0 \dots \alpha_i s_i \rangle$ is a position in g_{σ^*}, $\langle \alpha_0 s_0 \dots \alpha_i s_i \rangle \in \rho_{\sigma^*}$. But therefore $\langle \alpha_0 s_0 \dots \alpha_i s_i \dots \rangle$ is an infinite play against ρ_{σ^*}, a contradiction. Thus $\mu(p_\pi(\tau)) = 1$. ⊣

By our previous remarks the following are immediate corollaries of Theorem 1.2.

THEOREM 1.3 (AD+DC). Assume $\lambda \leq \kappa$ and that κ is a Suslin cardinal. Then the supercompact measure on $\wp_{\omega_1}(\lambda)$ is unique.

THEOREM 1.4 ($\mathrm{AD}_\mathbb{R}$). Assume $\lambda < \Theta$. Then the supercompact measure on $\wp_{\omega_1}(\lambda)$ is unique.

One can actually prove a slightly more general version of Theorem 1.4. Suppose S is a set. Let $\wp_{\omega_1}(S) = \{\sigma \subseteq S : \sigma \text{ is countable}\}$. A supercompact measure on $\wp_{\omega_1}(S)$ is a fine measure μ satisfying the following modified normality condition. Suppose $f\colon \wp_{\omega_1}(S) \to \wp_{\omega_1}(S)$ is such that for μ-almost all σ, $f(\sigma) \subseteq \sigma$. Then for some $a \in S$, $a \in f(\sigma)$ on a set of μ measure one.

If $S = \lambda$ this is just the usual notion of a supercompact measure. The modified normality condition is important in avoiding trivialities in the case of more general S. For instance if $S = \mathbb{R}$ then the usual notion of normality reduces to countable completeness.

THEOREM 1.5 ($\mathsf{AD}_{\mathbb{R}}$). Let S be an arbitrary set such that there is an onto map $g\colon \mathbb{R} \to S$. Then there is a supercompact measure on $\wp_{\omega_1}(S)$ and it is unique.

PROOF. For existence simply project the supercompact measure on $\wp_{\omega_1}(\mathbb{R})$ constructed by Solovay [Sol78B] or play the obvious real games.

The situation for uniqueness is identical to case of $S = \lambda$ except that in the proof of Theorem 1.2 one must work with "many valued" strategies. The modified normality condition is sufficient for the requirements of the proof. ⊣

§2. **Further Problems.** There are many questions left open regarding supercompact measures in the context of AD. We state two typical of these.

(1) Assume $\mathbf{V}{=}\mathbf{L}(\mathbb{R})$ and AD. How supercompact is ω_1? In particular is there a supercompact measure on $\wp_{\omega_1}(\lambda)$ for each $\lambda < \Theta$?

(2) Assume AD (or $\mathsf{AD}_{\mathbb{R}}$). Is $\mathbf{L}(\mathbb{R}, \mu)$ unique given $\mathbf{L}(\mathbb{R}, \mu) \models \mu$ is a supercompact measure on $\wp_{\omega_1}(\mathbb{R})$?

Assume $\mathsf{AD}_{\mathbb{R}}$ ($\mathsf{AD} + \mathbf{V}{\neq}\mathbf{L}(\mathbb{R})$ will suffice). Suppose $\lambda < \Theta^{\mathbf{L}(\mathbb{R})}$. Let μ be the supercompact measure on $\wp_{\omega_1}(\lambda)$. Let j denote the embedding corresponding to the ultrapower of the ordinals by μ. One can show that $j(\omega_1) \leq \delta$ for δ the first measurable cardinal past λ. In particular $j(\omega_1) < \Theta^{\mathbf{L}(\mathbb{R})}$. This suggests that μ should lie in $\mathbf{L}(\mathbb{R})$.

The following conjecture settles (1) positively. Fix, $\leq$, a prewellordering of $\mathbb{R}$ of length λ. Let $g\colon \mathbb{R} \to \lambda$ be the corresponding map. A λ-splitting tree, T, is a set of finite sequences of reals such that if $q \in T$ and $\alpha < \lambda$ then $q^\frown x \in T$ for some $x \in \mathbb{R}$, $g(x) = \alpha$.

CONJECTURE 2.1 (a strong coding lemma). Assume $\mathbf{V}{=}\mathbf{L}(\mathbb{R})$ and AD. Suppose $\leq$ is a prewellordering of $\mathbb{R}$ of length λ. Let T be a λ-splitting tree. Then there is a λ-splitting tree $T^* \subseteq T$ such that T^* is projective in $\leq$.

REFERENCES

HOWARD S. BECKER

[Bec81] AD *and the supercompactness of* $\aleph_1$, ***The Journal of Symbolic Logic***, vol. 46 (1981), pp. 822–841.

LEO A. HARRINGTON AND ALEXANDER S. KECHRIS

[HK81] *On the determinacy of games on ordinals*, ***Annals of Mathematical Logic***, vol. 20 (1981), pp. 109–154.

ALEXANDER S. KECHRIS AND YIANNIS N. MOSCHOVAKIS

[CABAL i] *Cabal Seminar* 76–77, Lecture Notes in Mathematics, vol. 689, Berlin, Springer-Verlag, 1978.

YIANNIS N. MOSCHOVAKIS

[Mos80] ***Descriptive Set Theory***, Studies in Logic and the Foundations of Mathematics, vol. 100, North-Holland, Amsterdam, 1980.

ROBERT M. SOLOVAY

[Sol78B] *The independence of* DC *from* AD, in Kechris and Moschovakis [CABAL i], pp. 171–184, reprinted in this volume.

DEPARTMENT OF MATHEMATICS
UNIVERSITY OF CALIFORNIA
BERKELEY, CALIFORNIA 94720-3840
UNITED STATES OF AMERICA
E-mail: woodin@math.berkeley.edu

THE EXTENDER ALGEBRA AND Σ^2_1-ABSOLUTENESS

ILIJAS FARAH

This note provides an introduction to Woodin's extender algebra and a proof (due to Steel and Woodin independently) of Woodin's Σ^2_1-absoluteness theorem, using the extender algebra, from a large cardinal assumption. Unlike the published accounts of this proof, the present account should be accessible to a set theorist familiar with forcing and basic large cardinals (*e.g.*, [Kun80, Kan94]). In particular, no familiarity with the inner model theory is required. A more comprehensive account of the extender algebra can also be found in [Ste10, §7.2] and [Nee04, §4] and the reader is invited to consult these excellent sources for more information and other applications. A strengthening of the Σ^2_1 absoluteness in terms of determinacy of games of length ω_1 was proved by Neeman [Nee07A] also using the extender algebra. In [Lar11], the extender algebra was used to prove the consistency of Woodin's Ω-conjecture.

Organization of the paper. In § 1 we introduce infinitary propositional logic $\mathcal{L}_{\delta,\gamma}$. In § 2 we review basics of elementary embeddings, Woodin cardinals and extenders. The extender algebra is introduced in § 3, where we also prove that it is powerfully δ-c.c. and introduce iteration trees and the iteration game. A variety of genericity iteration theorems is proved in § 4 and their applications to absoluteness are proved in § 5. Divergent models of AD^+ are constructed in § 6, and § 7 has no real content.

Acknowledgments. Over a period of fourteen years this paper evolved from a note not intended for publication to its present form. This subsection is appropriately long for a paper that took so long to write. My interest in the extender algebra was initiated by a conversation with John Steel in Oberwolfach in 2005 and motivated by (still open) Question 7.1. I am indebted to David Asperó, Phillip Doebler, Richard Ketchersid, Paul B. Larson, Ernest Schimmerling, Ralf Schindler, John Steel, W. Hugh Woodin, Martin Zeman and Stuart Zoble, and especially Menachem Magidor and Grigor Sargsyan, for explaining me the concepts presented below and pointing to flaws in the earlier versions of this note. I should also like to thank the anonymous referee for a most detailed and useful report. I am indebted to Grigor Sargsyan for

Partially supported by NSERC.

Large Cardinals, Determinacy, and Other Topics: The Cabal Seminar, Volume IV
Edited by A. S. Kechris, B. Löwe, J. R. Steel
Lecture Notes in Logic, 49

convincing me that the construction of divergent models of AD^+ was within ε of the content of the present paper as of January 2011.[1]

I should like to thank John Steel and Hugh Woodin for their kind permission to include their results in this note and to Joan Bagaria and John Steel for their encouragement to publish this paper. A part of this paper was prepared during my stay at the Mittag-Leffler Institute in September 2009. I should like to thank the staff of the Institute for providing a pleasant and stimulating atmosphere.

The definition of the extender algebra and every uncredited result presented here are due to W. H. Woodin, but some of the proofs and formulations, as well as the exposition, are mine. Mistakes, obscurities, and dwelling on trivialities are also all mine.

Notation and terminology. Our notation is mostly standard but inner model theorists may want to take note of two exceptions. First, π usually denotes a transitive collapse instead of its inverse. Second, to an ultrafilter $\mathcal{U}$ we associate quantifier $(\mathcal{U}x)$ so that $(\mathcal{U}x)\varphi(x)$ stands for "the set of x satisfying $\varphi(x)$ belongs to $\mathcal{U}$". All forcing considered here is set forcing.

§1. An infinitary propositional logic. The extender algebra is the Lindenbaum algebra of a certain theory in an infinitary propositional logic. The logic is described here and the theory will be described in § 3.

1.1. Logic $\mathcal{L}_{\delta,\gamma}$. For regular cardinals $\gamma \leq \delta$ we shall define the infinitary propositional logic $\mathcal{L}_{\delta,\gamma}$. The interesting cases are $\gamma = \omega$ and $\gamma = \delta$, but it will be easier to develop the basic theory in the two cases parallelly. Let $\mathcal{L}_{\delta,\gamma}$ be the propositional logic with γ variables a_ξ, for $\xi < \gamma$, which in addition to the standard propositional connectives $\vee, \wedge, \rightarrow, \leftrightarrow$, and $\neg$ allows infinitary conjunctions of the form $\bigwedge_{\xi<\kappa} \varphi_\xi$ and infinitary disjunctions of the form $\bigvee_{\xi<\kappa} \varphi_\xi$ for all $\kappa < \delta$.

In addition to the standard axioms and rules of inference for the finitary propositional logic, for each $\kappa < \delta$ and formulas φ_ξ, for $\xi < \kappa$ the logic $\mathcal{L}_{\delta,\gamma}$ has axioms $\vdash \bigvee_{\xi<\kappa} \neg\varphi_\xi \leftrightarrow \neg \bigwedge_{\xi<\kappa} \varphi_\xi$ and $\vdash \bigwedge_{\xi<\kappa} \varphi_\xi \rightarrow \varphi_\eta$, for every $\eta < \kappa$, as well as the infinitary rule of inference: from $\vdash \varphi_\xi$ for all $\xi < \kappa$ infer $\vdash \bigwedge_{\xi<\kappa} \varphi_\xi$. The provability relation for $\mathcal{L}_{\delta,\gamma}$ will be denoted by $\vdash_{\delta,\gamma}$ or simply $\vdash$ if δ and γ are clear from the context. Each proof in $\mathcal{L}_{\delta,\gamma}$ is a well-founded tree and the assertion that φ is provable in $\mathcal{L}_{\delta,\gamma}$ is upwards absolute between transitive models of ZFC. Since adding a new bounded subset of δ adds new formulas and new proofs to $\mathcal{L}_{\delta,\gamma}$, it is not obvious that the assertion that φ is not provable in $\mathcal{L}_{\delta,\gamma}$ is upwards absolute between transitive models of ZFC. This is, nevertheless, true: *cf.* Lemma 1.2.

[1] It wasn't.

Every $x \in \wp(\gamma)$ naturally defines a model for $\mathcal{L}_{\delta,\gamma}$ via $v_x(a_\xi) = \texttt{true}$ iff $x(\xi) = 1$ for $\xi < \gamma$. Define $A_\varphi = A_{\varphi,\delta,\gamma}$ via

$$A_\varphi = \{x \in \wp(\gamma) : x \models \varphi\}.$$

When $\gamma = \omega$ then these are the so-called ∞-**Borel** sets (*cf.*, *e.g.*, [Woo10, §9.1]). Note that the sets of the form A_φ for $\varphi \in \mathcal{L}_{\omega_1,\omega}$ are exactly the Borel sets.[2] Also note that $x \models \varphi$ is absolute between transitive models of ZFC containing x and φ.

1.2. Completeness of $\mathcal{L}_{\delta,\gamma}$. The following two lemmas are standard.

Lemma 1.1. (1) $\vdash \varphi$ implies $\models \varphi$ for every formula φ in $\mathcal{L}_{\delta,\gamma}$.
(2) $\delta > 2^\gamma$ implies that $\models \varphi$ but not $\vdash \varphi$ for some formula φ in $\mathcal{L}_{\delta,\gamma}$.

Proof. Clause (1) can be proved by recursion on the rank of the proof.

(2) For $x \subseteq \gamma$ let $\varphi_x = \bigvee_{\xi<\gamma} a^x_\xi$, where $a^x_\xi = a_\xi$ if $\xi \in x$ and $a^x_\xi = \neg a_\xi$ if $\xi \notin x$. Then $y \models \varphi_x$ if and only if $y = x$, and therefore the formula $\varphi = \bigwedge_{x\subseteq\gamma} \neg\varphi_x$ is not satisfiable. However, φ is satisfiable in every forcing extension in which there exists a new subset of γ. ⊣

Lemma 1.2. For every φ in $\mathcal{L}_{\delta,\gamma}$ the following are equivalent.
(1) $\vdash \varphi$.
(2) $A_\varphi = \wp(\gamma)$ in all generic extensions.
(3) $A_\varphi = \wp(\gamma)$ in the extension by $\mathrm{Col}(\omega,\kappa)$ for a large enough κ.

Proof. Clause (1) is upwards absolute and by recursion on the rank of the proof it easily implies (2). Also, (2) trivially implies (3).

Assume (1) fails for φ and let κ be the cardinality of the set of all subformulas of φ. We may assume a_ξ is a subformula of φ if and only if $\xi < \kappa$.

We claim that for any two formulas ψ_1 and ψ_2 in $\mathcal{L}_{\delta,\gamma}$ such that $\psi_1 \nvdash \varphi$ we have either $\psi_1 \wedge \psi_2 \nvdash \varphi$ or $\psi_1 \wedge \neg\psi_2 \nvdash \varphi$. Otherwise we have $\psi_1 \vdash \psi_2 \to \varphi$ and $\psi_1 \vdash \neg\psi_2 \to \varphi$ and therefore $\psi_1 \vdash \varphi$, a contradiction.

In the extension by $\mathrm{Col}(\omega,\kappa)$ enumerate all subformulas of φ as ψ_n, for $n \in \omega$, and also enumerate $\{a_\xi : \xi < \kappa\}$ as b_n, for $n \in \omega$. Recursively pick an increasing sequence of $\mathcal{L}_{\delta,\gamma}$-theories $\mathcal{T}_n$, for $n \in \omega$, satisfying the following requirements: (i) $\mathcal{T}_0 = \{\neg\varphi\}$ and each $\mathcal{T}_n$ is a finite consistent set of ground-model formulas in $\mathcal{L}_{\delta,\gamma}$. (ii) either $b_n \in \mathcal{T}_n$ or $\neg b_n \in \mathcal{T}_n$. (iii) if ψ_n is of the form $\bigwedge_{\xi<\lambda} \sigma_\xi$ for some λ then either $\psi_n \in \mathcal{T}_n$ or $\neg\sigma_\xi \in \mathcal{T}_n$ for some $\xi < \lambda$. (iv) if ψ_n is of the form $\bigvee_{\xi<\lambda} \sigma_\xi$ for some λ then either $\neg\psi_n \in \mathcal{T}_n$ or $\sigma_\xi \in \mathcal{T}_n$ for some $\xi < \lambda$.

By the above, if $\mathcal{T}_n$ satisfies the requirements then $\mathcal{T}_{n+1}$ as required can be chosen. After all $\mathcal{T}_n$ have been chosen define $x \subseteq \gamma$ by $x = \{\xi : a_\xi \in \bigcup_n \mathcal{T}_n\}$.

---

[2]Keep in mind that, in spite of the connection with Borel sets, this logic is **not** the 'usual' $\mathcal{L}_{\omega_1,\omega}$ [Kei71]. Our $\mathcal{L}_{\omega_1,\omega}$ happens to be the propositional fragment of the latter. This is really an accident since in each of the two notations 'ω' signifies a different constraint.

Then $x \models \neg\varphi$ can be proved by recursion on the rank of φ, using the fact that every infinitary conjunction and every infinitary disjunction appearing in φ is computed correctly. Therefore in this extension (3) fails. $\dashv$

Recall that the **Lindenbaum algebra** of a $\mathcal{L}_{\delta,\gamma}$-theory $\mathcal{T}$ is the Boolean algebra of all of equivalence classes of formulas in $\mathcal{L}_{\delta,\gamma}$ with respect to the equivalence relation $\sim_{\mathcal{T}}$ defined by $\varphi \sim_{\mathcal{T}} \psi$ if and only if $\mathcal{T} \vdash \varphi \leftrightarrow \psi$. We shall denote this algebra by $\mathcal{B}_{\delta,\gamma}/\mathcal{T}$ and consider its positive elements as a forcing notion (*cf.* Lemma 1.4 below).

LEMMA 1.3. For every $\mathcal{L}_{\delta,\gamma}$-theory $\mathcal{T}$ such that $\mathcal{B}_{\delta,\gamma}/\mathcal{T}$ has the δ-chain condition $\mathcal{B}_{\delta,\gamma}/\mathcal{T}$ is a complete Boolean algebra.

PROOF. Immediate, since both $\mathcal{B}_{\delta,\gamma}$ and $\mathcal{T}$ are δ-complete. $\dashv$

For $x \subseteq \gamma$ such that $x \models \mathcal{T}$ define an ultrafilter of $\mathcal{B}_{\delta,\gamma}/\mathcal{T}$ by (here $[\varphi]$ stands for the equivalence class of φ)

$$\Gamma_x = \{[\varphi] \in \mathcal{B}_{\delta,\gamma}/\mathcal{T} \,:\, x \models \varphi\}.$$

Note that for every generic $\Gamma \subseteq \mathcal{B}_{\delta,\gamma}/\mathcal{T}$ there is the unique $x \subseteq \gamma$ such that $\Gamma_x = \Gamma$, defined as $x = \{\xi \,:\, a_\xi \in \Gamma\}$.

LEMMA 1.4. Assume $\mathcal{T}$ is a theory in $\mathcal{L}_{\delta,\gamma}$ and M is a transitive model of a large enough fragment of ZFC such that $\{\delta, \gamma, \mathcal{T}\} \subseteq M$, and $\mathcal{B}_{\delta,\gamma}/\mathcal{T}$ has the δ-chain condition in M. Then for every $x \in \wp(\gamma)$ we have $x \models \mathcal{T}^M$ if and only if x is $\mathcal{B}_{\delta,\gamma}/\mathcal{T}$-generic over M.

PROOF. If x is $\mathcal{B}_{\delta,\gamma}/\mathcal{T}$-generic over M then a proof by induction on the complexity shows that $x \models \varphi$ for all $[\varphi] \in \Gamma$, hence $\{\varphi \in \mathcal{L}_{\delta,\gamma} \,:\, x \models \varphi\}$ includes $\mathcal{T}^M$.

Now assume $x \models \mathcal{T}^M$. Let φ_ξ $(\xi < \kappa)$ be a maximal antichain of $(\mathcal{B}_{\delta,\gamma}/\mathcal{T})^M$ that belongs to M. By the δ-chain condition $\kappa < \delta$ and therefore $\bigvee_{\xi<\kappa} \varphi_\xi$ is in $\mathcal{L}^M_{\delta,\gamma}$. By the maximality of the antichain we have $\mathcal{T}^M \models \bigvee_{\xi<\kappa} \varphi_\xi$. Therefore $x \models \bigvee_{\xi<\kappa} \varphi_\xi$. This implies $x \models \varphi_\xi$ for some $\xi < \kappa$ and Γ_x intersects the given maximal antichain. Since the antichain was arbitrary, we conclude x is generic. $\dashv$

§2. Elementary embeddings. In the present section we recall some basic facts about elementary embeddings and extenders.

2.1. Extenders I. We only sketch the bare minimum of the theory of extenders. For more details, *cf.* [MaS94] or [Kan94, §26]. An **extender** E is a set that codes an elementary embedding $j_E \colon \mathbf{V} \to M$. For every elementary embedding $j \colon \mathbf{V} \to M$ such that with $\kappa = \operatorname{crit}(j)$ we have $\kappa < \lambda < j(\kappa)$ there is an extender E in $\mathbf{V}_{\lambda+1}$ such that j_E and j coincide up to $\mathbf{V}_\lambda$ in the sense that $j(A) \cap \mathbf{V}_\lambda = j_E(A) \cap \mathbf{V}_\lambda$ for all $A \subseteq \mathbf{V}_\kappa$. The model M is constructed as a direct limit of ultrapowers of $\mathbf{V}$ and it is denoted by $\operatorname{Ult}(\mathbf{V}, E)$.

All extenders E used in this paper are such that $j_E(\kappa) > \lambda$, where $\kappa = \mathrm{crit}(E)$ and λ is the strength of E. Such extenders are called **short** extenders. Long extenders are needed to describe stronger large cardinal embeddings.

A **generator** of an elementary embedding $j : \mathbf{V} \to M$ is an ordinal ξ such that there are an inner model N and elementary embeddings i_1 and i_2 such that the diagram

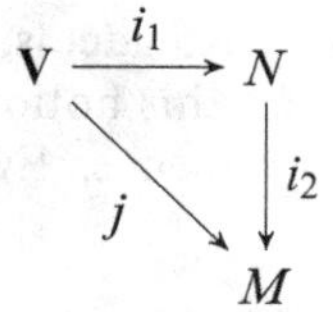

commutes and $\mathrm{crit}(i_2) = \xi$. For example, the critical point κ is the least generator and a counting argument shows that an elementary embedding such that $j(\kappa) \geq (2^\kappa)^+$ must have other generators. A **generator** of an extender E is a generator of j_E. The **strength** of E is the largest λ such that $\mathbf{V}_\lambda \subseteq M$, where $j_E : \mathbf{V} \to M$.

The only properties of extenders used in the present paper are (E1) and (E2).

(E1) If E is an extender with $\kappa = \mathrm{crit}(j_E)$ then $\mathrm{Ult}(N, E)$ can be formed for every model N such that $(\mathbf{V}_{\kappa+1})^N = \mathbf{V}_{\kappa+1}$. Also, $\mathrm{Ult}(N, E) \supseteq \mathbf{V}_\lambda$, where λ is the strength of E and whether or not ξ is a generator of E depends only on E and $\mathbf{V}_{\kappa+1}$.

For a proof of (E1), *cf.* [MaS94, Lemma 1.5], or note that this is immediate from Definition 2.3 below since all ultrafilters E_s concentrate on $[\kappa]^{<\omega}$. While (E1) is a property of all extenders, (E2) below is not. However, we shall consider only the extenders satisfying (E2).

(E2) The strength of E is greater than the supremum of the generators of E.

It is important to note that the strength depends only on E, and not on the model to which E was applied. In particular, if E is an extender in M with $\kappa = \mathrm{crit}(j_E)$, λ is the strength of E, and $M \cap \mathbf{V}_{\kappa+1} = N \cap \mathbf{V}_{\kappa+1}$ then we have $\mathrm{Ult}(M, E) \cap \mathbf{V}_\lambda = \mathrm{Ult}(N, E) \cap \mathbf{V}_\lambda$.

2.2. Woodin cardinals. If A is a set such that $j : \mathbf{V} \to M$ satisfies $\mathbf{V}_\lambda = (\mathbf{V}_\lambda)^M$ and $j(A) \cap \mathbf{V}_\lambda = A \cap \mathbf{V}_\lambda$ then we say j is an A, λ**-strong** embedding. This may differ from the standard definition of a λ-strong embedding but I shall consider only λ such that $\kappa + \lambda = \lambda$ for $\kappa = \mathrm{crit}(j)$, in which case there is no difference. If j_E is an A, λ-strong then we say E is A, λ-strong. A cardinal δ is a **Woodin cardinal** if for every $A \subseteq \mathbf{V}_\delta$ there is $\kappa < \delta$ such that there are A, λ-strong elementary embeddings with critical point κ for an arbitrarily large $\lambda < \delta$. We say that A **reflects** to κ. Note that the Woodinness of δ is witnessed by the extenders in $\mathbf{V}_\delta$, and therefore δ is Woodin in $\mathbf{V}$ if and only if it is

Woodin in $\mathbf{L}(\mathbf{V}_\delta)$. Moreover, it suffices to consider only the extenders that satisfy property (E2).

A cardinal κ is **λ-strong** if it is a critical point of an $\varnothing$, λ-strong elementary embedding $j\colon \mathbf{V} \to M$. In particular every Woodin cardinal δ is a limit of cardinals each of which is λ-strong for all $\lambda < \delta$.

2.3. Extenders II. A reader not interested in extenders per se may want to skip the rest of this section on the first reading and take (E1) and (E2) for granted. The actual definition of an extender is, strictly speaking, not necessary for our present purpose. However, this notion is central[3] in the theory and we include it for the reader's convenience. Every extender is of the form as described in Example 2.1.

EXAMPLE 2.1. Assume $j\colon \mathbf{V} \to M$ is an elementary embedding with $\operatorname{crit}(j) = \kappa$. Fix λ such that $\kappa \leq \lambda < j(\kappa)$. Typically, we take λ such that $M \supseteq \mathbf{V}_\lambda$. For $s \in [\lambda]^{<\omega}$ define $E_s \subseteq [\kappa]^m$ (where $m = |s|$) by

$$X \in E_s \text{ if and only if } s \in j(X).$$

Then $E(j, \lambda) = \langle E_s : s \in [\lambda]^{<\omega}\rangle$ is a (κ, λ)-extender. Ordinal λ is called the **length** of the extender E and is denoted by $\operatorname{lh}(E)$. In all of our applications $\operatorname{lh}(E)$ will equal the strength of E.

The following facts can be found *e.g.*, in [Ste10]. For an extender E we write

$$\kappa_E = \operatorname{crit}(E)$$

(the **critical point** of E, *i.e.*, the least ordinal moved by j_E) and

$$\lambda_E = \sup\{\eta : \mathbf{V}_\eta \subseteq (\mathbf{V}_\eta)^M\}$$

(the **strength** of E). Note that if $j\colon \mathbf{V} \to M$ is an elementary embedding such that $\operatorname{crit}(j) = \kappa$, the strength of j is λ, and λ is a strong limit cardinal, then the (κ, λ) extender E defined in Example 2.1 satisfies $\operatorname{Ult}(\mathbf{V}, E) \cap \mathbf{V}_\lambda = M \cap \mathbf{V}_\lambda$ and $j(A) \cap \mathbf{V}_\lambda = j_E(A) \cap \mathbf{V}_\lambda$ for all A and is therefore λ-strong. The assumption that λ is strong limit is needed to code subsets of $\wp(\alpha)$ by sets of ordinals for every $\alpha < \lambda$.

Fix a cardinal κ. For $m \in \omega$ and $s \subseteq m$ with $|s| = n$ consider the projection map $\pi = \pi_{m,s}\colon [\kappa]^m \to [\kappa]^s$ defined by

$$\pi(\langle \xi_i : i < m\rangle) = \langle \xi_i : i \in s\rangle.$$

More generally, if $s \subseteq t$ are finite sets of ordinals (listed in the increasing order) then the projection $\pi = \pi_{t,s}\colon [\kappa]^t \to [\kappa]^s$ is defined by

$$\pi(\langle \xi_i : i \in t\rangle) = \langle \xi_i : i \in s\rangle.$$

[3] After all, it is the *extender* algebra.

Recall that if $\mathcal{U}$ and $\mathcal{V}$ are ultrafilters on sets I and J, respectively, then we write $\mathcal{U} \leq_{\mathrm{RK}} \mathcal{V}$ if and only if there is $h\colon J \to I$ such that

$$X \in \mathcal{U} \text{ if and only if } h^{-1}(X) \in \mathcal{V}.$$

In this situation we say $\mathcal{U}$ **is Rudin-Keisler reducible to** $\mathcal{V}$ and that h is the **Rudin-Keisler reduction of** $\mathcal{U}$ **to** $\mathcal{V}$.

Respecting the notation commonly accepted in the theory of large cardinals, we denote an ultrapower of a structure M associated to an ultrafilter $\mathcal{U}$ by $\mathrm{Ult}(M, \mathcal{U})$. Its elements are the equivalence classes of $f \in M^I \cap M$,

$$[f]_{\mathcal{U}} = \{g \in M^I \cap M : (\mathcal{U}i) f(i) = g(i)\}$$

and the membership relation is defined by $[f]_{\mathcal{U}} \in [g]_{\mathcal{U}}$ if and only if $(\mathcal{U}i) f(i) \in g(i)$.

If $\mathcal{U}$ is $\aleph_1$-complete then $\mathrm{Ult}(M, \mathcal{U})$ is well-founded whenever M is well-founded, and we identify $\mathrm{Ult}(M, \mathcal{U})$ with its transitive collapse.

Assume $\mathcal{U}$ and $\mathcal{V}$ are ultrafilters on index-sets I and J, respectively, and $\mathcal{U} \leq_{\mathrm{RK}} \mathcal{V}$ is witnessed by a reduction $h\colon J \to I$. Then for any structure M we can define a map $j_h\colon M^I/\mathcal{U} \to M^J/\mathcal{V}$ by

$$j_h([f]_{\mathcal{U}}) = [f \circ h]_{\mathcal{V}}$$

In the following lemma M is any structure, $\mathcal{U}$ and $\mathcal{V}$ are arbitrary ultrafilters, and the proof is straightforward.

LEMMA 2.2. If $\mathcal{U} \leq_{\mathrm{RK}} \mathcal{V}$, h is the Rudin-Keisler reduction, and j_h is defined as above, then the diagram

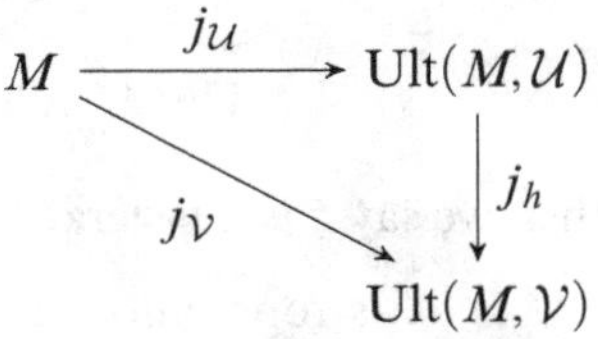

commutes and j_h is an elementary embedding. ⊣

For a finite set s and $i < |s|$ let s_i denote its ith element. As common in set theory, we start counting at 0.

DEFINITION 2.3. Assume $\kappa < \lambda$ are uncountable cardinals. A (κ, λ)**-extender** is $E\colon [\lambda]^{<\omega} \to \mathbf{V}_{\kappa+2}$ such that for all s and t in $[\lambda]^{<\omega}$ we have

(a) E_s is a nonprincipal κ-complete ultrafilter on $[\kappa]^{|s|}$.
(b) If $s \subseteq t$ then $\pi_{t,s}$ is a Rudin-Keisler reduction of E_s to E_t.
(c) **Normality**: If $s \in [\lambda]^{<\omega}$, $i < |s|$, and $f\colon [\kappa]^{|s|} \to \kappa$ is such that $f(u) < u_i$ for E_s many u, then there exist $\xi < s_i$ and j such that $f \circ \pi_{a\cup\{\xi\},a}(u) = u_j$ for $E_{s\cup\{\xi\}}$ many u.

(d) **Countable completeness**: if $s(n) \in [\lambda]^{<\omega}$ and $X(n) \in E_{s(n)}$ for all $n < \omega$ then there is increasing $h\colon \bigcup_n s(n) \to \kappa$ such that $h"s(n) \in X(n)$ for all n.

Note that, with the above notation for a (κ, λ)-extender, we have $\kappa_E = \kappa$ but not necessarily $\lambda_E = \lambda$. However, the length and strength of each extender in all of our applications will coincide.

If E is an extender, then the models

$$M_s = \mathrm{Ult}(\mathbf{V}, E_s)$$

are, by the κ-completeness of E_s, well-founded. By Lemma 2.2, they form a directed system under the embeddings (writing $j_{t,s}$ for $j_{\pi_{t,s}}$)

$$j_{t,s}\colon M_t \to M_s.$$

The direct limit of this system will be denoted by $\mathrm{Ult}(\mathbf{V}, E)$ and identified with its transitive collapse if it is well-founded.

It is not difficult to show that if E satisfies (a), (b) and (c) of Definition 2.3 then the corresponding ultrapower $\mathrm{Ult}(\mathbf{V}, E)$ is well-founded if and only if E satisfies (d) as well.

If E is a (κ, λ)-extender, let M be a model to which E can be applied and let $\kappa < \xi \le \lambda$. One defines $E\restriction\xi = \langle E_s : s \in [\xi]^{<\omega}\rangle$. Like in Lemma 2.2 one can define an elementary embedding i such that the diagram

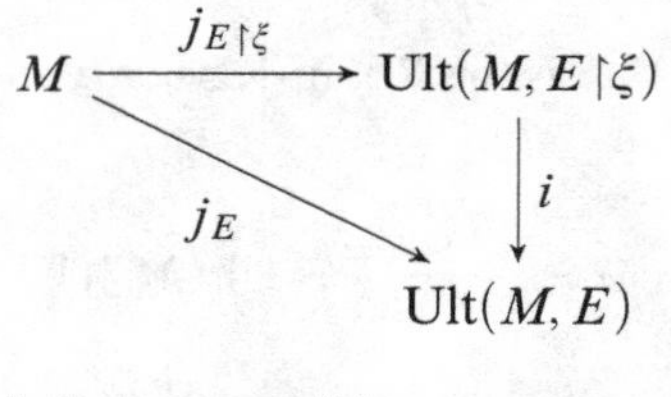

commutes. If $\mathrm{crit}(i) = \xi$ then we say ξ is a **generator** of E.

Lemma 2.4. If $j\colon \mathbf{V} \to M$ is a λ-strong embedding with $\mathrm{crit}(j) = \kappa$. Let $E = E(j, \lambda)$ be as in Example 2.1. Then

(1) E satisfies (E1) and (E2).

(2) For every $X \subseteq \kappa$ we have $j(X) \cap \lambda = j_E(X) \cap \lambda$.

In particular, if δ is a Woodin cardinal then there is a family $\vec{E} \subseteq \mathbf{V}_\delta$ of extenders satisfying (E1) and (E2) such that δ is Woodin in $\mathbf{L}[\vec{E}]$.

Proof. Clause (E1) is a consequence of the fact that $\mathrm{Ult}(N, E)$ depends only on $\mathbf{V}_{\kappa+1} \cap N$. Clearly every generator of a (κ, λ)-extender is $\le \lambda$, and therefore (E2) follows. $\dashv$

Lemma 2.5 below will not be needed elsewhere in the present paper. It is included only as an illustration that the countable completeness of the extenders is a necessary requirement for wellfoundedness of the ultrapower.

LEMMA 2.5. Assume κ is a measurable cardinal. Then there is $\lambda > \kappa$ and $E : [\lambda]^{<\omega} \to \mathbf{V}_{\kappa+2}$ such that

(a) E_s is a nonprincipal κ-complete ultrafilter on κ^s,

(b) If $s \subseteq t$ then $\pi_{t,s}$ is a Rudin-Keisler reduction of E_s to E_t

such that the ultrapower $\prod_E \mathbf{V}$ is ill-founded.

PROOF. Let $j : \mathbf{V} \to M$ be an elementary embedding with $\mathrm{crit}(j) = \kappa$. Let $\lambda = \sup\{n \in \omega : j^n(\kappa)\}$. For all X, n and k we have the following.

$$j^{n+k}(X) \cap j^k(\kappa) = j^k(j^n(X) \cap \kappa) = j^k(X \cap \kappa) = j^k(X) \cap j^k(\kappa).$$

Because of this for a finite $s \subseteq \lambda$ the following defines a subset E_s of $\kappa^{|s|}$:

$$X \in E_s \text{ if and only if } s \in j^n(X)$$

where n is such that $\max(s) < j^n(\kappa)$. Then E_s is clearly a κ-complete ultrafilter.

The map E, $[\lambda]^{<\omega} \ni s \mapsto E_s \in \mathbf{V}_{\kappa+1}$ satisfies (E1) and (E2). We can therefore form the direct limit of the ultrapowers $\prod_{E_s} \mathbf{V}$, $s \in [\lambda]^{<\omega}$. However, this ultrapower is not well-founded. Let $s(n) = \{j^i(\kappa) : i < n\}$ and $M_n = \mathrm{Ult}(\mathbf{V}, E_{s(n)})$. Then $j_{s(n),s(n+1)}(\kappa) > \kappa$, and therefore in the direct limit we have a decreasing ω-sequence of ordinals. ⊣

The problem with E defined in Lemma 2.5 is that the ultraproducts are iterated the 'wrong way'. Let us consider this example a little more closely. If $s(n) = \{j^m(\kappa) : m < n\}$ then $E_{s(n)}$ is the set of all $X \subseteq \kappa^n$ such that

$$(\mathcal{U}\xi_0)(\mathcal{U}\xi_1)\dots(\mathcal{U}\xi_{n-1})\langle \xi_{n-1}, \dots, \xi_1, \xi_0\rangle \in X$$

Then the set of all decreasing n-tuples of ordinals $< \kappa$ belongs to $E_{s(n)}$ for each n.

§3. The extender algebra.

Assume $\vec{E}$ is a family of extenders in $\mathbf{V}_{\delta+2}$. Typically, $\vec{E}$ will be a subset of $\mathbf{V}_\delta$ and it will witness that δ is a Woodin cardinal. Let $\mathcal{T}_{\delta,\gamma}(\vec{E})$ be the deductive closure in $\mathcal{L}_{\delta,\gamma}$ of all sentences of the form

$$\Psi(\vec{\varphi}, \kappa, \lambda) : \bigvee_{\xi<\kappa} \varphi_\xi \leftrightarrow \bigvee_{\xi<\lambda} \varphi_\xi$$

for a sequence $\vec{\varphi} = \langle \varphi_\xi : \xi < \delta\rangle$ in $\mathcal{L}_{\delta,\gamma}$ such that $\varphi_\xi \in \mathbf{V}_\kappa$ for all $\xi < \kappa$, ordinals $\kappa < \lambda$, and extender $\vec{E}$ with $\mathrm{crit}(E) = \kappa$ that is $\vec{\varphi}, \lambda$-strong.[4]

LEMMA 3.1. If $E \in \vec{E}$ and $\mathrm{crit}(E) = \kappa$ then for every $f : \kappa \to \mathcal{L}_{\delta,\gamma} \cap \mathbf{V}_\kappa$ we have $\mathcal{L}_{\delta,\gamma} \vdash \bigvee_{\xi<\kappa} f(\xi) \to j_E(f)(\kappa)$.

PROOF. If λ is the strength of E then with $\vec{\varphi} = \langle j_E(f)(\xi) : \xi < \lambda)\rangle$ we have that E is is $\vec{\varphi}, \lambda$-strong and $\Psi(\vec{\varphi}, \kappa, \lambda)$ implies the above formula. ⊣

[4]Early versions of this paper contained a nonstandard definition of theory $\mathcal{T}_{\delta,\gamma}(\vec{E})$. I have decided to adopt the standard, more flexible, definition.

In the following it may be worth emphasizing that x is assumed to belong to the same inner model as $\vec{E}$ (*cf.* Theorem 4.1).

LEMMA 3.2. *For every real x in $\mathbf{L}[\vec{E}]$ we have $x \models \mathcal{T}_{\delta,\omega}(\vec{E})$. In particular, $\mathcal{T}_{\delta,\omega}(\vec{E})$ is a consistent theory.*

PROOF. Fix a sequence $\vec{\varphi}$ that reflects to κ and this is witnessed by extenders in $\vec{E}$. We need to check $\Psi(\vec{\varphi}, \kappa, \lambda)$ for all $\lambda > \kappa$. We may assume $x \models \varphi_\xi$ for some ξ, since otherwise $\Psi(\vec{\varphi}, \kappa, \lambda)$ vacuously holds for all λ. Pick an extender $E \in \vec{E}$ such that j_E is $(\vec{\varphi}, \lambda)$-strong for some $\lambda > \xi$. Since x is a real it is not moved by any j and therefore by elementarity we have $x \models \varphi_\eta$ for some $\eta < \kappa$ and the conclusion follows. ⊣

Note that if a cardinal γ is δ-strong (or equivalently, in the terminology introduced above, if δ reflects to γ) and this is witnessed by extenders in $\vec{E}$ then $\gamma \not\models \mathcal{T}_{\delta,\gamma}(\vec{E})$, as can be seen by taking $\varphi_\xi(x)$ to be a_ξ if $\xi \in x$ and $\neg a_\xi$ if $\xi \notin x$. However, if $\operatorname{crit}(E) = \gamma$ then by the minimality of γ and elementarity we have $\gamma \models j_E(\mathcal{T}_{\delta\gamma}(\vec{E}))$.

The **extender algebra** with γ generators corresponding to $\vec{E}$ is the algebra

$$\mathcal{W}_{\delta,\gamma}(\vec{E}) = \mathcal{B}_{\delta,\gamma}/\mathcal{T}_{\delta,\gamma}(\vec{E}).$$

Most important instances of $\mathcal{W}_{\delta,\gamma}(\vec{E})$ are given by $\gamma = \omega$ and $\gamma = \delta$ but for convenience we develop the theory of $\mathcal{W}_{\delta,\gamma}(\vec{E})$ for an arbitrary $\gamma \leq \delta$.

LEMMA 3.3. *If δ is a Woodin cardinal and $\vec{E}$ is a system of extenders witnessing its Woodinness, then $\mathcal{W}_{\delta,\gamma}$ has the δ-chain condition and is therefore complete.*

PROOF. Pick a sequence $\vec{\varphi} = \{\varphi_\xi : \xi < \delta\}$ in $\mathcal{L}_{\delta,\gamma}$. We want to prove that $\{[\varphi_\xi] : \xi < \delta\}$ is not an antichain in $\mathcal{W}_{\delta,\gamma}$. The set

$$\mathbf{C} = \{\kappa < \delta : (\forall \xi < \kappa)\varphi_\xi \in \mathbf{V}_\kappa\}$$

is a club in δ. Using some reasonable coding of pairs in $\mathbf{V}_\delta$ by elements of $\mathbf{V}_\delta$ find $X \subseteq \mathbf{V}_\delta$ that codes the pair $\mathbf{C}, \vec{\varphi}$. Since δ is Woodin there exists an X, λ-strong extender $E \in \vec{E}$ such that with $\kappa = \operatorname{crit}(E)$ the set $[\kappa, \lambda) \cap \mathbf{C}$ is nonempty. By the elementarity of j_E this implies $\kappa \in \mathbf{C}$. Therefore Lemma 3.1 applies, and $[\varphi_\kappa]$ is compatible with $[\varphi_\xi]$ for some $\xi < \kappa$.

The completeness of $\mathcal{W}_{\delta,\gamma}$ now follows by Lemma 1.3. ⊣

A partial converse of Lemma 3.3, that if $\mathcal{W}_{\delta,\gamma}(\vec{E})$ has δ-c.c. and $\mathbf{V}_\delta^\#$ exists then δ is Woodin, was proved in [KZ06] and independently by Woodin. It is not known whether the converse of Lemma 3.3 is true.

The following lemma, which will play an important role in § 6, was proved in [Hjo97, Lemma 3.6] in the case when $\lambda = 2$. The general case requires no new ideas.

LEMMA 3.4 (Hjorth). if $\lambda < \delta$ then product of λ copies of $\mathcal{W}_{\delta,\gamma}$ has the δ-chain condition.

PROOF. Assume the contrary and let $A = \{\langle[\varphi_\xi(i)] : i < \lambda\rangle : \xi < \delta\}$ be an antichain in $(\mathcal{W}_{\delta,\gamma})^\lambda$. By argument as in the proof of Lemma 3.3 we can find κ such that $\varphi_\xi(i) \in \mathbf{V}_\kappa$ for all i and all $\xi < \kappa$ and there is an A, λ-strong extender E with critical point κ ($\lambda = \kappa + 1$ suffices).

Then $\mathcal{U} = \{X \subseteq \kappa : j_E(X) \ni \kappa\}$ is a normal κ-complete ultrafilter on κ. If there were $\xi < \eta < \kappa$ such that $[\varphi_\xi(i)]$ and $[\varphi_\eta(i)]$ are compatible for all $i < \lambda$, then A would not be an antichain.

Consider the map $f : [\kappa]^2 \to \lambda$ defined by

$$f(\{\xi, \eta\}) = \min\{i < \lambda : [\varphi_\xi(i)] \text{ and } [\varphi_\eta(i)] \text{ are incompatible}\}.$$

Since $\mathcal{U}$ is a normal ultrafilter we have $\mathcal{U} \to (\mathcal{U})^2_\lambda$ and there are $X \in \mathcal{U}$ and $i < \lambda$ such that f has constant value i on $[X]^2$. Let

$$\psi_\xi = \begin{cases} \varphi_\xi(i) & \text{if } \xi \in X \text{ or } \xi \geq \kappa, \\ \bot & \text{if } \xi \notin X \text{ and } \xi < \kappa. \end{cases}$$

Then $\psi_\kappa = j(\vec{\psi})_\kappa$. Łos's theorem implies that $[\psi_\kappa]$ is incompatible with $[\psi_\xi]$ for all $\xi < \kappa$. However, $\psi_\kappa = j(\vec{\psi})_\kappa$ and this contradicts Lemma 3.1. ⊣

The following curious fact (not needed in proofs of the main results of this note) was pointed out to me by Paul Larson.

LEMMA 3.5. Assume γ is any cardinal less than the least critical point of each extender in $\vec{E}$ and $\mathbb{P}$ is a forcing notion of cardinality γ. Then in $\mathcal{W}_{\delta,\gamma}(\vec{E})$ there is a condition p that forces that $\mathbb{P}$ is a regular subordering of $\mathcal{W}_{\delta,\gamma}(\vec{E})$.

THE FIRST PROOF OF LEMMA 3.5. Pick a bijection between $\mathbb{P}$ and γ and hence identify $\mathbb{P}$ with $\langle\gamma, \leq_{\mathbb{P}}\rangle$ for some partial ordering $\leq_{\mathbb{P}}$ on γ. Let the sentence φ be the conjunction of axioms expressing the following.

(a) The order on $\mathbb{P}$: $a_\xi \to a_\eta$ whenever $\xi \leq_{\mathbb{P}} \eta$,
(b) The incompatibility relation on $\mathbb{P}$: $\bigwedge_{\zeta<\gamma}(\neg(a_\zeta \to a_\eta) \vee \neg(a_\zeta \to a_\xi))$, if $\xi \perp_{\mathbb{P}} \xi$.
(c) For every maximal antichain $\mathcal{A}$ of $\mathbb{P}$, $\bigvee_{\xi\in\mathcal{A}} a_\xi$.

Then φ is in $\mathcal{L}_{\delta,\gamma}$ and since its size is below the least critical point of an extender in $\vec{E}$, it is consistent with $\mathcal{T}_{\delta\gamma}(\vec{E})$. Therefore we may take p to be (the equivalence class of) φ. ⊣

We shall give another proof of Lemma 3.5 after Theorem 4.2.

3.1. Iteration trees. We say that $T = (\zeta, \leq_T)$ is a **tree order** on an ordinal ζ if

(1) $T = (\zeta, \leq_T)$ is a tree with root 0,
(2) $\leq_T$ is coarser than $\leq$,
(3) every successor ordinal is a successor in $\leq_T$,

(4) if $\xi \leq \zeta$ is a limit ordinal then the set of $\leq_T$-predecessors of ξ is cofinal in ξ.

Consider a transitive model M of a large enough fragment of ZFC and a system $\vec{E}$ of extenders in M. We allow M to be a proper class. As a matter of fact, it is typically going to be a proper class. Nevertheless, we omit the (straightforward) nuisances involved in formalization of the notion of an iteration tree in ZFC.

An $\vec{E}$-**iteration tree** is a structure consisting of $\langle T, M_\eta, E_\xi : \eta \leq \zeta, \xi < \zeta\rangle$, together with a commuting system of elementary embeddings $j_{\xi\eta} \colon M_\xi \to M_\eta$ for $\xi \leq_T \eta$ such that

(5) $\leq_T$ is a tree order on ζ,
(6) E_ξ is an extender in $\vec{E}^{M_\xi}$,
(7) If $\kappa = \operatorname{crit}(E_\xi)$ then the immediate $\leq_T$-predecessor of $\xi + 1$ is the least ordinal η such that $M_\eta \cap \mathbf{V}_{\kappa+1} = M_\xi \cap \mathbf{V}_{\kappa+1}$,
(8) If $\xi + 1$ is the immediate $\leq_T$ successor of η then $M_{\xi+1} = \operatorname{Ult}(M_\eta, E_\xi)$, hence $j_{\eta,\xi+1} = j_{E_\xi}$ as computed with respect to M_η,
(9) If ξ is a limit ordinal then M_ξ is the direct limit of M_η, for $\eta <_T \xi$, and
(10) each M_ξ is well-founded.
(11) If $\xi < \eta$ and ζ is the length of E_ξ then $M_\xi \cap \mathbf{V}_\zeta = M_\eta \cap \mathbf{V}_\zeta$.

It is usually not required that an iteration tree satisfies condition (7), and the iteration trees satisfying this condition are called **normal** iteration trees. A straightforward induction shows that condition (11) is a consequence of the previous conditions.

If E_ξ was always applied to M_ξ, then we would have a **linear** iteration that is moreover **internal**—*i.e.*, each extender used in the construction belongs to the model to which it is applied. The wellfoundedness of such an iteration follows from a rather mild additional condition about the extenders. On the other hand, the choice of condition (7) is behind the power of the iteration trees (*cf.* the proof of Theorem 4.1). This condition also prevents the obstacle to wellfoundedness of the iteration exposed in Lemma 2.5 (*cf.* Lemma 3.6). It will be important that the extenders in $\vec{E}$ have the property (E2) (*cf.* § 2.1), that for every generator ξ of E we have $\operatorname{Ult}(\mathbf{V}, E) \cap \mathbf{V}_\xi = \mathbf{V}_\xi$, or in other words, that every extender is ξ-strong for each of its generators ξ.

Lemma 3.6. Assume $\langle T, M_\eta, E_\xi : \eta \leq \zeta, \xi < \zeta\rangle$, is an iteration tree such that every extender used in its construction satisfies (E2). Assume E_0 and E_1 are extenders used along the same branch of T and E_1 was used after E_0. Then $\kappa = \operatorname{crit}(E_1)$ is greater than the supremum of all generators of E_0.

Proof. Assume the contrary. Therefore κ is less or equal than some generator of E_0, and (E2) implies that κ is smaller than the strength of E_0. Let us first consider the case when E_1 was applied to $M_\beta = \operatorname{Ult}(M_\alpha, E_0)$ for some α along the branch. If ξ is the strength of E_0 then we have $M_\beta \cap \mathbf{V}_\xi = M_\alpha \cap \mathbf{V}_\xi$.

Therefore, since $\kappa < \xi$ and E_1 could be applied to M_β, it could be applied to M_α as well. This contradicts our assumption that E_1 was applied to M_β.

We may therefore assume E_1 was applied to a model M_γ that is a direct limit of other models on the branch, including $\mathrm{Ult}(M_\alpha, E_0)$ for some α. The above and an induction argument show that $\mathrm{crit}(j_{E_1})$ is greater than any generator of any extender used in the construction of this branch, including the generators of E_0. ⊣

Note that an iteration tree $\langle T, M_\eta, E_\xi : \eta \leq \omega_1, \xi < \omega_1\rangle$ has a branch of length ω_1 by (4). In Lemma 3.7 below, and elsewhere, we assume that all critical points of elementary embeddings $j_{\xi\eta}$ used in building an iteration tree are countable ordinals. Lemma 3.7 is an attempt to extract one of the key ideas from the proof of Theorem 4.1. It is part of the proof of the comparison lemma for mice, which at the level of generality we are dealing in now, is due to Martin and Steel [MaS94]. The preprint [Sch] was very helpful during the extraction of this lemma.

Lemma 3.7. Assume $\langle T, M_\eta, E_\xi : \eta \leq \omega_1, \xi < \omega_1\rangle$ is an iteration tree and $b \subseteq \omega_1$ is its cofinal branch. Assume $H \prec H_{(2^{\aleph_1})^+}$ is countable and it contains the iteration tree. Let $\bar{H}$ be its transitive collapse and let $\pi : H \to \bar{H}$ be the collapsing map.

With $\alpha = H \cap \omega_1$ we have the following.

(12) $\alpha \in b$ and M_α is the direct limit of M_ξ, $\xi \in b \cap \alpha$,

(13) $M_{\omega_1} \cap H$ is the direct limit of $\langle M_\xi \cap H, \xi \in b \cap \alpha\rangle$,

(14) π^{-1} and $j_{\alpha\omega_1}$ agree on $M_\alpha \cap \bar{H}$ and in particular
  (a) $\pi^{-1}[M_\alpha \cap \bar{H}]$ is included in $M_{\omega_1} \cap H$, and
  (b) $\mathrm{crit}(j_{\alpha\omega_1}) = \alpha$,

(15) If γ is the strength of the extender $E_{\alpha'}$ such that $\mathrm{Ult}(M_\alpha, E_{\alpha'})$ is the successor of M_α along b then $j_{\alpha\omega_1} = j_{E_{\alpha'}} \circ i$ for an elementary embedding i such that $i \restriction \mathbf{V}_\gamma$ is the identity,

(16) $M_{\omega_1} \cap \wp(\alpha + 1) = M_\alpha \cap \wp(\alpha + 1)$.

Proof. Clause (12) follows by (4) and (9). By elementarity in H it holds that M_{ω_1} is the direct limit of M_ξ, for $\xi \in b$, and therefore (13) follows. By applying this and (12), (14) follows as well. Let γ be as in (15). We have $j_{\alpha\omega_1} = j_{E_{\alpha'}} \circ i$ for some i. By Lemma 3.6 we have $\mathrm{crit}(i) \geq \gamma$ and (15) follows.

Clause (16) is a consequence of (14). ⊣

3.2. The iteration game. In what follows elements of an iteration tree will be proper classes instead of sets. The arguments can be formalized within ZFC by using the standard reflection and compactness devices. We leave out the well-known details. We define a two-player game of transfinite length ζ in which the players build an iteration tree, starting from a model M and a system of extenders $\vec{E}$ in M. Let $M_0 = M$. In his αth move player I picks an extender E_α in $\vec{E}^{M_\alpha}$ such that the strength of E_α is greater than the strength of E_β for all $\beta < \alpha$. Then the referee finds the minimal $\beta \leq \alpha$ such that (writing $\text{crit}(E)$ for the critical point of the elementary embedding j_E) $\mathbf{V}_{\text{crit}(E_\alpha)+1} \cap M_\beta = \mathbf{V}_{\text{crit}(E_\alpha)+1} \cap M_\alpha$. Hence M_β is the earliest model in the iteration to which E_α can be applied. Referee then defines

$$M_{\alpha+1} = \text{Ult}(M_\beta, E_\alpha),$$

with $j_{\beta\alpha+1}$ being the corresponding embedding. The referee also extends the tree order T by adding $\alpha + 1$ as an immediate successor to β. At a limit stage α player II picks a maximal branch $\langle M_\xi : \xi \in b\rangle$ of T such that b is cofinal in α and lets M_α be the direct limit of the system $\langle M_\xi, j_{\xi,\eta} : \xi < \eta \in b\rangle$. If M_α is well-founded then we identify it with its transitive collapse.

The first player who disobeys the rules loses. Assume both players obeyed the rules of the iteration game. If M_α is ill-founded then the game is over and player I wins. If all M_α are well-founded, then player II wins, and otherwise player I wins.

For definiteness, we call the above game the **$(\vec{E}, \zeta)$-iteration game in M**. We shall suppress $\vec{E}$ and M whenever they are clear from the context. An **$(\vec{E}, \zeta)$-iteration strategy** is a winning strategy for player II in the iteration game of length ζ. A pair $(M, \vec{E})$ is **$(\vec{E}, \zeta)$-iterable** if player II has a $\vec{E}, \zeta$-winning strategy. An **$(\vec{E}, \zeta)$-iteration** of M is an elementary embedding $j_{0\zeta} : M \to M_\zeta$ extracted from an $\vec{E}$-iteration tree on ζ. A model is **fully iterable** if it is $(\vec{E}, \zeta)$-iterable for every ordinal ζ.

Universally Baire sets of reals were introduced in [FMW92] and are defined in § 6.1.

THEOREM 3.8 (Martin & Steel; [MaS94]). Assume there exist n Woodin cardinals and a measurable above them all. For every $a \in \mathbb{R}$ and every $m \leq n$ there exists an inner model containing a and m Woodin cardinals, denoted by $M_m(a)$. It is $(\omega_1 + 1)$-iterable and its Woodin cardinals are countable ordinals in $\mathbf{V}$.

If there are class many Woodin cardinals then every $M_n(a)$ is fully iterable in every forcing extension and the iteration strategy is coded by a universally Baire set of reals. ⊣

The assumption that there are class many Woodin cardinals is not optimal. For the case when $n = 1$, *cf.* the proof of Theorem 5.6 below.

The following lemma will be needed in § 6.2.

LEMMA 3.9. Assume Σ is an ω_1-iteration strategy that is universally Baire and moreover that all sets projective in Σ are universally Baire. Then Σ can always be extended to a full iteration strategy.

PROOF. Since the statement "Σ is a winning iteration strategy" is projective in Σ it is forcing-absolute. One constructs an iteration strategy $\overline{\Sigma}$ for player II by induction on limit ordinals $\alpha \geq \omega_1$. The recursive hypothesis is that every play of the iteration game in which player II obeyed $\overline{\Sigma}$ has the following property. If κ is the cardinality of the resulting iteration tree, then in the extension by Lévy collapse $\mathrm{Col}(\omega, \kappa)$ of κ to ω tree T is the result of a play of an iteration game in which player II has obeyed Σ.

Let T be an iteration tree of height α resulting from an iteration game in which player II has obeyed the extension of Σ constructed so far. Go to a forcing extension by Lévy collapse $\mathrm{Col}(\omega, |\alpha|)$, and let $\mathbf{b}$ be the α-branch chosen by Σ.

We claim that for every $\beta < \alpha$ the condition whether $\beta \in \mathbf{b}$ is decided by the maximal condition. Otherwise, fix β and choose p_1 and p_2 such that $p_1 \Vdash \check{\beta} \in \dot{\mathbf{b}}$ and $p_2 \Vdash \check{\beta} \notin \dot{\mathbf{b}}$. Let $G_1 \subseteq \mathrm{Col}(\omega, \alpha)$ be a generic filter such that $p_1 \in G_1$. By using the homogeneity of $\mathrm{Col}(\omega, \alpha)$ in $\mathbf{V}[G_1]$ we can choose a generic filter G_2 such that $p_2 \in G_2$ and $\mathbf{V}[G_1] = \mathbf{V}[G_2]$. Therefore in $\mathbf{V}[G_1] = \mathbf{V}[G_2]$ we have that Σ does not choose a unique α-branch of T, a contradiction.

Therefore branch $\mathbf{b}$ is decided in the ground model. We can therefore extend strategy $\overline{\Sigma}$ by having player II choose $\mathbf{b}$ at the αth stage. Since well-foundedness is absolute for forcing extensions, this defines a winning $(\alpha + 1)$-iteration strategy. ⊣

§4. **Genericity iterations.** In the present section we formulate and prove results that make the extender algebra unique.

Assuming M is sufficiently iterable, we may talk about iteration strategies for player I. These are the strategies that, when played against player II's winning strategy, produce models (necessarily well-founded) with desirable properties.

THEOREM 4.1. Assume $(M, \vec{E})$ is (ω_1+1)-iterable and $\vec{E}$ witnesses a countable ordinal δ is a Woodin cardinal in M. Then for every $x \subseteq \omega$ there is a (well-founded) countable iteration $j : M \to M^*$ such that x is $j(\mathcal{W}_{\delta,\omega}(\vec{E}))$-generic over M^*.

PROOF. By Lemma 3.3 and Lemma 1.4 we only need to assure $x \models j(\mathcal{T}_{\delta,\omega}(\vec{E}))$. Define a strategy for player I for building an $\vec{E}$-iteration tree with $M_0 = M$ as follows. Assume $\langle T, M_\xi, E_\xi : \xi \leq \alpha\rangle$ has been constructed. If $x \models j_{0\alpha}(\mathcal{T}_{\delta,\omega}(\vec{E}))$ then $j_{0\alpha} : M \to M_\alpha$ is the required iteration and we stop. Otherwise, let λ be the minimal cardinal such that there are $\vec{\varphi}$, κ, and a $\vec{\varphi}, \lambda$-strong extender $E \in j_{0\alpha}(\vec{E})$ with $\mathrm{crit}(E) = \kappa$ such that $x \not\models \Psi(\vec{\varphi}, \kappa, \lambda)$. Fix such $\vec{\varphi}$, λ and E. We have $x \not\models \bigvee_{\xi<\kappa} \varphi_\xi$ and $x \models \bigvee_{\xi<\lambda} \varphi_\lambda$. and note that $\lambda < j_{0\alpha}(\delta)$. Then let player I play $E_\alpha = E$. Note that $j_{E_\alpha}(\kappa) \geq \lambda$.

This describes the iteration strategy for player I. We claim that if player II responds with his winning strategy then the process of building the iteration tree terminates at some countable stage. Assume otherwise. Let $\langle T, M_\xi, E_\xi : \xi < \omega_1\rangle$ be the resulting iteration tree and let $b \subseteq \omega_1$ be its cofinal branch such that M_{ω_1}, the direct limit of $\langle M_\xi : \xi \in b\rangle$, is well-founded.

Fix a countable $H \prec H_{(2^{\aleph_1})^+}$ containing everything relevant. Let $\alpha = H \cap \omega_1$ and let $\bar{H}$ be the transitive collapse of H, with $\pi^{-1}\colon \bar{H} \to \mathbf{H}(\vartheta)$ the inverse of the collapsing map. Since the iteration did not stop at stage α, we can consider the extender E applied to M_α along the well-founded ω_1-branch of T. Note that E may be different from E_α, the extender chosen in M_α during the run of the iteration game. Instead, E is equal to E_β for some $\beta \geq \alpha$.

By (15) of Lemma 3.7 we have that π^{-1} and $j_{\alpha\omega_1}$ agree on $M_\alpha \cap \bar{H}$ because i is the identity on $\mathbf{V}_\gamma$. In particular, $\operatorname{crit}(E) = \alpha$.

By the choice of the iteration, E is $\vec{\varphi}, \lambda$-strong for some $\vec{\varphi}$ and λ such that $\vec{\varphi}$ reflects to α but $x \not\models \bigvee_{\xi<\alpha} \varphi_\xi$ and $x \models \bigvee_{\xi<\lambda} \varphi_\xi$. By definition we have $\varphi_\eta \in M_\alpha \cap \mathbf{V}_\alpha$ for all $\eta < \alpha$, hence the formula $\bigvee_{\eta<\alpha} \varphi_\eta$ belongs to $\mathbf{V}_{\alpha+1}$. By (12) of Lemma 3.7 there are $\xi <_T \alpha$ and $\vec{\psi} \in M_\xi$ such that $\vec{\varphi} = j_{\xi\alpha}(\vec{\psi})$.

Since $j_{\alpha\omega_1}$ and j_E agree on $M_\alpha \cap \mathbf{V}_\lambda$, we have that $j_{\alpha\omega_1}(\vec{\varphi})$ implies $\bigvee_{\xi<\lambda} \varphi_\xi$, hence $x \models j_{\alpha\omega_1}(\vec{\psi})$ holds in H. Thus for some $\xi \in H \cap \omega_1$ we have $x \models \varphi_\xi$ and therefore $x \models \bigvee_{\xi<\alpha} \varphi_\xi$, a contradiction. $\dashv$

A number of extensions of Theorem 4.1 in different directions are in order (some of their obvious common generalization are being omitted).

Assume an extender E and a forcing $\mathbb{P}$ are such that for some κ we have $\operatorname{crit}(j_E) > \kappa$ and $\mathbb{P}$ is in $\mathbf{V}_\kappa$. Then E still defines an extender in the extension by $\mathbb{P}$ and this extender is λ-strong for every λ for which E is λ-strong. This is essentially a consequence of the Lévy-Solovay result that a measurable cardinal cannot be destroyed by a small forcing [Kan94]. The converse is also true [HW00]: in a forcing extension $\mathbf{V}[G]$ by forcing $\mathbb{P} \in \mathbf{V}_\kappa$ every λ-strong elementary embedding of $\mathbf{V}[G]$ into an inner model with critical point κ is a lift of a ground model λ-strong elementary embedding with critical point κ. This is a bit deeper than the corresponding result for elementary embeddings that are not necessarily λ-strong.

In the following lemma and elsewhere we shall slightly abuse the notation and denote the extender obtained from E in a forcing extension by the same letter.

Theorem 4.2. Assume $(M, \vec{E})$ is (ω_1+1)-iterable and $\vec{E}$ witnesses a countable ordinal δ is a Woodin cardinal in M. Assume moreover $\kappa < \min\{\operatorname{crit}(j_E) : E \in \vec{E}\}$ and $\mathbb{P} \in \mathbf{V}_\kappa \cap M$ is a forcing notion. If $G \subseteq \mathbb{P}$ in V is M-generic then for every $x \subseteq \omega$ there is a (well-founded) countable iteration $j\colon M[G] \to M^*[G]$ such that x is $j(\mathcal{W}_{\delta,\omega}(\vec{E}))$-generic over $M^*[G]$.

Proof. By the above discussion δ is still a Woodin cardinal in $M[G]$ as witnessed by $\vec{E}$. The proof is similar to the proof of Theorem 4.1. In his strategy, player I computes $\Psi(\vec{\varphi}, \kappa, \lambda)$ in $M[G]$ instead of M. Since $G \in V$, this describes an iteration strategy of player I in V. Thus player II can respond to it by using his winning iteration strategy and the other details of the proof are identical.

As a matter of fact, a simple proof shows that the full iterability of M implies the full iterability of $M[G]$ (essentially using the same strategy). The point is that, since $\mathbb{P}$ is small, the ultrapowers of M lift to ultrapowers of $M[G]$. Therefore this is a consequence of Theorem 4.1. $\dashv$

THE SECOND PROOF OF LEMMA 3.5. We now provide a different proof that if γ is any cardinal less than the least critical point of each extender in $\vec{E}$ and $\mathbb{P}$ is a forcing notion of cardinality γ then in $\mathcal{W}_{\delta,\gamma}(\vec{E})$ there is a condition p that forces that $\mathbb{P}$ is a regular subordering of $\mathcal{W}_{\delta,\gamma}(\vec{E})$. The present proof will also show that there exist condition $q \in \mathcal{W}_{\delta,\gamma}(\vec{E})$ and condition r in $\mathbb{P}$ that force that $\mathbb{P}$ is forcing-equivalent to $\mathcal{W}_{\delta,\gamma}(\vec{E})$. Assume $G \subseteq \mathbb{P}$ is generic over M. We may assume $\mathbb{P} = (\gamma, <_{\mathbb{P}})$ for some ordering $<_{\mathbb{P}}$ on γ. Since δ is a countable ordinal and $\gamma < \delta$, by using Theorem 4.1 we can find an iteration $j\colon M \to M^*$ such that G is $j(\mathcal{W}_{\delta,\gamma}(\vec{E}))$ generic over M^*. By our assumption γ is smaller than the critical point of j and therefore $j(\mathbb{P}) = \mathbb{P}$. Hence in M^* there is q_0 in $j(\mathcal{W}_{\delta,\gamma}(\vec{E}))$ and $r \in \mathbb{P}$ that forces $\mathbb{P}$ and $j(\mathcal{W}_{\delta,\gamma}(\vec{E}))$ are forcing equivalent. By the elementarity, this is true in M for $\mathbb{P}$ and $\mathcal{W}_{\delta,\gamma}(\vec{E})$. $\dashv$

Here is yet another variation of genericity iterations, also due to Woodin. Its proof was sketched in the appendix to [NZ00] (not included in the published version [NZ01]).

THEOREM 4.3. Assume $\mathbb{P}$ is a forcing notion of cardinality κ and $\dot{x}$ is a $\mathbb{P}$-name for a real. Assume $(M, \vec{E})$ is $(\kappa^+ + 1)$-iterable and $\vec{E}$ witnesses a countable ordinal δ is a Woodin cardinal in M. Then there is a (well-founded) iteration $j\colon M \to M^*$ of length $< \kappa^+$ such that $\mathbb{P}$ forces $\dot{x}$ is $j(\mathcal{W}_{\delta,\omega}(\vec{E}))$-generic over M^*.

PROOF OF THEOREM 4.3. has similar structure as the proof of Theorem 4.1. More precisely, player I and player II play the iteration game in which player II follows his winning strategy while player I at each move chooses a bad extender with the least possible strength. In the present situation an extender is bad if some condition in $\mathbb{P}$ forces that $\dot{x}$ violates an axiom associated with this extender. The game stops when $\mathbb{P}$ forces that $\dot{x}$ satisfies the theory corresponding to $\vec{E}$ and therefore that it is $j(\mathcal{W}_{\delta,\omega}(\vec{E}))$-generic over the iterate M^*.

Showing that the construction terminates before the κ^+th stage requires taking the elementary submodel H of cardinality κ and an appropriate analogue of Lemma 3.7. $\dashv$

In [NZ01] it was proved that if the forcing $\mathbb{P}$ is proper then one can find a countable genericity iteration. The authors of [NZ01] also pointed out that this is not necessarily true when $\mathbb{P}$ is only assumed to be semiproper.

4.1. Genericity iterations for subsets of ω_1. We finally turn to applications of the algebra $\mathcal{W}_{\delta,\delta}$ with δ generators.

THEOREM 4.4. Assume $(M, \vec{E})$ is (ω_1+1)-iterable and $\vec{E}$ witnesses a countable ordinal δ is a Woodin cardinal in M. Then for every $x \subseteq \omega_1$ there is a (well-founded) countable iteration $j: M \to M^*$ such that $x \cap j(\delta)$ is $j(\mathcal{W}_{\delta,\delta}(\vec{E}))$-generic over M^*.

Furthermore, we can assure $j(\delta) \in \mathbf{C}$ for any club $\mathbf{C} \subseteq \omega_1$ given in advance.

PROOF. The strategy for player I is identical to the strategy used in the proof of Theorem 4.1, and the proof of the latter shows that the iteration terminates at some countable stage at which $x \cap j(\delta) \models j(\mathcal{T}_{\delta,\delta}(\vec{E}))$.

Now we show how to assure $j(\delta) \in \mathbf{C}$. Let $c_\alpha \subseteq \omega$ be a real coding the αth element of $\mathbf{C}$ and let $\overline{\mathbf{C}}$ be a subset of ω_1 such that its intersection with the interval $[\omega\alpha, \omega(\alpha+1))$ codes c_α. Identify x with a set of even ordinals and $\overline{\mathbf{C}}$ with a set of odd ordinals and let y be the result. By applying the first part of this theorem to y find an iteration $j: M \to M^*$ such that $x \cap j(\delta)$ and $\overline{\mathbf{C}} \cap j(\delta)$ are both $j(\mathcal{W}_{\delta,\delta}(\vec{E}))$-generic over M^*. Let $G \subseteq [j(\mathcal{W}_{\delta,\delta}(\vec{E}))$ be the generic filter defined by $x \cap j(\delta)$ and $\overline{\mathbf{C}} \cap j(\delta)$. Since $c_\xi \in M^*[G]$ for all $\xi < j(\delta)$, we have that in $M^*[G]$ all ordinals below $j(\delta)$th element of $\mathbf{C}$ are collapsed to $\aleph_0$. By the $j(\delta)$-c.c. of $j(\mathcal{W}_{\delta,\delta}(\vec{E}))$, $j(\delta) = \aleph_1^{M^*[G]}$. Assume $\mathbf{C} \cap j(\delta)$ was bounded and pick ξ such that $\sup \mathbf{C} \cap j(\delta) < \xi < j(\delta)$. Then c_ξ codes an element of $\mathbf{C}$ greater than $j(\delta)$. But this means that $j(\delta)$ is collapsed to $\aleph_0$ in $M^*[G]$, contradicting the above. ⊣

The following theorem is an analogue of Theorem 4.1 for an arbitrary set of ordinals and it will be used in the proof of Theorem 5.6. More useful versions (Theorem 4.6 and Theorem 4.7) require stronger large cardinal assumptions.

THEOREM 4.5. Assume $(M, \vec{E})$ is fully iterable and $\vec{E}$ witnesses a countable ordinal δ is a Woodin cardinal in M. Then for every set of ordinals x there is a (well-founded) iteration $j: M \to M^*$ of length $< |x|^+$ such that x is $j(\mathcal{W}_{\delta,\delta}(\vec{E}))$-generic over M^*.

PROOF. Let $E_0 \in \vec{E}$ be an extender with minimal strength. Iterate E_0 to obtain an iteration $j_0: M \to M_0$ such that $x \subseteq j(\delta)$. Since this is a linear iteration, it is well-founded. By the minimality, the strength of every extender in $j(\vec{E})$ is greater than the strength of all iterates of E_0 used in j_0. Therefore for every iteration $i: M_0 \to M^*$ of $(M_0, j_0(\vec{E}))$ we have that $i \circ j_0: M \to M^*$ is an iteration of $(M, \vec{E})$, and therefore $(M_0, j_0(\vec{E}))$ is fully iterable.

From this point on the proof is analogous to the proof of Theorem 4.1. Define a strategy for player I for building an iteration tree starting with M_0 as follows. Assume $\langle T, M_\xi, E_\xi : \xi \leq \alpha \rangle$ has been constructed. If $x \models (j_{0\alpha} \circ j_0)(\mathcal{T}_{\delta,\delta}(\vec{E}))$ then $j_{0\alpha} \circ j_0: M \to M_\alpha$ is the required iteration and we stop. Otherwise, let λ be the minimal cardinal such that there are $\vec{\varphi}$, κ, and a $\vec{\varphi}, \lambda$-strong extender $E \in (j_{0\alpha} \circ j)(\vec{E})$ with $\mathrm{crit}(E) = \kappa$ such that $x \not\models \Psi(\vec{\varphi}, \kappa, \lambda)$. Fix such $\vec{\varphi}$, λ and

E. We have $x \not\models \bigvee_{\xi<\kappa} \varphi_\xi$ and $x \models \bigvee_{\xi<\lambda} \varphi_\lambda$. and note that $\lambda < (j_{0\alpha} \circ j_0)(\delta)$. Then let player I play $E_\alpha = E$. Note that $(j_{E_\alpha} \circ j_0)(\kappa) \geq \lambda$.

This describes the iteration strategy for player I. A proof that if player II responds with his winning strategy then the process of building the iteration tree terminates at some stage before $|x|^+$ is identical to the proof of Theorem 4.1, using the variant of Lemma 3.7 for an iteration tree of height $|x|^+$. ⊣

The iterability assumption of Theorem 4.6 below is almost the strongest known result of its kind. Its consistency modulo large cardinals was proved by Neeman [Nee02A]. If CH holds and $\{x_\xi : \xi < \omega_1\}$ is an enumeration of all reals we write

$$\mathbb{R}\restriction\gamma = \{x_\xi : \xi < \gamma\}.$$

The extender algebra with δ generators need not satisfy (3) of the following theorem, even when it collapses δ to $\aleph_1$.

Theorem 4.6. Assume $(M, \vec{E})$ is $(\omega_1 + 1)$-iterable and $\vec{E}$ witnesses δ is a Woodin limit of Woodin cardinals in M. Furthermore assume CH holds and A is a set of reals. Then there are a (well-founded) iteration $j : M \to M^*$ of countable length and an M^*-generic filter $G \subseteq j(\mathcal{W}_{\delta,\delta}(\vec{E}_0))$ such that the following hold:

(1) $\mathbb{R}^{M^*[G]} = \mathbb{R}\restriction j(\delta)$,
(2) $A \cap (\mathbb{R}\restriction j(\delta)) \in M^*[G]$, and
(3) Every real in $M^*[G]$ is M^*-generic for a forcing of cardinality $< j(\delta)$.
(4) There is $B \subseteq \omega_1$ such that $N = \mathbf{L}[M, B\restriction\alpha]$ satisfies $\mathbb{R}^{M^*[G]} = \mathbb{R}^N$

Proof. Let Σ be the winning strategy for player II in the iteration game for $M, \vec{E}$. It is naturally identified with a set of reals and therefore, using CH, with a subset of ω_1. Fix a subset B of ω_1 so that its intersection with ω codes M and B codes $\mathbb{R}$, A and Σ. Let $\mathbf{C}$ be the club of all $\alpha < \omega_1$ such $B\restriction\alpha$ is an elementary submodel of B.

Applying Theorem 4.4 find an iteration $j : M \to M^*$ such that (with $\alpha = j(\delta)$) we have that $B \cap \alpha$ is $j(\mathcal{W}_{\delta,\delta}(\vec{E}))$-generic over M^* and α is a limit point of club $\mathbf{C}$. Therefore for cofinally many $\xi < \alpha$ we have $B\restriction\xi \prec B$. Let G be the generic filter defined by $B\restriction\alpha$. Note that (2) is immediate. We shall now prove (4). Note that the converse inclusion is immediate.

Only the direct inclusion requires a proof. Let $\langle T, M_\eta, E_\xi : \eta \leq \alpha, \xi < \alpha\rangle$ be the iteration tree constructed during the course of finding M^*. Since the iteration strategy of player I is definable and $\Sigma\restriction\alpha$ is an elementary submodel of Σ, this tree (except possibly its final α-branch) can be constructed in N.

Now assume $\dot{y}$ is a nice $j(\mathcal{W}_{\delta,\delta}(\vec{E}))$-name for a real. By α-c.c. there is a Woodin cardinal $\gamma < \alpha$ in M^* such that all conditions in $\dot{y}$ use only generators less than γ. Let $\vec{E}(\gamma)$ be the extenders in $j(\vec{E})$ whose strength is below γ. Then $\vec{E}(\gamma)$ witnesses Woodinness of γ in M^*. Also, $G\restriction\gamma$ satisfies all axioms of theory

$\mathcal{T}(\vec{E}(\gamma))$ in M^*. Let $\beta < \alpha$ be large enough so that $M_\beta \cap \mathbf{V}_{\gamma+1} = M^* \cap \mathbf{V}_{\gamma+1}$ and therefore all axioms of $\mathcal{T}(\vec{E}(\gamma))$ are satisfied in M_β.

Therefore we have proved

(5) $G \restriction \gamma$ is generic for $\mathbb{P} = \mathcal{W}_{\gamma,\gamma}(\vec{E}_\gamma)$ over both M_β and M^*.

Since all conditions occurring in $\dot{y}$ belong to $\mathbb{P}$, the interpretation of $\dot{y}$ with respect to the filter in $\mathbb{P}$ defined by $G \restriction \gamma$ over M_β is definable from M and $G \restriction \alpha$ and therefore belongs to N. Since this interpretation coincides with the interpretation of $\dot{y}$ with respect to the filter in $j(\mathcal{W}_{\delta,\delta}(\vec{E}))$ defined by G, and since $\dot{y}$ was an arbitrary name for a real, the direct inclusion of (4) follows.

Since N is a definable inner model in $M^*[G]$, (1) follows.

It remains to prove (3). The elementarity of $\mathbb{R} \restriction \alpha$ in $\mathbb{R}$ implies that $j(\mathcal{W}_{\delta,\delta}(\vec{E}))$ collapses all cardinals below α to $\aleph_0$. By (5) each x_ξ, for $\xi < \gamma$, is generic over M^* for a small forcing. Since δ is a limit of Woodins all reals in $\mathbb{R} \restriction \alpha$ are generic over M^* for a small forcing notion. $\dashv$

The assumption of the existence of an $(\omega_1 + 1)$-iterable model with a measurable Woodin cardinal used in the following theorem is presently beyond reach of the inner model theory.

Theorem 4.7. Assume $(M, \vec{E})$ is (ω_1+1)-iterable and $\vec{E}$ witnesses a countable ordinal δ is a measurable Woodin cardinal in M. Then for every $x \subseteq \omega_1$ there is a (well-founded) ω_1-iteration $j\colon M \to M^*$ such that x is $j(\mathcal{W}_{\delta,\delta}(\vec{E}))$-generic over M^*.

Proof. This proof is an extension of the proof of Theorem 4.4. We need to assure $x \models j(\mathcal{T}_{\delta,\delta}(\vec{E}))$ for an ω_1-iteration j. Define a strategy of player I for building an iteration tree starting with $M_0 = M$ as follows. Assume $\langle T, M_\eta, E_\xi : \eta \leq \alpha, \xi < \alpha \rangle$ has been constructed.

(a) Assume there is a sequence $\vec{\varphi}$ in M_α that reflects to some κ but $x \not\models \Psi(\vec{\varphi}, \kappa, \lambda)$ for some λ satisfying $\lambda < j_{0\alpha}(\delta)$ and $\lambda > \sup_{\xi<\alpha}(\lambda_{E_\xi})$. Choose the minimal λ with this property, fix the appropriate $\vec{\varphi}$ and κ so that $x \not\models \Psi(\vec{\varphi}, \kappa, \lambda)$. Then let E_α be a $(\vec{\varphi}, \lambda)$-strong extender in M_α such that $\mathrm{crit}(E_\alpha) = \kappa$. Note that $j_{E_\alpha}(\kappa) \geq \lambda$.

(b) Now assume α is countable and $x \models j_{0\alpha}(\mathcal{T}_{\delta,\delta}(\vec{E}))$. Since δ is a measurable Woodin cardinal, use normal measure U on $j_{0\alpha}(\delta)$ to define $M_{\alpha+1} = \mathrm{Ult}(M_\alpha, U)$ and let $j_{\alpha,\alpha+1} : M_\alpha \to M_{\alpha+1}$.

This describes the iteration strategy for player I. Now consider the iteration tree formed when player II plays his winning strategy against the iteration strategy just defined for player I. By Theorem 4.4, the set C of all stages α in the iteration such that $x \models j_{0\alpha}(\mathcal{T}_{\delta,\delta}(\vec{E}))$ and $\alpha = j_{0\alpha}(\delta)$ is unbounded, and it is therefore a club. Note that $x \cap \alpha$ is $j_{0\alpha}(\mathcal{W}_{\delta,\delta}(\vec{E}))$-generic for all $\alpha \in C$.

By the choice of extenders and Lemma 3.6, for $\alpha \in C$ the critical points of embeddings constructed after αth stage will never drop below $j_{0\alpha}(\delta)$.

The critical sequence defines a club in ω_1 and an ω_1-iteration $\langle N_\xi, j_{\xi\eta} : \xi \leq \eta \leq \omega_1 \rangle$ such that $N_0 = M$, for $\xi < \eta$ the embedding $j_{\xi\eta} \colon N_\xi \to N_\eta$ has α_ξ as its critical point, and $x \cap \alpha_\xi \models j_{0\xi}(\mathcal{T}_{\delta,\delta}(\vec{E}))$ for all ξ. Since the critical points are increasing, this implies that $x \models j_{0\omega_1}(\mathcal{T}_{\delta,\delta}(\vec{E}))$, as required. ⊣

§5. **Absoluteness.** By "$\mathbf{M}_1$ exists" we denote the statement "There exists an inner model of the form $\mathbf{L}[\vec{E}]$ with a Woodin cardinal whose Woodinness is witnessed by extenders in $\vec{E}$". Similarly, if a is a real then "$\mathbf{M}_1(a)$ exists" denotes the statement "There exists an inner model of the form $\mathbf{L}[\vec{E}]$ with a Woodin cardinal whose Woodinness is witnessed by extenders in $\vec{E}$ and which contains a". Variants such as "M_n exists" or "M_1 exists and it is fully iterable" are interpreted similarly. The following lemma and its variations will be used tacitly in order to furnish the assumptions of genericity iteration theorems.

Lemma 5.1. Assume $\mathbf{M}_1$ exists and it is fully iterable. Then there exists a fully iterable inner model with a Woodin cardinal δ such that δ is a countable ordinal in $\mathbf{V}$.

Proof. Take a countable elementary submodel N of $\mathbf{M}_1$. Theorem 5.6 implies that N has a sharp. We can therefore find an iteration of N of length Ord whose resulting model is an inner model with a Woodin cardinal which is a countable ordinal in $\mathbf{V}$. The iteration strategy for N is obtained by copying the iteration strategy of $\mathbf{M}_1$ (*cf.* [Sar13A]). ⊣

5.1. Absoluteness in $\mathbf{L}(\mathbb{R})$. The following result and its proof are a prototype for the main result of this note, Theorem 5.9.

Theorem 5.2. Assume $\mathbf{M}_1(a)$ is fully iterable in all forcing extensions for all $a \in \mathbb{R}$. Then all $\utilde{\mathbf{\Sigma}}^1_3$ statements are forcing absolute.

Proof. A $\utilde{\mathbf{\Sigma}}^1_3$ statement $\varphi(a)$ with parameter $a \in \mathbb{R}$ is of the form $(\exists x)\psi(x, a)$ where ψ is Π^1_2. To φ we associate a sentence φ^* of $\mathbf{M}_1(a)$ (*cf.* Theorem 3.8) stating that there exists a forcing notion $\mathbb{P}$ forcing φ. We claim that φ holds (in $\mathbf{V}$) if and only if φ^* holds in $\mathbf{M}_1(a)$. This will suffice since the iterability of $\mathbf{M}_1(a)$ is not changed by forcing. If φ^* holds in $\mathbf{M}_1(a)$ then since $\mathbf{M}_1(a) \cap \mathbf{V}_{\delta+1}$ is countable we can find an $\mathbf{M}_1(a)$-generic filter $G \subseteq \mathbb{P}$ for $\mathbb{P}$ referred to in φ^*. If $\mathbf{M}_1(a)[G] \models \psi(x, a)$ then by Shoenfield's absoluteness theorem $\psi(x, a)$ holds in $\mathbf{V}$. For the converse implication, assume $\psi(x, a)$ holds in some forcing extension of $\mathbf{V}$ for some $x \in \mathbb{R}$. Since $\mathbf{M}_1(a)$ is fully iterable in all forcing extensions, apply Theorem 4.1 to find a countable iteration $j \colon \mathbf{M}_1(a) \to M^*$ such that x is generic over M^*. Then (again using Shoenfield) $M^* \models \varphi^*$, and by elementarity $\mathbf{M}_1(a) \models \varphi^*$. ⊣

The assumptions of Theorem 5.2 are far from optimal. By a result of Martin and Solovay [MS69], if κ is a measurable cardinal then $\utilde{\mathbf{\Sigma}}^1_3$ sentences are forcing absolute for forcing notions in $\mathbf{V}_\kappa$. As a matter of fact, all $\utilde{\mathbf{\Sigma}}^1_3$ sentences are absolute between all forcing extensions of $\mathbf{V}$ if and only if all sets have sharps (Martin & Steel, Woodin; *cf.*, *e.g.*, [Ste10] for terminology). One important fact about Theorem 5.2 and its extension, Theorem 5.4 below, is that its proof is susceptible to far-reaching generalizations. Also, Corollary 5.3 below is worth mentioning. In the language of Woodin's Ω-logic (*cf.* [Woo10, § 10], [Woo01, BCL06, Lar11]), it states that if the function $x \mapsto \mathbf{M}_1(x)^\#$ is universally Baire then it is an Ω-proof for all true $\mathbf{\Sigma}^1_3$ statements (*cf.* § 5.3). As common in inner model theory, by $\mathbf{M}_1(x)^\#$ we denote an iterable transitive model of ZFC that contains x and has a top measure (for more on mice, *cf.* [Sch01]).

COROLLARY 5.3. Assume N is a transitive model of a large enough fragment of ZFC that is closed under the map $x \mapsto \mathbf{M}_1(x)^\#$. Then N is $\mathbf{\Sigma}^1_3$-correct.

PROOF. For every real x in N, N can reconstruct $\mathbf{M}_1(x)$ and its iteration strategy. Therefore the proof of Theorem 5.2 applies inside N. ⊣

It should be noted that $\mathbf{M}_1$ is not necessarily $\mathbf{\Sigma}^1_3$ correct. For example, the fine-structural version of $\mathbf{M}_1$ has a Δ^1_3 well-ordering of the reals. Similarly, being closed under the sharps (*i.e.*, under the map $x \mapsto x^\#$) does not guarantee $\mathbf{\Sigma}^1_3$ correctness since $\mathbf{L}[\mu]$ has a Δ^1_3 well-ordering of the reals. Nevertheless, $\mathbf{M}_2$ is $\mathbf{\Sigma}^1_4$ correct (*cf.* [Ste95B]). Here we shall only show that a proof analogous to the above gives a weaker result.

THEOREM 5.4. Assume $\mathbf{M}_n(a)$ is fully iterable in all forcing extensions for every $a \in \mathbb{R}$. Then all $\mathbf{\Sigma}^1_{n+2}$ statements are forcing absolute.

PROOF. We first prove the case $n = 2$. A $\mathbf{\Sigma}^1_4$ sentence φ has the form $(\exists x)(\forall y)\psi(x, y, a)$ for some real parameter a and a Π^1_2 formula ψ. Let $\mathbf{M}_2(a)$ be the minimal model for two Woodin cardinals containing a fully iterable in all forcing extensions (Theorem 3.8), and let $\delta_0 < \delta_1$ be its Woodin cardinals.

To φ associate a sentence φ^* stating that there is a forcing $\mathbb{P}$ in $\mathbf{V}_{\delta_0+1}$ and a $\mathbb{P}$-name $\dot{x}$ for a real such that for every forcing $\dot{\mathbb{Q}} \in \mathbf{V}_{\delta_1+1}$ and a $\dot{\mathbb{Q}}$-name $\dot{y}$ for a real we have

$$\Vdash_{\mathbb{P}*\dot{\mathbb{Q}}} \psi(\dot{x}, \dot{y}, a).$$

We claim that φ holds in $\mathbf{V}$ if and only if φ^* holds in $\mathbf{M}_2(a)$.

Assume $\mathbf{M}_2(a) \models \varphi^*$. Find an $\mathbf{M}_2(a)$-generic $G \subseteq \mathbb{P}$ and let $x = \operatorname{int}_G(\dot{x})$. Let y be any real. Let $\vec{E}_1$ be the system of extenders witnessing δ_1 is Woodin in $\mathbf{M}_2(a)$. We may assume $\min\{\operatorname{crit}(j_E) : E \in \vec{E}_1\} > \delta_0$. Using the full iterability of $\mathbf{M}_2(a)$ and Theorem 4.2 find a well-founded iteration $j : \mathbf{M}_2(a)[G] \to M^*[G]$ such that y is $j(\mathcal{W}_{\delta,\omega}(\vec{E}_1))$-generic over $M^*[G]$. Then by elementarity we have $M^*[G][y] \models \psi(x, y, a)$ and by Shoenfield's absoluteness theorem $\psi(x, y, a)$ holds. Since y was arbitrary, we have proved φ holds in $\mathbf{V}$.

Now assume φ holds in $\mathbf{V}$ and let $x \in \wp(\omega)$ be such that $(\forall y)\psi(x, y, a)$. By Theorem 4.1 we can find an iteration $j_0 \colon \mathbf{M}_2(a) \to M^*$ such that x is $j_0(\mathcal{W}_{\delta,\omega}(\vec{E}_0))$-generic over M^*. Fix $y \in \wp(\omega)$. By Theorem 4.2 there is an iteration $j_1 \colon M^*[x] \to M^{**}[x]$ such that y is $(j_1 \circ j_0)(\mathcal{W}_{\delta,\omega}(\vec{E}_1))$-generic over $M^{**}[x]$. Then $\mathbf{V} \models \varphi(x, y, a)$ and by the Shoenfield's absoluteness theorem $M^{**}[x] \models \varphi(x, y, a)$. Since y was arbitrary, this shows φ^* holds in $\mathbf{M}_2(a)$.

This concludes proof of the theorem in the case when we have two Woodin cardinals. In general case, to a $\underset{\sim}{\mathbf{\Sigma}}^1_{n+2}$ formula $(\exists x_1)(\forall x_2)\dots\psi(x_1, \dots, x_n, a)$ with ψ being $\underset{\sim}{\mathbf{\Sigma}}^1_2$ one associates a formula φ^* of $\mathbf{M}_n(a)$ stating

$$(\exists \mathbb{P}_1 \in \mathbf{V}_{\delta_0+1})(\exists \dot{x}_1)(\forall \mathbb{P}_2 \in \mathbf{V}_{\delta_1+1})(\forall \dot{x}_2)\dots \Vdash_{\mathbb{P}_1 * \dot{\mathbb{P}}_2 * \dots * \dot{\mathbb{P}}_n} \psi(\dot{x}_1, \dots, x_n, a)$$

and proves that $\mathbf{V} \models \varphi$ is equivalent to $\mathbf{M}_n(a) \models \varphi^*$ as above. ⊣

Remark 5.5. The following was pointed out by Menachem Magidor. In Theorem 5.4 it is not sufficient to assume that there are n Woodin cardinals and a measurable above. Such an assumption can hold if $V = \mathbf{L}[\vec{E}]$ for a system of extenders $\vec{E}$. However a forcing extension of $\mathbf{L}[\vec{E}]$ in which a large enough cardinal is collapsed to $\aleph_0$ is of the form $\mathbf{L}[x]$, for a real x. Such an extension satisfies the projective statement "there exist a real x and a $\Delta^1_2(x)$ well-ordering of $\mathbb{R}$".

However, the existence of a proper class of Woodin cardinals gives such an absoluteness, and more (*cf.* [Lar04]).

By Remark 5.5 the existence of $\mathbf{M}_n(a)$ is a strictly weaker assumption than its iterability in all forcing extensions, needed in Theorem 5.2. We write $\mathbf{M}_n$ for $\mathbf{M}_n(0)$. Recall that a set of ordinals X **has a sharp** if there is a nontrivial elementary embedding of $\mathbf{L}[X]$ into itself.

The following result is also due to Woodin.

Theorem 5.6. The following are equivalent.

(1) $\mathbf{M}_1$ exists and it is fully iterable in all forcing extensions.
(2) Every set has a sharp and there is a proper class model with a Woodin cardinal.

Proof. (2) implies (1) is a difficult result using Mitchell-Steel constructions and we only prove the implication from (1) to (2). We only need to show that the full iterability of $\mathbf{M}_1$ implies that every set has a sharp. Assume the contrary, and let X be a set without a sharp. Then the Covering Lemma holds in $\mathbf{L}[X]$, hence the successor λ^+ of some singular cardinal λ such that $X \in \mathbf{V}_\lambda$ is correctly computed in $\mathbf{L}[X]$ [Kan94]. By Theorem 4.5 there is an iteration $j \colon \mathbf{M}_1 \to M^*$ such that X is generic over M^* and $j(\delta) < \lambda^+$. Then $M^*[X]$ correctly computes λ^+, since it includes $\mathbf{L}[X]$. On the other hand, by the chain condition of the extender algebra $j(\delta)$ remains a cardinal in $M^*[X]$. A contradiction. ⊣

I learned the above proof that (1) implies (2) from Ralf Schindler. This proof also shows that if $\mathbf{M}_{n+1}$ is fully iterable then for every set X there is an inner model including $\mathbf{L}[X]$ with n Woodin cardinals that has a sharp.

5.2. Absoluteness for $\mathbf{H}(\aleph_2)$. Theorem 5.9 below was proved independently by Steel and Woodin and a proof of its strengthening due to Neeman can be found in [Nee04]. Theorem 5.9 implies that the existence of a model with a measurable Woodin cardinal that is fully iterable in all forcing extensions would provide another proof of Woodin's Σ^2_1-absoluteness theorem ([Woo85]; *cf.* also [Doe10, Lar04, Far07]). The proof of this theorem will use the following standard forcing fact. The assumption that $\mathbf{V}_\delta \cap M$ is countable is used only to assure the existence of generic objects.

Lemma 5.7. Assume M is a transitive model of a large enough fragment of ZFC such that $M \cap \mathbf{V}_\delta$ is countable. Assume $\mathbb{P}$ and $\mathbb{Q}$ belong to $M \cap \mathbf{V}_\delta$ and $\mathbb{P}$ is a regular subalgebra of $\mathbb{Q}$. If $G \subseteq \mathbb{P}$ is M-generic, then there is an M-generic $H \subseteq \mathbb{Q}$ such that $G \in M[H]$. ⊣

Lemma 5.8. Assume $\mathbb{P}$ is a forcing notion in M with δ-c.c. of cardinality δ. If $j\colon M \to M^*$ is an elementary embedding with M^* a definable class in M and $\operatorname{crit}(j) = \delta$, then $\mathbb{P}$ is a regular subordering of $j(\mathbb{P})$ in M.

Proof. We may assume $\mathbb{P} \subseteq \mathbf{V}_\delta$. Let $\mathcal{A}$ be a maximal antichain in $\mathbb{P}$. By the δ-c.c. we have $\mathcal{A} \in \mathbf{V}_\delta$, and therefore $j(\mathcal{A}) = \mathcal{A}$. By the elementarity, $\mathcal{A}$ is a maximal antichain in $j(\mathbb{P})$ in M^*. Since being a maximal antichain is absolute, $\mathcal{A}$ is a maximal antichain of $j(\mathbb{P})$ in M. ⊣

Theorem 5.9. Assume there exists a model $\mathcal{M}_{\mathrm{mw}}$ with a countable ordinal δ that is a measurable Woodin cardinal in $\mathcal{M}_{\mathrm{mw}}$ which is fully iterable in all forcing extensions. Then to every Σ^2_1 statement φ we can associate a statement φ^* such that if $\mathbf{V} \models \varphi$ then $\mathcal{M}_{\mathrm{mw}} \models \varphi^*$ and if $\mathcal{M}_{\mathrm{mw}} \models \varphi^*$ and CH holds then $\mathbf{V} \models \varphi$.

Proof. The sentence φ is of the form $(\exists X \subseteq \mathbb{R})\psi(X)$ where ψ is a statement of $(\mathbf{H}(\aleph_1), X, \in)$. To it we associate φ^* stating that some condition in $\mathcal{W}_{\delta,\delta}$ forces φ and $|\check{\delta}| = \aleph_1$.

In order to prove that φ implies $\mathcal{M}_{\mathrm{mw}} \models \varphi^*$, assume $X \subseteq \mathbb{R}$ is such that $\psi(X)$ holds. Go to a forcing extension of V with the same reals that satisfies CH. Fix $Y \subseteq \omega_1$ that codes X and all reals. Using Theorem 4.7 find an iteration $j\colon \mathcal{M}_{\mathrm{mw}} \to \mathcal{M}$ of length ω_1 such that Y is $j(\mathcal{W}_{\delta,\delta}(\vec{E}))$-generic over $\mathcal{M}$. This forcing has $j(\delta)$-chain condition and it collapses all cardinals below $j(\delta)$ to ω. Therefore $\mathbb{R}^{\mathcal{M}_{\mathrm{mw}}[Y]} = \mathbb{R}$, hence $\mathcal{M}_{\mathrm{mw}}[Y] \models \psi(X)$.

We now assume CH and $\mathcal{M}_{\mathrm{mw}} \models \varphi^*$ and prove φ. In $\mathcal{M}_{\mathrm{mw}}$ fix a condition p in $\mathcal{W}_{\delta,\delta}$ and a name $\dot{X}$ such that p forces $\psi(\dot{X})$. By using CH we can enumerate $\mathbb{R}$ as r_ξ, for $\xi < \omega_1$. In this proof we shall write M_0 instead of $\mathcal{M}_{\mathrm{mw}}$. We shall now describe an iteration strategy for player I. Along with the iteration, player I constructs generic filters. After player I's strategy is described, we shall run it against player II's winning strategy for the iteration game and argue that the run of the game produces a well-founded iteration $j: M_0 \to M^*$ and $G \subseteq j(\mathcal{W}_{\delta,\delta}(\vec{E}))$ generic over M^* such that $p = j(p)$ is in G and that $M^*[G]$ contains all reals. All extenders E used in player I's strategy defined below will satisfy $\text{crit}(E) \geq \delta$, therefore assuring $j(p) = p$.

Since $M_0 \cap \mathbf{V}_{\delta+1}$ is countable, in V we can find a $G_0 \subseteq \mathcal{W}_{\delta,\delta}(\vec{E})$ generic over M_0 and containing p. Now use the extender in M_0 with critical point δ to find $j_0: M_0 \to M_1$. By Lemma 5.8, G_0 can be extended to a generic filter for $j_0(\mathcal{W}_{\delta,\delta}(\vec{E}))$.

We want to find a generic $G_1 \subseteq j_0(\mathcal{W}_{\delta,\delta}(\vec{E}))$ extending G_0 and such that $r_0 \in M_1[G_1]$. Let δ_0 be the least Woodin cardinal in M_1 greater than δ whose Woodinness is witnessed by (an initial segment of) $j_0(\vec{E})$. Let $\vec{E}_1$ consist of generators in $j_0(\vec{E})$ witnessing Woodinness of δ_0 whose critical points exceed δ. Now player I attempts to find an iteration $i_0: M_1 \to M_1^*$ using the extenders in $\vec{E}_1$ such that r_0 is generic over $i_0(\mathcal{W}_{\delta_0\omega}(\vec{E}_1))$. (Recall that player II continues playing his winning strategy for the iteration game corresponding to $\vec{E}$ from $\mathcal{M}_{\text{mw}}$, and therefore by Theorem 4.1 after countably many stages we shall assure that r_0 is generic.) Assume player I has succeeded in finding i_0. Since the critical point never drops below δ, this is an iteration of M_1 resulting in some M_1^*. Since p forces that $(i_0 \circ j_0)(\mathcal{W}_{\delta,\delta}(\vec{E}))$ collapses 2^{δ_0} to $\aleph_0$, $i_0(\mathcal{W}_{\delta,\delta}\omega(\vec{E}_1))$ can be embedded as a regular subalgebra of the former below p. We can therefore use Lemma 5.7 to find a generic $G_1 \subseteq (i_0 \circ j_0)(\mathcal{W}_{\delta,\delta}(\vec{E}))$ including G_0 such that $r_0 \in (i_0 \circ j_0)(M_1)[G_1]$.

We proceed in this manner. At the αth stage we have an iteration $j_\alpha^0: M_0 \to M_\alpha$ and G_α is a generic filter for $j_\alpha^0(\mathcal{W}_{\delta,\delta}(\vec{E})))$. Player I uses the extender in M_α with critical point $j_\alpha^0(\delta)$ to find $j_\alpha^1: M_\alpha \to M_\alpha^1$. Let δ_α be the least Woodin cardinal in M_α^1 above $j_\alpha^0(\delta)$. Choose a system of extenders $\vec{E}_{\alpha+1} \subseteq (j_\alpha^1 \circ j_\alpha^0)(\vec{E})$ which witnesses δ_α is Woodin and such that critical points of all extenders in $\vec{E}_{\alpha+1}$ exceed $j_\alpha^0(\delta)$. Using the genericity iteration theorem, player I finds an iteration j_α^2 of M_α^1 such that r_α is generic over $j_\alpha^2(\vec{E}_{\alpha+1})$. Using Lemma 5.8 to absorb this forcing in quotient, Find an iteration $j_\alpha^3: M_\alpha^1 \to M_\alpha^*$ with critical point $j_\alpha^0(\delta)$ such that the filter G_α can be extended to a generic ultrafilter $G_{\alpha+1}$ included in $j_\alpha(\mathcal{W}_{\delta,\delta}(\vec{E}))$, with $j_\alpha = j_\alpha^3 \circ j_\alpha^1 \circ j_\alpha^0$.

Fix the least Woodin cardinal in M_α^* above $j_\alpha^0(\delta)$. Using the algebra with ω generators on this cardinal and Theorem 4.1 find an iteration of $i_\alpha: M_\alpha^* \to M_\alpha^{**}$ that makes r_α generic for an algebra that is a regular subalgebra of $(i_\alpha \circ j_\alpha)(\mathcal{W}_{\delta,\delta}(\vec{E}))$. Again player I plays only the extenders E with $\text{crit}(E) \geq j_\alpha^0(\delta)$ so the responses of player II result in an iteration $i_\alpha: M_\alpha^* \to M_\alpha^{**}$, as required. By Lemma 5.8, we can find a generic filter $G_{\alpha+1} \subseteq (i_\alpha \circ j_\alpha)(\mathcal{W}_{\delta,\delta}(\vec{E}))$ extending G_α such that r_α belongs to $M^{**}_\alpha[G_{\alpha+1}]$.

At a limit stage of the construction player II chooses a maximal branch of the iteration tree constructed so far. By the δ-c.c., every maximal antichain of the image of $\mathcal{W}_{\delta,\delta}(\vec{E})$ in this model belongs to some earlier model. Therefore the direct limit of the G_ξ corresponding to the models on the branch is generic.

This describes a game which produces an iteration $j\colon M_0 \to M^*$ such that $j(\delta) = \aleph_1$ and a $G \subseteq j(\mathcal{W}_{\delta,\delta}(\vec{E}))$ generic over M^* and containing $j(p) = p$. By the choice of p and elementarity $\psi(X)$ holds in $M^*[G]$. Since the model $M^*[G]$ also contains all reals, $\psi(X)$ holds in $\mathbf{V}$. ⊣

Corollary 5.10 (Steel & Woodin). Assume there exists a fully iterable model $\mathcal{M}_{\rm mw}$ with a countable ordinal δ that is a measurable Woodin cardinal in $\mathcal{M}_{\rm mw}$. Then every Σ^2_1 statement true in some forcing extension of $\mathbf{V}$ is true in every forcing extension of $\mathbf{V}$ that satisfies CH. ⊣

5.3. A fairly complicated set of natural numbers. The following was pointed out by John Steel. Recall that if κ is a limit of Woodin cardinals then a set is κ-universally Baire if and only if it is $< \kappa$-homogeneously Suslin (*e.g.*, [Lar04, Theorem 3.3.13]). By Hom_∞ we denote the pointclass of all homogeneously Suslin sets. A sentence ψ is $\utilde{\Sigma}^2_1(\mathrm{Hom}_\infty)$ if there is $A \in \mathrm{Hom}_\infty$ and formula $\varphi(X, Y)$, where X and Y are second-order variables, such that ψ is of the following form:

$$(\exists X \in \mathrm{Hom}_\infty)(\mathbf{H}(\aleph_1), \in, A, X) \models \varphi.$$

By a result of Woodin [Ste09, Theorem 5.1], if there are class many Woodin cardinals then every $\utilde{\Sigma}^2_1(\mathrm{Hom}_\infty)$ sentence is absolute between forcing extensions. We shall need only the lightface version of this result.

Let $\mathcal{O}_\infty$ denote the set of all $\Sigma^2_1(\mathrm{Hom}_\infty)$ truths, with no real or Hom_∞ parameters. By identifying φ with its Gödel number $\ulcorner\varphi\urcorner$ we consider $\mathcal{O}_\infty$ as a set of natural numbers. By [Ste09, Theorem 5.1], if there exist class many Woodin cardinals $\mathcal{O}_\infty$ is invariant under forcing.

Corollary 5.11. Assume there exist class many Woodin cardinals. Let Γ be the set of all Σ^2_1 sentences φ that hold in forcing extensions that satisfy CH. If there exists a mouse with a measurable Woodin cardinal and Hom_∞ iteration strategy, then Γ is many-one reducible to $\mathcal{O}_\infty$.

Proof. By Theorem 5.9, truth of a sentence φ in Γ is equivalent to the existence of an iterable mouse with a Hom_∞ iteration strategy such that (φ holds in a forcing extension that satisfies CH$)^M$. ⊣

In the presence of class many Woodin cardinals, every generic absoluteness result proved using genericity iterations along the lines of Theorem 5.9 or Theorem 5.2 implies many-one reducibility of the relevant set Γ to $\mathcal{O}_\infty$, provided there are class many Woodin cardinals and the relevant mouse has a Hom_∞ iteration strategy. Proofs of Σ^2_1 absoluteness using stationary tower [Lar04] or Lévy collapse followed by forcing with a saturated ideal [Far07] do not seem to produce many-one reduction of the relevant set of sentences to $\mathcal{O}_\infty$.

Similarly, determinacy proofs produce many-one reductions of game-quantifier truths to $\mathcal{O}_\infty$. For example, Neeman's [Nee04] produces Hom_∞ strategies for open ω_1-games whose payoff set is $\utilde{\Pi}^1_1$ in the codes by transforming a Hom_∞ iteration strategy for the mouse. It therefore shows that $\Game_{\text{open-}\omega_1}\utilde{\Pi}^1_1$ truth is many-one reducible to $\mathcal{O}_\infty$.

Recall the definition of a proof in Woodin's Ω-logic (*cf.* [Woo10, § 10], [Woo01, BCL06, Lar11]). One first assumes there are class many Woodin cardinals, and therefore a set of reals is universally Baire if and only if it is homogeneously Suslin. If a set of reals A is universally Baire and M is a transitive model of a large enough fragment of ZFC then we say M is **A-closed** if for every $\mathbb{P} \in M$ and every $\mathbb{P}$-name τ for a real such that $\tau \in M$ the set $\{p \in \mathbb{P} : p \Vdash \tau \in A\}$ belongs to M. See [Woo10, Definition 10.141] or [BCL06, § 2.2] for more details. For production of A-closed models, *cf.* Lemma 6.2 and Lemma 7.2. An **Ω-proof** of φ is a universally Baire set A such that for every A-closed model M of a large enough fragment of ZFC we have $M \models \varphi$. If φ has an Ω-proof then we write $\vdash_\Omega \varphi$. Consider the set of all theorems of Ω-logic,

$$\mathcal{O}^\Omega = \{\varphi : \vdash_\Omega \varphi\}.$$

This $\Sigma^2_1(\mathrm{Hom}_\infty)$ set is easily seen to be many-one equivalent to $\mathcal{O}_\infty$.

Woodin's Ω-conjecture (*cf.* [Woo01, Lar11]) asserts that $\mathcal{O}^\Omega$ is the set of Gödel numbers of statements true in all forcing extensions. Therefore Ω-conjecture implies, and is essentially equivalent to, the assertion that all generic absoluteness comes via many-one reductions to $\mathcal{O}_\infty$.

§6. Divergent models of AD^+. In the present section we prove another previously unpublished theorem due to (surprise, surprise) Woodin. Recent sweeping results of Sargsyan added to the importance of this theorem by putting it in a chain of implications leading to dramatic lowering of the consistency strength of the axiom $\mathsf{AD}_\mathbb{R}$+"Θ is regular" (*cf.* [Sar13A] or [Woo10, § 1.6] for more details). I should like to thank Sargsyan for suggesting that I include this proof and for clarifying a number of details, in particular the ones sketched in § 6.2. I am indebted to Woodin for sketching the proof during a train ride from Oberwolfach in January 2011. Koellner's note [Koe] was quite helpful in reconstructing this proof.

Two inner models N_1 and N_2 of AD^+ are said to be **incompatible** if they have the same reals but $\wp(\mathbb{R})^{N_1} \not\subseteq \wp(\mathbb{R})^{N_2}$ and $\wp(\mathbb{R})^{N_2} \not\subseteq \wp(\mathbb{R})^{N_1}$. The terminology is justified by the fact that no inner model that includes both N_1 and N_2 can satisfy Wadge determinacy.

THEOREM 6.1. Assume there exists Woodin limit of Woodin cardinals. Then in some forcing extension of $\mathbf{V}$ there are incompatible AD^+ models.

The remainder of this section contains the proof of Theorem 6.1.

6.1. Universally Baire sets and absoluteness. A set of reals A is **κ-universally Baire** if there are trees S and T on $\omega \times \lambda$ for some cardinal λ such that S projects to A, T projects to $\mathbb{R}\setminus A$, and in every forcing extension by $\mathbb{P} \in \mathbf{V}_\kappa$ trees S and T project to complementary sets of reals. Set A is **universally Baire** if it is κ-universally Baire for all κ. In forcing extension we use these trees as a code for A and identify set A with the projection of T. We therefore define $A^{\mathbf{V}[G]}$ to be $\mathrm{p}[T]^{\mathbf{V}[G]}$ for a generic filter $G \subseteq \mathbb{P}$. In [Woo10] set $A^{\mathbf{V}[G]}$ is denoted A_G. Good sources for universally Baire sets are [Lar04] and [Woo10, §10].

Given a κ-universally Baire set A we shall need a sufficiently closed with respect to A countable transitive model of a large enough fragment of ZFC.

LEMMA 6.2. Assume A is κ-universally Baire and $\vartheta > |\mathbf{V}_\kappa|$. Expand the language of ZFC by adding a predicate for A. Then for club many $M \prec \mathbf{H}(\vartheta)$ the transitive collapse $\bar{M}$ of M satisfies the following.

(1) $\bar{M}$ is **A-correct**: $A^{\bar{M}} = \mathbb{R}^{\bar{M}} \cap A$, and
(2) $\bar{M}$ is **A-absolute**: If $\mathbb{P}$ is a forcing notion in $(\mathbf{V}_{\bar{\kappa}})^{\bar{M}}$ and $G \subseteq \mathbb{P}$ is $\bar{M}$-generic, then $A^{\bar{M}[G]} = \mathbb{R}^{\bar{M}[G]} \cap A$.

PROOF. Let M be an elementary submodel of a large enough $\mathbf{H}(\vartheta)$ containing trees S and T that witness κ-universal Baireness of A and let $\bar{M}$ be its transitive closure. We also denote images of S, T, and κ under the transitive collapse by $\bar{S}$, $\bar{T}$, and $\bar{\kappa}$. Let $A^{\bar{M}}$ denote the projection of $\bar{S}$ in $\bar{M}$. In $\bar{M}$ interpret A to be the projection of $\bar{T}$. Since the reals are not moved by the collapsing map, the elementarily of M implies (1) and (2). ⊣

The model $\bar{M}$ in the conclusion of Lemma 6.2 is **A-closed** (*cf.* § 5.3).

If A is a set of reals and there is a nontrivial elementary embedding of $\mathbf{L}(A,\mathbb{R})$ into itself then $(A,\mathbb{R})^\#$ is the theory of the first ω many indiscernibles for $\mathbf{L}(A,\mathbb{R})$ with parameters from $\mathbb{R}$ and predicate for A. It is naturally identified with a set of reals.

Theorem 6.3 is due to Woodin. Its variant for projective sets was proved in [Woo82]. A very similar result also appears in [Woo83, Theorem 1.13] (also compare [Woo83, Theorem 1.14] with Corollary 6.4 below) and a variant of part (2) for $\mathbf{L}(\mathbb{R})$ and proper forcing was proved in [NZ01].

Let us remind the reader that, while large cardinals imply that for a pointclass Γ all sets of reals in $\mathbf{L}(\Gamma,\mathbb{R})$ are universally Baire, these sets are not universally Baire in $\mathbf{L}(\Gamma,\mathbb{R})$. A forcing notion $\mathbb{P}$ is **weakly proper** if every countable set of ordinals in the extension is included in a ground-model countable set. In particular every proper forcing is weakly proper.

THEOREM 6.3. Suppose A is a set of reals such that all sets in $\mathbf{L}(A,\mathbb{R})$ are κ-universally Baire and $(A,\mathbb{R})^\#$ exists.

(1) Then for every forcing notion $\mathbb{P} \in \mathbf{V}_\kappa$ and V-generic filter $G \subseteq \mathbb{P}$ there is a generic elementary embedding $j_G \colon \mathbf{L}(A,\mathbb{R})^{\mathbf{V}} \to \mathbf{L}(A,\mathbb{R})^{\mathbf{V}[G]}$.

(2) If in addition AD and DC hold in $\mathbf{L}(A,\mathbb{R})$ and $\mathbb{P}$ is weakly proper then the canonical j_G fixes all ordinals.

PROOF. (1) This is essentially [Woo10, Theorem 2.30], where analogous statement was proved under the assumption that all sets in $\mathbf{L}(A,\mathbb{R})$ were weakly homogeneously Suslin. In this case the assumption that $(A,\mathbb{R})^\#$ exists is automatic. The present theorem is proved using appropriately modified versions of [Woo10, Lemmas 2.27–2.29], but we shall sketch a proof for the convenience of a reader. If $B \subseteq \mathbb{R}$ is in $\mathbf{L}(A,\mathbb{R})$ then we have trees T and S such that $\mathrm{p}[T] = B$ and $\mathrm{p}[S] = \mathbb{R}\setminus B$ and $\mathrm{p}[T]$ and $\mathrm{p}[S]$ cover $\mathbb{R}$ in forcing extension by $\mathbb{P}$. We therefore let $j_G(B) = \mathrm{p}[T]^{\mathbf{V}[G]}$. Note that $(A,\mathbb{R})^\#$ is a countable union of sets in $\mathbf{L}(A,\mathbb{R})$ (the nth set codes the theory of the first n indiscernibles), and is therefore universally Baire. If U and V are trees witnessing universal Baireness of $(A,\mathbb{R})^\#$ then an application of Lemma 6.2 easily shows that $\mathrm{p}[U]^{\mathbf{V}[G]}$ still codes the first order diagram of

$$(\mathbf{H}(\aleph_1)^{\mathbf{V}[G]}, \in, j_G(C))$$

where C ranges over universally Baire sets in $\mathbf{L}(A,\mathbb{R})$. Therefore map $j_G \colon (\mathbf{H}(\aleph_1), \in, A)^{\mathbf{V}} \to (\mathbf{H}(\aleph_1), \in, A)^{\mathbf{V}[G]}$ is elementary and so is its canonical extension $\bar{j}_G \colon \mathbf{L}(A,\mathbb{R})^{\mathbf{V}} \to \mathbf{L}(A,\mathbb{R})^{\mathbf{V}[G]}$.

(2) Let j_G be as defined in the proof of (1). The fact that j_G does not move ordinals is essentially [Woo10, Theorem 10.63]. Proof that j_G does not move ordinals relies on a result of Steel [Woo10, Theorem 3.40]. ⊣

An another way to prove variants of Theorem 6.3 (1) in case when the generic filter of $\mathbb{P}$ is determined by a single real x_G is to construe $\mathbf{L}(A,\mathbb{R})^{\mathbf{V}[G]}$ as a generic ultrapower of $\mathbf{L}(A,\mathbb{R})^{\mathbf{V}}$ associated with

$$\mathcal{U}(x_G) = \{B \in (\wp(\mathbb{R}) \cap \mathbf{L}(A,\mathbb{R}))^{\mathbf{V}} : x_G \in B\}.$$

By using universal Baireness of sets in $\mathbf{L}(A,\mathbb{R})$, $\mathcal{U}(x_G)$ is a generic ultrafilter in $(\wp(\mathbb{R}) \cap \mathbf{L}(A,\mathbb{R}))^{\mathbf{V}}$. In general case one can construe $\mathbf{L}(A,\mathbb{R})^{\mathbf{V}[G]}$ as a direct limit of ultrapowers of this sort (*cf.* [Woo10, p. 752]).

By a result of Woodin [Woo10] there exists a semiproper forcing $\mathbb{P}$ such that the generic embedding as in Theorem 6.3 necessarily moves ordinals.

COROLLARY 6.4. Assume A satisfies the assumption of Theorem 6.3 (2) for an uncountable cardinal κ. In addition assume $\mathbb{P}$ is a forcing notion of cardinality $\aleph_1$ such that every real added by $\mathbb{P}$ is added by a countable regular subordering. Then the identity map is an elementary embedding of $\mathbf{L}(\Gamma,\mathbb{R})^{\mathbf{V}}$ into an inner model N that contains all reals.

PROOF. If $\kappa > \aleph_1$ and $\mathbb{P}$ is weakly proper then this is an immediate consequence of Theorem 6.3, but it is exactly the case when $\kappa = \aleph_1$ that we shall need. By recursion we can extract a regular subordering $\mathbb{Q}$ of $\mathbb{P}$ that is an increasing ω_1-chain of countable regular suborderings and such that the quotient $\mathbb{P}/\mathbb{Q}$ does not add any new reals. We can write $\mathbb{Q}$ as a direct limit of regular suborderings $\mathbb{Q}_\xi$, for $\xi < \omega_1$, each of which is forcing-equivalent to $\mathbb{C}$.

By iterating Theorem 6.3 ω_1 times we can conclude that $\mathbf{L}(A,\mathbb{R})^{\mathbf{V}}$ is an elementary submodel of $N = \mathbf{L}(A,\mathbb{R})^{\mathbf{V}^{\mathbb{Q}}}$. After forcing by quotient $\mathbb{P}/\mathbb{Q}$ no reals are added and therefore $\mathbf{L}(A,\mathbb{R})^{\mathbf{V}^{\mathbb{P}}} = \mathbf{L}(A,\mathbb{R})^{\mathbf{V}^{\mathbb{Q}}}$. This model contains all reals of the extension and therefore satisfies the conclusion of the corollary. ⊣

In the following lemma we assume $\mathbb{P}$ is a forcing notion with the following properties:

(1) $|\mathbb{P}| = \delta$,
(2) $\mathbb{P}$ collapses δ to $\aleph_1$,
(3) for every $\mathbb{P}$-name $\dot{x}$ for a real there exists a regular subordering $\mathbb{Q}$ of $\mathbb{P}$ such that $\dot{x}$ is a $\mathbb{Q}$-name and $|\mathbb{Q}| < \delta$,
(4) $\mathbb{P} \times \mathbb{P}$ has δ-chain condition.

LEMMA 6.5. Assume $\mathbb{P}$ is as above and $\dot{B}$ is a $\mathbb{P}$-name for a set of reals such that $\mathbb{P}$ forces AD+DC hold in $\mathbf{L}(\dot{B},\mathbb{R})^{\mathbf{V}^{\mathbb{P}}}$, all sets in $\mathbf{L}(\dot{B},\mathbb{R})$ are δ-universally Baire, and $(\dot{B},\mathbb{R})^{\#}$ exists. Then $\mathbb{P} \times \mathbb{P}$ forces that there is an elementary embedding which fixes all reals and all ordinals from $\mathbf{L}(\dot{B},\mathbb{R})^{\mathbf{V}^{\mathbb{P}}}$ into an inner model of $\mathbf{V}^{\mathbb{P}\times\mathbb{P}}$ that contains all reals.

PROOF. Since every real added by $\mathbb{P}$ is added by a small forcing and since $\mathbb{P}$ collapses its own cardinality to $\aleph_1$, this is an immediate consequence of Corollary 6.4. ⊣

6.2. An iterable model with Woodin limit of Woodin cardinals. As the final preparation for the proof of Theorem 6.1 we outline some properties of Neeman's mouse. Assume $W = \mathbf{L}[\vec{E}]$ is an inner model with cardinal δ which is a Woodin limit of Woodin cardinals as constructed in [Nee02A]. Then W has the following properties.

(5) W is an extender model, $W = \mathbf{L}[\vec{E}]$, where $\vec{E}$ witnesses Woodinness of both δ and cofinally many cardinals below δ.
(6) For every δ-universally Baire set of reals A in W the model $\mathbf{L}(A,\mathbb{R})$ satisfies AD^+.
(7) The transitive collapse of every countable $X \prec (\mathbf{V}_{\delta+1})^W$ is fully iterable with a δ-universally Baire iteration strategy.

Clause (5) is a result of [MiS94]. Clause (6) is a result of Neeman [Nee02B]. Clause (7) can be proved by using methods from [Ste08A] where similar statements were proved.

We shall need conditions (10) and (11) given below, while (8) and (9) will not be used outside of the present subsection. Let us call a countable model of the form $\mathbf{L}[\vec{E}]$ that contains a Woodin limit of Woodins **premouse**. Every premouse M has the canonical relative constructibility well-ordering $<_M$ of the reals. We shall refer to an iterable premouse as a **mouse**.

Recall that Γ_{uB} denotes the class of all δ-universally Baire sets of reals. A sentence ψ is $\Sigma_1^2(\Gamma_{\mathrm{uB}})$ if there is $A \in \Gamma_{\mathrm{uB}}$ and formula $\varphi(X, Y)$, where X and Y are second-order variables, such that ψ is of the following form:

$$(\exists X \in \Gamma_{\mathrm{uB}})(H(\aleph_1), \in, A, X) \models \varphi.$$

Under our assumption that δ is a limit of Woodin cardinals $<\delta$-homogeneously Suslin is equivalent to δ-universally Baire and this pointclass coincides with $\Sigma_1^2(\mathrm{Hom}_{<\delta})$ considered in § 5.3.

(8) If M and N are $(\omega_1 + 1)$-iterable mice then either $\mathbb{R}^M \subseteq \mathbb{R}^N$ and $<_M$ is an initial segment of $<_N$ or $\mathbb{R}^N \subseteq \mathbb{R}^M$ and $<_N$ is an initial segment of $<_M$.

(9) The set of $(\omega_1 + 1)$-iterable mice is $\Sigma_1^2(\Gamma_{\mathrm{uB}})$.

Given M and N as in (8), a standard comparison argument (that can be reconstructed from the proof of Theorem 4.1[5]) shows that M and N have countable iterations $j : M \to M^*$ and $k : N \to N^*$ such that M^* is an initial segment of N^* or N^* is an initial segment of M^*. In either case, the well-orderings $<_{M^*}$ and $<_{N^*}$ coincide. Since the reals of a mouse and their well-ordering are unaffected by iteration, (8) follows.

Now we prove (9). A set Σ is an ω_1-iteration strategy if and only if for all countable iteration trees T obtained when player II obeys Σ all branches chosen by Σ result in well-founded models. Therefore the assertion that Σ is an ω_1-iteration strategy is projective in Σ. Therefore stating that a countable transitive model M is of the form $\mathbf{L}[\vec{E}]$ and is ω_1-iterable with a δ-universally Baire iteration strategy is $\Sigma_1^2(\Gamma_{\mathrm{uB}})$.

By Lemma 3.9, every such model is $\omega_1 + 1$-iterable, and this completes the proof of (9).

(10) There is a $\Sigma_1^2(\Gamma_{\mathrm{uB}})$ good well-ordering of the reals in W.

(11) For every real r there is a δ-universally Baire set A such that r is ordinal definable in $\mathbf{L}(A, \mathbb{R})$.

By (7) every real belongs to an $\omega_1 + 1$-iterable mouse with a δ-universally Baire strategy. Since canonical well-orderings of the reals in such mice cohere by (8), we can define well-ordering by letting $x < y$ if there exists an $\omega_1 + 1$-iterable mouse M with a δ-universally Baire iteration strategy such that $x <_M y$. By (9) this is a $\Sigma_1^2(\Gamma_{\mathrm{uB}})$ statement.

[5]Actually the proof of Theorem 4.1 was based on the proof of Comparison Lemma.

We finally prove (11). Let M be an ω_1+1-iterable mouse containing real x and let Σ be its δ-universally Baire strategy. Let α be such that x is the αth real in $<_M$. By (8), x is the αth real in every mouse N such that $x \in N$ and Σ is an ω_1+1-iteration strategy for N. We want to show that x is ordinal-definable from α in $\mathbf{L}(\Sigma\restriction\omega_1, \mathbb{R})$. Since $\Sigma\restriction\omega_1$ need not be δ-universally Baire in $\mathbf{L}(\Sigma\restriction\omega_1, \mathbb{R})$, Lemma 3.9 does not apply. Nevertheless, $\Sigma\restriction\omega_1$ can be extended to an ω_1+1-iteration strategy as follows. Since $\mathbf{L}(\Sigma\restriction\omega_1, \mathbb{R})$ is a model of AD, $\aleph_1$ is a measurable cardinal in it. In particular it has the tree property and after an iteration game of length ω_1 player II can choose an ω_1 branch of the iteration tree. As a direct limit of well-founded models of length ω_1, the model corresponding to this branch is well-founded.

6.3. Proof of Theorem 6.1. Assume that $\mathbf{V}$ is Neeman's model described in § 6.2 in which δ is a Woodin limit of Woodin cardinals and continue numbering of formulas started in § 6.2. Let M be the transitive collapse of an elementary submodel X of $\mathbf{H}(\lambda)$ for a large enough $\lambda > \kappa$. An initial segment of M is an initial segment of $X \cap \mathbf{V}_{\kappa+1}$ and it therefore satisfies (7). This iteration strategy clearly gives an iteration strategy for M. Pick a Cohen real c over M. Now fix a δ-universally Baire set A as in (11) so that c is ordinal definable in $\mathbf{L}(A, \mathbb{R})$. Since W satisfies CH, both A and $A^\#$ can be identified with subsets of ω_1.

Let $\mathbb{R}\restriction\alpha$ be the first α reals in the well-ordering of the reals as in (10) and write $B\restriction\alpha = B \cap (\mathbb{R}\restriction\alpha)$ for $B \subseteq \mathbb{R}$. By applying Theorem 4.6 to a $C \subseteq \omega_1$ that codes $A, A^\#$ and trees witnessing $\aleph_1$-universal Baireness of all sets in $\mathbf{L}(A, \mathbb{R})$, we can find an iteration $j\colon M \to M^*$ such that with $\alpha = j(\delta)$ the following hold:

(12) $\mathbf{L}(C\restriction\alpha, \mathbb{R}\restriction\alpha) \prec \mathbf{L}(C, \mathbb{R})$,
(13) the pair $(C\restriction\alpha, \mathbb{R}\restriction\alpha)$ is $j(\mathcal{W}_{\delta,\delta})$-generic over M^*,
(14) every real in the extension is added by a small forcing,
(15) the reals of the forcing extension from (13) are equal to $\mathbb{R}\restriction\alpha$.

We also assure that α is large enough to have $c \in \mathbb{R}\restriction\alpha$. Then for generic $G \subseteq j(\mathcal{W}_{\delta,\delta})$ as in (13) we have that in $M^*[G]$ real c is ordinal definable over a model of the form $\mathbf{L}(B, \mathbb{R})$ for some $\aleph_1$-universally Baire set B, and this model contains all reals of the extension by (15). This model is a model of AD^+ as witnessed by $A^\#$. Note that, since M and M^* have the same reals, c is Cohen over M^*.

By elementarity between M and M^* we can fix a condition p in $(\mathcal{W}_{\delta,\delta})^M$ which forces that there exist a real c, an $\aleph_1$-universally Baire set B, and a countable ordinal α such that

(16) c is the αth real in the well-ordering of ordinal definable reals in $\mathbf{L}(B, \mathbb{R})$,
(17) $\mathbf{L}(B, \mathbb{R}) \models \mathsf{AD}^+$,
(18) every new real belongs to a forcing extension by a forcing of cardinality $< \delta$,
(19) c does not belong to any dense G_δ set coded in M,

Let $G_1 \times G_2$ be a $\mathcal{W}_{\delta,\delta} \times \mathcal{W}_{\delta,\delta}$-generic filter over M below (p, p). We therefore have an $\aleph_1$-universally Baire set B_1 and real c_1 in $M[G_1]$ such that in $\mathbf{L}(\mathbb{R}, B_1)^{M[G_1]}$ conditions (16) to (19) hold. We also have a universally Baire set B_2 and real c_2 in $M[G_2]$ such that (16) to (19) hold with the same α.

By Hjorth's Lemma 3.4 product $\mathcal{W}_{\delta,\delta} \times \mathcal{W}_{\delta,\delta}$ has δ-c.c. and $\mathcal{W}_{\delta,\delta}$ satisfies the assumptions of Lemma 6.5. By applying Lemma 6.5 twice, we find elementary embeddings

$$j_1 \colon \mathbf{L}(\mathbb{R}, B_1)^{M[G_1]} \to \mathbf{L}(\mathbb{R}, B_1')^{M[G_1][G_2]}$$

and

$$j_2 \colon \mathbf{L}(\mathbb{R}, B_2)^{M[G_1]} \to \mathbf{L}(\mathbb{R}, B_2')^{M[G_2][G_1]}.$$

Both j_1 and j_2 fix all reals and all ordinals.

Now assume that in $M[G_1][G_2]$ there are no divergent models of AD^+. This applies to images of j_1 and j_2. Therefore one of these models includes the other, and in this model both c_1 and c_2 are the αth ordinal definable real in the well-ordering of ordinal-definable reals. We therefore must have that $c_1 = c_2$. This implies $c_1 \in M$, contradicting the fact that c_1 is a Cohen real over M.

§7. Other applications.

A positive answer to the following (modulo sufficient large cardinals) was conjectured by John Steel (*cf.* also [Woo02]).

QUESTION 7.1. Assume φ is a Σ^2_2 sentence such that $\mathsf{CH}+\varphi$ holds in some forcing extension. Is it true that φ holds in every forcing extension that satisfies $\diamondsuit$?

Several partial positive results were proved in [DS13] and in [FKLM08]. By a result of Woodin (*cf.* [KLZ10]), if there is a measurable Woodin cardinal then there is a forcing $\mathbb{P}$ that simultaneously forces every Σ^2_2 sentence φ that holds in some forcing extension satisfying CH. The forcing $\mathbb{P}$ is the iteration of the collapse of $2^{\aleph_0}$ to $\aleph_1$ and an another forcing notion. It is not known whether the collapse of $2^{\aleph_0}$ to $\aleph_1$ alone suffices for this conclusion.

The extender algebra has many other applications but it is time to finish (*cf.* [Mil95, p. 155]). For example, in [Lar11] a proof that the Ω-conjecture is true in many inner models (for example, Neeman's model briefly described in § 6.2) was presented. One of the key points in the proof is Lemma 7.2 below. If A is a universally Baire set of reals and M is a transitive model of a large enough fragment of ZFC, we say that M is ***A*-closed** if for every generic extension $M[G]$ of M we have that $A \cap M[G] \in M[G]$. This condition is closely related to the conclusion of Lemma 6.2, the only difference being that in the present situation the language of ZFC is not expanded by adding a predicate for A. Note that the assumption of the following lemma is a consequence of clause (7) in § 6.2.

LEMMA 7.2. Assume that δ is a Woodin cardinal and $\kappa > \delta$ is an inaccessible cardinal such that the transitive collapse of every elementary submodel $H \prec \mathbf{V}_\kappa$ is fully iterable with a universally Baire iteration strategy. Then for every universally Baire set of reals A there is a fully iterable A-closed transitive model. ⊣

The proof of Lemma 7.2 is identical to the proof of Lemma 6.2. Its conclusion, together with an application of Theorem 4.1, implies Ω-conjecture (*cf.* [Lar11, Theorem 9.2]).

REFERENCES

Joan Bagaria, Neus Castells, and Paul B. Larson
[BCL06] *An Ω-logic primer*, ***Set theory*** (Joan Bagaria and Stevo Todorčević, editors), Trends in Mathematics, Birkhäuser, Basel, 2006, pp. 1–28.

Philipp Doebler
[Doe10] *Stationary set preserving $\mathcal{L}$-forcings and their applications*, ***Ph.D. thesis***, Westfälische Wilhelms-Universität Münster, 2010.

Philipp Doebler and Ralf Schindler
[DS13] *The extender algebra and vagaries of Σ^2_1 absoluteness*, ***Münster Journal of Mathematics***, vol. 6 (2013), pp. 117–166.

Ilijas Farah
[Far07] *A proof of the Σ^2_1-absoluteness theorem*, ***Advances in logic. Papers from the North Texas Logic Conference held at the University of North Texas, Denton, TX, October 8–10, 2004*** (Yi Zhang Su Gao, Steve Jackson, editor), Contemporary Mathematics, vol. 425, American Mathematical Society, 2007, pp. 9–22.

Ilijas Farah, Richard O. Ketchersid, Paul B. Larson, and Menachem Magidor
[FKLM08] *Absoluteness for universally Baire sets and the uncountable II*, ***Computational Prospects of Infinity. Part II. Presented Talks*** (Chitat Chong, Qi Feng, Theodore A. Slaman, W. Hugh Woodin, and Yue Yang, editors), World Scientific, 2008, pp. 163–192.

Qi Feng, Menachem Magidor, and W. Hugh Woodin
[FMW92] *Universally Baire sets of reals*, in Judah et al. [JJW92], pp. 203–242.

Joel David Hamkins and W. Hugh Woodin
[HW00] *Small forcing creates neither strong nor Woodin cardinals*, ***Proceedings of the American Mathematical Society***, vol. 128 (2000), no. 10, pp. 3025–3029.

Gregory Hjorth
[Hjo97] *Some applications of coarse inner model theory*, ***The Journal of Symbolic Logic***, vol. 62 (1997), no. 2, pp. 337–365.

H. Judah, W. Just, and W. Hugh Woodin
[JJW92] ***Set Theory of the Continuum***, MSRI publications, vol. 26, Springer-Verlag, 1992.

Akihiro Kanamori
[Kan94] ***The Higher Infinite. Large Cardinals in Set Theory from Their Beginnings***, Perspectives in Mathematical Logic, Springer-Verlag, Berlin, 1994.

Akihiro Kanamori and Matthew Foreman
[KF10] ***Handbook of Set Theory***, Springer-Verlag, 2010.

ALEXANDER S. KECHRIS, DONALD A. MARTIN, AND YIANNIS N. MOSCHOVAKIS
[CABAL iii] *Cabal Seminar* 79–81, Lecture Notes in Mathematics, vol. 1019, Berlin, Springer-Verlag, 1983.

H. JEROME KEISLER
[Kei71] ***Model Theory for Infinitary Logic***, Studies in Logic and the Foundations of Mathematics, vol. 62, North-Holland, 1971.

RICHARD O. KETCHERSID, PAUL B. LARSON, AND JINDŘICH ZAPLETAL
[KLZ10] *Regular embeddings of the stationary tower and Woodin's Σ^2_2 maximality theorem*, ***The Journal of Symbolic Logic***, vol. 75 (2010), pp. 711–727.

RICHARD O. KETCHERSID AND STUART ZOBLE
[KZ06] *On the extender algebra being complete*, ***Mathematical Logic Quarterly***, vol. 52 (2006), no. 6, pp. 531–533.

PETER KOELLNER
[Koe] *Incompatible* AD^+ *models*, preprint.

KENNETH KUNEN
[Kun80] ***Set Theory: An Introduction to Independence Proofs***, North-Holland, 1980.

PAUL B. LARSON
[Lar04] ***The Stationary Tower: Notes on a Course by W. Hugh Woodin***, University Lecture Series, vol. 32, American Mathematical Society, Providence, RI, 2004.
[Lar11] *Three days of Ω-logic*, ***Annals of the Japan Association for Philosophy of Science***, vol. 19 (2011), pp. 57–86.

DONALD A. MARTIN AND ROBERT M. SOLOVAY
[MS69] *A basis theorem for Σ^1_3 sets of reals*, ***Annals of Mathematics***, vol. 89 (1969), pp. 138–160.

DONALD A. MARTIN AND JOHN R. STEEL
[MaS94] *Iteration trees*, ***Journal of the American Mathematical Society***, vol. 7 (1994), no. 1, pp. 1–73.

ARNOLD W. MILLER
[Mil95] ***Descriptive Set Theory and Forcing: How to prove theorems about Borel sets the hard way***, Lecture Notes in Logic, vol. 4, Springer-Verlag, Berlin, 1995.

WILLIAM J. MITCHELL AND JOHN R. STEEL
[MiS94] ***Fine Structure and Iteration Trees***, Lecture Notes in Logic, vol. 3, Springer-Verlag, Berlin, 1994.

ITAY NEEMAN
[Nee02A] *Inner models in the region of a Woodin limit of Woodin cardinals*, ***Annals of Pure and Applied Logic***, vol. 116 (2002), no. 1-3, pp. 67–155.
[Nee02B] *Optimal proofs of determinacy II*, ***Journal of Mathematical Logic***, vol. 2 (2002), no. 2, pp. 227–258.
[Nee04] ***The Determinacy of Long Games***, de Gruyter Series in Logic and its Applications, vol. 7, Walter de Gruyter, Berlin, 2004.
[Nee07A] *Games of length ω_1*, ***Journal of Mathematical Logic***, vol. 7 (2007), no. 1, pp. 83–124.

ITAY NEEMAN AND JINDŘICH ZAPLETAL
[NZ00] *Proper forcing and* $\mathbf{L}(\mathbb{R})$, preprint, arXiv:0003027v1, 2000.
[NZ01] *Proper forcing and* $\mathbf{L}(\mathbb{R})$, ***The Journal of Symbolic Logic***, vol. 66 (2001), no. 2, pp. 801–810.

Grigor Sargsyan

[Sar13A] *Descriptive inner model theory*, ***The Bulletin of Symbolic Logic***, vol. 19 (2013), no. 1, pp. 1–55.

Ernest Schimmerling

[Sch01] *The ABC's of mice*, ***The Bulletin of Symbolic Logic***, vol. 7 (2001), no. 4, pp. 485–503.

[Sch] *Notes on Woodin's extender algebra*, preprint, undated.

John R. Steel

[Ste95B] *Projectively wellordered inner models*, ***Annals of Pure and Applied Logic***, vol. 74 (1995), no. 1, pp. 77–104.

[Ste08A] *Derived models associated to mice*, ***Computational Prospects of Infinity. Part II. Presented Talks*** (Chitat Chong, Qi Feng, Theodore A. Slaman, W. Hugh Woodin, and Yue Yang, editors), World Scientific, 2008, pp. 105–193.

[Ste09] *The derived model theorem*, ***Logic Colloquium '06. Proc. of the Annual European Conference on Logic of the Association for Symbolic Logic held at the Radboud University, Nijmegen, July 27–August 2, 2006*** (S. Barry Cooper, Herman Geuvers, Anand Pillay, and Jouko Väänänen, editors), Lecture Notes in Logic, vol. 19, Association for Symbolic Logic, 2009, pp. 280–327.

[Ste10] *An outline of inner model theory*, in Kanamori and Foreman [KF10], pp. 1595–1684.

W. Hugh Woodin

[Woo82] *On the consistency strength of projective uniformization*, ***Proceedings of the Herbrand Symposium. Logic Colloquium '81. Held in Marseille, July 16–24, 1981*** (Jacques Stern, editor), Studies in Logic and the Foundations of Mathematics, vol. 107, North-Holland, Amsterdam, 1982, pp. 365–384.

[Woo83] *Some consistency results in* ZFC *using* AD, in Kechris et al. [Cabal iii], pp. 172–198, reprinted in this volume.

[Woo85] Σ^2_1*-absoluteness*, handwritten note, May 1985.

[Woo01] *The* Ω *conjecture*, ***Aspects of Complexity. Minicourses in Algorithmics, Complexity and Computational Algebra. Proceedings of the Workshop on Computability, Complexity, and Computational Algebra held in Kaikoura, January 7–15, 2000*** (Rod Downey and Denis Hirschfeldt, editors), de Gruyter Series in Logic and its Applications, vol. 4, de Gruyter, Berlin, 2001, pp. 155–169.

[Woo02] *Beyond* Σ^2_1 *absoluteness*, ***Proceedings of the International Congress of Mathematicians, Vol. I. Plenary Lectures and Ceremonies. Held in Beijing, August 20–28, 2002*** (Beijing) (Tatsien Li, editor), Higher Education Press, 2002, pp. 515–524.

[Woo10] ***The Axiom of Determinacy, Forcing Axioms, and the Nonstationary Ideal***, revised ed., de Gruyter Series in Logic and its Applications, vol. 1, Walter de Gruyter, Berlin, 2010.

DEPARTMENT OF MATHEMATICS AND STATISTICS
YORK UNIVERSITY
N520 ROSS
4700 KEELE STREET
TORONTO, ONTARIO M3J 1P3
CANADA
E-mail: ifarah@mathstat.yorku.ca

PART VIII: OTHER TOPICS

ON VAUGHT'S CONJECTURE

JOHN R. STEEL

§0. **Introduction.** We have two purposes in the present paper. The first is to survey some of the descriptive set theory and model theory directly relevant to Vaught's conjecture. This is done in §1. The survey is a bit distorted by our second purpose, which is to prove a special case of the conjecture. That is done in §2, where we show that if φ is an $\mathcal{L}_{\omega_1\omega}$ sentence all of whose models are trees (*i.e.*, partially ordered sets so that the set of predecessors of any element is linearly ordered) then φ has either countably many or $2^{\aleph_0}$ countable models. (This extends earlier work of Marcus, Miller, and Rubin [Mar80, Mil77, Rub77, Rub74].)

It is the author's pleasure to acknowledge his debt to Arnold Miller, both for many conversations on this topic of a general nature, and for some of the key ideas involved in the proof in case 2 of Theorem 2.1.

§1. **Background.**

1.1. The conjecture. All languages in this paper will be countable. If $\mathcal{L}$ is a countable list of symbols, then $\mathcal{L}_{\omega_1\omega}$ is the set of formulae built up from symbols of $\mathcal{L}$ using countable conjunctions and disjunctions as well as the usual logical operations.

Vaught's conjecture is the statement: for any sentence φ in $\mathcal{L}_{\omega_1\omega}$, φ has either countably many or $2^{\aleph_0}$ countable models (up to isomorphism of course).

Note that countable structures can be coded by reals. (For us, a real is an element of ${}^{\omega}2$, the Cantor space.) The strong Vaught conjecture states: For any φ in $\mathcal{L}_{\omega_1\omega}$, either φ has countably many countable models, or there is a perfect $P \subseteq {}^{\omega}2$ so that any two distinct elements of P code non-isomorphic models of φ.

One reason to concentrate on the strong conjecture is to avoid trivialities involving the Continuum Hypothesis. Vaught's conjecture is provable in ZF+CH, and provably equivalent in ZF+$\neg$CH to the strong conjecture (*cf.* Theorem 1.1). The strong conjecture, restricted to φ in $\mathcal{L}_{\omega\omega}$, is provably equivalent in ZF to a Σ^1_2 sentence. So the strong conjecture is probably decidable in ZF.

Large Cardinals, Determinacy, and Other Topics: The Cabal Seminar, Volume IV
Edited by A. S. Kechris, B. Löwe, J. R. Steel
Lecture Notes in Logic, 49

Closely related to Vaught's conjecture is the ordinal spectrum conjecture, of Barwise. Recall that if $x \in {}^{\omega}2$, then ω_1^x is the least ordinal not recursive in x; equivalently, ω_1^x is the least $\mu > \omega$ so that $\mathbf{L}_\mu[x]$ is admissible. Let $\omega_1^{\varnothing} = \omega_1^{\mathrm{CK}}$. If $\mathfrak{A}$ is a countable structure, then

$$\omega_1^{\mathfrak{A}} = \min\{\omega_1^x : x \in {}^{\omega}2 \text{ and } x \text{ codes } \mathfrak{A}\}.$$

(In the notation of [Bar76], $\omega_1^{\mathfrak{A}} = \max(\omega_1^{\mathrm{CK}}, \mathrm{o}(\mathfrak{A}^+)))$. The ordinal spectrum conjecture states: Let φ be a sentence in $\mathcal{L}_{\omega_1\omega} \cap \mathbf{L}_{\omega_1^{\mathrm{CK}}}$. Then either $\{\omega_1^{\mathfrak{A}} : \mathfrak{A} \models \varphi\} = \{\omega_1^{\mathrm{CK}}\}$ or $\{\omega_1^{\mathfrak{A}} : \mathfrak{A} \models \varphi\} = \{\mu < \aleph_1 : \mu \text{ is admissible}\}$. We prove this conjecture for trees in §2.

1.2. Effective descriptive set theory. Some information about Vaught's conjecture can be obtained simply from the fact that the relation of coding isomorphic structures is a Σ_1^1 equivalence relation on ${}^{\omega}2$. The first result in this direction is due to Silver. We say that an equivalence relation E on ${}^{\omega}2$ has perfectly many classes if there is $P \subseteq {}^{\omega}2$ such that P is perfect and for all $x \neq y \in P$, we have $\neg xEy$.

THEOREM 1.1 (Silver; *cf.* [Har, Sil]). Let E be a $\boldsymbol{\Pi}_1^1$ equivalence relation on ${}^{\omega}2$. Then there are either $\leq \aleph_0$ or perfectly many E-equivalence classes.

COROLLARY 1.2 (Burgess; [Bur74]). Let E be a $\boldsymbol{\Sigma}_1^1$ equivalence relation on ${}^{\omega}2$. Then there are either $\leq \aleph_1$ or perfectly many E-equivalence classes.

PROOF. For notational simplicity let E be Σ_1^1 lightface. Thus xEy if and only if T_{xy} is well-founded, where T_{xy} is a tree on ω depending recursively on x and y. For $\mu < \aleph_1$, let

$$xE_\mu y \quad \text{if and only if} \quad \neg(|T_{xy}| < \mu).$$

Here $|T_{xy}|$ is the ordinal rank of T_{xy}, or $\aleph_1$ if T_{xy} is not well-founded. Note that each E_μ is Borel, $E_\mu \subseteq E_\nu$ if $\nu < \mu$, and $E = \bigcap_{\mu<\aleph_1} E_\mu$.

If μ is admissible, then E_μ is an equivalence relation. For example, to see that E_μ is transitive, suppose not and let $\nu < \mu$ be such that $S \neq \varnothing$, where

$$S = \{\langle x, y, z\rangle : xE_\mu y \text{ and } yE_\mu z \text{ and } \neg xE_\nu z\}.$$

Pick an $r \in {}^{\omega}2$ so that $\omega_1^r = \mu$. Then S is $\Sigma_1^1(r)$, so by Gandy's Basis Theorem we have $\langle x, y, z\rangle \in S$ so that $\omega_1^{\langle x,y,z\rangle} \leq \mu$. But notice that for all x, y, and ν, if $\nu = \omega_1^{x,y}$ and $xE_\nu y$, then xEy. Thus xEy and yEz and $\neg xE_\nu z$, a contradiction.

We can now show that Corollary 1.2 is true in any countable model M of a sufficiently large fragment of ZF, thereby proving the corollary. For let M be such. Let N be a $\mathbb{P}$-generic extension of M, where $\mathbb{P}$ is the Levy algebra for collapsing $\aleph_1^M$ to ω. If $M \models$"E has at least $\aleph_2$ classes", then $N \models$"$E_{\aleph_1^M}$ has at least $\aleph_1$ classes". (Since $\aleph_1^N = \aleph_2^M$, and inequivalent reals in M remain so in N.) By Theorem 1.1 in N, $N \models$"$E_{\aleph_1^M}$ has perfectly many classes". Thus

$N \models$"E has perfectly many classes". By Shoenfield's Absoluteness Theorem, $M \models$"E has perfectly many classes". ⊣

REMARK 1.3. In [Bur74], Burgess has given a purely combinatorial (forcing-free) proof of an improvement of Corollary 1.2. A similar improvement is proved model-theoretically in [HM77]. In [Bur78], Burgess has shown that the classes of a Σ_1^1 equivalence relation with $\aleph_1$ classes are the "levels" of some $\Delta_2^1(0^\#)$ prewellordering of ${}^\omega 2$ (provided $0^\#$ exists).

COROLLARY 1.4. Let E be a $\underset{\sim}{\Sigma}_1^1$ equivalence relation, and $S \subseteq {}^\omega 2$ be $\underset{\sim}{\Sigma}_1^1$. Then there are either $\leq \aleph_1$ or perfectly many classes of E represented by elements of S.

PROOF. Let $S = \operatorname{ran}(f)$, where $f : {}^\omega 2 \to {}^\omega 2$ is Borel. Let xE^*y if and only if $f(x)Ef(y)$. Apply Corollary 1.2 to E^*. ⊣

We say a class C of $\mathcal{L}$ structures is $\mathrm{PC}_{\mathcal{L}_{\omega_1\omega}}$ if and only if there is a (countable) language $\mathcal{L}' \supseteq \mathcal{L}$, and $\sigma \in \mathcal{L}'_{\omega_1\omega}$ and list $\vec{R}$ of relation symbols in $\mathcal{L}' - \mathcal{L}$ so that $C = \{\mathfrak{A} : \mathfrak{A} \models \exists \vec{R} \sigma\}$. (A class C is $\mathrm{EC}_{\mathcal{L}_{\omega_1\omega}}$ if and only if $C = \{\mathfrak{A} : \mathfrak{A} \models \varphi\}$ for some $\varphi \in \mathcal{L}_{\omega_1\omega}$.)

COROLLARY 1.5 (Morley; [Mor70]). Let C be $\mathrm{PC}_{\mathcal{L}_{\omega_1\omega}}$. Then C has either $\leq \aleph_1$ or perfectly many countable members, up to isomorphism.

1.3. Invariant descriptive set theory. For simplicity, let $\mathcal{L}$ have one binary relation symbol only. Consider the space $2^{\omega\times\omega}$ with the product topology induced by the discrete topology on $\omega \times \omega$. A set $C \subseteq 2^{\omega\times\omega}$ is invariant if and only if for all R, S, we have that if $R \in C$ and $(\omega, R) \cong (\omega, S)$, then $S \in C$.

THEOREM 1.6. Let $C \subseteq 2^{\omega\times\omega}$. Then

(a) C is $\underset{\sim}{\Sigma}_1^1$ and invariant if and only if there is a C^* which is $\mathrm{PC}_{\mathcal{L}_{\omega_1\omega}}$ and $C = \{R : (\omega, R) \in C^*\}$;

(b) C is $\underset{\sim}{\Delta}_1^1$ and invariant if and only if there is a C^* which is $\mathrm{EC}_{\mathcal{L}_{\omega_1\omega}}$ and $C = \{R : (\omega, R) \in C^*\}$.

PROOF. (a) is immediate. (b) follows directly from (a) and the Interpolation Theorem for $\mathcal{L}_{\omega_1\omega}$ [Kei71, p. 19]. ⊣

REMARK 1.7. A lightface version of Theorem 1.6 can be proved. Also, there is a "level by level" version of Theorem 1.6(b).

In view of Theorem 1.6, it is natural to ask which theorems of classical descriptive set theory hold in the more general setting of invariant descriptive set theory. Vaught's conjecture states that the classical theorem on the cardinality of Borel sets generalizes to an invariant version.

Invariant descriptive set theory has been investigated in [Vau74].

1.4. Counterexamples. Life is sometimes difficult.

EXAMPLE 1.8. Let xEy if and only if $\omega_1^x = \omega_1^y$. Then E is a Σ_1^1 equivalence relation, and each E-equivalence class is Borel. (By Theorem 1.13, isomorphism has the latter property as well.) However, E has $\aleph_1$ but not perfectly many equivalence classes. There appear to be no definability-theoretic grounds on which to distinguish between E and isomorphism.

EXAMPLE 1.9 (Friedman). There is a $\mathrm{PC}_{\mathcal{L}_{\omega_1\omega}}$ class with exactly $\aleph_1$ countable members. *E.g.*, let $C := \{\langle A, <\rangle : \text{there is an } \omega\text{-model } M \text{ of } \mathsf{KP} \text{ and } \langle A, <\rangle \cong \langle \mathrm{Ord}^M, <\rangle\}$. H. Friedman has shown in [Fri73] that the countable members of C are precisely the orderings α or $\alpha + \alpha \cdot \eta$, where $\alpha < \aleph_1$ is admissible. (Here η is the order type of the rationals.) So C has exactly $\aleph_1$ models, but C doesn't have perfectly many models (*cf.* Remark 1.16).

EXAMPLE 1.10 (Kunen). There is a $\mathrm{PC}_{\mathcal{L}_{\omega\omega}}$ class with exactly $\aleph_1$ countable models. In order to see this, let $C = \{\langle A, <\rangle : \langle A, <\rangle \text{ is a linear order and for all } a, b \in A \text{ there is an automorphism } F \text{ of } \langle A, <\rangle \text{ such that } F(a) = b\}$. Let $\mathbb{Z}$ be the ordered set of integers, and let

$$
\begin{aligned}
Z^0 &:= \{0\};\\
Z^{\alpha+1} &:= Z^\alpha \times \mathbb{Z}, \text{ ordered lexicographically;}\\
Z^\lambda &:= \text{direct limit of } \langle Z^\beta : \beta < \lambda\rangle, \text{ for } \lambda \text{ limit.}
\end{aligned}
$$

Then one can show that $\langle A, <\rangle \in C$ if and only if there is an α such that $\langle A, <\rangle \cong Z^\alpha$ or $\langle A, <\rangle = Z^\alpha \cdot \eta$. (As a first step, notice every order in C is either dense or discrete.) Thus C has $\aleph_1$ countable models. Again, C doesn't have perfectly many models (*cf.* Remark 1.16).

Examples 1.9 & 1.10 indicate a failure of the analogy between classical and invariant descriptive set theory. They also indicate an important restriction: a proposed proof of Vaught's conjecture must differentiate between $\mathrm{EC}_{\mathcal{L}_{\omega_1\omega}}$ and $\mathrm{PC}_{\mathcal{L}_{\omega_1\omega}}$ classes. The work of §2 may be of interest in this regard.

1.5. Some model theory for $\mathcal{L}_{\omega_1\omega}$. [Bar76, Kei71] are references for this material.

1.5.1. *Scott sentences.* For $\varphi \in \mathcal{L}_{\omega_1\omega}$ we define $\mathrm{qr}(\varphi)$, the quantifier rank of φ, by induction on φ:

(i) $\mathrm{qr}(\varphi) = 0$ for φ atomic;

(ii) $\mathrm{qr}(\neg\varphi) = \mathrm{qr}(\varphi)$;
$\mathrm{qr}(\bigwedge \Phi) = \mathrm{qr}(\bigvee \Phi) = \sup\{\mathrm{qr}(\varphi) : \varphi \in \Phi\}$;

(iii) $\mathrm{qr}(\forall v\varphi) = \mathrm{qr}(\exists v\varphi) = \mathrm{qr}(\varphi) + 1$

Let $\mathfrak{A} \equiv_\nu \mathfrak{B}$ if and only if $\mathfrak{A}$ and $\mathfrak{B}$ satisfy the same sentence of quantifier rank $\leq \nu$.

We use "$|\mathfrak{A}|$" to denote the universe of $\mathfrak{A}$.

For any $\mathfrak{A}$ there is a least ordinal ν so that for all n, all $\vec{a} \in |\mathfrak{A}|^n$ and all $\vec{b} \in |\mathfrak{A}|^n$, we have

$$(\mathfrak{A}, \vec{a}) \equiv_\nu (\mathfrak{A}, \vec{b}) \text{ implies } (\mathfrak{A}, \vec{a}) \equiv_{\nu+1} (\mathfrak{A}, \vec{b}).$$

We call ν the Scott rank of $\mathfrak{A}$, and write "$\nu = \mathrm{sr}(\mathfrak{A})$". Note that if M is admissible, then $\{\langle \mathfrak{A}, \nu\rangle \in M : \mathrm{sr}(\mathfrak{A}) = \nu\}$ is Δ over M. (However, $\mathfrak{A} \in M$ does not imply $\mathrm{sr}(\mathfrak{A}) \in M$.)

Let $\mathfrak{A}$ be given. For $\vec{a} \in |\mathfrak{A}|^n$ and $\vec{v} = \langle v_1 \cdots v_n\rangle$ we define a formula $\varphi^\alpha_{\vec{a}}(\vec{v})$ by induction on α:

$$\varphi^0_{\vec{a}}(\vec{v}) := \bigwedge \{\vartheta(\vec{v}) : \vartheta \text{ is atomic or negation of atomic and } \mathfrak{A} \models \vartheta[\vec{a}]\};$$

$$\varphi^{\alpha+1}_{\vec{a}} := \bigwedge_{b \in |\mathfrak{A}|} \exists v_{n+1} \varphi^\alpha_{\vec{a},b}(\vec{v}, v_{n+1}) \wedge \forall v_{n+1} \bigvee_{b \in |\mathfrak{A}|} \varphi^\alpha_{\vec{a},b}(\vec{v}, v_{n+1});$$

$$\varphi^\lambda_{\vec{a}}(\vec{v}) := \bigwedge_{\beta<\lambda} \varphi^\beta_{\vec{a}}(\vec{v}), \text{ for } \lambda \text{ limit.}$$

Let ν be least so that

$$\mathfrak{A} \models \varphi^\nu_\varnothing \wedge \bigwedge_n \bigwedge_{\vec{a} \in |\mathfrak{A}|^n} \forall \vec{v} (\varphi^\nu_{\vec{a}}(\vec{v}) \Rightarrow \varphi^{\nu+1}_{\vec{a}}(\vec{v})).$$

Then $\nu = \mathrm{sr}(\mathfrak{A})$, and the displayed sentence is called the canonical Scott sentence of $\mathfrak{A}$ and denoted by "$\mathrm{CSS}(\mathfrak{A})$".

THEOREM 1.11. Let $\mathfrak{A}$ and $\mathfrak{B}$ be countable. Then $\mathfrak{A} \cong \mathfrak{B}$ if and only if $\mathcal{L} \models \mathrm{CSS}(\mathfrak{A})$.

REMARK 1.12. Let C be a $\mathrm{PC}_{\mathcal{L}_{\omega_1\omega}}$ class with at least $\aleph_1$ non-isomorphic countable models of Scott rank $< \nu$, where $\nu < \aleph_1$. By Theorem 1.11 there are $\aleph_1$ $\equiv_{\nu+\omega}$-classes of countable elements of C. Since $\equiv_{\nu+\omega}$ is a Borel equivalence, Theorem 1.1 and the proof of Corollary 1.4 yield perfectly many models in C.

THEOREM 1.13 (Nadel; [Nad71]). For $\mathfrak{A}$ countable, $\mathrm{sr}(\mathfrak{A}) \leq \omega_1^{\mathfrak{A}}$.

PROOF. Notice that $\{\langle \vec{a}, \vec{b}\rangle : \exists \nu\ ((\mathfrak{A}, \vec{a}) \not\equiv_\nu (\mathfrak{A}, \vec{b}))\}$ is definable by a Σ_1 positive inductive definition over $\mathbf{L}_{\omega_1^x}[x]$, whenever $x \in {}^\omega 2$ codes $\mathfrak{A}$. Such a definition terminates in $\leq \omega_1^x$ steps. ⊣

THEOREM 1.14 (Sacks). Suppose C is $\mathrm{PC}_{\mathcal{L}_{\omega_1\omega}}$ and for all $\mathfrak{A} \in C$, we have $\mathrm{sr}(\mathfrak{A}) < \omega_1^{\mathfrak{A}}$. Then C has either $\leq \aleph_0$ or perfectly many countable models.

PROOF. View countable elements of C as reals. By hypothesis then for every $x \in {}^\omega 2$, if $x \in C$, then there is an e such that $\{e\}^x$ is a well-order and $|\{e\}^x| = \mathrm{sr}(x)$. This is of the form $\forall x \exists e P(x, e)$ where P is $\underset{\sim}{\Pi}^1_1$. Let $f : {}^\omega 2 \to \omega$ be $\underset{\sim}{\Delta}^1_1$ so that for all x, we have $P(x, f(x))$. Then $\{\{f(x)\}^x : x \in {}^\omega 2\}$ is a $\underset{\sim}{\Sigma}^1_1$ set of well-orders, hence there is a $\nu < \aleph_1$ such that for all x, we have

$|\{f(x)\}^x| < \nu$. But then $\mathrm{sr}(x) < \nu$, whenever $x \in C$. The remark after Theorem 1.11 completes the proof. ⊣

There are $\mathfrak{A}$ so that $\mathrm{sr}(\mathfrak{A}) = \omega_1^{\mathfrak{A}}$. *E.g.*, let $\mathfrak{A} = \langle \omega_1^{\mathrm{CK}} + \omega_1^{\mathrm{CK}} \cdot \eta, < \rangle$ Then $\mathrm{sr}(\mathfrak{A}) = \omega_1^{\mathfrak{A}} = \omega_1^{\mathrm{CK}}$.

There is a game theoretic characterization of $\equiv_\nu$, due to Benda and Karp. Namely, $\mathfrak{A} \equiv_\nu \mathfrak{B}$ if and only if player II has a winning strategy in the following game: On the ith move player I plays an ordinal $\nu_i < \nu_{i-1}$ (where $\nu_0 = \nu$) and an element of one of the structures; player II then plays an element of the other structure. So eventually player I plays $\nu_n = 0$. Let $a_1 \cdots a_n$ be the first n elements of $|\mathfrak{A}|$ played, and $b_1 \cdots b_n$ the corresponding elements of $|\mathfrak{B}|$ played. Then player II wins if and only if $\langle \mathfrak{A}, a_1 \cdots a_n \rangle \equiv_0 \langle \mathfrak{B}, b_1 \cdots b_n \rangle$.

REMARK 1.15. This characterization provides a trivial proof of the following. Suppose $\mathfrak{A}$ and $\mathfrak{B}$ are $\mathcal{L}$ structures, where for simplicity $\mathcal{L}$ has one binary relation symbol R. Let $\mathcal{P}$ and $\mathcal{Q}$ be partitions of $|\mathfrak{A}|$ and $|\mathfrak{B}|$, and $\pi \colon \mathcal{P} \leftrightarrow \mathcal{Q}$ so that for all $A \in \mathcal{P}$, we have that $\langle A, R^{\mathfrak{A}} \rangle \equiv_\nu \langle \pi(A), R^{\mathfrak{B}} \rangle$. Suppose that whenever $\langle A, R^{\mathfrak{A}}, a \rangle \equiv_0 \langle \pi(A), R^{\mathfrak{B}}, c \rangle$ and $\langle B, R^{\mathfrak{A}}, b \rangle \equiv_0 \langle \pi(B), R^{\mathfrak{B}}, d \rangle$ and $A \neq B$, then $R^{\mathfrak{A}}(a, b)$ if and only if $R^{\mathfrak{B}}(c, d)$. Then $\mathfrak{A} \equiv_\nu \mathfrak{B}$. This remark is used repeatedly in §2.

REMARK 1.16. Let C be $\mathrm{PC}_{\mathcal{L}_{\omega_1\omega}}$. Then C has perfectly many models if and only if there is $\nu < \aleph_1$ such that C has uncountably many models of Scott rank $< \nu$. The "if" part follows from Theorems 1.11 and 1.1. For the "only if" direction, let $P \subseteq {}^{\omega}2$ be a perfect set of codes for nonisomorphic models in C. Regard P itself as a real. For uncountably many $x \in P$ then $\omega_1^x \leq \omega_1^P$. For such x, $\mathrm{sr}(x) \leq \omega_1^P$ by Theorem 1.13.

1.5.2. *Prime models.* A fragment of $\mathcal{L}_{\omega_1\omega}$ is a set of formulae closed under the finitary logical operations, substitution of terms, and subformulae. For 1.17 through 1.21, fix a countable fragment F and a complete theory T in F.

DEFINITION 1.17. Let $\varphi(\vec{v}) \in F$. Then $\varphi(\vec{v})$ is **complete over** T if and only if $\forall \psi(\vec{v}) \in F$ ("$\forall \vec{v}(\varphi(\vec{v}) \to \psi(\vec{v}))$"$\in T$ or "$\forall \vec{v}(\varphi(\vec{v}) \to \neg\psi(\vec{v}))$"$\in T$).

DEFINITION 1.18. Let $\Sigma \subseteq F$. Then Σ is an **n-type of** T if and only if there is an $\mathfrak{A}$ such that $\mathfrak{A} \models T$ and there is $\vec{a} \in |\mathfrak{A}|^n$ such that $\Sigma = \{\vartheta(\vec{v}) : \mathfrak{A} \models \vartheta[\vec{a}]\}$). We say that $\vec{a}$ **realizes Σ in** $\mathfrak{A}$. A type Σ is **principal** over T if and only if there is a $\vartheta \in \Sigma$ such that ϑ is complete over T.

Let $\mathrm{Th}(\mathfrak{A}) = \{\varphi \in \mathcal{L}_{\omega_1\omega} : \mathfrak{A} \models \varphi\}$.

DEFINITION 1.19. The structure $\mathfrak{A}$ is **prime over** F if and only if every n-type realized in $\mathfrak{A}$ is principal over $\mathrm{Th}(\mathfrak{A}) \cap F$.

Let "$\mathfrak{A} \preccurlyeq_F \mathfrak{B}$" mean that there is an embedding of $\mathfrak{A}$ into $\mathfrak{B}$ which preserves all formulae of F. "$\mathfrak{A} \prec_F \mathfrak{B}$" means the inclusion map is such an embedding.

THEOREM 1.20. Suppose that for all n, there are $\leq \aleph_0$ n-types over T. Then T has a countable model $\mathfrak{A}$, prime over F, which is unique up to isomorphism. Moreover, for all $\mathfrak{B}$ if $\mathfrak{B} \models T$, then $\mathfrak{A} \preccurlyeq_F \mathfrak{B}$.

In §2, the hypothesis of Theorem 1.20 will usually be available, due to Theorem 1.21.

THEOREM 1.21. Let C be $\mathrm{PC}_{\mathcal{L}_{\omega_1\omega}}$ without perfectly many models. Then $\{\Sigma \subseteq F$: there are $\mathfrak{A} \in C$ and n such that Σ is an n-type realized in $\mathfrak{A}\}$ is countable.

PROOF. Otherwise there are at least $\aleph_1$ $(\nu+1)$-equivalence classes in C, where $\nu = \sup\{\mathrm{qr}(\varphi) : \varphi \in F\}$. This contradicts the remark after Theorem 1.11.

Our use for prime models in §2 stems from the observation of M. Nadel that if $\mathfrak{A}$ is prime over F, then $\mathrm{sr}(\mathfrak{A}) \leq \sup\{\mathrm{qr}(\varphi) : \varphi \in F\}$.

These concepts yield a slightly different proof of [HM77, Theorem 1]. Say that $\varphi \in \mathcal{L}_{\omega_1\omega}$ is a **counterexample** if φ has at least $\aleph_1$ but not perfectly many countable models. A counterexample φ is **minimal** if for all $\vartheta \in \mathcal{L}_{\omega_1\omega}$, either $\varphi \wedge \vartheta$ or $\varphi \wedge \neg\vartheta$ has $\leq \aleph_0$ countable models. ⊣

THEOREM 1.22. Let φ be a counterexample. Then there is a minimal counterexample ψ so that $\models \psi \to \varphi$.

PROOF. Let φ be given, and suppose no such ψ exists. It is easy to construct by induction on the length of s for $s \in 2^{<\omega}$, structures $\mathfrak{A}_s$ and fragments F_s so that $\varphi \in F_\varnothing$ and

(i) $F_s \subseteq F_t$ if $s \subseteq t$;
(ii) $\mathfrak{A}_{s\frown i}$ is prime over F_s, $i \in \{0,1\}$, and $\mathfrak{A}_{s\frown 0} \not\equiv_{F_s} \mathfrak{A}_{s\frown 1}$;
(iii) $\mathrm{Th}(\mathfrak{A}_{s\frown i}) \cap F_s$ has $\aleph_1$ models, and $\mathrm{Th}(\mathfrak{A}_{s\frown i}) \cap F_s = \mathrm{Th}(\mathfrak{A}_{s\frown i\frown j}) \cap F_s$, for $i, j \in \{0,1\}$;
(iv) $\mathfrak{A}_s \models \varphi$.

Then by (ii), (iii), and Theorem 1.20, $\mathfrak{A}_{s\frown i} \preccurlyeq \mathfrak{A}_{s\frown i\frown j}$ for $i, j \in \{0,1\}$. For $x \in {}^{\omega}2$ let $\mathfrak{A}_x$ be the direct limit of $\langle \mathfrak{A}_{x\restriction n} : n \in \omega\rangle$. Then $x \neq y$ implies $\mathfrak{A}_x \not\equiv \mathfrak{A}_y$, a contradiction. ⊣

§2. Trees.

2.1. Some Lemmas. We now descend to a much more concrete level. A structure $\mathfrak{A} = \langle A, <\rangle$ is a **tree** if and only if $<$ partially orders A so that for all $a \in A$, $\{b : b < a\}$ is linearly ordered by $<$. The main goal of §2 is to prove the following theorem.

THEOREM 2.1. Let $\varphi \in \mathcal{L}_{\omega_1\omega}$. If every model of φ is a tree, then φ has either $\leq \aleph_0$ or perfectly many countable models.

Throughout the proof of Theorem 2.1 we shall be computing the Scott ranks of various structures. In general the computation is done simply by writing down sentences which characterize the structure and observing their quantifier rank. The next lemma contains a typical example of such a computation.

If $\mathfrak{A}$ is a tree, then for $a \in |\mathfrak{A}|$ let

$$C_a = \{b \in |\mathfrak{A}| : \exists c(c \leq^{\mathfrak{A}} b \wedge c \leq^{\mathfrak{A}} a)\}$$

and

$$\mathfrak{C}_a = \langle C_a, <^{\mathfrak{A}} \restriction C_a \times C_a \rangle.$$

The $\mathfrak{C}_a$'s partition $\mathfrak{A}$ into its maximal connected subtrees. The structure $\mathfrak{A}$ is connected if and only if there is an $a \in |\mathfrak{A}|$ such that $|\mathfrak{A}| = C_a$.

LEMMA 2.2. Suppose $\mathfrak{A}$ is a tree and $\mathrm{sr}(\mathfrak{A}) = \omega_1^{\mathfrak{A}}$. Then there is an $a \in |\mathfrak{A}|$ such that $\mathrm{sr}(\mathfrak{C}_a) = \omega_1^{\mathfrak{A}}$.

PROOF. Assume not. Let $x \in {}^{\omega}2$ code $\mathfrak{A}$, $\omega_1^x = \omega_1^{\mathfrak{A}}$. Notice that the map $a \mapsto \mathfrak{C}_a$ is Δ over $\mathbf{L}_{\omega_1^x}[X]$. So then is the map $a \mapsto \mathrm{sr}(\mathfrak{C}_a)$. By assumption this map is total, so the boundedness principle for admissible sets yields a $\nu < \omega_1^{\mathfrak{A}}$ so that for all $a \in |\mathfrak{A}|$; $\mathrm{sr}(\mathfrak{C}_a) < \nu$. We show $\mathrm{sr}(\mathfrak{A}) < \nu + \omega \cdot 2$, a contradiction. ⊣

The direct method for doing this is to produce for each $\vec{a} \in |\mathfrak{A}|^n$ a sentence ψ with $\mathrm{qr}(\psi) < \nu + \omega \cdot 2$ so that for all countable $\mathfrak{B}$ and all $\vec{b} \in |\mathfrak{B}|^n$, we have

$$(\mathfrak{B}, \vec{b}) \cong (\mathfrak{A}, \vec{a}) \text{ if and only if } (\mathfrak{B}, \vec{b}) \models \psi.$$

So let $\vec{a} \in |\mathfrak{A}|^n$ be given. For simplicity, let $n = 1$, $\vec{a} = \langle a \rangle$. Let $S = \{\mathrm{CSS}(\mathfrak{C}_b) : b \in |\mathfrak{A}|\}$, and for $\varphi \in S$ let K_φ be the number of distinct $\mathfrak{C}_b$'s so that $\mathfrak{C}_b \models \varphi$. Let ϑ_φ say that exactly K_φ maximal connected subtrees satisfy φ; *e.g.*, if $K_\varphi = \omega$

$$\vartheta_\varphi = \bigwedge_{i \leq K_\varphi} \exists v_1 \cdots v_i \Big(\bigwedge_{k,j \leq i} \neg \exists u(u \leq v_k \wedge u \leq v_j) \wedge \bigwedge_{j \leq i} \varphi^{\exists u(u \leq v_0 \wedge u \leq v_j)} \Big).$$

(Here $\varphi^{\vartheta(v_0)}$ is φ with all quantifiers relativized to $\vartheta(v_0)$.)

Let $\vartheta = \mathrm{CSS}(\langle \mathfrak{C}_a, a \rangle)$. The desired sentence ψ is given by:

$$\psi = \vartheta^{\exists u(u \leq v_0 \wedge u \leq \underline{a})} \wedge \bigwedge_{\varphi \in S} \vartheta_\varphi \wedge \forall x \Big(\bigvee_{\varphi \in S} \varphi^{\exists u(u \leq v_0 \wedge u \leq x)} \Big).$$

In the future we'll indicate proofs like that of Lemma 2.2 in an elliptical fashion. The next lemma will be our other main tool for evaluating Scott ranks. Its proof is more subtle than that of Lemma 2.2.

Let $\mathfrak{A}$ be a tree. For $a <^{\mathfrak{A}} b$, let

$$[a, b]^{\mathfrak{A}} = \{c : a \leq^{\mathfrak{A}} c \wedge \neg(b <^{\mathfrak{A}} c)\}.$$

Thus when $\mathfrak{A}$ is a linear order, $[a, b]^{\mathfrak{A}}$ is just a closed interval. We also use "$[a, b]^{\mathfrak{A}}$" to denote the structure $\langle [a, b]^{\mathfrak{A}}, <^{\mathfrak{A}}, a, b \rangle$.

LEMMA 2.3. Let $\mathfrak{A}$ be a connected tree, and L a maximal linearly ordered subset of $|\mathfrak{A}|$. Let λ be a limit ordinal so that for all $a, b \in L$, we have that $\mathrm{sr}([a,b]^{\mathfrak{A}}) < \lambda$ and there are fewer than perfectly many structures λ-equivalent to $\langle \mathfrak{A}, L\rangle$. Then $\mathrm{sr}(\langle \mathfrak{A}, L\rangle) \leq \lambda \cdot 3$.

PROOF. Let $\lambda, \mathfrak{A}$, and $L \subseteq |\mathfrak{A}|$ be as in the hypotheses. Suppose $\mathrm{sr}(\langle \mathfrak{A}, L\rangle) > \lambda \cdot 3$. Let F be the smallest fragment of $\mathcal{L}_{\omega_1\omega}$ so that whenever $\vec{a} \in |\mathfrak{A}|^n$ and $\alpha < \lambda \cdot 2$ then $\varphi^{\alpha}_{\vec{a}} \in F$. ($\mathcal{L}$ is the language of $\langle \mathfrak{A}, L\rangle$, and $\varphi^{\alpha}_{\vec{a}}$ is computed in the sense of $\langle \mathfrak{A}, L\rangle$.) By our hypotheses and Theorem 1.21, $\mathrm{Th}(\langle \mathfrak{A}, L\rangle) \cap F$ has a model $\langle \mathfrak{B}, M\rangle$ prime over F. Since $\sup\{\mathrm{qr}(\psi) : \psi \in F\} = \lambda \cdot 2$, $\mathrm{sr}(\langle \mathfrak{B}, M\rangle) \leq \lambda \cdot 2$. But by the choice of F, $\langle \mathfrak{B}, M\rangle \equiv_{\lambda\cdot 2} \langle \mathfrak{A}, L\rangle$. Thus $\mathrm{sr}(\langle \mathfrak{B}, M\rangle) = \lambda \cdot 2$. We obtain a contradiction by showing $\langle \mathfrak{A}, L\rangle \cong \langle \mathfrak{B}, M\rangle$.

It is enough to get an F-elementary embedding j of $\langle \mathfrak{B}, M\rangle$ into $\langle \mathfrak{A}, L\rangle$ so that j"M is cofinal in L in both directions. For suppose we had such a j. Let $i \mapsto b_i$ map I order-preservingly and cofinally into M, where I is one of the ordered sets ω, ω^*, or $\omega^* + \omega$. Say for definiteness that $I = \omega$. For $i \in \omega$ let

$$\pi_i \colon [b_i, b_{i+1}]^{\mathfrak{B}} \cong [j(b_i), j(b_{i+1})]^{\mathfrak{A}}.$$

Such a π_i exists because j is F-elementary and $\mathrm{sr}([j(b_i), j(b_{i+1})]^{\mathfrak{A}}) < \lambda$. Now for $b \in |\mathfrak{B}|$ let $\pi(b) = \pi_i(b)$ where $b \in [b_i, b_{i+1}]^{\mathfrak{B}}$. Then $\pi \colon \langle \mathfrak{B}, M\rangle \cong \langle \mathfrak{A}, L\rangle$, as desired. (That $\mathrm{dom}(\pi) = |\mathfrak{B}|$ and $\mathrm{ran}(\pi) = |\mathfrak{A}|$ follows from the maximality of L in $\mathfrak{A}$, the connectedness of $\mathfrak{A}$, and the corresponding properties of M and $\mathfrak{B}$.)

Now by Theorem 1.20, there is an F-elementary map j of $\langle \mathfrak{B}, M\rangle$ into $\langle \mathfrak{A}, L\rangle$. If we weave into the usual construction of j (*cf.* [Kei71, p. 61]) steps designed to make $\mathrm{ran}(j)$ as large as possible, we can arrange that for all $a \in (|\mathfrak{A}| - \mathrm{ran}(j))$ there is $\vec{b} \in \mathrm{ran}(j)^n$ such that $\langle a, \vec{b}\rangle$ satisfies in $\langle \mathfrak{A}, L\rangle$ no complete formula of $\mathrm{Th}(\langle \mathfrak{A}\rangle, L) \cap F$. We show now that if j has this property, then j"M is cofinal in L.

Let $R = \{a \in |\mathfrak{A}| : \exists b, c \in j\text{"}M\ (a \in [b,c]^{\mathfrak{A}})\}$. Let

$$\mathfrak{C} = \langle (|\mathfrak{A}| - R) \cup \mathrm{ran}(j), <^{\mathfrak{A}}, L\rangle.$$

Then $\mathfrak{C} \prec_F \langle \mathfrak{A}, L\rangle$. In fact, if $\vec{c} \in |\mathfrak{C}|^n$, then $\langle \mathfrak{A}, L, \vec{c}\rangle \cong \langle \mathfrak{C}, \vec{c}\rangle$ (the proof being like the proof that a cofinal embedding yields an isomorphism).

Let F' be the smallest fragment of $\mathcal{L}_{\omega_1\omega}$ containing each $\varphi^{\alpha}_{\vec{a}}$ for $\alpha < \lambda$ and $\vec{a} \in |\mathfrak{A}|^n$, and let $\langle \mathfrak{B}', M'\rangle \prec_{F'} \langle \mathfrak{B}, M\rangle$ with $\mathrm{sr}(\langle \mathfrak{B}', M'\rangle) = \lambda$. Define

$$\mathfrak{C}' = \langle (|\mathfrak{A}| - R) \cup \mathrm{ran}(j \restriction |\mathfrak{B}'|), L, <^{\mathfrak{A}}\rangle.$$

It is convenient to have an endpoint for L, so pick $a \in L \cap \mathrm{ran}(j \restriction |\mathfrak{B}'|)$. Then either $\mathrm{sr}(\langle \{b \in |\mathfrak{A}| : a \leq^{\mathfrak{A}} b\}, L, <^{\mathfrak{A}}\rangle) \geq \lambda \cdot 3$ or $\mathrm{sr}(\langle \{b \in |\mathfrak{A}| : a \not\leq^{\mathfrak{A}} b\}, L, <^{\mathfrak{A}}\rangle) \geq \lambda \cdot 3$. (Otherwise, by direct computation, as in Lemma 2.2, $\mathrm{sr}(\langle \mathfrak{A}, L\rangle) \leq \lambda \cdot 3$.) We assume the former alternative; the proof in the other case is the same. It follows that there is a $b \in L$ so that for all $c \in j$"M, we have that $c <^{\mathfrak{A}} b$. For otherwise j"M is cofinal upward in

L, and j would yield an isomorphism of $\langle\{b \in |\mathfrak{A}| : a \leq^{\mathfrak{A}} b\}, L, <^{\mathfrak{A}}\rangle$ with $\langle\{b \in |\mathfrak{B}| : j^{-1}(a) \leq^{\mathfrak{B}} b\}, M, <^{\mathfrak{B}}\rangle$. which is impossible since the latter structure has Scott rank $\leq \lambda \cdot 2$.

Now by Remark 1.15 we have $\langle\mathfrak{C}, a, b\rangle \equiv_\lambda \langle\mathfrak{C}', a, b\rangle$. Since $\mathfrak{C} \prec_F \langle\mathfrak{A}, L\rangle$, $\mathrm{sr}([a,b]^{\mathfrak{C}}) < \lambda$, whence we have an isomorphism π from $[a,b]^{\mathfrak{C}}$ onto $[a,b]^{\mathfrak{C}'}$. Clearly π preserves L and π cannot map $\mathrm{ran}(j)$ onto $\mathrm{ran}(j \restriction |\mathfrak{B}'|)$, for then we have an isomorphism of $\langle\{b \in |\mathfrak{B}| : j^{-1}(a) \leq b\}, <^{\mathfrak{B}}, M\rangle$ onto $\langle\{b \in |\mathfrak{B}'| : j^{-1}(a) \leq b\}, <^{\mathfrak{B}'}, M\rangle$, which is impossible since the first structure has Scott rank $\lambda \cdot 2$ and the second has Scott rank λ. There are now two cases.

Case 1. There is a $c \in \mathrm{ran}(j) \cap [a,b]^{\mathfrak{C}}$ such that $\pi(c) \notin \mathrm{ran}(j \restriction |\mathfrak{B}'|)$.

Using Remark 1.15 we have

$$\langle [a,b]^{\mathfrak{C}'}, \pi(c)\rangle \equiv_\lambda \langle [a,b]^{\mathfrak{C}}, \pi(c)\rangle.$$

Since the latter structure has Scott rank $< \lambda$ (recall $\mathfrak{C} \prec_F \langle\mathfrak{A}, L\rangle$), we have an isomorphism σ from the former to the latter. Clearly σ preserves L. Then $\sigma \circ \pi$ is an automorphism of $\langle [a,b]^{\mathfrak{C}}, <^{\mathfrak{C}}, L\rangle$ taking c to $\pi(c)$, where $c \in \mathrm{ran}(j)$ and $\pi(c) \notin \mathrm{ran}(j)$.

Since $\pi(c) \notin \mathrm{ran}(j)$, the construction of j yields $\vec{d} \in \mathrm{ran}(j)^n$ so that $\langle \pi(c), \vec{d}\rangle$ satisfies in $\langle\mathfrak{A}, L\rangle$, hence in $\mathfrak{C}$, no complete formula of $\mathrm{Th}(\langle\mathfrak{A}, L\rangle) \cap F$. But $\sigma \circ \pi$ extends trivially to an automorphism of $\mathfrak{C}$ taking some $\vec{e} \in \mathrm{ran}(j)^n$ to $\vec{d}$. Thus $\langle c, \vec{e}\rangle$ satisfies in $\mathfrak{C}$, hence in $\langle\mathfrak{A}, L\rangle$, no complete formula of $\mathrm{Th}(\langle\mathfrak{A}, L\rangle) \cap F$. Since $c \in \mathrm{ran}(j)$, this is a contradiction.

Case 2. There is a $c \in [a,b]^{\mathfrak{C}} - \mathrm{ran}(j)$ such that $\pi(c) \in \mathrm{ran}(j \restriction |\mathfrak{B}'|)$.

The proof in this case is similar to that in Case 1. The diagram below illustrates the proof in the case $\mathfrak{A}$ is a linear order (so that $L = |\mathfrak{A}|$ and $M = |\mathfrak{B}|$).

In the case that $\mathfrak{A}$ is a linear order, Lemma 2.3 has a simple formulation, which would become still simpler if the hypothesis on the number of models $\equiv_\lambda \mathfrak{A}$ could be dropped. We conjecture that this can be done. ⊣

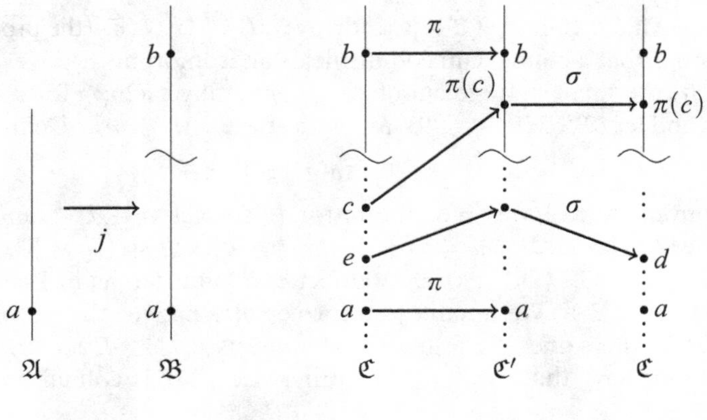

2.2.

Proof of Theorem 2.1. Let $\varphi \in \mathcal{L}_{\omega_1\omega}$ be a counterexample to Theorem 2.1. Then by Theorem 1.14 (essentially) we have a structure $\mathfrak{A} \models \varphi$ so that $\text{qr}(\varphi) < \omega_1^{\mathfrak{A}}$ and $\text{sr}(\mathfrak{A}) = \omega_1^{\mathfrak{A}}$. Let $\lambda = \text{qr}(\varphi) + \omega$. Our plan is to construct for each $x \subseteq \omega$ a structure $\mathfrak{A}_x \equiv_\lambda \mathfrak{A}$ so that $x \neq y$ implies $\mathfrak{A}_x \not\cong \mathfrak{A}_y$. The real x is coded onto the isomorphism type of $\mathfrak{A}_x$ by manipulating the Scott ranks of "pieces" of $\mathfrak{A}_x$.

We may assume $\mathfrak{A}$ is connected. For by Lemma 2.2, $\text{sr}(\mathfrak{C}_a) = \omega_1^{\mathfrak{C}_a}$ for some $a \in |\mathfrak{A}|$. But then the proof to follow would produce perfectly many structures $\equiv_\lambda \mathfrak{C}_a$, and thus by Remark 1.15 perfectly many structures $\equiv_\lambda \mathfrak{A}$.

For $a \in |\mathfrak{A}|$ let $\mathfrak{A}_a = \langle \{b : a \leq^{\mathfrak{A}} b\}, <^{\mathfrak{A}} \rangle$. Then for some $a \in |\mathfrak{A}|$, $\text{sr}(\mathfrak{A}_a) = \omega_1^{\mathfrak{A}}$. For pick any $a \in |\mathfrak{A}|$. If $\text{sr}(\mathfrak{A}_a) < \omega_1^{\mathfrak{A}}$, then consider $\mathfrak{B} = \langle \{c \in |\mathfrak{A}| : \neg(a <^{\mathfrak{A}} c)\}, <^{\mathfrak{A}} \rangle$. By direct computation, $\text{sr}(\mathfrak{B}) = \omega_1^{\mathfrak{A}}$. Let $L = \{c : c \leq^{\mathfrak{A}} a\}$, so that L is a maximal linearly ordered subset of $|\mathfrak{B}|$ definable in $\mathfrak{B}$ (from a). We may assume there are fewer than perfectly many models $\equiv_\lambda \mathfrak{B}$, as otherwise Remark 1.15 would bring our proof to an early end. Thus there are fewer that perfectly many models of $\equiv_\lambda \langle \mathfrak{B}, L \rangle$. If for all $b, c \in L$, we have $\text{sr}([b, c]^{\mathfrak{B}} < \omega_1^{\mathfrak{A}})$, then by boundedness and Lemma 2.3 $\text{sr}(\langle \mathfrak{B}, L \rangle) < \omega_1^{\mathfrak{A}}$, so $\text{sr}(\mathfrak{B}) < \omega_1^{\mathfrak{A}}$. So we have $\text{sr}([b, c]^{\mathfrak{B}}) = \omega_1^{\mathfrak{A}}$ for some $b, c \in L$. But then $\text{sr}(\mathfrak{A}_b) = \omega_1^{\mathfrak{A}}$, as desired.

The proof now splits into two cases. In Case 1, $\mathfrak{A}$ behaves like the graph of a unary function; in Case 2, $\mathfrak{A}$ behaves like a linear order.

Case 1. For all $a \in |\mathfrak{A}|$, if $\text{sr}(\mathfrak{A}_a) = \omega_1^{\mathfrak{A}}$, then there are $b, c \in |\mathfrak{A}|$ such that $\text{sr}(\mathfrak{A}_b) = \text{sr}(\mathfrak{A}_c) = \omega_1^{\mathfrak{A}}$, $a <^{\mathfrak{A}} b$, $a <^{\mathfrak{A}} c$, and b and c are incomparable.

Proof. Find sequences $\langle b_i : i \in \omega \rangle$ and $\langle a_i : i \in \omega \rangle$ so that for each i, $b_i < b_{i+1}$, $b_i < a_i$, a_i and b_{i+1} are incomparable, and $\text{sr}(\mathfrak{A}_{b_i}) = \text{sr}(\mathfrak{A}_{a_i}) = \omega_1^{\mathfrak{A}}$. If $c \in |\mathfrak{A}|$, let $|\mathfrak{B}_c| = \{d : c \leq d$ or there is $e < c$ such that for all i, $e \not\leq b_i$ and $d \in [e, c]\}$ and let $\mathfrak{B}_c = \langle |\mathfrak{B}_c|, <^{\mathfrak{A}} \rangle$.

Let $\langle \lambda_i : i \in \omega \rangle$ be a strictly increasing sequence of limit ordinals with limit $\omega_1^{\mathfrak{A}}$, and $\lambda < \lambda_0$. Let $x \subseteq \omega$ be infinite. For each i, choose $\mathfrak{C}_i \prec_{\lambda_k} \mathfrak{B}_{a_i}$ so that $\text{sr}(\mathfrak{C}_i) = \lambda_k$, where k is the ith element of x. (Such a $\mathfrak{C}_i$ can be found by Theorems 1.20 and 1.21; by Remark 1.15 we can assume $\mathfrak{B}_{a_i}$ has fewer than perfectly many λ-equivalent models.)

For $a \notin \bigcup_i |\mathfrak{B}_{a_i}|$, let $\mathfrak{C}_a \prec_\lambda \mathfrak{B}_a$ and $\text{sr}(\mathfrak{C}_a) = \lambda$. Choose the $\mathfrak{C}_a$'s so that $\mathfrak{B}_a = \mathfrak{B}_b$ implies $\mathfrak{C}_a = \mathfrak{C}_b$. Finally, let

$$\mathfrak{A}_x = \left\langle \{c : \exists i\ (c \leq^{\mathfrak{A}} b_i)\} \cup \bigcup_i |\mathfrak{C}_i| \cup \bigcup_a |\mathfrak{C}_a|, <^{\mathfrak{A}} \right\rangle.$$

By Remark 1.15, $\mathfrak{A}_x \equiv_\lambda \mathfrak{A}$. To see that x can be recovered from the isomorphism type of $\mathfrak{A}_x$, notice that there is some i such that $c \leq b_i$ if and only if $\text{sr}((\mathfrak{A}_x)_c) = \omega_1^{\mathfrak{A}}$. Thus $\{c : \exists i\ (c \leq b_i)\}$ can be recovered. But then

$\langle \gamma_k : k \in x \rangle$, and thus x, can be recovered. Since there is $\nu < \aleph_1$ such that for all x, $\mathrm{sr}(\mathfrak{A}_x) < \nu$, Theorem 1.1 yields perfectly many models of φ. ⊣ (Case 1)

Case 2. Otherwise.

Proof. We have an $a \in |\mathfrak{A}|$ so that $\mathrm{sr}(\mathfrak{A}_a) = \omega_1^{\mathfrak{A}}$ and $M = \{b \in |\mathfrak{A}_a| : \mathrm{sr}(\mathfrak{A}_b) = \omega_1^{\mathfrak{A}}\}$ is linearly ordered by $<^{\mathfrak{A}}$. We claim that there is some b such that $\mathrm{sr}([a,b]^{\mathfrak{A}}) = \omega_1^{\mathfrak{A}}$. For suppose first that for all $c \in M$, we have $c < b$. Then $\mathrm{sr}(\mathfrak{A}_b) < \omega_1^{\mathfrak{A}}$, and so by direct computation $\mathrm{sr}([a,b]^{\mathfrak{A}}) = \omega_1^{\mathfrak{A}}$. If no upper bound for M exists, then M is a maximal linearly ordered subset of $|\mathfrak{A}_a|$. A boundedness argument yields a $\nu < \omega_1^{\mathfrak{A}}$ so that $\mathrm{sr}(\mathfrak{A}_b) < \nu$ whenever $b \in |\mathfrak{A}_a| - M$. Thus M is definable in $\mathfrak{A}_a$ by a formula of $\mathrm{qr} < \omega_1^{\mathfrak{A}}$. We may assume there are fewer than perfectly many models $\equiv_\lambda \mathfrak{A}_a$. If for all $b \in M$, we have $\mathrm{sr}([a,b]^{\mathfrak{A}}) < \omega_1^{\mathfrak{A}}$, then boundedness and Lemma 2.3 imply $\mathrm{sr}(\mathfrak{A}_a) < \omega_1^{\mathfrak{A}}$, a contradiction. Thus we have a b so that $\mathrm{sr}([a,b]^{\mathfrak{A}}) = \omega_1^{\mathfrak{A}}$.

Our efforts will now be bent toward finding perfectly many models $\equiv_\lambda [a,b]^{\mathfrak{A}}$; by Remark 1.15 this will complete the proof.

Let $L = \{c : a \le c \le b\}$. For $c, d \in L$ let $c \sim d$ if and only if $\mathrm{sr}([c,d]^{\mathfrak{A}}) < \omega_1^{\mathfrak{A}}$. It is easy to check that $\sim$ is an equivalence relation partitioning L into segments. For $d \in L$, let $C_d = \{c : \exists e\, (e \sim d \wedge (c \in [e,d]^{\mathfrak{A}} \vee c \in [d,e]^{\mathfrak{A}}))\}$. We call C_d a component. Clearly $e \nsim d$ if and only if $C_e \cap C_d = \varnothing$.

We claim the components are densely ordered (where $C_d < C_e$ if and only if $d < e$ and $d \nsim e$). For suppose $d < e$ and C_d and C_e are immediately adjacent components. Then for all $c \in [d,e] \cap \mathbf{L}$, we have $\mathrm{sr}([c,e]^{\mathfrak{A}}) < \omega_1^{\mathfrak{A}}$ or $\mathrm{sr}([d,c]^{\mathfrak{A}}) < \omega_1^{\mathfrak{A}}$. By boundedness this is true with "ν" in place of "$\omega_1^{\mathfrak{A}}$" for some $\nu < \omega_1^{\mathfrak{A}}$. But then by Lemma 2.3 each of

$$\Big\langle \bigcup_{\substack{c\sim d\\ d<c}} [d,c]^{\mathfrak{A}}, <^{\mathfrak{A}}, L \Big\rangle$$

and

$$\Big\langle \bigcup_{\substack{c\sim e\\ c<e}} [c,e]^{\mathfrak{A}}, <^{\mathfrak{A}}, L \Big\rangle$$

has Scott rank $< \omega_1^{\mathfrak{A}}$. Let P be the set of points in $[d,e]$ appearing in neither of the above structures. Then by 2.2 essentially, $\mathrm{sr}(P, <^{\mathfrak{A}}) < \omega_1^{\mathfrak{A}}$. By direct computation it follows that $\mathrm{sr}([d,e]^{\mathfrak{A}} < \omega_1^{\mathfrak{A}})$, a contradiction.

Our plan is to juggle the Scott ranks of the components, roughly as we did in Case 1. There are two subcases according to the distribution of these Scott ranks in $[a,b]^{\mathfrak{A}}$.

Subcase 1. For all $\nu < \omega_1^{\mathfrak{A}}$, $\{C_d : \mathrm{sr}(\langle C_d, <^{\mathfrak{A}}, L \rangle) > \nu\})$ is a dense set of components.

PROOF. Recall λ is given; we are to find perfectly many models $\equiv_\lambda [a,b]^{\mathfrak{A}}$. Let $\langle C_i : i < \omega \rangle$ be a strictly increasing sequence of components so that $\mathrm{sr}(\langle C_i, <^{\mathfrak{A}}, L\rangle) \geq \lambda + \omega \cdot 2$. Let $x \in {}^{\omega}2$ be given. For each i, choose $C_i' \subseteq C_i$ so that $\langle C_i', <^{\mathfrak{A}}, L\rangle \prec_\lambda \langle C_i, <^{\mathfrak{A}}, L\rangle$ and $\mathrm{sr}(\langle C_i', <^{\mathfrak{A}}, L\rangle) = \lambda + \omega + \omega \cdot x(i)$. If C is a component so that $\mathrm{sr}(\langle C, <^{\mathfrak{A}}, L\rangle) \leq \lambda + \omega \cdot 2$ and $C \notin \{C_i : i < \omega\}$ let $C' \subseteq C$ be so that $\langle C', <^{\mathfrak{A}}, L\rangle \prec_\lambda \langle C, <^{\mathfrak{A}}, L\rangle$ and $\mathrm{sr}(\langle C', <^{\mathfrak{A}}, L\rangle) \leq \lambda$. Otherwise for C a component, let $C' = C$. (The existence of the C's follows from Remark 1.15, Theorems 1.20 and 1.21, and the assumption that there are fewer than perfectly many models $\equiv_\lambda [a,b]^{\mathfrak{A}}$.)

Let

$$A_x = \bigcup\{C' : C \text{ a component}\} \cup \{c \in [a,b]^{\mathfrak{A}} : c \text{ is a member of no components}\},$$

and let $\mathfrak{A}_x = \langle A_x, <^{\mathfrak{A}}\rangle$.

By Remark 1.15, $\mathfrak{A}_x \equiv_\lambda \mathfrak{A}$. To see that $x \neq y$ implies $\mathfrak{A}_x \not\cong \mathfrak{A}_y$, it suffices to show that for all $c, d \in L$, we have $\mathrm{sr}([c,d]^{\mathfrak{A}_x}) < \omega_1^{\mathfrak{A}}$ if and only if $\mathrm{sr}([c,d]^{\mathfrak{A}}) < \omega_1^{\mathfrak{A}}$. The "if" direction is clear. Now suppose $\mathrm{sr}([c,d]^{\mathfrak{A}_x}) = \nu < \omega_1^{\mathfrak{A}}$ and $\mathrm{sr}([c,d]^{\mathfrak{A}}) = \omega_1^{\mathfrak{A}}$. By the subcase hypothesis and Lemma 2.3 we can find $i, j \in [c,d]^{\mathfrak{A}} \cap L$ so that $\omega_1^{\mathfrak{A}} > \mathrm{sr}([i,j]^{\mathfrak{A}}) > \nu + \lambda + \omega \cdot 2$. But then by construction of $\mathfrak{A}_x$, $[i,j]^{\mathfrak{A}} = [i,j]^{\mathfrak{A}_x}$. Then $\mathrm{sr}([c,d]^{\mathfrak{A}_x}) < \mathrm{sr}([i,j]^{\mathfrak{A}_x})$, which is impossible.

To obtain perfectly many $\mathfrak{A}_x$'s, note that there is a fixed $z \in {}^{\omega}2$ so that for all x, $\mathfrak{A}_x \in \mathbf{L}_{\omega_1^{\langle x,z\rangle}}[x,z]$. But then Theorem 1.13 and the existence of perfectly many x so that $\omega_1^{\langle x,z\rangle} = \omega_1^z$ give the desired conclusion (*cf.* Remark 1.16). ⊣ (Subcase 1)

SUBCASE 2. Otherwise.

PROOF. Let $\nu < \omega_1^{\mathfrak{A}}$ be so that the components of Scott rank $\geq \nu$ are not dense. By moving a and b we can assume there are no such components.

Pick an increasing sequence $\langle c_i : i < \omega\rangle$ so that for all i, both $a < c_i < b$ and $c_i \nsim c_{i+1}$. Thus $\mathrm{sr}([c_i, c_{i+1}]^{\mathfrak{A}}) = \omega_1^{\mathfrak{A}}$. Let $x \in {}^{\omega}2$. Choose $\mathfrak{C}_i \prec_\lambda [c_{2i}, c_{2i+1}]^{\mathfrak{A}}$ so that $\mathrm{sr}(\mathfrak{C}_i) = \nu + \lambda + \omega + \omega \cdot x(i)$. Let $A_x = \bigcup_i |\mathfrak{C}_i| \cup ([a,b]^{\mathfrak{A}} - \bigcup_i [c_{2i}, c_{2i+1}]^{\mathfrak{A}})$, and let $\mathfrak{A}_x = \langle A_x, <^{\mathfrak{A}}\rangle$.

Again, $\mathfrak{A}_x \equiv_\lambda [a,b]^{\mathfrak{A}}$ by Remark 1.15. To show that there are perfectly many $\mathfrak{A}_x$'s, one proceeds along the lines of Subcase 1. ⊣ (Subcase 2)

⊣ (Case 2)

The proof of Theorem 2.1 is now complete. ⊣ (Theorem 2.1)

2.3. Remarks and corollaries. Two corollaries of Theorem 2.1 have been proven earlier and by different methods.

COROLLARY 2.4 (Miller & Marcus; [Mil77, Mar80] for $\mathcal{L}_{\omega\omega}$). Let φ be an $\mathcal{L}_{\omega_1\omega}$ sentence all of whose models are graphs of unary functions. Then φ has $\leq \aleph_0$ or perfectly many models.

Corollary 2.5 (Rubin; [Rub77, Rub74]). Let φ be an $\mathcal{L}_{\omega_1\omega}$ sentence all of whose models are linear orders. Then φ has $\leq \aleph_0$ or perfectly many models.

The proof of Theorem 2.1 is somewhat simpler than the earlier proofs of these corollaries. In particular, in the special case of trees anyway, there is no advantage in restricting one's attention to $\mathcal{L}_{\omega\omega}$. The machinery of Scott sentences and admissible sets leads naturally into $\mathcal{L}_{\omega_1\omega}$.

It would be natural to attempt to extend Theorem 2.1 from trees to arbitrary partial orders. However, Arnold Miller has shown Vaught's conjecture for partial orders is equivalent to the full conjecture, and his proof shows that partial orders are really no simpler in this context than arbitrary binary relations. What classes lie between trees and partial orders we leave to the reader's imagination.

There are effective refinements of Theorem 2.1, for which we need the following lemma.

Lemma 2.6. Let $\varphi \in \mathcal{L}_{\omega_1\omega} \cap \mathbf{L}_{\omega_1^{\mathrm{CK}}}$ and suppose for all $\mathfrak{A}$, if $\mathfrak{A} \models \varphi$, then $\mathrm{sr}(\mathfrak{A}) < \omega_1^{\mathrm{CK}}$. Then (a) if φ has $\leq \aleph_0$ countable models, then every countable model of φ is isomorphic to an element of $\mathbf{L}_{\omega_1^{\mathrm{CK}}}$; (b) if φ has $> \aleph_0$ countable models, then $\{\omega_1^{\mathfrak{A}} : \mathfrak{A} \models \varphi\} = \{\mu : \omega < \mu < \aleph_1 \text{ and } \mu \text{ is admissible}\}$.

Proof. By boundedness let $\nu < \omega_1^{\mathrm{CK}}$ be so that $\mathfrak{A} \models \varphi$ implies $\mathrm{sr}(\mathfrak{A}) < \nu$. Adopt a reasonable coding of formulae of $\mathcal{L}_{\omega_1\omega}$ with quantifier rank $< \nu$ by reals, so that each formula gets exactly one code. The set of codes will be a lightface Δ^1_1 set of reals. Let $D = \{x :$ there is an $\mathfrak{A}$ such that $\mathfrak{A} \models \varphi$ and x codes $\mathrm{CSS}(\mathfrak{A})\}$. By [Mak77, 2.14], D is Δ^1_1. (D is clearly Σ^1_1. But $D = \{x :$ there is an $\mathfrak{A} \in \mathbf{L}_{\omega_1^x}[x]$ such that $\mathfrak{A} \models \varphi$ and x codes $\mathrm{CSS}(\mathfrak{A})\}$, so D is Π^1_1.)

(a) If φ has $\leq \aleph_0$ models, D is countable, hence $D \subseteq \mathbf{L}_{\omega_1^{\mathrm{CK}}}$. Since $\mathfrak{A}$ has an isomorphic copy inside any admissible set which thinks $\mathrm{CSS}(\mathfrak{A})$ is countable, (a) is proved.

(b) If D is uncountable, then D has members of every hyperdegree above Kleene's ϑ (*cf.* [Mar76]). Since our coding can be taken so that any admissible set is closed under the map $\mathfrak{A} \mapsto$ code for $\mathrm{CSS}(\mathfrak{A})$ (for $\mathfrak{A} \models \varphi$!), this implies $\{\omega_1^{\mathfrak{A}} : \mathfrak{A} \models \varphi\}$ is cofinal in $\aleph_1$. By Gandy's Basis theorem relative to any x so that $\omega_1^x = \mu$, we have an $\mathfrak{A} \models \varphi$ such that $\omega_1^{\mathfrak{A}} = \mu$. ⊣

Corollary 2.7. Let $\varphi \in \mathcal{L}_{\omega_1\omega} \cap \mathbf{L}_{\omega_1^{\mathrm{CK}}}$. Suppose φ has $\leq \aleph_0$ countable models, all of which are trees. Then every model of φ is isomorphic to an element of $\mathbf{L}_{\omega_1^{\mathrm{CK}}}$.

Proof. If the hypothesis of Lemma 2.6 is not satisfied, then by Gandy's Basis theorem, there is an $\mathfrak{A}$ such that $\mathfrak{A} \models \varphi$ and $\mathrm{sr}(\mathfrak{A}) = \omega_1^{\mathfrak{A}} = \omega_1^{\mathrm{CK}}$. The proof of Theorem 2.1 then yields perfectly many models of φ. ⊣

COROLLARY 2.8. *Let $\varphi \in \mathcal{L}_{\omega_1\omega} \cap \mathbf{L}_{\omega_1^{\mathrm{CK}}}$, and suppose every model of φ is a tree. Then either $\{\omega_1^{\mathfrak{A}} : \mathfrak{A} \models \varphi\} \subseteq \{\omega_1^{\mathrm{CK}}\}$ or $\{\omega_1^{\mathfrak{A}} : \mathfrak{A} \models \varphi\} = \{\mu : \omega < \mu < \aleph_1$ and μ is admissible$\}$.*

PROOF. If the hypothesis of Lemma 2.6 is not satisfied, then there is an $\mathfrak{A}$ such that $\mathfrak{A} \models \varphi$ and $\mathrm{sr}(\mathfrak{A}) = \omega_1^{\mathfrak{A}} = \omega_1^{\mathrm{CK}}$. But then the proof of Theorem 2.1 implies $\{\omega_1^{\mathfrak{A}} : \mathfrak{A} \models \varphi\}$ is cofinal in $\aleph_1$. Thus $\{\omega_1^{\mathfrak{A}} : \mathfrak{A} \models \varphi\} = \{\mu : \omega < \mu < \aleph_1$ and μ is admissible$\}$. ⊣

§3. Determinateness. We wish to conclude my mentioning some consequences of the hypothesis that all $\underset{\sim}{\mathbf{\Sigma}}^1_1$ games are determined which shed some light on Theorem 1.14 and the counterexamples of §1.4. We omit proofs.

If we rephrase Theorem 1.14 in the language of invariant descriptive set theory, we obtain the following statement: Let $C \subseteq {}^{\omega}2$ be $\underset{\sim}{\mathbf{\Sigma}}^1_1$ and invariant, and suppose that for all $x \in C$ there is an $\alpha < \omega_1^x$ such that $\{y : y \cong x\}$ is $\underset{\sim}{\mathbf{\Sigma}}^0_\alpha$; then C has $\leq \aleph_0$ or perfectly many classes.

THEOREM 3.1. *Assume all $\underset{\sim}{\mathbf{\Sigma}}^1_1$ games are determined. Let $C \subseteq {}^{\omega}2$ be $\underset{\sim}{\mathbf{\Sigma}}^1_1$ and invariant, and suppose that for all $x \in C$, $\{y : y \cong x\}$ is $\underset{\sim}{\mathbf{\Sigma}}^0_{\omega_1^x+2}$; then C has $\leq \aleph_0$ or perfectly many classes.*

By Theorems 1.11 and 1.13, $\{y : y \cong x\}$ is always $\underset{\sim}{\mathbf{\Pi}}^0_{\omega_1^x+2}$, and so the examples of §1.4 show that Theorem 3.1 cannot be improved.

THEOREM 3.2. *Assume all $\underset{\sim}{\mathbf{\Sigma}}^1_1$ games are determined. Let C be a $\mathrm{PC}_{\mathcal{L}_{\omega_1\omega}}$ class with uncountably many but not perfectly many countable models. Then there is $\langle \mathfrak{A}_\nu : \nu < \aleph_1 \rangle$ so that*

(i) *for all ν, $\mathfrak{A}_\nu \in C$;*
(ii) *$\{\mathfrak{B} : \exists \nu(\mathfrak{B} \cong \mathfrak{A}_\nu)\}$ is $\mathrm{PC}_{\mathcal{L}_{\omega_1\omega}}$;*
(iii) *let $\alpha_\nu = \omega_1$; then $\nu \mapsto \alpha_\nu$ is strictly increasing and continuous;*
(iv) *$\mathrm{sr}(\mathfrak{A}_\nu) = \alpha_\nu$;*
(v) *$\mathfrak{A}_\nu \preccurlyeq_{\alpha_\nu} \mathfrak{A}_\mu$ if $\nu < \mu$;*
(vi) *(Saturation) suppose $\langle \mathfrak{A}_\mu, b_0 \cdots b_n \rangle \equiv_{\alpha_\nu} \langle \mathfrak{A}_\nu, j(b_0) \cdots j(b_n) \rangle$, where $\nu \leq \mu$. Then j can be extended to an α_ν-elementary embedding of $\mathfrak{A}_\mu$ into $\mathfrak{A}_\nu$.*

We shall describe only the game which enters into the proofs of Theorems 3.1 and 3.2. Let $\mathrm{WO} = \{x \in {}^{\omega}2 : x$ codes a well-order of $\omega\}$, and for $x \in \mathrm{WO}$ let $|x| =$ length of well-order coded by x. Let C be $\mathrm{PC}_{\mathcal{L}_{\omega_1\omega}}$ with at least $\aleph_1$ but not perfectly many models. Consider the following Solovay-style game: player I plays x, player II plays y. Player I loses unless $x \in \mathrm{WO}$. If $x \in \mathrm{WO}$, then player II loses unless y codes a model $\mathfrak{A} \in C$ so that $|x| \leq \mathrm{sr}(\mathfrak{A})$. By a standard argument, player I can have no winning strategy in this game, so by determinateness player II has a winning strategy.

The existence of such a strategy, together with forcing arguments like those of [Har78], implies Theorems 3.1 and 3.2.

REFERENCES

Jon Barwise

[Bar76] ***Admissible Sets and Structures. An approach to definability theory***, Perspectives in Mathematical Logic, Springer-Verlag, 1976.

John P. Burgess

[Bur74] *Infinitary Languages and Descriptive Set Theory*, ***Ph.D. thesis***, University of California at Berkeley, 1974.

[Bur78] *Equivalences generated by families of Borel sets*, ***Proceedings of the American Mathematical Society***, vol. 69 (1978), no. 2, pp. 323–326.

Harvey Friedman

[Fri73] *Countable models of set theories*, ***Cambridge Summer School in Mathematical Logic (held in Cambridge, England, August 1–21, 1971)*** (A. R. D. Mathias and H. Rogers, editors), Lecture Notes in Mathematics, vol. 337, Springer-Verlag, 1973, pp. 539–573.

Victor Harnik and Michael Makkai

[HM77] *A tree argument in infinitary model theory*, ***Proceedings of the American Mathematical Society***, vol. 67 (1977), no. 1, pp. 309–314.

Leo A. Harrington

[Har78] *Analytic determinacy and* $0^{\#}$, ***The Journal of Symbolic Logic***, vol. 43 (1978), no. 4, pp. 685–693.

[Har] *A powerless proof of a theorem of Silver*, unpublished notes, undated.

H. Jerome Keisler

[Kei71] ***Model Theory for Infinitary Logic***, Studies in Logic and the Foundations of Mathematics, vol. 62, North-Holland, 1971.

Michael Makkai

[Mak77] *An "admissible" generalization of a theorem on countable* Σ^1_1 *sets of reals with applications*, ***Annals of Mathematical Logic***, vol. 11 (1977), no. 1, pp. 1–30.

Leo Marcus

[Mar80] *The number of countable models of a theory of one unary function*, ***Fundamenta Mathematicae***, vol. 108 (1980), no. 3, pp. 171–181.

Donald A. Martin

[Mar76] *Proof of a conjecture of Friedman*, ***Proceedings of the American Mathematical Society***, vol. 55 (1976), no. 1, p. 129.

Arnold W. Miller

[Mil77] *Some Problems in Set Theory and Model Theory*, ***Ph.D. thesis***, University of California at Berkeley, 1977.

Michael Morley

[Mor70] *The number of countable models*, ***The Journal of Symbolic Logic***, vol. 35 (1970), pp. 14–18.

Mark E. Nadel

[Nad71] *Model Theory in Admissible Sets*, ***Ph.D. thesis***, University of Wisconsin, 1971.

Mati Rubin

[Rub74] *Thoeries of linear order*, ***Israel Journal of Mathematics***, vol. 17 (1974), pp. 392–443.

[Rub77] *Vaught's conjecture for linear orderings*, ***Notices of the American Mathematical Society***, vol. 24 (1977), p. A 390.

Jack H. Silver
[Sil] $\underset{\sim}{\Pi}_1^1$ *equivalence relations*, unpublished notes, undated.

Robert L. Vaught
[Vau74] *Invariant sets in topology and logic*, ***Fundamenta Mathematicae***, vol. 82 (1974), pp. 269–293.

DEPARTMENT OF MATHEMATICS
UNIVERSITY OF CALIFORNIA
BERKELEY, CALIFORNIA 94720-3840
UNITED STATES OF AMERICA
E-mail: steel@math.berkeley.edu

CAPACITIES AND ANALYTIC SETS

CLAUDE DELLACHERIE

Introduction. In these notes I have tried to convince the logician reader that the notion of capacity, far from being merely a "refinement for analyst" of a measure (several examples more and more deviating from the expected behavior of a measure are given in § 1), lies in the heart of the theory of analytic sets. Therefore I chose in § 2 to *define* the analytic sets (in a more general setting than the usual Polish frame) with the help of capacities, as I did in [Del80]: that allows, for example, to get the separation theorem as a natural consequence of the separation of two disjoint compact sets by open sets. On the way is defined a large class of monotone set operations closely related to capacities, stable under composition and preserving the property to be analytic. In § 3, going back into the Polish frame, I consider under the name of caliber a natural and useful extension of the notion of capacity (there is a kind of analogy between the couple compact/analytic and the couple capacity/caliber) and under the name of analytic operation the related notion of monotone set operation. Finally, in § 4, I give an extension to capacities of Suslin's theorem on uncountable analytic sets, improving a result of [Del72]. Applications to Hausdorff measure theory are succinctly given in almost all sections. For a more comprehensive treatment (and also for substantial bibliographies) the reader may consult my already quoted former works.

I am very happy to thank here Martin and Moschovakis for their hospitality (I hope they will not forget I am still the best in ping-pong). It is also a great pleasure, and a duty, to say my indebtedness to Miss Sarbadhikari who has written carefully all of these notes from a confused version in frenglish jargon.

§1. Introduction to the notion of capacity.

1.1. Definition and basic properties. We use $\mathbb{R}_+$ and $\overline{\mathbb{R}}_+$ to mean the non-negative real numbers and the extended nonnegative real numbers, respectively. For any set E, $\wp(E)$ denotes the power set of E. By $\mathcal{K}$ and $\mathcal{K}_\sigma$, we denote the class of compact and σ-compact subsets of Hausdorff spaces respectively.

DEFINITION 1.1. A **capacity** I on a Hausdorff space E is a function on $\wp(E)$ into $\overline{\mathbb{R}}_+$ such that

Large Cardinals, Determinacy, and Other Topics: The Cabal Seminar, Volume IV
Edited by A. S. Kechris, B. Löwe, J. R. Steel
Lecture Notes in Logic, 49

(1) $A \subseteq B$ implies $I(A) \leq I(B)$, *i.e.*, I is non-decreasing.
(2) If $A_n \nearrow A$ then $I(A_n) \nearrow I(A)$, *i.e.*, if $A_1 \subseteq A_2 \subseteq \cdots$ and $\bigcup_n A_n = A$, $I(A_1) \leq I(A_2) \leq \cdots$ and $\sup_n I(A_n) = I(A)$. We express this by "I is going up".
(3) For compact K, $I(K)$ is finite and if $I(K) < t$, then there exists an open $U \supseteq K$ such that $I(U) < t$, *i.e.*, $I(K) = \inf\{I(U) : U \supseteq K \text{ and } U \text{ is open}\}$. We express this by saying "I is right continuous over the compacts".

REMARK 1.2. If E is compact metrisable, (3) can be replaced by

(3′) For compact K, $I(K)$ is finite and if $K_1 \supseteq K_2 \supseteq \cdots$ are compact sets with $\bigcap_n K_n = K$, then $I(K_1) \geq I(K_2) \geq \cdots$ and $\inf_n I(K_n) = I(K)$, *i.e.*, I is going down on compacts. In symbols, $K_n \searrow K \Rightarrow I(K_n) \searrow I(K)$.

In general, however, (1), (2), (3) $\Rightarrow$ (1), (2), (3′) but not conversely. For example, let $E = {}^{\omega}\omega$, with the product of discrete topologies, and let $I(A) = 0$ if A is contained in some $\mathcal{K}_\sigma$-subset of E and $I(A) = 1$ otherwise. The capacity I satisfies (1), (2), (3′) but not (3).

In the rest of this section, we take E, F to be Polish spaces.

EXAMPLES OF CAPACITIES.

1.1 Let m be a nonnegative, σ-additive measure on the Borel σ-field $\mathcal{B}(F)$ of a Polish space F. Extend m to an exterior measure on $\wp(F)$ so that for any $A \subseteq F$, $m(A) = \inf\{m(B) : B \supseteq A,\ B \in \mathcal{B}(F)\}$. Then m is a capacity.
1.2 Suppose J is a capacity on F and f a continuous map from E into F. Define I on $\wp(E)$ by $I(A) = J(f(A))$. Then I is a capacity.

In particular, we can take J to be an exterior measure on $\wp(F)$. The induced capacity I on $\wp(E)$ is not, in general, an exterior measure. However, if f is injective, I is also an exterior measure.

DEFINITION 1.3. Let I be a capacity on E. A subset A of E is called **I-capacitable** if $I(A) = \sup\{I(K) : K \subseteq A,\ K \in \mathcal{K}\}$. The set A is called **universally capacitable** if it is I-capacitable for all capacities I on E.

REMARK 1.4. Using Example 1.2, it is easy to see that if $A \subseteq E$ is universally capacitable and $f : E \to F$ a continuous map, then $f(A)$ is universally capacitable. As a matter of fact, this is almost the only known stability property of the class of universally capacitable sets.

We now prove the main theorem of this section. It is a somewhat stronger version of Choquet's capacitability theorem. This version is due to Sion.

THEOREM 1.5 (Choquet, Sion; [Cho55, Cho59, Sio63]). Any Borel subset of a Polish space is universally capacitable. More generally, any analytic subset of a Polish space (*i.e.*, a continuous image of a Borel subset of a Polish space) is universally capacitable.

PROOF. Note that the analytic subsets of Polish spaces are just the continuous images of ${}^{\omega}\omega$ with the product of discrete topologies (*cf.* [Kur58, §35, I]). Hence in view of the above remark, it is enough to prove that ${}^{\omega}\omega$ is universally capacitable.

Let I be a capacity on ${}^{\omega}\omega$ with $I({}^{\omega}\omega) > t$. We shall show that there exists a compact set $K \subseteq {}^{\omega}\omega$ such that $I(K) \geq t$.

Put $A_n = \{1, 2, \ldots, n\} \times {}^{\omega}\omega$. Then $A_n \nearrow {}^{\omega}\omega$. Hence for some n_1, $I(A_{n_1}) > t$. Repeating this argument, we can find natural numbers $n_1, n_2, \ldots, n_k, \ldots$ such that $I(\{1, \ldots, n_1\} \times \cdots \times \{1, \ldots, n_k\} \times {}^{\omega}\omega) > t$ for all k. Put $K = \prod_k \{1, \ldots, n_k\}$. As I is right continuous on compacts, $I(K) \geq t$. ⊣

REMARK 1.6. We shall see later that there exist coanalytic sets (*i.e.*, complements of analytic sets in Polish spaces) which are not universally capacitable. On the other hand, assuming the continuum hypothesis, it is possible to construct a nonanalytic (even nonprojective) universally capacitable set. This construction is due to Martin and will be given in an appendix to this section.

REMARK 1.7. The same proof shows that any subset of a Hausdorff space which is a continuous image of ${}^{\omega}\omega$ (*i.e.*, any Suslin space in the sense of Bourbaki [Bou75]) is universally capacitable. In the next section, we shall prove the following more general result: Any $\mathcal{K}$-analytic set is universally capacitable.

DEFINITION 1.8. Two capacities on a set E are called **equivalent** if they agree on compact sets (equivalently on universally capacitable sets or on Borel sets or on analytic sets).

The next theorem implies the first principle of separation for analytic sets, as we shall see later.

THEOREM 1.9. *Let I be a capacity on E. There exists a greatest capacity J on E equivalent to I. The capacity J is given by $J(A) = \inf\{I(B) : B \supseteq A,\ B \text{ Borel}\}$.*

PROOF. Since I and J agree on Borel sets, it is enough to check that J is a capacity.

Clearly, J satisfies (1) and (3). To verify (2), let $A_n \nearrow A$ and let $B_n \supseteq A_n$ be Borel sets such that $I(B_n) = J(A_n)$. Without loss of generality, we can take $B_n = \bigcap_{m \geq n} B_m$ so that $B_n \nearrow B$ (say). Then $I(B) \geq J(A) \geq J(A_n) = I(B_n)$ for all n. But $I(B_n) \nearrow I(B)$, hence $J(A_n) \nearrow J(A)$. ⊣

COROLLARY 1.10. *If A is a universally capacitable subset of E, then $\sup\{I(K) : K \in \mathcal{K} \text{ and } K \subseteq A\} = I(A) = \inf\{I(B) : B \supseteq A \text{ and } B \text{ is Borel}\}$. In particular, there exist Borel sets B_1 and B_2 such that $B_1 \subseteq A \subseteq B_2$ and $I(B_1) = I(A) = I(B_2)$.*

Note however that $I(B_2 - B_1) \neq 0$ in general.

1.2. Further examples of capacities. In this section, we give several examples of capacities together with some of their properties. By E and F, we denote compact metrisable spaces.

EXAMPLE 1.11. Let m be an exterior measure on E. Define I on $\wp(E \times F)$ by $I(A) = m[\Pi_E(A)]$, where Π_E denotes projection to E. Then I is a capacity. The I-capacitable sets are characterized by the following:

THEOREM 1.12. *A is I-capacitable iff for all $\varepsilon > 0$, there exists a Borel set $B_\varepsilon \subseteq \Pi_E(A)$ and a Borel map $f_\varepsilon \colon B_\varepsilon \to F$ such that*

(1) $I(A) \leq m(B_\varepsilon) + \varepsilon$, *and*

(2) *graph* $f_\varepsilon \subseteq A$.

This theorem may be proved using the capacitability theorem, *i.e.*, Theorem 1.5.

REMARK 1.13. This type of capacity is often met in the theory of stochastic processes.

EXAMPLE 1.14. Let m be as above and $L \subseteq E \times F$ be compact. We define a capacity I_L on F by

$$I_L(A) = m[\Pi_E(L \cap (E \times A))].$$

Concerning this kind of capacity, we have the following deep result of Mokobodzki [Mok78].

THEOREM 1.15.

(i) *There exists a measure λ on F satisfying "$\lambda(K) = 0 \Rightarrow I_L(K) = 0$ for all compact $K \subseteq F$" iff for m-almost all $x \in E$, $L(x) = L \cap (\{x\} \times F)$ is at most countable.*

(ii) *There exists a measure λ on F satisfying "for all $\varepsilon > 0$, there is a $\delta > 0$ such that $\lambda(K) < \delta \Rightarrow I_L(K) < \varepsilon$ for any compact $K \subseteq F$" iff for m-almost all $x \in E$, $L(x)$ is finite.*

DEFINITION 1.16. A capacity I on F is called **strongly subadditive** or alternating of order two of for all compact $K_1, K_2 \subseteq F$,

$$I(K_1 \cup K_2) + I(K_1 \cap K_2) \leq I(K_1) + I(K_2). \qquad (*)$$

Such a capacity is interesting because it can be constructed from a function, satisfying suitable conditions, defined on the compact sets as the following theorem shows.

THEOREM 1.17 (Choquet; [Cho55]). *Let J be a function on the compact subsets of F into the nonnegative real numbers satisfying*

(1) *For every compact $K \subseteq F$ and every $t > J(K)$, there exists an open subset U of F, $U \supseteq K$ such that if $L \subseteq U$ is compact, $J(L) < t$.*

(2) $J(K_1 \cup K_2) + J(K_1 \cap K_2) \leq J(K_1) + J(K_2)$.

Define I as follows:

If U is open, let $I(U) = \sup\{I(K) : K \subseteq U, K \in \mathcal{K}\}$. For any $A \subseteq F$, put $I(A) = \inf\{I(U) : U \supseteq A, U \text{ open}\}$.

Then I is a strongly subadditive capacity agreeing with J on the compact sets. (Compare with the Caratheodory extension theorem in measure theory). Moreover, it is clear that any strongly subadditive capacity is equivalent to such an I. In particular, for such an I and a universally capacitable set A, we have $\sup\{I(K) : K \subseteq A \text{ and } K \in \mathcal{K}\} = I(A) = \inf_{\{} I(U) : U \supseteq A \text{ and } U \text{ is open}\}$.

Remark 1.18. A capacity alternating of order two is a "weak" version of a capacity alternating of order infinity. The latter is characterized by a series of inequalities among which $(*)$ occurs. A typical example of such a capacity is I_L in Example 1.14. As a matter of fact, Choquet proved that any capacity alternating of order infinity is equivalent to I_L form some L, E and m. These capacities occur in potential theory. In particular, the Newtonian capacity is one such.

Example 1.19. Let $\mathcal{M}^1(F)$ be the set of all probability measures on F with the weakest topology making the maps $m \to m(K)$ upper semicontinuous for all compact $K \subseteq F$. It can be proved that $\mathcal{M}^1(F)$ is compact and metrizable.

Let $L \subseteq \mathcal{M}^1(F)$ be compact. For $A \subseteq F$, define $I_L(A) = \sup_{m \in L} m(A)$ where, as usual, $m(A)$ stands for the exterior measure of A induced by m. Then I_L is a capacity on F.

Using the Hahn–Banach theorem, it is possible to prove that any strongly subadditive capacity on F is equivalent to some I_L. The converse is, however, not true as can be seen from the next theorem.

Theorem 1.20 (Preiss; [Pre73]). If $B \subseteq F$ is a Borel set which is not in $\mathcal{K}_\sigma$, then there exists a compact subset L of $\mathcal{M}^1(F)$ such that

(1) $I_L(B) = 0$, and

(2) $\inf\{I_L(U) : U \supseteq B \text{ and } U \text{ is open}\} > 0$.

Example 1.21. Let (E, d) be a compact metric space and $h\colon \mathbb{R}_+ \to \mathbb{R}_+$ a continuous nondecreasing function with $h(t) > 0$ if $t > 0$. For any $\varepsilon > 0$, we define Λ^h_ε as follows:

$$\Lambda^h_\varepsilon(\varnothing) = 0.$$

For $A \neq \varnothing$,

$$\Lambda^h_\varepsilon(A) = \inf\Big\{\sum_{n\geq 1} h(\delta(F_n)) : \{F_n\}_{n\geq 1} \text{ is a countable cover of } A \text{ by closed sets with } \delta(F_n) \leq \varepsilon \text{ for all } n\Big\},$$

where $\delta(F_n)$ stands for the diameter of F_n.

If $\varepsilon \searrow 0$, $\Lambda^h_\varepsilon(A) \nearrow \Lambda^h(A)$ (say). The function Λ^h is called the h-Hausdorff measure on E. The function Λ^h, restricted to the Borel subsets of E is certainly a nonnegative, σ-additive measure but it is not, in general, bounded or even σ-finite. For example, let $h(t) \equiv 1$. Then

$$\Lambda^h(A) = \begin{cases} \text{cardinality of } A & \text{if } A \text{ is finite,} \\ \infty & \text{otherwise.} \end{cases}$$

So, in general Λ^h, Λ^h_ε are not capacities although they are pointwise limits of increasing sequences of capacities.

However, if we take $\varepsilon \geq \delta(E)$, Λ^h_ε, also denoted by Λ^h_∞, is a capacity. Note that taking $\varepsilon \geq \delta(E)$ is the same as removing all restrictions on $\delta(F_n)$ in the definition of Λ^h_ε.

THEOREM 1.22 (Davies & Rogers; [DR69]). There exists a triplet (E, d, h) as in the example such that for any measure m on E, there is a partition of E into two Borel sets B_1, B_2 with $m(B_1) = 0$ and $\Lambda^h_\infty(B_2) = \Lambda^h(B_2) = 0$.

EXAMPLE 1.23. Let $\mathcal{K}(E)$ be the set of compact subsets of E equipped with the topology induced by the Hausdorff metric. Note that this is the weakest topology such that the restriction of any capacity on E to compact sets, considered as a function on $\mathcal{K}(E)$, is upper semicontinuous.

Let L be a compact subset of $\mathcal{K}(E)$ such that $K \in L \Rightarrow K$ is finite. Define I by

$$I(A) = \begin{cases} 1 & \text{if there is some } K \in L \text{ with } K \subseteq A, \\ 0 & \text{otherwise.} \end{cases}$$

It is easy to check that I is a capacity. If $K \in L \Rightarrow$ cardinality of $K = 1$, then I is a capacity alternating of order infinity. In general, however, it is not even strongly subadditive. For example, suppose $\{a_1, a_2\} \in L$, $\{a_1\} \notin L$, $\{a_2\} \notin L$, $a_1 \neq a_2$. Then $I(\{a_1\}) = 0$, $I(\{a_2\}) = 0$, but $I(\{a_1\} \cup \{a_2\}) = 1$.

On the other hand, any capacity on E with values in $\{0, 1\}$ is the limit of an increasing sequence of such capacities.

1.3. Theorem of Separation. We now prove the theorem of separation for universally capacitable sets, using the capacitability theorem.

THEOREM 1.24. Let E be a Polish space and A_1, A_2 two disjoint subsets of E such that $A_1 \times A_2$ is universally capacitable in $E \times E$. There exist Borel subsets B_1, B_2 of E such that $B_1 \supseteq A_1$, $B_2 \supseteq A_2$ and $B_1 \cap B_2 = \varnothing$.

PROOF. Without loss of generality, take $A_1, A_2 \neq \varnothing$. For $A \subseteq E \times E$, define $\boxed{A} = \Pi_1(A) \times \Pi_2(A)$, Π_1, Π_2 denoting projections to the first and second coordinate, respectively. Then $\boxed{A}$ is the smallest rectangle containing A.

Put

$$I(A) = \begin{cases} 1 & \text{if } \boxed{A} \text{ meets the diagonal,} \\ 0 & \text{otherwise.} \end{cases}$$

Note that I is a capacity and $I(A_1 \times A_2) = 0$. Thus there is a Borel set $C_1 \supseteq A_1 \times A_2$ such that $I(C_1) = 0$. Put $C_2 = \boxed{C_1}$. Then C_2 is analytic, hence universally capacitable, and $I(C_2) = 0$. Repeating the process, we can find C_n, $n \geq 1$, such that $A_1 \times A_2 \subseteq C_1 \subseteq C_2 \subseteq \cdots$ and, for all n, C_{2n-1} is Borel and C_{2n} is an analytic rectangle of null capacity. Hence $C = \bigcup_n C_n$ is a Borel rectangle of null capacity. Put finally, $B_1 = \Pi_1(C)$ and $B_2 = \Pi_2(C)$. ⊣

REMARK 1.25. It is known that there exists a pair (A_1, A_2) of coanalytic sets which are disjoint but cannot be separated by Borel sets. For such a pair, $A_1 \times A_2$ is a coanalytic set which is not universally capacitable. On the other hand Shochat and Busch proved assuming (PD) that any projective set is capacitable for any capacity alternating of order infinity [Sho72, Bus73].

Appendix. We denote by E a fixed perfect Polish space. We give first a lemma (due to Martin) to show the inability of the notion of capacity to grasp the notion of "topological thickness".

LEMMA 1.26. Let I be a capacity on E and A an I-capacitable subset of E. Then there exist a meager $\mathcal{K}_\sigma$ set $L \subseteq A$ such that $I(A) = I(L)$.

PROOF. Clearly it is sufficient to prove that if A is compact and $I(A) > t$, then there is a nowhere dense compact set $K \subseteq A$ such that $I(K) \geq t$. Note that for this it is enough to prove:

If A is compact, B open and $I(A) > t$, then there is an open set $B' \subseteq B$ such that $I(A - B') > t$. $(*)$

To see this, assume $(*)$ and let $B_1, B_2, \ldots$ be an open base for E. By induction on n, get open sets $B_1', B_2', \ldots$ such that for all n, $B_n' \subseteq B_n$ and $I(A - \bigcup_{i=1}^n B_i') > t$. Put $K = A - \bigcup_n B_n'$. The set K is clearly nowhere dense and compact. Since I is going down on compacts, $I(K) \geq t$.

Now we prove $(*)$. Let A be compact, B open and $I(A) > t$. Let $x \in B$ and $B \supseteq U_1 \supseteq U_2 \supseteq \cdots$ where each U_i is open and $\bigcap_i U_i = \{x\}$. Now as $U_i(A - (U_i - \{x\})) = A$, $I(A - (U_i - \{x\})) > t$ for some i. Put $B' = U_i - \{x\}$ for such an i. ⊣

THEOREM 1.27. Assuming CH, there exists a nonprojective, universally capacitable subset of E.

PROOF. Without loss of generality, take E to be the Cantor set. Restrict the capacities on E to the set of universally capacitable subsets of E. Then we get c many capacities, say $\{I_i : i < c\}$. Enumerate the projective subsets of E as $\{P_i : i < c\}$.

We now construct, by transfinite induction, two families $\{A_i : i < c\}$, $\{B_i : i < c\}$ of subsets of E such that

(a) Each A_i is a meager $\mathcal{K}_\sigma$ set, each G_i is a comeager G_δ set.
(b) $\{A_i : i < c\}$ is an increasing family and $\{B_i : i < c\}$ is a decreasing family.
(c) For each i, $A_i \subseteq B_i$ and $I_i(A_i) = I_i(B_i)$.
(d) For each i, either $A_i - P_i \neq \varnothing$ or $P_i - B_i \neq \varnothing$.

Then any subset H of E such that $\bigcup_i A_i \subseteq H \subseteq \bigcup_i B_i$ will be universally capacitable and not equal to any P_i.

Suppose A_i, B_i have been constructed for $i < j < c$. Let L_j be a meager $\mathcal{K}_\sigma$ set contained in $B'_j = \bigcap_{i<j} B_i$ such that $I_j(L_j) = I_j(B'_j)$. This is possible since by CH, B'_j is a countable intersection of G_δ sets and is hence a G_δ and therefore a universally capacitable set. Put $A'_j = L_j \cup \bigcup_{i<j} A_i$. By CH, A'_j is a meager $\mathcal{K}_\sigma$ set.

If $A'_j - P_j \neq \varnothing$ or $P_j - B'_j \neq \varnothing$, take $A_j = A'_j$, $B_j = B'_j$. If $A'_j \subseteq P_j \subseteq B'_j$, then either $B'_j - P_j \neq \varnothing$ or $P_j = B'_j \neq A'_j$ (since B'_j is comeager and A'_j is meager). In the first case, pick a point $x \in (B'_j - P_j)$ and put $A_j = A'_j \cup \{x\}$, $B_j = B'_j$. In the second case pick some $x \in (P_j - A'_j)$ and put $A_j = A'_j$ and $B_j = B'_j - \{x\}$. ⊣

§2. Multicapacities, capacitary operations and $\mathcal{K}$-analytic sets. In this section E, F, with or without suffices, denote nonempty Hausdorff spaces.

2.1. Multicapacities.

Definition 2.1. If $\{E_n\}_{n\geq 1}$ is a sequence (finite or countably infinite) of spaces, a **multicapacity on** $\prod E_n$ (*i.e.*, $E_1 \times E_2 \times \cdots$) is a function I on $\prod \wp(E_n)$ into $\overline{\mathbb{R}}_+$ which is

(1) globally increasing, *i.e.*, $A_n \subseteq B_n$ for all n implies $I(A_1, A_2, \dots) \leq I(B_1, B_2, \dots)$,
(2) separately going up, *i.e.*, if for some n, $A_n^m \nearrow A_n$, then $I(A_1, A_2, \dots, A_{n-1}, A_n^m, A_{n+1}, \dots) \nearrow I(A_1, A_2, \dots, A_{n-1}, A_n, A_{n+1}, \dots)$,
(3) globally right continuous over compact sets, *i.e.*, if $K_1, K_2, \dots$ are compact, then $I(K_1, K_2, \dots)$ is finite and if $I(K_1, K_2, \dots) < t$, then there exist open sets $U_1, U_2, \dots, K_n \subseteq U_n \subseteq E_n$ and $U_n = E_n$ for all but finitely many n, such that $I(U_1, U_2, \dots) < t$.

Note that in (3), for any open set V in $\prod E_n$ which contains $\prod K_n$, there must exist a $\prod U_n$ such that $\prod K_n \subseteq \prod U_n \subseteq V$.

Example 2.2. Any capacity is a multicapacity with one argument.

Example 2.3. Consider the capacities I_L in Example 1.14 and 1.19 of 1. Remover the restriction L is compact and let $I(L, A) = I_L(A)$. The functions I thus obtained are multicapacities with two arguments.

EXAMPLE 2.4. Let $E_n = E$ for all n. Let

$$I(A_1, A_2, \dots) = \begin{cases} 0 & \text{if } \bigcap_n A_n = \varnothing, \\ 1 & \text{otherwise.} \end{cases}$$

I is a multicapacity. (The number of arguments may be taken to be finite or countably infinite).

DEFINITION 2.5. If I is a multicapacity on $\prod E_n$, a sequence $\{A_n\}_{n\geq 1}$, $A_n \subseteq E_n$, is **I-capacitable** if $I(A_1, A_2, \dots) = \sup\{I(K_1, K_2, \dots) : K_n \subseteq A_n,\ K_n \in \mathcal{K}\}$. The sequence $\{A_n\}_{n\geq 1}$ is called **universally capacitable** if it is I-capacitable for any multicapacity I having $\{A_n\}$ as a sequence of arguments.

THEOREM 2.6. Any sequence $\{A_n\}_{n\geq 1}$ of $\mathcal{K}_\sigma$-sets is universally capacitable.

PROOF. For each n, let $\{K_n^m\}_{m\geq 1}$ be a sequence of compact sets with $K_n^m \nearrow A_n$. Let I be any multicapacity with $I(A_1, A_2, \dots) > t$. By (2) of the definition of multicapacity, we can find $m_1, m_2, \dots$ such that for all n, $I(K_1^{m_1}, \dots, K_n^{m_n}, A_{n+1}, A_{n+2}, \dots) > t$. Hence by (1) $I(K_1^{m_1}, \dots, K_n^{m_n}, E_{n+1}, E_{n+2}, \dots) > t$. Finally by (3), $I(K_1^{m_1}, \dots, K_n^{m_n}, K_{n+1}^{m_{n+1}}, \dots) \geq t$. ⊣

2.2. Capacitary operations. In this section, we are going to consider what is a "good change of variables" in a multicapacity.

DEFINITION 2.7. If $\{E_n\}$ is a sequence (finite or infinite) of spaces and F is a space, a **capacitary operation** on $\prod E_n$ with values in F is a map $J\colon \prod \wp(E_n) \to \wp(F)$ which is

(1) globally increasing (with the obvious meaning),
(2) separately going up (with the obvious meaning),
(3) globally right continuous on compacts, *i.e.*,
   (a) if $K_1, K_2, \dots$ are compact, then so is $J(K_1, K_2, \dots)$ and
   (b) if $V \subseteq F$ is open and $J(K_1, K_2, \dots) \subseteq V$, then there exist open sets $U_1, U_2, \dots$ in $E_1, E_2, \dots$ such that $K_n \subseteq U_n$, $U_n = E_n$ for all but finitely many n and $J(U_1, U_2, \dots) \subseteq V$.

Note that for a fixed $y \in F$, if we define

$$J_y(A_1, A_2, \dots) = \begin{cases} 1 & \text{if } y \in J(A_1, A_2, \dots), \\ 0 & \text{otherwise,} \end{cases}$$

then J_y is a multicapacity on $\prod E_n$ with values in $\{0, 1\}$.

The next theorem, although easy, is very important.

THEOREM 2.8 (Composition). If $J^i, i = 1, 2, \dots$, are capacitary operations on $\prod_j E_j^i$ with values in E^i and if I is a multicapacity (or capacitary operation) on $\prod_i E^i$, then the composition $I(J^1(A_1^1, A_2^1, \dots), J^2(A_1^2, A_2^2, \dots), \dots)$ is a multicapacity (or capacitary operation) on $\prod_{(i,j)} E_j^i$.

EXAMPLE 2.9. f is a continuous map from $\prod E_n$ into F. For $A_n \subseteq E_n, n = 1, 2, \dots$ put $J(A_1, A_2, \dots) = f(\prod A_n)$. Then J is a capacitary operation.

EXAMPLE 2.10. If $E_n = E$ for all n, then $J(A_1, A_2, \dots) = \bigcap_n A_n$ is a capacity operation.

EXAMPLE 2.11. If $E_n = E$ for all n, $\{A_n\} \to \bigcup_n A_n$ is not, in general, a capacity operation. However, add an argument $A_0 \subseteq N = \{1, 2, \dots\}$ and put $J(A_0, A_1, A_2, \dots) = \bigcup_{n \in A_0} A_n$. Then J is a capacitary operation and $\bigcup_{n \geq 1} A_n = J(N, A_1, A_2, \dots)$.

EXAMPLE 2.12. Let $A_1 \subset E \times F$, $A_2 \subseteq E$, $A_3 \subseteq F$ and put $J(A_1, A_2, A_3, \dots) = \Pi_F[A_1 \cap (A_2 \times A_3)]$. Then J is a capacity operation by the composition theorem.

Note that if A_1 is the graph of a function $f: E \to F$ and $A_3 = F$, then $J(A_1, A_2, A_3) = f(A_2)$. If A_1 is the graph of a function $f: F \to E$ and $A_3 = F$, then $J(A_1, A_2, A_3) = f^{-1}(A_2)$.

EXAMPLE 2.13. Let $E = {}^\omega\omega$ and $f: {}^\omega\omega \to \mathcal{K}(F)$ (where $\mathcal{K}(F)$ is the family of compact subsets of F) be upper semicontinuous, *i.e.*, if $f(\sigma) \subseteq V \subseteq F$ for some $\sigma \in {}^\omega\omega$ and open V, then there is an open subset U of ${}^\omega\omega$ such that $\sigma \in U$ and $\tau \in U \Rightarrow f(\tau) \subseteq V$ for all τ. Now for any $A \subseteq {}^\omega\omega$, put $J(A) = \bigcup_{\sigma \in A} f(\sigma)$. Then J is a capacitary operation.

EXAMPLE 2.14. In this example we shall see that the Suslin operation, although it is not a capacitary operation, can be obtained from a capacitary operation.

Let $E_s = E$ for any finite sequence s of positive integers. If $A_s \subseteq E$, $\mathcal{A}(\dots A_s \dots)$ is defined to be $\bigcup_{\sigma \in {}^\omega\omega} \bigcap_{s \prec \sigma} A_s$ where $s \prec \sigma$ means "s is an initial segment of σ". Add an argument $A_0 \subseteq {}^\omega\omega$ and put $J(A_0, \dots, A_s, \dots) = \bigcup_{\sigma \in A_0} \bigcap_{s \prec \sigma} A_s$. Using the composition theorem, it is easy to prove that J is a capacitary operation. Note that $\mathcal{A}(\dots A_s \dots) = J({}^\omega\omega, \dots, A_s, \dots)$.

2.3. $\mathcal{K}$-analytic sets. Consider a capacitary operation J. We know that if $A_1, A_2, \dots$ are all compact, then so is $J(A_1, A_2, \dots)$. However, since J is only separately (and not globally) going up, we do not know what kind of a set $J(A_1, A_2, \dots)$ is when the A_n's are σ-compact (and there are infinitely many arguments). We now introduce a definition for such sets.

DEFINITION 2.15. A subset A of a space F is called $\mathcal{K}$-analytic if there is a sequence of spaces $\{E_n\}_{n \geq 1}$, a capacitary operation J on $\prod E_n$ with values in $\wp(F)$ and $\mathcal{K}_\sigma$-sets $H_n \subseteq E_n$ such that $J(H_1, H_2, \dots) = A$.

Note that in the above definition, if we take $E_n = H_n$ for all n and restrict J to $\prod H_n$, then J takes values in $\wp(A)$. Thus the property of being $\mathcal{K}$-analytic is intrinsic.

THEOREM 2.16 (Invariance). If J is a capacitary operation and $\{A_n\}$ a sequence of $\mathcal{K}$-analytic arguments for J, then $J(A_1, A_2, \dots)$ is $\mathcal{K}$-analytic.

PROOF. This is immediate from the composition theorem. ⊣

From this theorem and our examples we get the stability properties of the class of $\mathcal{K}$-analytic sets.

COROLLARY 2.17. The class of $\mathcal{K}$-analytic sets contains all $\mathcal{K}_\sigma$ sets and is closed under countable unions, countable intersections and Suslin operations.

PROOF. It is enough to note that in Example 2.11, N is a $\mathcal{K}_\sigma$ set and in Example 2.14, ${}^\omega\omega$ is a $\mathcal{K}_{\sigma\delta}$ (and hence $\mathcal{K}$-analytic) set in the compact metrisable space $\overline{N}^N$, $\overline{N}$ being the natural one-point compactification of N. ⊣

COROLLARY 2.18. Any closed subset of a $\mathcal{K}$-analytic space A is $\mathcal{K}$-analytic.

PROOF. Suppose $A = J(H_1, H_2, \dots)$ where $H_1, H_2, \dots$ are $\mathcal{K}_\sigma$ sets and J a capacitary operation. If X is a closed subset of A, define for $A_i \subseteq H_i$, $i = 1, 2, \dots$, $J_X(A_1, A_2, \dots) = J(A_1, A_2, \dots) \cap X$. Clearly, J_X is a capacitary operation into A and $X = J_X(H_1, H_2, \dots)$. ⊣

COROLLARY 2.19.

(a) Any Borel subset of a Polish space is $\mathcal{K}$-analytic.
(b) If E, F are Polish and $f : E \to F$ is a Borel map, then the direct image of $\mathcal{K}$-analytic subsets of E and the inverse image of $\mathcal{K}$-analytic subsets of F are $\mathcal{K}$-analytic.

PROOF. (a) By Corollary 2.17, any Borel subset of a compact metrisable space is $\mathcal{K}$-analytic. As any Polish space can be imbedded as a G_δ subset in a compact metrisable space, the result follows.

(b) Use Example 2.12 and the fact that graph f is Borel in $E \times F$.

Actually, a subset of a Polish space is $\mathcal{K}$-analytic iff it is "classical" analytic, *i.e.*, a continuous image of ${}^\omega\omega$. This is an easy consequence of the following

THEOREM 2.20. A subset A of F is $\mathcal{K}$-analytic iff there is an upper semi-continuous map $f : {}^\omega\omega \to \mathcal{K}(F)$ (the compact subsets of F) such that $A = \bigcup_{\sigma \in {}^\omega\omega} f(\sigma)$, *i.e.*, iff it is "$\mathcal{K}$-analytic" in the sense of Frolik.

REMARK 2.21. By a result of Jayne [Jay76], this implies that our $\mathcal{K}$-analytic sets are precisely the "$\mathcal{K}$-analytic" sets defined by Choquet, *i.e.*, continuous images of $\mathcal{K}_{\sigma\delta}$ subsets of compact spaces.

PROOF OF THEOREM 2.20. The "if" part follows by Example 2.13. For the "only if" part, let $A = J(H_1, H_2, \dots)$ where J is a capacitary operation and $H_1, H_2, \dots$ are $\mathcal{K}_\sigma$ sets. For each n, let $K_n^m \nearrow H_n$, where $K_n^1, K_n^2, \dots$ are compact. By Theorem 1.5, for $y \in F$, we have $J_y(H_1, H_2, \dots) = \sup J_y(K_1^{m_1}, K_2^{m_2}, \dots)$. Hence $J(H_1, H_2, \dots) = \bigcup_{\sigma \in {}^\omega\omega} J(K_1^{\sigma_1}, K_2^{\sigma_2}, \dots)$ where

$\sigma = (\sigma_1, \sigma_2, \dots)$. Let $f(\sigma) = J(K_1^{\sigma_1}, K_2^{\sigma_2}, \dots)$. As J is right continuous on the compacts, f is upper semicontinuous. ⊣ (Theorem 2.20)

This completes the proof. ⊣ (Corollary 2.19)

REMARK 2.22. Actually, in our definition of a $\mathcal{K}$-analytic set A, it is always possible to take $E_n = H_n = N$ for each n so that our definition really reduces to Frolik's definition. To see this, we argue as follows.

Let $A = J(H_1, H_2, \dots)$ where J is a capacitary operation and $H_1, H_2, \dots$ are $\mathcal{K}_\sigma$ sets. Note that $H_n = J_n(N, K_1^n, K_2^n, \dots)$ where J_n is a capacitary operation and $K_1^n, K_2^n, \dots$ are compact sets. Now define $I_n(H) = J_n(H, K_1^n, K_2^n, \dots), H \subseteq N$. It is easy to see that I_n is a capacitary operation. Put $I(A_1, A_2, \dots) = J(I_1(A_1), I_2(A_2), \dots)$. Note $A = I(N, N, \dots)$ and I is a capacitary operation.

2.4. Approximation properties of $\mathcal{K}$-analytic sets.

THEOREM 2.23 (Capacitability). If I is a multicapacity or a capacitary operation and $\{A_n\}$ is a sequence of $\mathcal{K}$-analytic arguments for I, then $I(A_1, A_2, \dots) = \sup\{I(K_1, K_2, \dots) : K_n \subseteq A_n,\ K_n \in \mathcal{K}\}$.

PROOF. *Case* 1: I is a multicapacity.

For each n, let $H_1^n, H_2^n, \dots$ be $\mathcal{K}_\sigma$ sets and J^n a capacitary operation such that $A_n = J^n(H_1^n, H_2^n, \dots)$. Let $J(M_i^n : i \geq 1,\ n \geq 1) = I(J^1(M_1^1, M_2^1, \dots), J^2(M_1^2, M_2^2, \dots), \dots)$ for any arguments M_i^n. Then by the composition theorem J is a multicapacity. Note that $I(A_1, A_2, \dots) = J(H_i^n : i \geq 1,\ n \geq 1)$.

Now suppose $I(A_1, A_2, \dots) > t$. By Theorem 1.5, there exist compact sets $K_i^n \subseteq H_i^n$ such that $J(K_i^n,\ i \geq 1,\ n \geq 1) \geq t$. Let $K_n = J^n(K_1^n, K_2^n, \dots)$. Then K_n is compact, $K_n \subseteq J^n(H_1^n, H_2^n, \dots) = A_n$ and $I(K_1, K_2, \dots) = J(K_i^n,\ i \geq 1,\ n \geq 1) \geq t$.

Case 2: I is a capacitary operation.

Note that for each y, I_y is a multicapacity and use Case 1. ⊣

DEFINITION 2.24. A subset B of a space F is called **$\mathcal{G}$-Borel** if it belongs to the smallest class of subsets containing open sets and closed under countable unions and countable intersections.

Note that a $\mathcal{G}$-Borel set is always Borel. If F is Polish, the converse is true. However, in general a Borel set, or even a compact set, need not be $\mathcal{G}$-Borel. Again, in a Polish space, every $\mathcal{K}$-analytic (equivalently "classical" analytic) is $\mathcal{G}$-Borel but in general a $\mathcal{G}$-Borel set need not be $\mathcal{K}$-analytic.

THEOREM 2.25 (Borel approximation). If I is a multicapacity or a capacitary operation on $\prod E_n$ and if $\{A_n\}$ is a sequence of $\mathcal{K}$-analytic arguments for I, then $I(A_1, A_2, \dots) = \inf\{I(B_1, B_2, \dots) : B_n \supseteq A_n \text{ and } B_n \text{ is } \mathcal{G}\text{-Borel in } E_n\}$.

PROOF. As in Theorem 2.23, it is enough to consider the case when I is a multicapacity. Define J on $\prod \wp(E_n)$ by $J(H_1, H_2, \dots) = \inf\{I(B_1, B_2, \dots) : B_n \supseteq H_n,\ B_n \text{ is } \mathcal{G}\text{-Borel in } E_n\}$.

Since I is right continuous on compacts, J and I agree if $H_1, H_2, \ldots$ are compact. Hence to show that they agree when $H_1, H_2, \ldots$ are $\mathcal{K}$-analytic, it is enough to show that J is a multicapacity (capacitability theorem).

Clearly, J satisfies (1) and (3). We now show J satisfies (2).

Suppose $H_1^m \nearrow H_1$. We can choose $\mathcal{G}$-Borel sets B_n^m such that $B_1^m \supseteq H_1^m$, $B_n^m \supseteq H_n$ for $n \geq 2$ and $J(H_1^m, H_2, \ldots) = I(B_1^m, B_2^m, \ldots)$. Let $B_n = \bigcap_m B_n^m$ for $n \geq 2$. Then $J(H_1^m, H_2, \ldots) = I(B_1^m, B_2, \ldots)$. Replacing B_1^m by $\bigcap_{p \geq m} B_1^p$ if necessary, we suppose $\{B_1^m\}$ to be an increasing sequence, say $B_1^m \nearrow B_1$. Now $J(H_1, H_2, \ldots) \geq J(H_1^m, H_2, \ldots) = I(B_1^m, B_2, \ldots)$ and $I(B_1^m, B_2, \ldots) \nearrow I(B_1, B_2, \ldots) \geq J(H_1, H_2, \ldots)$. Thus $J(H_1^m, H_2, \ldots) \nearrow J(H_1, H_2, \ldots)$. ⊣

As corollaries, we get several separation theorems.

Corollary 2.26. If I is a multicapacity (capacitary operation) on $\prod E_n$ and $I(A_1, A_2, \ldots) = o(\varnothing)$ where $A_1, A_2, \ldots$ are $\mathcal{K}$-analytic, then there exist $\mathcal{G}$-Borel subsets B_n of E_n, $n = 1, 2, \ldots$ such that $B_n \supseteq A_n$ and $I(B_1, B_2, \ldots) = o(\varnothing)$.

Proof. By the theorem, the result is clearly true when I is a multicapacity. Let I be a capacitary operation with values in F. Compose I with the capacity J on F such that

$$J(H) = \begin{cases} 0 & \text{if } H = \varnothing, \\ 1 & \text{otherwise.} \end{cases}$$

Get $B_n \supseteq A_n$, $\mathcal{G}$-Borel in E_n, such that $J(I(B_1, B_2, \ldots)) = 0$. Then $I(B_1, B_2, \ldots) = \varnothing$. ⊣

Corollary 2.27.

(a) (Extension of Novikov's separation theorem.) If $A_1, A_2, \ldots$ are $\mathcal{K}$-analytic subsets of E such that $\bigcap_n A_n = \varnothing$, then there exist $\mathcal{G}$-Borel subsets $B_1, B_2, \ldots$ of E such that $B_n \supseteq A_n$ and $\bigcap_n B_n = \varnothing$.

(b) (Extension of Liapunov's separation theorem.) Suppose for each finite sequence s of natural numbers, A_s is a $\mathcal{K}$-analytic subset of E such that $\bigcup_{\sigma \in {}^\omega\omega} \bigcap_{s \prec \sigma} A_s = \varnothing$. Then there exist $\mathcal{G}$-Borel subsets B_s of E, $B_s \supseteq A_s$, such that $\bigcup_{\sigma \in {}^\omega\omega} \bigcap_{s \prec \sigma} B_s = \varnothing$.

Proof. These are particular cases of Corollary 2.26. ⊣

Definition 2.28. A subset H of E is $\mathcal{F}$-Suslin if $H = \bigcup_{\sigma \in {}^\omega\omega} \bigcap_{s \prec \sigma} F_s$, where the F_s's are closed in E. The set H is called $\mathcal{K}$-Suslin if the F_s's are compact.

It follows from Example 2.14 that every $\mathcal{K}$-Suslin set is $\mathcal{K}$-analytic and easily, from Frolik's definition, that every $\mathcal{K}$-analytic set is $\mathcal{F}$-Suslin. Clearly, if E is compact or Polish every $\mathcal{F}$-Suslin set is $\mathcal{K}$-analytic but this is not necessarily true for general E. Nevertheless, we do have the following stronger version of Corollary 2.26.

COROLLARY 2.29. Suppose I is a capacitary operation on $\prod E_n$ with values in F, for $n \geq 1$, $A_n \subseteq E_n$ is $\mathcal{K}$-analytic and $A' \subseteq F$ is $\mathcal{F}$-Suslin. If $I(A_1, A_2, \dots) \cap A' = \varnothing$, then there exist $\mathcal{G}$-Borel subsets B_n of E_n such that $B_n \supseteq A_n$ and $I(B_1, B_2, \dots) \cap A' = \varnothing$.

In particular if $A \subseteq F$ is $\mathcal{K}$-analytic and $A \cap A' = \varnothing$, then there exist a $\mathcal{G}$-Borel subset B of F such that $B \supseteq A$ and $B \cap A' = \varnothing$.

PROOF. The last statement clearly follows from the first. We now prove the first statement. First note that if J is a capacitary operation with values in F and H is a closed subset of F, then $(X_1, X_2, \dots) \to J(X_1, X_2, \dots) \cap H$ is a capacitary operation.

Write A' as $\bigcup_{\sigma \in {}^\omega\omega} \bigcap_{s \prec \sigma} H_s$, H_s being closed in F.

If $X_o \subseteq {}^\omega\omega$ and for each sequence s of natural numbers, $J_s(X_1^s, X_2^s, \dots)$ is a capacitary operation, then by the composition theorem $(X_0, X_n^s : n \geq 1$, s is a finite sequence of natural numbers$) \to \bigcup_{\sigma \in X_0} \bigcap_{s \prec \sigma} [J_s(X_1^s, X_2^s, \dots) \cap H_s]$ is a capacitary operation. Now for all s, take $J_s = I$ and $X_n^s = A_n$. Applying Corollary 2.26, we get $\mathcal{G}$-Borel sets $B_n^s \supseteq A_n$ such that $\bigcup_{\sigma \in {}^\omega\omega} \bigcap_{s \prec \sigma} [I(B_1^s, B_2^s, \dots) \cap H_s] = \varnothing$. Let $B_n = \bigcap_s B_n^s$. ⊣

COROLLARY 2.30. Suppose I is a capacitary operation on $\prod E_n$ with values in F and let $A_n \subseteq E_n, n = 1, 2, \dots$ be $\mathcal{K}$-analytic. For every $\mathcal{G}$-Borel subset B of F such that $I(A_1, A_2, \dots) \subseteq B$, there exist $\mathcal{G}$-Borel subsets B_n of E_n such that $A_n \subseteq B_n$ and $I(B_1, B_2, \dots) \subseteq B$.

PROOF. The set $F - B$ belongs to the smallest family of subsets of F containing closed sets and closed under countable unions and intersections. Hence $F - B$ is $\mathcal{F}$-Suslin. Now we use Corollary 2.29 with $A' = F - B$. ⊣

2.5. An application of the separation theorem. Let E be a compact metrisable space with card $E \geq 2$ and $\Omega \subseteq E^{\mathbb{R}_+}$ the space of all right continuous maps from $\mathbb{R}_+$ into E. Since the elements of Ω are completely determined by their values on the rationales, we can suppose $\Omega \subseteq E^{\mathbb{Q}_+}$ and give it the relative topology. It can be shown that Ω is coanalytic but not Borel in $E^{\mathbb{Q}_+}$.

Let $X_t, t \in \mathbb{R}_+$, be defined on Ω into E by $X_t(\omega) = \omega(t)$. Let $\mathcal{B}$ be the smallest σ-field on Ω making $\{X_t : t \in \mathbb{R}_+\}$ measurable. Then $\mathcal{B}$ can be shown to equal the Borel σ-field on Ω.

PROPOSITION 2.31.

(1) The map $X: \mathbb{R}_+ \times \Omega \to E$ defined by $X(t, \omega) = \omega(t)$ is measurable.
(2) The map $\vartheta: \mathbb{R}_+ \times \Omega \to \Omega$ given by $\vartheta(t, \omega) = \vartheta_t(\omega)$ is measurable where $\vartheta_t(\omega)(s) = \omega(t + s)$.

THEOREM 2.32. Let $\{m_x : x \in E\}$ be a family of probability measures on Ω such that for all $B \in \mathcal{B}$, the map $x \to m_x(B)$ is Borel measurable. There exists a subset Ω_0 of Ω such that

(1) $\vartheta_t(\Omega_0) \subseteq \Omega_0$ for all $t \in \mathbb{R}_+$

(2) $m_x(\Omega - \Omega_0) = 0$ for all $x \in E$
(3) Ω_0 is a Borel subset of $E^{\mathbb{Q}_+}$.

PROOF. Let $\mathcal{M}^1(E^{\mathbb{Q}_+})$ be as described in 1, Example 2.11. Define I on $\wp(\mathcal{M}^1(E^{\mathbb{Q}_+})) \times \wp(E^{\mathbb{Q}_+})$ by $I(H_1, H_2) = \sup_{m \in H_1} m(H_2)$. Note that I is a multicapacity. Let $A = \{m_x : x \in E\} \subseteq \mathcal{M}^1(E^{\mathbb{Q}_+})$, where we extend m_x to $E^{\mathbb{Q}_+}$ by taking $m_x(E^{\mathbb{Q}_+} - \Omega) = 0$. The map $x \to m_x$ is Borel. Hence A is analytic and $I(A, E^{\mathbb{Q}_+} - \Omega) = \sup_{x \in E} m_x(E^{\mathbb{Q}_+} - \Omega) = 0$. Find Borel sets B_1, B_2 such that $A \subseteq B_1$, $E^{\mathbb{Q}_+} - \Omega \subseteq B_2$, and $I(B_1, B_2) = 0$. Put $\Omega_1 = E^{\mathbb{Q}_+} - B_2$. Then $\Omega_1 \subseteq \Omega$ and Ω_1 is a Borel set satisfying $m_x(\Omega - \Omega_1) = 0$ for all x. Let Ω_2 be the smallest set $\supseteq \Omega_1$ such that $\vartheta_t(\Omega_2) \subseteq \Omega_2$ for all $t \in \mathbb{R}_+$. Clearly Ω_2 is analytic and $\Omega_2 \subseteq \Omega$. Let Ω_3 be a Borel subset of $E^{\mathbb{Q}_+}$ such that $\Omega_2 \subseteq \Omega_3 \subseteq \Omega$. Clearly $m_x(\Omega - \Omega_3) = 0$ for all x. In general, we can get for all $n \geq 1$, $\Omega_n \subseteq \Omega$ such that $\Omega_n \subseteq \Omega_{n+1}$, Ω_n is Borel if n is odd, $\vartheta(\Omega_n) \subseteq \Omega_n$ for all t if n is even and $m_x(\Omega - \Omega_n) = 0$ for all n. Put $\Omega_0 = \bigcup_n \Omega_n$. ⊣

§3. Multicalibers and analytic operations. In this section E, F (with or without indices) are taken to be Polish spaces. Thus $\mathcal{K}$-analytic sets are just the "classical" analytic sets. Here we consider operations obtained from multicapacities (or capacitary operations) by

(a) identifying some arguments and
(b) replacing some other arguments by fixed analytic sets (called analytic parameters).

3.1. Calibers. Suppose I is a multicapacity on $\prod(E_n : n \in M)$, where M is a countable set. Let $\varnothing \neq M_1 \subseteq M$ and suppose $E_n =$ a fixed space E for $n \in M_1$. For each $n \in M - M_1$, let $A_n \subseteq E_n$ be analytic.

We say $(I, M_1, \{A_n : n \in M - M_1\})$ is an **analytic representation** of some map $J : \wp(E) \to \overline{\mathbb{R}}_+$ if

(a) J is nondecreasing and
(b) for every analytic $X \subseteq E$, $J(X) = I(X_1, X_2, \dots)$ where

$$X_n = \begin{cases} X & \text{for } n \in M_1, \\ A_n & \text{if } n \notin M_1. \end{cases}$$

REMARK 3.1. It is possible to take $M = M_1$.

DEFINITION 3.2. A **caliber** J on E is a nondecreasing map $J : \wp(E) \to \overline{\mathbb{R}}_+$ which has an analytic representation.

EXAMPLE 3.3. Let E be compact metrisable. Define J on $\wp(E)$ by

$$J(X) = \begin{cases} 0 & \text{if } X \text{ is countable,} \\ 1 & \text{otherwise.} \end{cases}$$

Then J is a caliber.

PROOF. Let $E_1 = \mathcal{M}_1(E)$ be the space of probability measure on E with the natural topology (*cf.* 1, Example 2.11), $E_2 = E$. Define I on $\wp(E_1) \times \wp(E_2)$ by $I(X_1, X_2) = \sup_{m \in X_1} m(X_2)$, where $m(X_2)$ is the outer measure of X_2 induced by m. Then I is a multicapacity. Let $D \subseteq \mathcal{M}^1(E)$ be the set of diffuse probabilities (*i.e.*, probabilities taking the value zero on singletons). Then $(I, \{2\}, D)$ is an analytic representation of J. To see this, it is enough to note that if X_2 is uncountable analytic, there is a diffuse probability measure m with $m(X_2) = 1$. This is true because the Cantor set carries such a measure, namely the product measure of the measures μ on $\{0, 1\}$ with $\mu(\{0\}) = \mu(\{1\}) = \frac{1}{2}$, and X_2 contains a homeomorph of the Cantor set. ⊣

Note that $J(X) = I(D, X)$ need not hold when X is not analytic.

EXAMPLE 3.4. Let $\{I_n\}$ be a sequence of capacities on E. Then $J_1(X) = \sup_n I_n(X)$ and $J_2(X) = \inf_n I_n(X)$ are calibers.

The proof is similar to Examples 1.14 & 1.19.

EXAMPLE 3.5. More generally, suppose for each finite sequence of integers s, I_s is a capacity on E. For $X \subseteq E$, put $J(X) = \sup_{\sigma \in {}^\omega\omega} \inf_{s \prec \sigma} I_s(X)$. Then J is a caliber.

REMARK 3.6. It is not known if every caliber agrees on analytic arguments with a J of this form. However, it is possible to prove the following weaker fact:

Call I a **primitive caliber** if it has an analytic representation with no parameters (*i.e.*, $M = M_1$). Every caliber on E agrees on analytic arguments with some J given by $J(X) = \sup_{\sigma \in {}^\omega\omega} \inf_{s \prec \sigma} I_s(X)$ where I_s is a primitive caliber on E for every finite sequence of integers s.

Following Cenzer and Mauldin, call a class $\mathcal{C}$ of subsets of a space E a **Π^1_1-monotone class** iff there is an analytic subset T of E^N such that $X \in \mathcal{C} \leftrightarrow X^N \cap T = \varnothing$. Examples of Π^1_1-monotone classes are

(a) $\mathcal{C} = \{X : \operatorname{card} X \leq 1\}$,
(b) $\mathcal{C} = \{X : X \text{ is relatively compact, } i.e., \overline{X} \text{ is compact}\}$, and
(c) $\mathcal{C} = \{X : X \text{ is nowhere dense}\}$.

EXAMPLE 3.7. Let $\mathcal{C}$ be a Π^1_1-monotone class. Let

$$J(X) = \begin{cases} 0 & \text{if } X \in \mathcal{C}, \\ 1 & \text{otherwise.} \end{cases}$$

Then J is a caliber.

PROOF. Let I be the multicapacity on $E^N \times E \times E \times \cdots$ defined by

$$I(H, X_1, X_2, \dots) = \begin{cases} 0 & \text{if } H \cap \prod X_n = \varnothing, \\ 1 & \text{otherwise.} \end{cases}$$

Then $J(X) = I(H, X_1, X_2, \dots)$ where $H = T$, the analytic set in the definition of $\mathcal{C}$, and $X_1 = X_2 = \dots = X$. ⊣

EXAMPLE 3.8. In this example, we take E to be compact metrisable and $\mathcal{K}(E)$ to be the set of closed subsets of E, endowed with the natural topology. Thus $\mathcal{K}(E)$ is compact metrisable. If J is a caliber on E and $\overline{\mathcal{C}} = \{K \in \mathcal{K}(E) : J(K) = 0\}$, then it is easy to show that $\overline{\mathcal{C}}$ is coanalytic in $\mathcal{K}(E)$. Trivially, $\overline{\mathcal{C}}$ is also hereditary, *i.e.*, if $K_1 \subseteq K_2$ are in $\mathcal{K}(E)$ and $K_2 \in \overline{\mathcal{C}}$, then $K_1 \in \overline{\mathcal{C}}$.

Conversely, let $\overline{\mathcal{C}}$ be nay coanalytic hereditary subset of $\mathcal{K}(E)$. For $X \in \wp(E)$, define

$$J(X) = \begin{cases} 0 & \text{if } \overline{X} \in \overline{\mathcal{C}}, \\ 1 & \text{otherwise.} \end{cases}$$

Then J is a caliber.

PROOF. The class $\mathcal{C} = \{X : J(X) = 0\}$ is Π^1_1-monotone. To see this, take $T \subseteq E^N$ to be all sequences $\{x_n\}$ such that the closure of $\{x_1, x_2, \dots\} \notin \overline{\mathcal{C}}$. Then T is analytic and $X \in \mathcal{C} \leftrightarrow X^N \cap T = \varnothing$. ⊣

3.2. Multicalibers and analytic operations. We now generalize the notion of analytic representation for a caliber.

Let I be a multicapacity (or capacitary operation with values in F) on $\prod_{n \in M} E_n$, where M is countable. Let $\{M_k\}$ be a sequence (finite or infinite) of nonempty, pairwise disjoint subsets of M. Suppose

(a) for $n \in M_k$, $E_n = E^k$, a fixed space depending only on k and
(b) $A_n \subseteq E_n$ are analytic for $n \in M - \bigcup_k M_k$ (note that this set may be empty).

We say $(I, \{M_k\}, \{A_n\}_{n \in M - \bigcup_k M_k})$ is an **analytic representation** of some map J from $\prod_k \wp(E_k)$ into $\overline{\mathbb{R}}_+$ (or $\wp(F)$) if J is globally nondecreasing and for $X_1, X_2, \dots$ analytic subsets of $E^1, E^2, \dots$ respectively, $J(X_1, X_2, \dots) = I(Y_1, Y_2, \dots)$ where

$$Y_n = \begin{cases} X_k & \text{if } n \in M_k, \\ A_n & \text{if } n \in M - \bigcup_k M_k. \end{cases}$$

DEFINITION 3.9. A **multicaliber on** $\prod E^k$ (or an **analytic operation on** $\prod E^k$ **with values in** F) is a nondecreasing map J from $\prod \wp(E^k)$ into $\overline{\mathbb{R}}_+$ (or $\wp(F)$) which admits an analytic representation.

EXAMPLE 3.10. The Suslin operation is an analytic operation (*cf.* 2). It is obtained by taking one parameter $= {}^{\omega}\omega$ in a capacitary operation.

EXAMPLE 3.11. $X \mapsto$ closure of X is an analytic operation.

PROOF. Let, for $n \geq 1$, $X_n, Y_n \subseteq E$ and $Z_n \subseteq E \times E$. Define $I(\{X_n, Y_n, Z_n : n \geq 1\}) = \bigcap_n \Pi[(X_n \times Y_n) \cap Z_n]$ where Π denotes projection to the second coordinate. By the composition theorem, I is a capacitary operation. Put $J(X) = I(\{X_n, Y_n, Z_n : n \geq 1\})$ where for $n \geq 1$,

(a) $X_n = X$,
(b) $Y_n = E$,
(c) $Z_n = \{(x, y) : \text{distance}(x, y) < \frac{1}{n} \text{ in the metric of } E\}$.

Clearly $J(X)$ is an analytic operation and $J(X) = \overline{X}$. ⊣

REMARK 3.12. The operation $X \to$ interior of X is not an analytic operation but $X \to$ interior of $\overline{X}$ is one.

EXAMPLE 3.13. Let R be an analytic equivalence relation on E. Then $J(A) = \{x : (\exists y \in A)xRy\}$ is an analytic operation.

In general, multicalibers and analytic operations are not "separately going up" or "right continuous over compact sets". However they have many of the regularity properties of multicapacities and capacitary operations.

For example we have

THEOREM 3.14 (Theorem of Invariance). An analytic operation applied to analytic arguments results in an analytic set. Moreover, any analytic set is the image of N under some analytic operation with one argument.

THEOREM 3.15 (Theorem of Composition). The composition of multicalibers and analytic operations yields multicalibers or analytic operations. Identification of arguments in a multicaliber (or an analytic operation) yields a multicaliber (or an analytic operation).

THEOREM 3.16 (Theorem of Capacitability). Here we do not have approximation from below by compact sets. However, we do have approximation from below by $\mathcal{K}_\sigma$ sets, *i.e.*, if I is a multicaliber (or an analytic operation) and $X_1, X_2, \ldots$ is a sequence of analytic arguments, then $I(X_1, X_2, \ldots) = \sup\{I(X_1, X_2, \ldots) : K_n \subseteq X_n,\ K_n \text{ are } \mathcal{K}_\sigma \text{ sets}\}$.

PROOF. We prove the problem for a caliber J. Proof in the other cases are similar. Let I be a multicapacity and $A_1, A_2, \ldots$ analytic sets such that for any analytic argument X, $J(X) = I(A_1, A_2, \ldots, X_1, X_2, \ldots)$ where $X = X_1 = X_2 = \cdots$. Given $\varepsilon > 0$, we can find by the capacitability theorem for multicapacities, compact sets $K_1, K_2, \ldots \subseteq X$ such that $J(X) < I(A_1, A_2, \ldots, K_1, K_2, \ldots) + \varepsilon$. Let $K = \bigcup_n K_n$. Clearly $K \subseteq X$, K is a $\mathcal{K}_\sigma$ set and $J(X) < J(K) + \varepsilon$. ⊣

THEOREM 3.17. If J is a multicaliber (or an analytic operation) and $A_1, A_2, \ldots$ a sequence of analytic arguments for J, then $J(A_1, A_2, \ldots) = \inf\{J(B_1, B_2, \ldots) : A_n \subseteq B_n,\ B_n \text{ Borel for } n \geq 1\}$.

THEOREM 3.18 (Theorem of Separation). If J is a multicaliber (or an analytic operation) and $A_1, A_2, \ldots$ is a sequence of analytic arguments for

J with $J(A_1, A_2, \dots) = 0$ (or $\varnothing$), then there exist Borel sets $B_1, B_2, \dots$ with $A_n \supseteq B_n$ for all n and $J(B_1, B_2, \dots) = 0$ (or $\varnothing$).

3.3. An application of the separation theorem.

DEFINITION 3.19. We call a nonnegative function f defined on a Polish space an **analytic function** if for every $t \in \mathbb{R}_+$, $\{x : f(x) \geq t\}$ is analytic (equivalently for every $t \in \mathbb{R}_+$, $\{x : f(x) \geq t\}$ is analytic).

The point (2) of the next theorem is an extension of a recent result of Cenzer and Mauldin on Π^1_1-monotone classes.

THEOREM 3.20. Let J be a caliber on E with $J(\varnothing) = 0$ and $A \subseteq E \times F$ an analytic set. Then

(1) The function on F defined by $y \to J[A(y)]$, where $A(y)$ is the section of A with respect to $y \in F$, is analytic.
(2) If $J[A(y)] = 0$ for each $y \in F$, then there exists a Borel set B in $E \times F$ such that $A \subseteq B$ and $J[B(y)] = 0$ for each $y \in F$.

PROOF. We can suppose E and F to be compact metrisable since otherwise, we can imbed them as G_δ subsets of compact metrisable spaces $\widetilde{E}, \widetilde{F}$ respectively and extend J to $\widetilde{E}$ by $\widetilde{J}(X) = J(X \cap E)$. The function $\widetilde{J}$ is a caliber on $\widetilde{E}$. This simplifies the proof to some extent.

Fix $t \in \mathbb{R}_+$. We shall prove that the operation J_t on $E \times F$ with values in $\wp(F)$ given by

$$J_t(H) = \{y : J[H(y)] > t\}$$

is analytic.

Now (1) follows from the invariance theorem and (2) from the separation theorem. ⊣

Define an operation Γ on $(E \times F) \times \overline{\mathbb{R}}_+$ into $F \times \overline{\mathbb{R}}_+$ as follows:

For $X \subseteq \overline{\mathbb{R}}_+$, let $i(X)$ denote the smallest interval containing $\{0\}$ and X. Let $H \subseteq E \times F$. Put $\Gamma(H, X) = \{(y, v) \in F \times \overline{\mathbb{R}}_+ : \exists u \in i(X)(v \leq J[H(y)] + u)\}$, in other words, $\Gamma(H, X)$ is obtained by adding the interval $i(X)$ above the graph of $y \to J(H, y)$. It is not difficult to see that Γ is a capacitary operation. (Note that if H is compact, then $y \to J[H(y)]$ is an upper semicontinuous function as J is right continuous on compact sets and hence $\{(y, v) : J[H(y)] \geq v\}$ is compact.)

Finally we have $J_t(H) = \bigcap_n \Pi_F[\Gamma(H, \varnothing) \cap F \times [t + \frac{1}{n}, \infty]]$ so that J_t is an analytic operation.

When J is a caliber, we look at its analytic representation and work with the corresponding multicapacity.

REMARKS.

(a) In part (2), we can replace "$J[A(y)] = 0$" by "$J[A(y)] \leq t$" and "$J[B(y)] = 0$" by "$J[B(y)] \leq t$" for any t.

To prove this, replace the caliber J by the caliber J' where

$$J'(X) = \begin{cases} J(X) - t & \text{if } J(X) \geq t, \\ 0 & \text{otherwise.} \end{cases}$$

(b) If the caliber J satisfies $J(X) = J(\overline{X})$ then in part (2) we can obtain a B with its sections closed. Note that this is not trivial!

(c) In his recent work, Louveau proved another extension of the Cenzer-Mauldin result which is "disjoint" from ours but nevertheless deeper [Lou79]. In any case, it is more interesting to Descriptive Set Theorists.

We finish with an application of Theorem 3.20 to the theory of Hausdorff measure (*cf.* 1).

Suppose m is a (finite) measure on E and B is a Borel subset of $E \times F$. Then by Fubini's theorem $y \to m[B(y)]$ is a Borel function on F.

Now assume E is a compact metric space and replace m by some Hausdorff measure Λ^h. It is not generally true that for any Borel B, $y \to \Lambda^h[B(y)]$ is Borel. For example, let Λ^h be the counting measure. Then, $\{y : \Lambda^h[B(y)] > 0\}$ is the projection of B into F. Nevertheless, since Λ^h is the limit of an increasing sequence of capacities and hence is a caliber, we get

COROLLARY 3.21. Suppose E is a compact metric space and Λ^h a Hausdorff measure on E. If A is an analytic subset of a product space $E \times F$, then the map $y \to \Lambda^h[A(y)]$ is an analytic function on F.

§4. Thick and thin sets with respect to a capacity. In this section, we take E to be a compact metrisable space.

4.1. Definition and examples. Suppose I is a capacity on E such that

(a) $I(\varnothing) = 0$,
(b) for all $A \in \wp(E)$, $I(A) = \inf\{I(B) : B \supseteq A \text{ and } B \text{ is Borel}\}$, and
(c) $I(A_1) = 0$ and $I(A_2) = 0$ implies $I(A_1 \cup A_2) = 0$.

Condition (a) is imposed to avoid trivialities. Condition (b) is also not very important since we generally work with analytic A for which it is anyway true. Condition (c), together with the fact that I is a capacity, implies that the class $\mathcal{N}$ of subsets of E of null capacity is a σ-ideal *i.e.*, it is closed under taking of subsets and countable unions. We call a member of $\mathcal{N}$ a **null set**.

Suppose we have fixed I.

DEFINITION 4.1. An analytic subset A of E is called **thin** if for any family $\{A_\gamma : \gamma \in \Gamma\}$ of disjoint analytic subsets of A, A_γ is a null set for all but countably many $\gamma \in \Gamma$. A subset of E is called **thin** if it is contained in some thin analytic set. Call a subset **thick** if it is not thin.

Let

$$\widetilde{J}(A) = \begin{cases} 0 & \text{if } A \text{ is thin,} \\ 1 & \text{otherwise.} \end{cases}$$

The main result of this section is that $\widetilde{J}$ is a caliber which is "going up". We begin with an easy proposition which implies that $\widetilde{J}$ is "going up".

PROPOSITION 4.2. The class $\mathcal{M}$ of the thin sets is a σ-ideal containing $\mathcal{N}$.

PROOF. The only nontrivial part is the proof of the fact that the countable union of thin analytic sets is thin. Let $\{A_n : n \geq 1\}$ be thin analytic sets and $A = \bigcup_n A_n$. Suppose $\{A_\gamma : \gamma \in \Gamma\}$ is an uncountable family of disjoint analytic subsets of A with $I(A_\gamma) > 0$ for all $\gamma \in \Gamma$. Since $\mathcal{N}$ is a σ-ideal, for each γ there is an integer $n(\gamma) \geq 1$ such that $I(A_{n(\gamma)} \cap A_\gamma) > 0$. As Γ is uncountable, there is some n_0 such that $n(\gamma) = n_0$ for uncountably many values of γ. Thus A_{n_0} is thick. Contradiction! ⊣

In the next section, we show that any σ-ideal ζ such that $\mathcal{N} \subseteq \zeta \subseteq \mathcal{M}$ is completely determined by its compact elements. Such an ζ is called a **σ-ideal of thin sets**. For the time being we assume this result. We now give some examples.

EXAMPLE 4.3. If $I(\varnothing) = 0$ and $I(A) = 1$ for $A \neq \varnothing$, then $\mathcal{N} = \{\varnothing\}$ and $\mathcal{M}$ is the class of countable sets.

EXAMPLE 4.4. Let I be a capacity, alternating of order infinity. It can be deduced from Mokobodzki's theorem (*cf.* 1) that I has a unique decomposition of the form $I = I_1 + I_2$ where I_1, I_2 are alternating of order infinity, every I_1 thin set is I_1-null and E is I_2 thin. In the case of the classical potential theory with I as the Newtonian capacity, we have $I_1 = I$ and $I_2 = 0$.

EXAMPLE 4.5. Suppose I is the capacity Λ^h in 1, Example 2.12. It is easy to check that $\Lambda^h_\infty(A) = 0$ iff $\Lambda^h(A) = 0$. A set A is said to be σ-finite (for Λ^h) if there exists a sequence $\{A_n : n \geq 1\}$ such that $A \subseteq \bigcup_n A_n$ and $\Lambda^h(A_n) < \infty$ for each n. Since Λ^h is a σ-additive measure on the Borel σ-field of E, and since for any $A \subseteq E$, $\Lambda^h(A) = \inf\{\Lambda^h(B) : B \supseteq A \text{ and } B \text{ is Borel}\}$, it is clear that any σ-finite set is thin.

The converse is true in some cases, for example if E is a compact subset of an Euclidean space. In the general case the problem is probably still open.

In any case, the class of σ-finite sets is a σ-ideal of thin sets.

EXAMPLE 4.6. Suppose Ω is the family of right continuous maps from $\mathbb{R}_+$ into E and $\{m_x : x \in E\}$ is the family of probability measures over Ω constructed from a Hunt Borel semigroup (for notation, *cf.* the end of 2; the reader is not obliged to know the definition of a Hunt Borel semigroup in order to understand something of this example). For each $x \in E$, m_x is what we know about the stochastic process $(X_t)_{t\in\mathbb{R}_+}$ when $X_0(\omega) = x$ for each $\omega \in \Omega$. For $x \in E$, define a capacity I_x on E by

$$I_x(A) = m_x\left(\{\omega \in \Omega : X_t(\omega) \in A \text{ for some } t \in \mathbb{R}_+\}\right)$$

(The "going down on compacts" is a consequence of the fact that we are dealing with a Hunt semigroup.)

Since $x \to I_x(A)$ is an analytic function when A is analytic, we can define a capacity I_μ for each probability measure μ on E by $I_\mu(A) = \int I_x(A)d\mu(x)$. An analytic set A is called **polar** if $m_x(\{\omega : X_t(\omega) \in A \text{ for some } t > 0\}) = 0$, for every $x \in E$; in other words, it is a set which is (almost) never met by the process. To simplify matters, we suppose there is a probability measure λ on E such that A is polar iff A is I_λ-null. This condition is in fact satisfied by a large class of Markov processes.

We call an analytic set **semipolar**[1] if $m_x(\{\omega : X_t(\omega) \in A \text{ for uncountably many } t\}) = 0$ for every $x \in E$; in other words, it is a set which is (almost) only countably met by the process.[2] Using Mokobodzki's theorem, it is possible to prove that an analytic set A is semipolar iff A is I_λ-thin and, for each $y \in A$, the set $\{y\}$ is semipolar.

4.2. Thickness of a set. Let I be a capacity on E satisfying (a), (b), (c) at the beginning of the section. For an analytic subset A of E, the **thickness** of A is defined as $J(A) = \text{lub}\{t \geq 0 : \text{there exists an uncountable family } \{A_\gamma : \gamma \in \Gamma\}$ of disjoint analytic subsets of A such that $I(A_\gamma) \geq t$ for all $\gamma \in \Gamma\}$.

The **thickness** of an arbitrary $C \subseteq E$ is defined by $J(C) = \inf\{J(A) : A \text{ analytic}, A \supseteq C\}$.

PROPOSITION 4.7.

(1) The thickness function J is nondecreasing and going up.
(2) If A is analytic with $J(A) > t$, then there exist disjoint compact sets K_0 and K_1 such that $J(A \cap K_i) > t$, for $i = 0, 1$.

PROOF. The proof of (1) is similar to Proposition 4.2.

To prove (2), first note that by the capacitability theorem applied to I, there exists an uncountable family $\{K_\gamma : \gamma \in \Gamma\}$ of disjoint compact subsets of A, with $I(K_\gamma) > t$ for each $\gamma \in \Gamma$.

Now $\mathcal{K}(E)$ is a compact metrisable space and $\{K_\gamma : \gamma \in \Gamma\}$ is an uncountable subset of it. Hence this set has two distinct members K_α and K_β which are condensation points of it (*i.e.*, any neighborhood of K_α or K_β in $\mathcal{K}(E)$ contains an uncountable number of K_γ's). To finish the proof, take K_0 and K_1 to be disjoint compact neighborhoods of K_α and K_β in E and note that for any neighborhood U of a compact set K in E, $\{L \in \mathcal{K}(E) : L \subseteq U\}$ is a neighborhood of K in $\mathcal{K}(E)$. ⊣

---

[1]This is not the classical definition of a semipolar set in potential theory. However it is equivalent although this is difficult to prove.

[2]The class of semipolar sets is a σ-ideal of I_λ-thin sets.

The main step in the proof that J is a caliber is a generalization of the classical Suslin theorem on uncountable analytic sets. We state it as a lemma.

LEMMA 4.8. Let $\mathcal{C}$ be a map from $\wp(E)$ into $\{0, 1\}$ such that

(a) $\mathcal{C}$ is nondecreasing and going up.
(b) if A is analytic and $\mathcal{C}(A) = 1$, then there exist two disjoint compact sets K_0, K_1 such that $\mathcal{C}(A \cap K_i) = 1$ for $i = 0, 1$.

Then if A is analytic and $\mathcal{C}(A) = 1$, there exist an upper semicontinuous map K from the Cantor space $\{0, 1\}^N$ into $\mathcal{K}(E)$ such that

(1) the $K(\sigma)$ are disjoint,
(2) $\bigcup_\sigma K(\sigma)$ is compact and contained in A, and
(3) for any σ and any open U such that $K(\sigma) \subseteq U$, $\mathcal{C}(U) = 1$.

PROOF. Let A be an analytic set with $\mathcal{C}(A) = 1$. We can write $A = \vartheta(H_1, H_2, \dots)$ where ϑ is an analytic operation and each H_n is a K_σ set. As we have seen in 2, we can suppose $H_n = N$ for all n. Let $\underline{m} = \{1, 2, \dots, m\}$.

We define, for each finite sequence s of 0's and 1's a $K_s \in \mathcal{K}(E)$ and for each natural number n, another natural number m_n such that the following holds.

(1) For all s, t, if t is an extension of s, then $K_t \subseteq K_s$.
(2) If s, t are incompatible, $K_t \cap K_s = \varnothing$.
(3) If $n = \ell(s)$, *i.e.*, the length of s, and $A_n = \vartheta(\underline{m}_1, \dots, \underline{m}_n, N, \dots, N, \dots)$ then $\mathcal{C}(A_n \cap K_s) = 1$ and $K_s \subseteq \overline{A}_n$.

Suppose we have constructed K_s and m_n for all s and n. Put $K(\sigma) = \bigcap_n K_{\sigma|n}$ where $\sigma|n$ denotes the finite sequence $\sigma(1)\sigma(2)\dots\sigma(n)$. Then it results from (2) that the $K(\sigma)$ are disjoint and from (3) that for all σ, $K(\sigma) \subseteq \vartheta(\underline{m}_1, \underline{m}_2, \dots, \underline{m}_n, \dots)$, ϑ being right continuous on compacts, so that $K(\sigma) \subseteq A$. Now from (1) and (3) it follows that for any σ and any open $U \supseteq K(\sigma)$ there is some n such that $K_{\sigma|n} \subseteq U$. Thus $\mathcal{C}(U) = 1$ and K is upper semicontinuous, in particular $\bigcup_\sigma K_\sigma$ is compact. (Note that this can also be deduced from (2).)

We now proceed with the construction of the K_s and m_n by induction on $\ell(s)$.

First choose m_1 so that $\mathcal{C}(A_1) = 1$ where $A_1 = \vartheta(\underline{m}_1, N, N, \dots)$. This is clearly possible. Since A_1 is analytic there exist, by hypothesis, two disjoint compact sets L_0 and L_1 such that $\mathcal{C}(A_1 \cap L_0) = \mathcal{C}(A_1 \cap L_1) = 1$.

Put $K_0 = \overline{A_1 \cap L_0}$, $K_1 = \overline{A_1 \cap L_1}$.

Now suppose K_s and m_n have been constructed for $\ell(s) \leq p$ and $n \leq p$. Choose m_{p+1} such that if $A_{p+1} = \vartheta(\underline{m}_1, \dots, \underline{m}_p, \underline{m}_{p+1}, N, \dots)$, then $\mathcal{C}(A_{p+1} \cap K_s) = 1$ for all s with $\ell(s) \leq p$. This is possible since by hypothesis, $\mathcal{C}(A_p \cap K_s) = 1$ for each such s and there are only finitely many such s. Now A_{p+1} is analytic, so for a fixed s of length p, we can find disjoint compact sets L_{s0}, L_{s1} such that

$$\mathcal{C}(A_{p+1} \cap K_s \cap L_{s0}) = \mathcal{C}(A_{p+1} \cap K_s \cap L_{s1}) = 1.$$

Put $K_{s0} = \overline{A_{p+1} \cap K_s \cap L_{s0}}$, $K_{s1} = \overline{A_{p+1} \cap K_s \cap L_{s1}}$.
This completes the construction. ⊣

PROPOSITION 4.9. Let A be analytic with $J(A) > t$. Then there exists an upper semicontinuous map K from $\{0, 1\}^N$ into $\mathcal{K}(E)$ such that each $K(\sigma) \subseteq A$ and $I(K(\sigma)) > t$ for all σ. In particular, $\bigcup_\sigma K(\sigma)$ is a compact subset of A and $J(\bigcup_\sigma K(\sigma)) > t$.

PROOF. Choose $t' > t$ such that $J(A) > t'$ and put, for any $H \subseteq E$,

$$\mathcal{C}(H) = \begin{cases} 1 & \text{if } J(H) > t', \\ 0 & \text{otherwise.} \end{cases}$$

The map $\mathcal{C}$ satisfies the conditions of the lemma. Taking K as in the lemma, we have for any σ and any open $U \supseteq K(\sigma)$, $J(U) > t'$ so that $I(U) > t'$. Since I is a capacity, by right continuity, $I(K(\sigma)) \geq t'$. ⊣

THEOREM 4.10. The thickness J is a caliber.

PROOF. Let $\mathcal{D} \subseteq \mathcal{M}^1(\mathcal{K}(E))$ consist of all probability measures λ on $\mathcal{K}(E)$ such that

(1) λ is diffuse
(2) there exists a compact subset Λ of $\mathcal{K}(E)$ made of disjoint compact subsets of E such that $\lambda(\Lambda) = 1$.

So we have $\lambda \in \mathcal{D}$ iff $(\forall K \in \mathcal{K}(E))(\lambda(\{K\}) = 0)$ and $(\exists \Lambda \in \mathcal{K}(\mathcal{K}(E)))$ $((\lambda(\Lambda) = 1)$ and $(K_1 \in \Lambda$ and $K_2 \in \Lambda$ and $K_1 \neq K_2 \Rightarrow K_1 \cap K_2 = \varnothing))$.

It is easy to prove that $\mathcal{D}$ is a G_δ subset of the compact metrisable space $\mathcal{M}^1(\mathcal{K}(E))$.

We claim that for an analytic subset A of E, $J(A) > t$ iff $(\exists \lambda \in \mathcal{D})\lambda(\{K \in \mathcal{K}(E) : I(K \cap A) > t\}) > 0$.

In Theorem 3.20 of 3, replacing F by $\mathcal{K}(E)$, J by I and A by $(A \times \mathcal{K}(E)) \cap \{(x, K) \in E \times \mathcal{K}(E) : x \in K\}$, we see that $\{K \in \mathcal{K}(E) : I(K \cap A) > t\}$ is an analytic subset of $\mathcal{K}(E)$.

Now to prove the "if" part, let $\lambda\{K \in \mathcal{K}(E) : I(K \cap A) > t\} > 0$ and Λ be the compact subset of $\mathcal{K}(E)$ corresponding to λ as in (2); then $\lambda(\{K \in \Lambda : I(K \cap A) > t\}) = \lambda(\{K \in \mathcal{K}(E) : I(K \cap A) > t\}) > 0$. Since λ is diffuse, the set $\{K \in \Lambda : I(K \cap A) > t\}$ is uncountable and so we have $J(A) > t$.

For the converse, let $K : \{0, 1\}^N \to \mathcal{K}(E)$ be as in the preceding theorem and let μ be the image under K of the Lebesgue measure on $\{0, 1\}^N$. Then μ is a diffuse probability measure on $\mathcal{K}(E)$ and $\mu(M) = 1$ where $M = \{K(\sigma) : \sigma \in \{0, 1\}^N\}$. Now M is the image of $\{0, 1\}^N$ under the Borel map K, hence M is analytic (in fact it is Borel, K being injective). Thus we can find a compact subset Λ of $\mathcal{K}(E)$ such that $\Lambda \subseteq M$ and $\mu(\Lambda) > 0$. Finally, let λ be the probability measure $\frac{1}{\mu(\Lambda)}\mu|_\Lambda$.

So we have proved that for analytic A

$$J(A) > t \leftrightarrow (\exists\lambda \in \mathcal{D})\lambda(\{K \in \mathcal{K}(E) : I(K \cap A) > t\}) > 0.$$

It is easy to see that

$$(\exists\lambda \in \mathcal{D})\lambda(\{K \in \mathcal{K}(E) : I(K \cap A) > t\}) > 0 \leftrightarrow$$
$$(\exists\lambda \in \mathcal{D})\lambda(\{K \in \mathcal{K}(E) : I(K \cap A) > t\}) = 1.$$

For any fixed $t \in \mathbb{R}_+$, using the proof of Theorem 3.20 in 3 and the composition theorem, we have $A \to \{K : I(K \cap A) > t\}$ is an analytic operation.

Using the composition theorem again, we see that $A \to \sup_{\lambda\in\mathcal{D}} \lambda(\{K : I(K \cap A) > t\})$ is a caliber with values in $\{0, 1\}$. Finally we get that, for fixed t, the function

$$J_t(A) = \begin{cases} 1 & \text{if } J(A) > t, \\ 0 & \text{otherwise} \end{cases}$$

is a caliber. To finish the proof, note that for fixed A, $t \to J_t(A)$ is decreasing and that $J(A) = \int_0^\infty J_t(A)dt$ and approximate this integral by an increasing sequence of Riemann sums. ⊣

Corollary 4.11. If we set

$$\widetilde{J}_t(A) = \begin{cases} 1 & \text{if } A \text{ is thick}, \\ 0 & \text{otherwise}, \end{cases}$$

then $\widetilde{J}$ is a caliber.

Proof. Put $\widetilde{J} = J_0$ where J_t is defined as in the theorem. ⊣

Structure of the σ-ideals of thin sets. Recall that σ-ideal ζ of subsets of E is called a **σ-ideal of thin sets** if $\mathcal{N} \subseteq \zeta \subseteq \mathcal{M}$ where $\mathcal{N}$ is the σ-ideal consisting of all null sets (*i.e.*, sets of null capacity with respect to I) and $\mathcal{M}$ is the σ-ideal consisting of all thin sets.

Theorem 4.12. Let ζ be a σ-ideal of thin sets. An analytic subset A of E belongs to ζ iff A is of the form $N \cup (\bigcup_n K_n)$ where K_n is a sequence of disjoint compact sets belonging to ζ and N is a null set.

Proof. The "if" part is trivial. For the "only if" let $A \in \zeta$. Consider a maximal family of disjoint compact subsets of A with a strictly positive capacity. Since A is thin, such a family is countable, say $\{K_n : n \geq 1\}$. Now $N = A - \bigcup_n K_n$ is analytic and does not contain a nonnull compact set. By the capacitability theorem, N is a null set. ⊣

Theorem 4.13. Let ζ be a σ-ideal of thin sets. An analytic subset A of E belongs to ζ iff every compact subset of A belongs to ζ.

PROOF. The "only if" part is trivial. For the "if" part, suppose first A is thick. Then by Proposition 4.9, A contains a compact set which is thick and hence does not belong to ζ. Next suppose A is thin. Applying Theorem 4.12 to $\mathcal{M}$, we find $A = N \cup (\bigcup_n K_n)$ where N is null and K_n is a thin compact set for $n \geq 1$. Now if every compact subset of A belongs to ζ, applying Theorem 4.12 to ζ, we get $A \in \zeta$. ⊣

REFERENCES

NICOLAS BOURBAKI

[Bou75] ***Eléments de mathématique. Topologie générale***, 3ème ed., Hermann, Paris, 1975.

DOUGLAS R. BUSCH

[Bus73] *Some Problems Connected with the Axiom of Determinacy*, ***Ph.D. thesis***, Rockefeller University, 1973.

GUSTAVE CHOQUET

[Cho55] *Theory of capacities*, ***Annales de l'Institut Fourier***, vol. 5 (1955), pp. 131–295.

[Cho59] *Forme abstraite du théorème de capacitabilité*, ***Annales de l'Institut Fourier***, vol. 9 (1959), pp. 83–89.

ROY O. DAVIES AND C. AMBROSE ROGERS

[DR69] *The problem of subsets of finite positive measure*, ***Bulletin of the London Mathematical Society***, vol. 1 (1969), pp. 47–54.

CLAUDE DELLACHERIE

[Del72] ***Ensembles analytiques, capacités, mesures de Hausdorff***, Lecture Notes in Mathematics, vol. 295, Springer-Verlag, Heidelberg, 1972.

[Del80] *Un cours sur les ensembles analytiques*, ***Analytic sets (Developed from lectures given at the London Mathematical Society Instructional Conference on Analytic Sets, University College London, July 1978)*** (C. A. Rogers, J. E. Jayne, Claude Dellacherie, Flemming Topsøe, Jørgen Hoffmann-Jørgensen, Donald A. Martin, Alexander S. Kechris, and A. H. Stone, editors), Academic Press, 1980, pp. 183–316.

JOHN E. JAYNE

[Jay76] *Structure of analytic Hausdorff spaces*, ***Mathematika***, vol. 23 (1976), pp. 208–211.

KAZIMIERZ KURATOWSKI

[Kur58] ***Topologie. Vol. I***, 4ème ed., Monografie Matematyczne, vol. 20, Państwowe Wydawnictwo Naukowe, Warsaw, 1958.

ALAIN LOUVEAU

[Lou79] *Familles séparantes pour les ensembles analytiques*, ***Comptes Rendus de l'Académie des Sciences***, vol. 288 (1979), pp. 391–394.

GABRIEL MOKOBODZKI

[Mok78] *Ensembles à coupes dénombrables et capacités dominées par une mesure*, ***Séminaire de Probabilités XII***, Lecture Notes in Mathematics, vol. 649, Springer-Verlag, Heidelberg, 1978, pp. 492–511.

DAVID PREISS

[Pre73] *Metric spaces in which Prohorov's theorem is not valid*, ***Zeitschrift für Wahrscheinlichkeitstheorie und Verwandte Gebiete***, vol. 27 (1973), pp. 109–116.

DAVID DAWSON SHOCHAT
[Sho72] *Capacitability of* $\sum_2^1$ *sets*, ***Ph.D. thesis***, University of California at Los Angeles, 1972.

MAURICE SION
[Sio63] *On capacitability and measurability*, ***Annales de l'Institut Fourier***, vol. 13 (1963), pp. 88–99.

MATHÉMATIQUES, SITE COLBERT
FACULTÉ DES SCIENCES
UNIVERSITÉ DE ROUEN
76821 MONT SAINT AIGNAN
FRANCE
E-mail: Claude.Dellacherie@univ-rouen.fr

MORE SATURATED IDEALS

MATTHEW FOREMAN

§1. In this paper we prove three theorems relating the consistency of huge cardinals with the consistency of saturated ideals on regular cardinals and with the consistency of model theoretic transfer properties.

We prove:

THEOREM. Con(ZFC+"there is a 2-huge cardinal") implies Con(ZFC+"for all $m, n \in \omega$ with $m > n$, $(\aleph_{m+1}, \aleph_m) \twoheadrightarrow (\aleph_{n+1}, \aleph_n)$").

THEOREM. Con(ZFC+"there is a huge cardinal") implies Con(ZFC+"for all $n \in \omega$, there is a normal, $\aleph_n$-complete, $\aleph_{n+1}$-saturated ideal on $\aleph_n$"+"there is a normal, $\aleph_{\omega+1}$-complete, $\aleph_{\omega+2}$-saturated ideal on $\aleph_{\omega+1}$").

The theorem above contains all the new ideas necessary to prove the following theorem:

THEOREM. Con(ZFC+"there is a huge cardinal") implies Con(ZFC+"every regular cardinal κ carries a κ^+-saturated ideal").

We now make some definitions: Let $\mathcal{L}$ be a countable language with a unary predicate U. An $\mathcal{L}$-structure $\mathfrak{A}$ is said to have **type** (κ, λ) if and only if $|\mathfrak{A}| = \kappa$ and $U^{\mathfrak{A}} = \lambda$. If $\kappa \geq \kappa'$, $\lambda \geq \lambda'$ we say that $(\kappa, \lambda) \twoheadrightarrow (\kappa', \lambda')$ if and only if every structure $\mathfrak{A}$ of type (κ, λ) has an elementary substructure $\mathcal{B} \prec \mathfrak{A}$ of type (κ', λ').

An ideal $\mathcal{I} \subseteq \mathcal{P}$ is set to be **α-complete** if and only if whenever $\{X_\gamma : \gamma < \beta\} \subseteq \mathcal{I}$ and $\beta < \alpha$, $\bigcup_{\gamma<\beta} X_\gamma \in \mathcal{I}$. A set $A \subseteq \wp(\kappa)$ is said to be **positive** if $A \notin \mathcal{I}$. The ideal $\mathcal{I}$ is **normal** if and only if for every positive set $A \subseteq \kappa$ and every regressive function f defined on A there is a $\beta \in \kappa$ such that $\{\alpha : f(\alpha) = \beta\}$ is positive. The ideal $\mathcal{I}$ is said to be **λ-saturated** if and only if it is normal and $\wp(\kappa)/\mathcal{I}$ has the λ-chain condition. (We shall never consider "non-normal" saturated ideals.) There is an extensive literature on saturated ideals. (*Cf.* [KM78, Kun78].)

Let $j: V \to M$ be an elementary embedding from V into a transitive class M. Let κ_0 be the critical point of j (*i.e.*, the first ordinal moved by j), and let $\kappa_{i+1} = j(\kappa_i)$. We shall call j an n-huge embedding and κ_0 an n-huge

Large Cardinals, Determinacy, and Other Topics: The Cabal Seminar, Volume IV
Edited by A. S. Kechris, B. Löwe, J. R. Steel
Lecture Notes in Logic, 49

cardinal if and only if M is closed under κ_n-sequences. (This means that if $\langle x_a : \alpha < \kappa_n\rangle \subseteq M$ then $\langle x_a : a < \kappa_n\rangle \in M$.) An almost n-huge embedding is an embedding j as above such that M is closed under $< \kappa_n$-sequences, *i.e.*, if $\beta < \kappa_n$ and $\{x_\alpha : \alpha < \beta\} \subseteq M$ then $\{x_\alpha : \alpha < \beta\} \in M$. We shall use the notation "crit(j)" for the critical point of j.

If $\kappa_0 \leq \lambda \leq \kappa_1$ and j is at least an almost huge embedding, then j induces a supercompact measure on $\wp_\kappa(\lambda)$. In particular, j induces a normal measure on κ.

PROPOSITION 1.1. Let j be an n-huge embedding. Let U be the measure on κ induced by j. Then there is a set A of measure one for U such that for all α and $\beta \in A$, there is an almost n-huge embedding $j_{\alpha,\beta}$ such that the critical point of $j_{\alpha,\beta}$ is α and $j_{\alpha,\beta}(\alpha) = \beta$.

PROOF. This is a routine reflection argument. ⊣ (Proposition 1.1)

We now precisely state the theorems we shall prove:

THEOREM 1.2. Con(ZFC+ there is a sequence of huge embeddings $\langle j_n : n \in \omega\rangle$ with j_n (critical point of j_n) = critical point of j_{n+1}) implies Con(ZFC+ for all $m, n \in \omega$, $m > n$ implies $(\aleph_{m+1}, \aleph_m) \twoheadrightarrow (\aleph_{n+1}, \aleph_n)$).

THEOREM 1.3. Con(ZFC+there is a sequence of almost huge embeddings $\langle j_n : n \in \omega\rangle$ with j_n (critical point of j_n) = critical point of j_{n+1}) implies Con(ZFC+ for all $n \in \omega$ there is a normal, $\aleph_n$-complete, $\aleph_{n+1}$-saturated ideal on $\aleph_n$).[1]

THEOREM 1.4. Con(ZFC+there is a huge cardinal) implies Con(ZFC + $\aleph_{\omega+1}$ carries a normal $\aleph_{\omega+2}$-saturated ideal and for all $n \in \omega$, $\aleph_n$ carries an $\aleph_{n+1}$-saturated ideal).

We shall assume that the reader is familiar with iterated forcing. (*Cf.* [Bau83] for an excellent exposition.) All of our partial orderings $\mathbb{P}$ will have a unique greatest element, $1_{\mathbb{P}}$. Our notion of "support" will be the standard one and if p is a condition in an iteration we shall write "supp p" for its support. For the inverse limit of a system $\{\mathbb{P}_i : i \in I\}$ we shall write $\varprojlim\{\mathbb{P}_i : i \in I\}$. We shall also use the notion of support for products of partial orderings. If $\{\mathbb{Q}_i : i \in I\}$ is a collection of partial orderings, we let $\prod\langle \mathbb{Q}_i : i \in I\rangle = \{f : f$ is a function with domain I and for all $i \in I$ $f(i) \in \mathbb{Q}_i\}$. The product $\prod\langle \mathbb{Q}_i : i \in I\rangle$ is ordered coordinatewise. If $p \in \prod\langle \mathbb{Q}_i : i \in I\rangle$, then $\operatorname{supp} p = \{i : p(i) \neq 1_{\mathbb{Q}_i}\}$. If $K \subseteq \wp(I)$ is an ideal then $\prod_{\operatorname{supp} p \in K}\langle \mathbb{Q}_i : i \in I\rangle = \{f \in \prod\langle \mathbb{Q}_i : i \in I\rangle : \operatorname{supp} f \in K\}$.

If $\langle \mathbb{Q}_i : i \in \omega\rangle$ is a sequence of terms such that $\mathbb{Q}_{i+1} \in V^{\mathbb{Q}_0 * \mathbb{Q}_1 * \cdots * \mathbb{Q}_i}$, we write $\mathop{*}_{i=0}^{n} \mathbb{Q}_i$ for the finite iteration $\mathbb{Q}_0 * \mathbb{Q}_1 * \cdots * \mathbb{Q}_n$. If S is a partial ordering

[1] We note that the model used to prove Theorem 1.2 also satisfies the conclusion of Theorem 1.3.

with a uniform definition, we shall use $S^{\mathbb{P}}$ to denote the partial ordering S defined in $V^{\mathbb{P}}$. To simplify notation, we shall write $\mathbb{P} * \bar{S}$ to mean $\mathbb{P} * S^{\mathbb{P}}$.

If $\mathbb{P}$ is a partial ordering we shall use $\mathcal{B}(\mathbb{P})$ to denote the canonical complete boolean algebra obtained from $\mathbb{P}$. If $\varphi(\dot{t}_1, \dots, \dot{t}_n)$ is a formula in the forcing language of $\mathbb{P}$, we shall use the notation $p \wedge q$, and $p \vee q$ to be the meet and join of p and q in $\mathcal{B}(\mathbb{P})$. Similarly $\neg p$ will denote the complement of $p \in \mathcal{B}(\mathbb{P})$. If $b \in \mathcal{B}(\mathbb{P})$, we shall say that p "decides" b (in symbols $p \parallel b$) if and only if $p \leq b$ or $p \leq \neg b$. We shall write $p \Vdash b$ if $p \leq b$.

We define $C(\kappa, \gamma) = \langle \{p : p \colon \kappa \to \gamma, |p| < \kappa\}, \supseteq \rangle$. $C(\kappa, \gamma)$ is the partial ordering appropriate for making γ have cardinality κ. We shall call this the Levy collapse. Similarly, we shall define the Silver collapse $S(\kappa, \lambda)$ by $p \in S(\kappa, \lambda)$ if and only if

(a) $p \colon \lambda \times \kappa \to \lambda$
(b) $|p| \leq \kappa$
(c) there is a $\xi < \kappa$, $\operatorname{dom} p \subseteq \lambda \times \xi$
(d) for all $\alpha < \kappa, \gamma < \lambda, p(\gamma, \alpha) < \gamma$

$S(\kappa, \lambda)$ is ordered by reverse inclusion: Standard arguments show that for inaccessible λ, $S(\kappa, \lambda)$ makes λ into κ^+.

If $\gamma' < \gamma$ and $\kappa < \gamma'$, $C(\kappa, \gamma')$ is a subset of $C(\kappa, \gamma)$. If $p \in C(\kappa, \gamma)$, we define $p \cap C(\kappa, \gamma')$ to be q, where $\operatorname{dom} q = \{\alpha < \kappa : p(\alpha) < \gamma'\}$ and for each $\alpha \in \operatorname{dom} q$, $q(\alpha) = p(\alpha)$. It is easy to see that $p \cap C(\kappa, \gamma') \in C(\kappa, \gamma')$. If $p \in C(\kappa, \gamma)$, we define $\sup p$ to be $\sup[\{p(\alpha) + 1 : \alpha \in \operatorname{dom} p\}]$.

We shall write $\vec{\alpha}$ for a finite sequence of ordinals. If $\alpha_0, \alpha_1, \dots, \alpha_n$ are mentioned in connection with $\vec{\alpha}$ we shall assume that $\vec{\alpha} = (\alpha_0, \alpha_1, \dots, \alpha_n)$.

If κ and λ are cardinals, with $\kappa < \lambda$ and $x, y \in \wp_\kappa(\lambda) = \{z \subseteq \lambda : |z| < \kappa\}$ then we write $x \prec y$ if and only if $x \subseteq y$ and the order type of x is less the the order type of $y \cap \kappa$. If $x \in \wp_\kappa(\lambda)$, let $\operatorname{crit}(x) =$ order type of $x \cap \kappa$.

If $\mathcal{B}$ and $\mathcal{C}$ are complete Boolean algebras and $\pi \colon \mathcal{B} \to \mathcal{C}$ is an order preserving function, π is called a **projection map** if whenever $G \subseteq \mathcal{B}$ is generic, $\pi''G \subseteq \mathcal{C}$ is generic.

Let $\mathbb{P}$ and $\mathbb{Q}$ be partial orderings. If $i \colon \mathbb{P} \to \mathbb{Q}$ is one to one, order and incompatibility preserving function with the property that whenever $A \subseteq \mathbb{P}$ is a maximal antichain, $i''A \subseteq \mathbb{Q}$ is a maximal antichain, then we say that i is a **neat embedding** from $\mathbb{P}$ into $\mathbb{Q}$, and write $i \colon \mathbb{P} \hookrightarrow \mathbb{Q}$. Standard theory says that if $i \colon \mathbb{Q} \hookrightarrow \mathbb{P}$ is a neat embedding then there is a projection map $\pi \colon \mathcal{B}(\mathbb{Q}) \to \mathcal{B}(\mathbb{P})$ such that for all $p \in \mathbb{P}$, $\pi(i(p)) = p$.

If $e \colon \mathbb{P} \hookrightarrow \mathbb{Q}$ and $S \in V^{\mathbb{Q}}$ is a partial ordering we define $\mathbb{P} *_e S$ to be the iteration with amalgam $e''\mathbb{Q}$: conditions consist of pairs (p, τ) where $p \in \mathbb{P}$ and τ is a $\mathbb{Q}$-term for an element of S. Let $(p, \tau) \leq (p', \tau')$ if and only if $p \leq_{\mathbb{P}} p'$ and $\pi(p) \Vdash \tau \leq \tau'$. (*Cf.* [For82] for more on this definition.)

We shall say that a partial ordering $\mathbb{P}$ has the **κ-chain condition** (is κ-c.c.) if and only if there is no antichain in $\mathbb{P}$ of size κ. The partial ordering $\mathbb{P}$

is κ-closed if and only whenever $\langle p_\alpha : \alpha < \beta\rangle$ is a decreasing sequence of conditions with $\beta < \kappa$, there is a $q \in \mathbb{P}$ such that for all $\alpha < \beta, q < p_\alpha$.

If $\mathbb{P}$ is a partial ordering and $p \in \mathbb{P}$, then $\mathbb{P}/p$ is defined to be $\{q : q \leq p\}$.

§2. In this section we prove Theorems 1.2 & 1.3. We begin by mentioning theorems of Magidor and Kunen that we shall use extensively in this paper.

THEOREM (Kunen; [Kun78]). Let $j: V \to M$ be a huge embedding. Suppose $\mathbb{P} * \mathbb{Q}$ is a forcing notion such that

(a) $\mathbb{P}$ is κ_0-c.c.,
(b) $\mathbb{Q}$ is κ_0-closed in $V^{\mathbb{P}}$,
(c) $|\mathbb{P}| = \kappa_0$, $|\mathbb{P} * \mathbb{Q}| = \kappa_1$,
(d) there is a projection map $\pi: \mathcal{B}(j(\mathbb{P}) * j(\mathbb{Q})) \to \mathcal{B}(\mathbb{P} * \mathbb{Q})$ and a $q \in j(\mathbb{P}) * j(\mathbb{Q})$ such that for all $r \in j(\mathbb{P}) * j(\mathbb{Q})$, $r \leq q$ implies $j(\pi(r)) \geq r$

then in $V^{\mathbb{P}*\mathbb{Q}}$:

(i) κ_0 carries a normal, κ_0-complete, $j(\kappa_0)$-saturated ideal, and
(ii) $(j(\kappa_0), \kappa_0) \twoheadrightarrow (\kappa_0, < \kappa_0)$.

Magidor proved that to get a normal κ_0-complete, $j(\kappa_0)$-saturated ideal on κ_0, it is enough to have j be almost huge and replace (d) by:

(d′) There is a neat embedding $i: \mathcal{B}(\mathbb{P}' * \mathbb{Q}) \to \mathcal{B}(j(\mathbb{P}'))$.

We shall need the following preservation lemmas.

LEMMA 2.1. Let λ be a regular cardinal. Suppose $\mathbb{P}$ is λ-c.c. and $\mathbb{Q}$ is λ-closed. (We do not rule out $\mathbb{P} = \varnothing$.)

(a) If $\kappa < \lambda$ and κ carries a normal λ-saturated ideal in $V^{\mathbb{P}}$, then κ carries a normal λ-saturated ideal in $V^{\mathbb{P}\times\mathbb{Q}}$.
(b) If $\lambda' < \lambda$, $\gamma \leq \kappa \leq \lambda$ and $(\lambda, \lambda') \twoheadrightarrow (\kappa, \gamma)$ in V, then $(\lambda, \lambda') \twoheadrightarrow (\kappa, \gamma)$ in $V^{\mathbb{Q}}$.

PROOF. Since $\mathbb{P}$ is λ-c.c., $\mathbb{Q}$ is (λ, ∞)-distributive in $V^{\mathbb{P}}$. Thus, if $\mathcal{I}$ is a normal λ-saturated ideal on κ in $V^{\mathbb{P}}$, $\mathcal{I}$ remains a normal ideal in $(V^{\mathbb{P}})^{\mathbb{Q}}$. Suppose $\mathcal{I}$ is no longer λ-saturated in $V^{\mathbb{P}\times\mathbb{Q}}$. Let $\langle \sigma_\alpha : \alpha < \lambda\rangle$ be terms in $V^{\mathbb{P}\times\mathbb{Q}}$ for an antichain. We shall build a descending sequence $\langle q_\alpha : a < \lambda\rangle \subseteq \mathbb{Q}$ in V with the property that there is a term $\tau_\alpha \in V^{\mathbb{P}}$ such that $\|q_\alpha \Vdash_{\mathbb{Q}} \sigma_\alpha = (\tau_\alpha)^{V^{\mathbb{P}}}\|_{\mathbb{P}} = 1$. We do this by induction on α. Let $q_{-1} = 1_{\mathbb{Q}}$.

Assume we have defined $\{q_\beta : \beta < \alpha\}$. Let $q_\alpha^{-1} \leq \{q_\beta : \beta < \alpha\}$. There is such a q_α^{-1} because $\alpha < \lambda$ and $\mathbb{Q}$ is λ-closed. We shall simultaneously build $\{q_\alpha^\gamma : \gamma < \gamma_\alpha\} \subseteq \mathbb{Q}$ and a maximal antichain $\{p_\gamma : \gamma < \gamma_\alpha\} \subseteq \mathbb{P}$. These will have the property that for each γ there is a term $\tau_\alpha^\gamma \in V^{\mathbb{P}}$ such that $(p_\gamma, q_\alpha) \Vdash_{\mathbb{P}\times\mathbb{Q}} \sigma_a = (\tau_\alpha^\gamma)^{V^{\mathbb{P}}}$. Assume that we have built $\langle q_\alpha^\gamma : \gamma < \xi\rangle$ and $\langle p_\gamma : \gamma < \xi\rangle$ with the above property. Note that $\xi < \lambda$ since $\mathbb{P}$ has the λ-chain condition. If $\langle p_\gamma : \gamma < \xi\rangle$ is not a maximal antichain, pick p_ξ, q_ξ^α such that

$q^\alpha_\xi \le \langle q^\alpha_\gamma : \gamma < \xi\rangle$ and for some $\tau^\xi_\alpha \in V^{\mathbb{P}}$, $(p_\xi, q^\xi_\alpha) \Vdash_{\mathbb{P}\times\mathbb{Q}} (\tau^\xi_\alpha)^{V^{\mathbb{P}}} = \sigma_\alpha$ and p_ξ is incompatible with each p_γ, $\gamma < \xi$.

Since $\mathbb{P}$ has the λ-c.c., for some $\xi < \lambda$, $\{p_\gamma : \gamma < \xi\}$ is a maximal antichain. Let $q_\gamma \le \{q^\gamma_\alpha : \gamma < \xi\}$ and $\tau_\alpha \in V^{\mathbb{P}}$ be such that $p_\gamma \Vdash \tau^\gamma_\alpha = \tau_\alpha$. Then $\|q_\alpha \Vdash_{\mathbb{Q}} (\tau^{V^{\mathbb{P}}}_\alpha) = \sigma_\alpha\|_{\mathbb{P}} = 1$.

In $V^{\mathbb{P}}$, $\langle q_\alpha : \alpha < \gamma\rangle$ identifies a sequence $\langle \tau_\alpha : \alpha < \lambda\rangle \subseteq (\wp(\kappa))^{V^{\mathbb{P}}}$. If $\alpha < \beta < \lambda$, then $q_\beta \Vdash_{\mathbb{Q}} \tau_\alpha \cap \tau_\beta \in \mathcal{I}$ and $\tau_\alpha, \tau_\beta \notin \mathcal{I}$. But "$\tau_\alpha \cap \tau_\beta \in \mathcal{I}$" and "$\tau_\alpha \notin \mathcal{I}$" is absolute between $V^{\mathbb{P}}$ and $V^{\mathbb{P}\times\mathbb{Q}}$. Hence in $V^{\mathbb{P}}$, $\langle \tau_\alpha : \alpha < \lambda\rangle$ is an antichain of size λ in $\wp(\kappa)/\mathcal{I}$, a contradiction.

(b) Let $\mathfrak{A} = \langle \lambda ; \lambda', f_i\rangle_{i\in\omega}$ be a fully skolemized structure of type (λ, λ') in $V^{\mathbb{Q}}$. Using the λ-closure of $\mathbb{Q}$, in V we can find a sequence $\langle p_\alpha : \alpha < \lambda\rangle \subseteq \mathbb{Q}$ and a structure $\mathfrak{A}' = \langle \lambda ; \lambda', f'_i\rangle_{i\in\omega}$ such that f'_i are defined on all of λ and for each $\vec{\beta} \in \lambda^{<\omega}$ and each i there is an $\alpha < \lambda$ $p_\alpha \Vdash f_i(\vec{\beta}) = f'_i(\vec{\beta})$. Let $\mathcal{B} \prec \mathfrak{A}'$ be of type (κ, γ). Since $\kappa < \lambda$ and λ is regular there is an α such that for all $\beta \in \mathcal{B}$, $p_\alpha \Vdash f_i(\vec{\beta}) = f'_i(\vec{\beta})$. But then $p_\alpha \Vdash \mathcal{B}$ is closed under all of the f_i's. Hence $p_\alpha \Vdash \mathcal{B} \prec \mathfrak{A}$. ⊣ (Lemma 2.1)

A partial ordering $\mathbb{P}$ will be called **κ-centered** if and only if there is a collection $\{C_\alpha : \alpha < \kappa\}$ of disjoint subsets of $\mathbb{P}$ such that

(a) $\mathbb{P} = \bigcup_{\alpha<\kappa} C_\alpha$.

(b) If $p_1, \dots, p_n \in C_\alpha$ then $\bigwedge_{i=1}^n p_i \ne 0$.

In particular, if $|\mathbb{P}| \le \kappa$, $\mathbb{P}$ is κ-centered.

LEMMA 2.2. Let $\mathcal{I}$ be a κ^{++}-saturated ideal on κ^+. Suppose $\mathbb{P}$ is a κ-centered partial ordering. Then $\mathcal{I}$ generates a normal κ^+-complete, κ^{++}-saturated ideal in $V^{\mathbb{P}}$. (Laver proved stronger results unknown to the author at the time this work was done. *Cf.* [Lav82A] and [Lav82B].)[2]

PROOF. Fix a centering $\langle C_\gamma : \gamma < \kappa\rangle$ of $\mathbb{P}$. Then $\mathbb{P}$ is manifestly κ^+-c.c. Hence $\mathcal{I}$ remains κ^+-complete and normal. We need to see that $\mathcal{I}$ is κ^{++}-saturated. Let $\langle \dot{X}_\alpha : \alpha < \kappa^{++}\rangle$ be a term of an antichain of size κ^{++}. Let $\dot{Y}_{\alpha,\beta} = \dot{X}_\alpha \cap \dot{X}_\beta$. Then $\dot{Y}_{\alpha,\beta} \in V^{\mathbb{P}}$ and $\|\dot{Y}_{\alpha,\beta} \in \bar{\mathcal{I}}\| = 1$. (Here we are using $\bar{\mathcal{I}}$ to stand for the ideal generated by $\mathcal{I}$ in $V^{\mathbb{P}}$.) Hence for each pair α, β

$$\|\exists Y \in V(Y \in \mathcal{I} \text{ and } Y_{\alpha,\beta} \subseteq Y)\| = 1.$$

Since $\mathbb{P}$ is κ^+-c.c. and $\mathcal{I}$ is κ^+-complete, there is a sequence in V, $\langle Y_{\alpha,\beta} : \alpha < \beta < \kappa^{++}\rangle \subseteq \mathcal{I}$ and $\|X_\alpha \cap X_\beta \subseteq Y_{\alpha,\beta}\| = 1$. For each X_α and each $\gamma < \kappa$, let $X_{\alpha,\gamma} = \{\xi :$ there is a $p \in C_\gamma, p \Vdash \xi \in X_\alpha\}$. Since $\|X_\alpha \subseteq \bigcup_{\gamma<\kappa} X_{\alpha,\gamma}\| = 1$ and $\mathcal{I}$ is κ^+-complete, there must be some γ_α with $X_{\alpha,\gamma_\alpha} \notin \mathcal{I}$. By the pigeon-hole principle, there is a set $S \subseteq \kappa^{++}$, $S \in V$ with $|S| = \kappa^{++}$ such that for all $\alpha, \beta \in S$, $\gamma_\alpha = \gamma_\beta$. Fix such a set S and such a $\gamma = \gamma_\alpha$. We claim that if $\alpha, \beta \in S$, then $X_{\alpha,\gamma} \cap X_{\beta,\gamma} \subseteq Y_{\alpha,\beta}$.

[2] *Note* (2018): This remark is in the original paper.

Let $\xi \in X_{\alpha,\gamma} \cap X_{\alpha,\beta}$. Since C_γ is a collection of compatible elements we can find a $q \in C_\gamma$, $q \Vdash \xi \in X_{\alpha,\gamma} \cap X_{\beta,\gamma}$. But then $q \Vdash \xi \in \check{Y}_{\alpha,\beta}$. Hence $\xi \in Y_{\alpha,\beta}$. We now derive our contradiction: $\langle X_{\alpha,\gamma} : \alpha \in S\rangle$ is a set of positive elements of $\wp(\kappa^+)$ with respect to $\mathcal{I}$ and if $\alpha, \beta \in S$, $\alpha \neq \beta$, $X_{\alpha,\gamma} \cap X_{\beta,\gamma} \subseteq Y_{\alpha,\beta} \in \mathcal{I}$. Hence $\langle X_{\alpha,\gamma} : \alpha \in S\rangle$ are an antichain of size κ^{++} in the ground model, a contradiction. $\dashv$ (Lemma 2.2)

The following lemma is standard (*Cf.* [KM78]).

LEMMA 2.3. Suppose $\lambda' < \kappa' \leq \lambda < \kappa$ are regular cardinals.
(a) If $\mathbb{P}$ has the λ'-c.c. and $(\kappa, \lambda) \twoheadrightarrow (\kappa', \lambda')$ then in $V^{\mathbb{P}}$, $(\kappa, \lambda) \twoheadrightarrow (\kappa', \lambda')$.
(b) If $\mathbb{P}$ has the λ-c.c. and κ carries a κ^+-saturated ideal $\mathcal{I}$, then in $V^{\mathbb{P}}$, $\mathcal{I}$ is a κ^+-saturated ideal.

Let $\langle j_n : n \in \omega\rangle$ be a sequence of huge embeddings. Let $\kappa_n = \text{crit}(j_n)$ and suppose $j_n(\kappa_n) = \kappa_{n+1}$. Suppose $\langle \mathbb{R}_n : n \in \omega\rangle$ is a sequence of terms for partial orderings with $\mathbb{R}_{n+1} \in V^{\ast_{i=0}^{n} \mathbb{R}_i}$ such that

(a) $\ast_{i=0}^{n} \mathbb{R}_i$ has the κ_n-chain condition and in $V^{\ast_{i=0}^{n} \mathbb{R}_i}$, $\mathbb{R}_{n+1}$ is κ_n-closed
(b) In $V^{\ast_{i=0}^{n} \mathbb{R}_i}$, for all $i \leq n$ $\kappa_i = \aleph_{i+2}$
(c) In $V^{\ast_{i=0}^{n} \mathbb{R}_i}$, $\mathbb{R}_{n+1}$ has cardinality κ_{n+1}.
(d) For each n, there is a projection map $\pi : j_n(\ast_{i=0}^{n} \mathbb{R}_i) \to \ast_{i=0}^{n} \mathbb{R}_i \ast \mathbb{R}_{n+1}$ and a condition $q \in j_n(\ast_{i=0}^{n} \mathbb{R}_i) \ast j_n(\mathbb{R}_{n+1})$ such that for all $r \leq q$, $r \in j_n(\ast_{i=0}^{n} \mathbb{R}_i) \ast j_n(\mathbb{R}_{n+1})$, $j_n(\pi(r)) \geq r$. (Thus, if $\mathbb{P}' = \ast_{i=0}^{n} \mathbb{R}_i$ and $\mathbb{Q} = \mathbb{R}_n$, then $\mathbb{P}'$ and $\mathbb{Q}$ satisfy the hypothesis (d) in Kunen's theorem.)

Let $\mathbb{P} = \varprojlim \langle \ast_{i=0}^{n} \mathbb{R}_i : n \in \omega\rangle$. By Kunen's Theorem, in $V^{\ast_{i=0}^{n} \mathbb{R}_i}$ there is κ_{n+1}-saturated ideal on κ_n and $(\kappa_{n+1}, \kappa_n) \twoheadrightarrow (\kappa_n, \kappa_{n-1})$. (Let $\kappa_{-1} = \aleph_1$.) But $\mathbb{P}/\ast_{i=0}^{n} \mathbb{R}_i$ is κ_{n+1}-closed. Hence by Lemma 2.1 in $V^{\mathbb{P}}$, for all $m, n \in \omega$, $m > n > 0$, $(\aleph_{m+1}, \aleph_m) \twoheadrightarrow (\aleph_{n+1}, \aleph_n)$ and $\aleph_m$ carries a normal $\aleph_m$-complete $\aleph_{m+1}$-saturated ideal.

By Lemmas 2.2 & 2.3, in $V^{\mathbb{P} \ast C(\aleph_0, \aleph_1)}$ for all $m, n \in \omega$, $m > n$, $(\aleph_{m+1}, \aleph_m) \twoheadrightarrow (\aleph_{n+1}, \aleph_n)$ and $\aleph_m$ carries an $\aleph_m$-complete $\aleph_{m+1}$-saturated ideal.

We shall now concentrate our attention on building the $\mathbb{R}$'s. The construction given here is considerably simpler than our original one at the expense of introducing somewhat more technicality.[3] The next few definitions and lemmas explore these technicalities.

[3] *Note* (2018): This wording is from the original published manuscript.

LEMMA 2.4. Let $\mathbb{P}$ and $\mathbb{Q}$ be partial orderings. Suppose $\pi\colon \mathbb{P} \to \mathbb{Q}$ has the properties:

(1) $p_1 \leq p_2$ implies $\pi(p_1) \leq \pi(p_2)$, and
(2) for all $p \in \mathbb{P}$ there is a $q \leq \pi(p)$ such that for all $q' \leq q$ there is a $p' \leq p$ such that $\pi(p') \leq q'$.

$$\begin{array}{ccc}
\mathbb{P} & & \mathbb{Q} \\
\hline
p & \xrightarrow{\pi} & \pi(p) \\
 & & \rotatebox{90}{$\leq$} \\
 & & q \\
\rotatebox{90}{$\leq$} & & \rotatebox{90}{$\leq$} \\
 & & q' \\
 & & \rotatebox{90}{$\leq$} \\
p' & \xrightarrow{\pi} & \pi(p')
\end{array}$$

Then π is a projection map.

PROOF. This is standard. ⊣ (Lemma 2.4)

The following was discovered independently by Avraham:

DEFINITION 2.5 (Laver). Let $\mathbb{P}$ be a partial ordering and $\mathbb{Q} \in V^{\mathbb{P}}$ be a term for partial ordering. We define the **termspace** partial ordering $\mathbb{Q}^*$ to be the following partial order:

$$\operatorname{dom}\mathbb{Q}^* = \{\tau : \tau \in V^{\mathbb{P}}, \|\tau \in \mathbb{Q}\| = 1 \text{ and for all } \tau' \in V^{\mathbb{P}}, \text{ if } \operatorname{rank}(\tau') < \operatorname{rank}(\tau) \text{ then } \|\tau' = \tau\|_{\mathbb{P}} \neq 1\}.$$

For $\tau, \tau' \in \operatorname{dom}\mathbb{Q}^*$,

$$\tau \leq_{\mathbb{Q}^*} \tau' \quad \text{if and only if} \quad \|\tau \leq_{\mathbb{Q}} \tau'\|_{\mathbb{P}} = 1.$$

(Technically we want to take the universe of $\mathbb{Q}^*$ to be equivalence classes of terms modulo the relation $\tau_1 \sim \tau_2$ if and only if $\|\tau_1 = \tau_2\|_{\mathbb{P}} = 1$. In practice we ignore this distinction.)

The main use of the termspace partial ordering is summed up by:

LEMMA 2.6.

(a) Let $\mathbb{Q} \in V^{\mathbb{P}}$ be a term for a partial ordering. There is a projection map $\pi\colon \mathbb{P} \times \mathbb{Q}^* \to \mathbb{P} * \mathbb{Q}$
(b) Let $\langle \mathbb{Q}_i : i \in \omega \rangle \subseteq V^{\mathbb{P}}$ be terms for partial orderings. Then there is a projection map

$$\pi\colon \mathbb{P} \times \prod_{i\in\omega}(\mathbb{Q}_i)^* \to \mathbb{P} * \Big(\prod_{i\in\omega}\mathbb{Q}_i\Big)^{V^{\mathbb{P}}}$$

(c) If $\langle \mathbb{Q}_i : i \in \omega \rangle$, $\langle \mathbb{P}_i : i \in \omega \rangle$ are two sequences of partial orderings and for each $i \in \omega$ $\varphi_i : \mathbb{Q}_i \to \mathbb{P}_i$ is a neat embedding, then there is a $\varphi : \prod_{i \in \omega} \mathbb{Q}_i \to \prod_{i \in \omega} \mathbb{P}_i$ extending each φ_i.

PROOF. (a) Define π as follows: If $(p, \tau) \in \mathbb{P} \times \mathbb{Q}^*$, let $\pi(p, \tau) = (p, \tau) \in \mathbb{P} * \mathbb{Q}$. We want to apply Lemma 2.4. It is enough to see that if $(p, \tau) \in \mathbb{P} \times \mathbb{Q}^*$ and $(q, \sigma) \in \mathbb{P} * \mathbb{Q}$ with $(q, \sigma) \leq_{\mathbb{P}*\mathbb{Q}} (p, \tau)$ there is a $\sigma' \in \mathbb{Q}^*$ such that $(q, \sigma') \leq_{\mathbb{P}\times\mathbb{Q}^*} (p, \tau)$ and $(q, \sigma') \leq_{\mathbb{P}*\mathbb{Q}} (q, \sigma)$. Let σ' be a term of minimal rank such that $\|\sigma' = \sigma\|_{\mathbb{P}} \geq q$ and $\|\sigma' = \tau\|_{\mathbb{P}} \geq \neg q$. Then $q \Vdash \sigma' \leq \sigma$, hence $(q, \sigma') \leq_{\mathbb{P}*\mathbb{Q}} (q, \sigma)$. Also $\neg q \leq \|\sigma' = \tau\|$, hence $q \vee (\neg q) \leq \|\sigma' \leq \tau\|$ so $\|\sigma' \leq \tau\| = 1$ and $(q, \sigma') \leq_{\mathbb{P}\times\mathbb{Q}^*} (p, \tau)$ as desired.

The proof of (b) is similar and (c) is standard. ⊣ (Lemma 2.6)

If $\mathbb{Q}$ is in $V^{\mathbb{P}}$ we shall use the notation $A(\mathbb{P} ; \mathbb{Q})$ for the termspace partial ordering.

We now make a definition we shall use to get an easy hold on the chain condition of the termspace partial ordering.

DEFINITION 2.7.

(a) Let $\mathbb{P}$ and $\mathbb{Q}$ be partial orderings. $\mathbb{Q}$ is said to be **representable** in $\mathbb{P}$ if and only if there is an incompatibility preserving map $i : \mathbb{Q} \to \mathbb{P}$. (In other words: if q and q' are incompatible in $\mathbb{Q}$, $i(q)$ and $i(q')$ are incompatible in $\mathbb{P}$.)

(b) Let $\mathbb{P}$ and S be partial orderings. Define $\mathcal{F}(\mathbb{P}, S)$ as the partial ordering with domain

$$\{f : f : \mathbb{P} \to S \text{ and f is order preserving}\}$$

The ordering of $\mathcal{F}(\mathbb{P}, S)$ is $f \leq g$ if and only if for all $p \in \mathbb{P}$ $f(p) \leq g(p)$.

If $\mathbb{Q}$ is representable in $\mathbb{P}$ then the chain condition of $\mathbb{P}$ gives an upper bound on the chain condition for $\mathbb{Q}$.

EXAMPLE 2.8. (a) Let $\mathbb{P}$ and S be partial orderings and $\prod_{p \in \mathbb{P}} S$ be the product of $|\mathbb{P}|$ copies of S, one for each element of $\mathbb{P}$. Then $\mathcal{F}(\mathbb{P}, S)$ is representable in $\prod_{p \in \mathbb{P}} S$, namely map $f \in \mathcal{F}(\mathbb{P}, S)$ to $g \in \prod_{p \in \mathbb{P}} S$ where g has the value $f(p)$ on the pth copy of S (so g is pointwise equal to f).

(b) Let $\mathbb{P}$ be a κ-c.c. partial ordering. Let $S^{\mathbb{P}}(\kappa, \lambda)$ be the Silver collapse of λ to κ^+ defined in $V^{\mathbb{P}}$. Let $S(\kappa, \lambda)$ be the Silver collapse of λ to κ^+ defined in V. Then $(S^{\mathbb{P}}(\kappa, \lambda))^*$ is representable in $\mathcal{F}(\mathbb{P} ; S(\kappa, \lambda))$: define a map $i : (S^{\mathbb{P}}(\kappa, \lambda))^* \to \mathcal{F}(\mathbb{P} ; S(\kappa, \lambda))$ by $i(\tau) = f_\tau$ where

$$f_\tau(p) = \{(\alpha, \beta, \gamma) : p \Vdash \tau(\alpha, \beta) = \gamma.\}$$

Using the κ-c.c. of $\mathbb{P}$ we see that $|f_\tau(p) \leq \kappa|$. Further, $p \Vdash f_\tau(p) \in S^{\mathbb{P}}(\kappa, \lambda)$, so there is a $\xi < \kappa$ such that $\operatorname{dom} f_\tau(p) \subseteq \lambda \times \xi$. Hence $f_\tau(p) \in S(\kappa, \lambda)$. If $p \leq_{\mathbb{P}} p'$ and $p' \Vdash \tau(\alpha, \beta) = \gamma$ then $p \Vdash \tau(\alpha, \beta) = \gamma$. Thus f_τ is order preserving and $f_\tau \in \mathcal{F}(\mathbb{P}, S)$.

We must see that i preserves incompatibility. Let $\tau, \tau' \in (S^{\mathbb{P}}(\kappa, \lambda))^*$, τ and τ' incompatible. Then there is a $p \in \mathbb{P}$,

$$p \Vdash_{\mathbb{P}} \tau \text{ and } \tau' \text{ are incompatible in } S^{\mathbb{P}}(\kappa, \lambda).$$

Pick $p' \leq p, \gamma \neq \gamma', p' \Vdash \tau(\alpha, \beta) = \gamma$ and $p' \Vdash \tau'(\alpha, \beta) = \gamma'$. Then $f_\tau(p')$ is incompatible with $f_{\tau'}(p')$, and hence f_τ is incompatible with $f_{\tau'}$.

(In fact if $|\mathbb{P}| = \kappa$ it is possible to see that $(S^{\mathbb{P}}(\kappa, \lambda))^* \cong \{f : f : \mathbb{P} \to S(\kappa, \lambda)$, f is order preserving and there is a $\xi < \kappa$ $\bigcup_{p \in \mathbb{P}} \operatorname{dom} f(p) \subseteq \lambda \times \xi\}$. We leave this to the reader.)

If $\mathbb{Q} \in V^{\mathbb{P}}$ and $\|\mathbb{Q}$ is λ-closed$\|_{\mathbb{P}} = 1$ and $\langle \tau_\alpha : \alpha < \beta \rangle$ $(\beta < \lambda)$ is a descending sequence in $\mathbb{Q}^*$ then $\|$there is a $\tau \in \mathbb{Q}$, $\tau \leq \tau_\alpha$ for all $\alpha\|_{\mathbb{P}} = 1$. Let τ be a term for an element of $\mathbb{Q}$ such that for each α $\|\tau \leq \tau_\alpha\|_{\mathbb{P}} = 1$. Then $\tau \leq_{\mathbb{Q}^*} \inf_{\alpha<\beta} \tau_\alpha$. Hence $\mathbb{Q}^*$ is λ-closed.

Lemma 2.9. Let $\mathbb{R} \in V^{\mathbb{P}*\mathbb{Q}}$ be a term for a partial ordering. Then $A(\mathbb{P} ; A(\mathbb{Q} ; \mathbb{R})) \cong A(\mathbb{P} * \mathbb{Q} ; \mathbb{R})$.

Proof. If $\tau \in A(\mathbb{P} ; A(\mathbb{Q} ; \mathbb{R})))$ let $\pi(\tau) \in A(\mathbb{P} * \mathbb{Q} ; \mathbb{R})$ be a term τ^* such that $\|\tau^* = ((\tau)^{V^{\mathbb{P}}})^{V^{\mathbb{Q}}}\|_{\mathbb{P}*\mathbb{Q}} = 1$. We verify that π is an isomorphism. π is clearly order preserving. Suppose $\tau, \sigma \in A(\mathbb{P} ; S(\mathbb{Q} ; \mathbb{R}))$, $\|\tau = \sigma\|_{\mathbb{P}} \neq 1$. Let $p \in \mathbb{P}$, $p \Vdash$"$\tau \neq \sigma$ in $A(\mathbb{Q} ; \mathbb{R})$". Pick $q \in V^{\mathbb{P}}$, $\|q \in \mathbb{Q}\| \geq p$, with $p \Vdash$"$q \Vdash (((\tau)^{V^{\mathbb{P}}})^{V^{\mathbb{Q}}}) \neq ((\sigma)^{V^{\mathbb{P}}})^{V^{\mathbb{Q}}}$." Then $(p, q) \in \mathbb{P}*\mathbb{Q}$ and $(p, q) \Vdash \tau^* \neq \sigma^*$. Let $\tau^* \in V^{\mathbb{P}*\mathbb{Q}}$, $\|\tau^* \in \mathbb{R}\|_{\mathbb{P}*\mathbb{Q}} = 1$. In $V^{\mathbb{P}}$, there is a term τ for an element of $V^{\mathbb{P}*\mathbb{Q}}$ such that $\big\| \|\tau = \tau^*\|_{\mathbb{Q}} = 1 \big\|_{\mathbb{P}} = 1$. Then $\pi(\tau) = \tau^*$. Thus π is one-to-one, onto and order preserving. ⊣ (Lemma 2.9)

The following lemma is standard:

Lemma 2.10. Let λ be a measurable cardinal. Let $\alpha < \lambda$ and $\langle \mathbb{P}_\gamma : \gamma < \lambda \rangle$ be a sequence of λ-c.c. partial orderings. Then $\prod_{\alpha\text{-supports}} \{\mathbb{P}_\gamma : \gamma < \lambda\}$ has the λ-c.c. (The term "α-supports" refers to the ideal of subsets of λ with cardinality $\leq \alpha$.)

We shall be interested in constructing partial orderings with nice embedding properties. We make the following definition, which is similar to one that will appear in [FL88].

Our partial orderings will be in the form $\mathbb{R}(\kappa, \lambda)$ for all regular $\kappa > \omega$ and all measurable $\lambda > \kappa$. The partial orderings $\mathbb{R}(\kappa, \lambda)$ are given by a uniform definition in the next few paragraphs.

If λ is greater than the first measurable above κ, the partial ordering $\mathbb{R}(\kappa, \lambda)$ is the product $\prod_{n \in \omega} S^n(\kappa, \lambda)$, where the partial orderings $S^n(\kappa, \lambda)$ are defined by induction on n.

Assume that we have defined $\mathbb{R}(\beta, \alpha)$ for all regular $\beta, \omega < \beta < \alpha$ and all inacessible $\alpha < \lambda$. We now define $\mathbb{R}(\beta, \lambda)$.

Case 1: λ is the first measurable above β. Let $\mathbb{R}(\beta, \lambda) = S(\beta, \lambda) =$ the Silver collapse of λ to β^+.

Case 2: otherwise.

By induction on n, we define $S^n(\alpha, \lambda)$ for all regular $\alpha, \beta \leq \alpha < \lambda$. We shall have a uniform definition of $S^n(\alpha, \lambda)$ no matter which model we define it in, hence we define it only in the ground model. Then $(S^n(\alpha, \lambda))^{V^{\mathbb{R}(\beta,\alpha)}}$ is the partial ordering constructed in $V^{\mathbb{R}(\beta,\alpha)}$ using this definition.

If $n = 0$:

$$\text{Let } S^0(\alpha, \lambda) = S(\alpha, \lambda) = \text{ the Silver collapse of } \lambda \text{ to } \alpha^+.$$

Assume that we have defined $S^n(\alpha, \lambda)$ for all regular α.

For $n + 1$:

Let $S^{n+1}(\alpha, \lambda) =$

$$\prod_{\alpha\text{-supports}} \{A(\mathbb{R}(\alpha, \gamma) \,;\, S^n(\gamma, \lambda)) \,:\, \alpha < \gamma < \lambda \text{ and } \gamma \text{ is measurable}\}.$$

Lemma 2.11. For all regular $\beta > \omega$ and all measurable λ, $\mathbb{R}(\beta, \lambda)$ is λ-c.c. and has cardinality λ.[4]

Proof. By Lemma 2.10, it is enough to see that each $S^n(\beta, \lambda)$ has the λ-c.c. We show by induction on λ and n, that for all measurable $\alpha_1, \dots, \alpha_m$ $A(\overset{m-1}{\underset{i=1}{*}} \mathbb{R}(\alpha_i, \alpha_{i+1}) \,;\, S^n(\alpha_m, \lambda))$ has the cardinality λ and the λ-c.c.

Assume that this is true for all $\lambda' < \lambda$. We first consider the case $n = 0$. By Example 2.8 we know that $A(\overset{m-1}{\underset{i=1}{*}} \mathbb{R}(\alpha_i, \alpha_{i+1}) \,;\, S^0(\alpha_m, \lambda))$ is representable in $\mathcal{F}(\overset{m-1}{\underset{i=1}{*}} \mathbb{R}(\alpha_i, \alpha_{i+1}) \,;\, S^0(\alpha_m, \lambda))$. This in turn is representable in the product of α_m copies of $S^0(\alpha_m, \lambda)$. By Lemma 2.10, this has the λ-c.c.

Assume that $A(\overset{m-1}{\underset{i=1}{*}} \mathbb{R}(\alpha_i, \alpha_{i+1}) \,;\, S^n(\alpha_m, \lambda))$ is λ-c.c. for all $\alpha_1, \dots, \alpha_m$ measurable. We want for each $\alpha_1, \dots, \alpha_m$ measurable, $A(\overset{m-1}{\underset{i=1}{*}} \mathbb{R}(\alpha_i, \alpha_{i+1}) \,;\, S^{n+1}(\alpha_m, \lambda))$ is λ-c.c.

Fix a sequence $\alpha_1, \dots, \alpha_m$. In $V^{\overset{m-1}{\underset{i=1}{*}} \mathbb{R}(\alpha_i, \alpha_{i+1})}$, $S^{n+1}(\alpha_m, \lambda) =$

$$\prod_{\alpha_m\text{-supports}} \{A(\mathbb{R}(\alpha_m, \beta) \,;\, S^n(\beta, \lambda)) \,:\, \alpha_m < \beta < \lambda \text{ and } \beta \text{ is measurable}\}.$$

Since $\overset{m-1}{\underset{i=1}{*}} \mathbb{R}(\alpha_i, \alpha_{i+1})$ has the α_m-c.c., if $\tau \in A(\overset{m-1}{\underset{i=1}{*}} \mathbb{R}(\alpha_i, \alpha_{i+1}) \,;\, S^{n+1}(\alpha_m, \lambda))$ there is a set $D \subseteq \lambda$ lying in V such that $|D| = \alpha_m$ and $\|\operatorname{supp} \tau \subseteq D\|_{\overset{m-1}{\underset{i=1}{*}} \mathbb{R}(\alpha_i, \alpha_{i+1})} = 1$. For each $\beta \in D$, we get a term τ_β such that $\|\tau(\beta) =$

[4] In fact we could have replaced the requirement that λ be measurable by asking that λ be Mahlo throughout, but we leave this to the reader.

$\tau_\beta\|_{\overset{m-1}{\underset{i=1}{*}}\mathbb{R}(\alpha_i,\alpha_{i+1})} = 1$ and $\tau_\beta \in A(\overset{m-1}{\underset{i=1}{*}}\mathbb{R}(\alpha_i, \alpha_{i+1})\ ;\ A(\mathbb{R}(\alpha_m, \beta)\ ;\ S^n(\beta, \lambda)))$. It is easy to verify that the map $\tau \mapsto \langle \tau_\beta : \beta \in D\rangle$ is an isomorphism between $A(\overset{m-1}{\underset{i=1}{*}}\mathbb{R}(\alpha_i, \alpha_{i+1})\ ;\ S^{n+1}(\alpha_m, \lambda))$ and $\prod_{\alpha_m\text{-supports}}\{A(\overset{m-1}{\underset{i=1}{*}}\mathbb{R}(\alpha_i, \alpha_{i+1})\ ;\ A(\mathbb{R}(\alpha_m, \beta)\ ;\ S^n(\beta, \lambda)))\ \alpha_m < \beta < \lambda$ and β is measurable$\}$. By Lemma 2.9,

$$A(\overset{m-1}{\underset{i=1}{*}}\mathbb{R}(\alpha_i, \alpha_{i+1})\ ;\ A(\mathbb{R}(\alpha_m, \beta)\ ;\ S^n(\beta, \lambda))) \cong A(\overset{m-1}{\underset{i=1}{*}}\mathbb{R}(\alpha_i, \alpha_{i+1}) * \mathbb{R}(\alpha_m, \beta)\ ;\ S^n(\beta, \lambda)).$$

By the induction hypothesis for n, $A(\overset{m-1}{\underset{i=1}{*}}\mathbb{R}(\alpha_i, \alpha_{i+1}) * \mathbb{R}(\alpha_m, \beta)\ ;\ S^n(\beta, \lambda))$ has the λ-c.c. and thus by 2.10 $A(\overset{m-1}{\underset{i=1}{*}}\mathbb{R}(\alpha_i, \alpha_{i+1})\ ;\ S^{n+1}(\alpha_m, \lambda))$ has λ-c.c.

⊣ (Lemma 2.11)

We now establish the properties we need to satisfy the conditions of Kunen's Theorem:

LEMMA 2.12. For each $\kappa > \omega$, κ regular and each λ measurable:
(a) $\prod_{\substack{n\in\omega\\ n\geq 1}} S^n(\kappa, \lambda)$ is κ^+-closed.
(b) $\mathbb{R}(\kappa, \lambda) = \prod_{n\in\omega} S^n(\kappa, \lambda)$ is κ-closed.
(c) If α is measurable and $\mathrm{id}\colon \mathbb{R}(\kappa, \alpha) \hookrightarrow \mathbb{R}(\kappa, \lambda)$ then there is a map φ extending id such that

$$\varphi\colon \mathbb{R}(\kappa, \alpha) \times \prod_{n\in\omega} A(\mathbb{R}(\kappa, \alpha)\ ;\ S^n(\alpha, \lambda)) \hookrightarrow \mathbb{R}(\kappa, \lambda).$$

(d) If α is measurable and $\mathrm{id}\colon \mathbb{R}(\kappa, \alpha) \hookrightarrow \mathbb{R}(\kappa, \lambda)$ then there is a map ψ extending id,

$$\psi\colon \mathcal{B}(\mathbb{R}(\kappa, \alpha) * \mathbb{R}(\alpha, \lambda)) \hookrightarrow \mathcal{B}(\mathbb{R}(\kappa, \lambda)).$$

PROOF. (a) We show by induction on $n > 0$ that for all regular α, $S^n(\alpha, \lambda)$ is α^+-closed.

$n = 1$: $S^1(\alpha, \lambda) = \prod_{\alpha\text{-supports}}\{A(\mathbb{R}(\alpha, \beta)\ ;\ S^0(\beta, \lambda))\ \alpha < \beta < \lambda$ and β is measurable$\}$

By our remarks preceding Lemma 2.9, each $A(\mathbb{R}(\alpha, \beta)\ ;\ S^0(\beta, \lambda))$ is β-closed and hence α^+-closed. Thus $\prod_{\alpha\text{-supports}}\{A(\mathbb{R}(\alpha, \beta)\ ;\ S^0(\beta, \lambda))\ \alpha < \beta < \lambda$ and β is measurable$\}$ is α^+-closed.

Assume that for all regular β, $S^n(\beta, \lambda)$ is β^+-closed and $n \geq 1$. Then, $A(\mathbb{R}(\alpha, \beta)\ ;\ S^n(\beta, \lambda))$ is β^+-closed and hence α^+-closed. Again this implies $\prod_{\alpha\text{-supports}}\{A(\mathbb{R}(\alpha, \beta)\ ;\ S^0(\beta, \lambda))\ \alpha < \beta < \lambda$ and β is measurable$\}$ is α^+-closed.

(b) $\mathbb{R}(\kappa, \lambda) = S^0(\kappa, \lambda) \times \prod_{n\geq 1} S^n(\kappa, \lambda)$ where $S^0(\kappa, \lambda)$ is the Silver collapse of λ to κ^+. The Silver collapse is κ-closed and by (a), $\prod_{n\geq 1} S^n(\kappa, \lambda)$ is κ^+-closed. Hence the product is κ-closed.

(c) Suppose id: $\mathbb{R}(\kappa,\alpha) \hookrightarrow \mathbb{R}(\kappa,\lambda)$. By the definition of $S^{n+1}(\kappa,\lambda)$, if α is measurable then $A(\mathbb{R}(\kappa,\alpha)\ ;\ S^n(\alpha,\lambda))$ is a factor of $S^{n+1}(\kappa,\lambda)$. Let φ_n be the map from $A(\mathbb{R}(\kappa,\alpha)\ ;\ S^n(\alpha,\lambda))$ to $S^{n+1}(\kappa,\lambda)$ that sends $p \in A(\mathbb{R}(\kappa,\alpha)\ ;\ S^n(\alpha,\lambda))$ to the element $q \in S^{n+1}(\kappa,\lambda)$ that is 1 on each factor of $S^{n+1}(\kappa,\lambda)$ except $A(\mathbb{R}(\kappa,\alpha)\ ;\ S^n(\alpha,\lambda))$ where it is equal to p.[5] The product, $\prod_{n\in\omega} A(\mathbb{R}(\kappa,\alpha)\ ;\ S^n(\alpha,\lambda))$ can be embedded in $\prod_{n\in\omega} S^n(\kappa,\lambda)$ by the map

$$\langle p_n\ :\ n\in\omega\rangle \overset{\varphi_\omega}{\longmapsto} \langle q_n\ :\ n\in\omega\rangle$$

where $q_0 = 1$, $q_{n+1} = \varphi_n(p_n)$.

By Lemma 2.6(c), φ_ω is a neat embedding from $\prod_{n\in\omega} A(\mathbb{R}(\kappa,\alpha)\ ;\ S^n(\alpha,\lambda))$ into $\prod_{n\in\omega} S^n(\kappa,\lambda)$.

Each

$$\begin{aligned} S^{n+1}(\kappa,\lambda) = &\prod_{\kappa\text{-supports}} \{A(\mathbb{R}(\kappa,\beta)\ ;\ S^n(\beta,\lambda))\ :\ \kappa<\beta<\alpha \text{ and } \beta \text{ is measurable.}\} \\ &\times \prod_{\kappa\text{-supports}} \{A(\mathbb{R}(\kappa,\beta)\ ;\ S^n(\beta,\lambda))\ :\ \alpha\le\beta<\lambda \text{ and } \beta \text{ is measurable }\}. \end{aligned}$$

Let

$$S_0^{n+1}(\kappa,\lambda) = \prod_{\kappa\text{-supports}} \{A(\mathbb{R}(\kappa,\beta)\ ;\ S^n(\beta,\lambda)) : \kappa<\beta<\alpha \text{ and } \beta \text{ is measurable.}\},$$

$$S_1^{n+1}(\kappa,\lambda) = \prod_{\kappa\text{-supports}} \{A(\mathbb{R}(\kappa,\beta)\ ;\ S^n(\beta,\lambda))\ :\ \alpha\le\beta<\lambda \text{ and } \beta \text{ is measurable}\}.$$

By rearranging our product we get

$$\mathbb{R}(\kappa,\lambda) = S^0(\kappa,\lambda) \times \prod_{\substack{n\in\omega\\ n\ge 1}} S_0^n(\kappa,\lambda) \times \prod_{\substack{n\in\omega\\ n\ge 1}} S_1^n(\kappa,\lambda).$$

For $n\ge 1$, $\mathrm{id}''S^n(\kappa,\alpha) \subseteq S_0^n(\kappa,\lambda)$. Hence, if id: $\mathbb{R}(\kappa,\alpha)\hookrightarrow\mathbb{R}(\kappa,\lambda)$ then

$$\mathrm{id}\colon \mathbb{R}(\kappa,\alpha) \hookrightarrow S^0(\kappa,\lambda) \times \prod_{1\le n\in\omega} S_0^n(\kappa,\lambda)$$

By the definition of φ_n, $\varphi_n'' A(\mathbb{R}(\kappa,\alpha)\ ;\ S^n(\alpha,\lambda)) \subseteq S_1^n(\kappa,\lambda)$. So,

$$\varphi_\omega'' \prod_{n\in\omega} A(\mathbb{R}(\kappa,\alpha)\ ;\ S^n(\alpha,\lambda)) \subseteq \prod_{n\in\omega} S_1^n(\kappa,\lambda).$$

Thus we have a map $\varphi = \mathrm{id}\times\varphi_\omega$

$$\begin{aligned} \varphi\colon \mathbb{R}(\kappa,\alpha) \times \Big[\prod_{n\in\omega} A(\mathbb{R}(\kappa,\alpha)\ ;\ S^n(\alpha,\lambda))\Big] \hookrightarrow& \\ \Big(S^0(\kappa,\lambda) \times \prod_{1\le n\in\omega} S_0^n(\kappa,\lambda)\Big) &\times \Big[\prod_{1\le n\in\omega} S_1^n(\kappa,\lambda)\Big]\Big). \end{aligned}$$

[5] In essence φ_n is the identity map of $A(\mathbb{R}(\kappa,\alpha)\ ;\ S^n(\alpha,\lambda))$ to its factor in $S^{n+1}(\kappa,\lambda)$.

Hence

$$\varphi\colon \mathbb{R}(\kappa,\alpha)\times\Big(\prod_{n\in\omega} A(\mathbb{R}(\kappa,\alpha)\ ;\ S^n(\alpha,\lambda))\Big)\hookrightarrow \mathbb{R}(\kappa,\lambda).$$

(d) By Lemma 2.6(b), we have a neat embedding

$$\psi_1\colon \mathcal{B}(\mathbb{R}(\kappa,\alpha)*\mathbb{R}(\alpha,\lambda))\hookrightarrow \mathcal{B}\Big(\mathbb{R}(\kappa,\alpha)\times\prod_{n\in\omega} A(\mathbb{R}(\kappa,\alpha)\ ;\ S^n(\alpha,\lambda))\Big)$$

By (c), we have

$$\psi_2\colon \mathcal{B}\Big(\mathbb{R}(\kappa,\alpha)\times\prod_{n\in\omega} A(\mathbb{R}(\kappa,\alpha)\ ;\ S^n(\alpha,\lambda))\Big)\hookrightarrow \mathcal{B}(\mathbb{R}(\kappa,\lambda))$$

Composing we get:

$$\psi=\psi_1\circ\psi_2\colon \mathcal{B}(\mathbb{R}(\kappa,\alpha)*\mathbb{R}(\alpha,\lambda))\hookrightarrow \mathcal{B}(\mathbb{R}(\kappa,\lambda)).$$

This completes the definition and verification of the basic properties of the $\mathbb{R}$'s. ⊣ (Lemma 2.12)

Returning to the proofs of Theorems 1.2 & 1.3:
Let

$$\mathbb{P}=\varinjlim\ \langle \mathbb{R}(\aleph_1,\kappa_0)*\overline{\mathbb{R}(\kappa_0,\kappa_1)}*\overline{\mathbb{R}(\kappa_1,\kappa_2)}*\cdots*\overline{\mathbb{R}(\kappa_{n-1},\kappa_n)}\ :\ n\in\omega\rangle.$$

To see in $V^{\mathbb{P}}$:

(a) for all $m>1$, $\aleph_m$ carries an $\aleph_{m+1}$-saturated ideal
(b) for all $m>n\geq 1$, $(\aleph_{m+1},\aleph_m)\twoheadrightarrow(\aleph_{n+1},\aleph_n)$

It is enough to see that $\mathbb{P}'=\mathbb{R}(\aleph_1,\kappa_0)*\cdots*\bar{\mathbb{R}}(\kappa_{n-1},\kappa_n)$ and $\mathbb{Q}=\mathbb{R}^{\mathbb{P}'}(\kappa_n,\kappa_{n+1})$ together with the embedding j_n satisfy the conditions for Kunen's Theorem. Conditions (a), (b) and (c) follow directly from Lemma 2.11 and the remarks before Lemma 2.9.

Since $\mathbb{P}'$ is κ_n-c.c., $j_n\colon \mathbb{P}'\to j(\mathbb{P}')$ is a neat embedding. Since $\mathbb{P}'\subseteq V_{\kappa_n}$, $j_n\restriction\mathbb{P}'=\mathrm{id}$ and hence $\mathrm{id}\colon \mathbb{P}'\hookrightarrow j(\mathbb{P}')$. Unraveling this we get

$$\begin{aligned}\mathrm{id}\colon \mathbb{R}(\aleph_1,\kappa_0)*\bar{\mathbb{R}}(\kappa_0,\kappa_1)*\cdots*\bar{\mathbb{R}}(\kappa_{n-1},\kappa_n)&\\ \hookrightarrow \mathbb{R}(\aleph_1,\kappa_0)*\bar{\mathbb{R}}(\kappa_0,\kappa_1)*\cdots*\bar{\mathbb{R}}(\kappa_{n-1},\kappa_{n+1}).&\end{aligned}$$

By Lemma 2.12 there is a generic object $G*H\subseteq\mathbb{P}'*\mathbb{Q}$ in $V^{j_n(\mathbb{P}')}$. In $V^{j(\mathbb{P}')}$, $H\cong H_0\times H_{>0}$, where $H_0\subseteq S^0(\kappa_n,\kappa_{n+1})$ and $H_{>0}\subseteq\prod_{k>0}S^k(\kappa_n,\kappa_{n+1})$. Still in $V^{j(\mathbb{P}')}$: $j''(H)=j''(H_0)\times j''(H_{>0})\subseteq S^0(\kappa_{n+1},\kappa_{n+2})\times\prod_{k>0}S^k(\kappa_{n+1},\kappa_{n+2})$ and $j''H$ has cardinality κ_{n+1}. By Lemma 2.12, $\prod_{k>0}S^k(\kappa_{n+1},\kappa_{n+2})$ is κ_{n+1}^+-closed and hence there is a $q_{>0}\in\prod_{k>0}S^k(\kappa_{n+1},\kappa_{n+2})$ such that for all $p\in j''H_{>0}$, $q_{>0}\leq p$. Let $q_0=\bigcup_{\tau\in H_0}(j(\tau))^{V^{j(\mathbb{P}')}}$. We show that $q_0\in S^0(\kappa_{n+1},\kappa_{n+2})$. Clearly q_0 is a function satisfying from $\kappa_{n+2}\times\kappa_{n+1}$ all but the cardinality conditions for being an element of $S^0(\kappa_{n+1},\kappa_{n+2})$. To see that it satisfies these, we note that for each $\tau \in (S^0(\kappa_n, \kappa_{n+1}))^{V^{\mathbb{P}'}}$ there is a $\xi < \kappa_n$ such that $|\mathrm{dom}\,\tau \subseteq \kappa_{n+1} \times \xi|_{\mathbb{P}'} = 1$. Hence, $\mathrm{dom}\, q_0 \subseteq \kappa_{n+2} \times \kappa_n$. Further,

$$|\mathrm{dom}\, q_0| = \left| \bigcup_{\tau \in H_0} \mathrm{dom}\, j(\tau)^{V^{j(\mathbb{P}')}} \right| = \kappa_{n+1} \times \kappa_{n+1} = \kappa_{n+1}.$$

Hence, $q_0 \in (S^0(\kappa_{n+1}, \kappa_{n+2}))^{V^{j(\mathbb{P}')}}$. Let

$$q = (q_0, q_{>0}) \in S^0(\kappa_{n+1}, \kappa_{n+2}) \times \prod_{k>0} S^k(\kappa_{n+1}, \kappa_{n+2}).$$

It is easy to check that in $V^{j(\mathbb{P})}$, if $p \in H$, $j(p) \geq q$. From this we conclude that $(1, q) \in j(\mathbb{P}') * \mathbb{R}(\kappa_{n+1}, \kappa_{n+2})$ functions as a master condition and satisfies (d) of Kunen's Theorem.

Forcing with $\mathbb{P} \times C(\aleph_0, \aleph_1)$ we get a model as in Theorem 1.2. This completes the proof of Theorem 1.2. For Theorem 1.3 we remark that if our original sequence $\langle j_n : n \in \omega \rangle$ was a sequence of almost huge embeddings then the $\mathbb{R}'s$ satisfy the Magidor conditions and our proof would go through to get the saturated ideals.(Although apparently not the strong transfer principles.)

We end this section with some problems. First, not much is known about transfer principles that go "infinite distances". A problem along this line might be:

Open Problem 2.13. It is consistent to have $(\aleph_{\omega+1}, \aleph_\omega) \twoheadrightarrow (\aleph_1, \aleph_0)$?

Magidor has remarked that if the GCH holds then $(\aleph_{\omega+1}, \aleph_\omega) \not\twoheadrightarrow (\aleph_{m+1}, \aleph_m)$ for $m > 0$. Along these lines, the author has shown:

Con(ZFC+"there is a 2-huge cardinal") implies
Con(ZFC+"there is a supercompact cardinal κ with $(\kappa^+, \kappa) \twoheadrightarrow (\aleph_1, \aleph_0)$").

This might be relevant to this problem.

In this paper we have "gap one" transfer principles. In [For82] it was shown that Con(ZFC+"there is a 2-huge cardinal") implies Con(ZFC+$(\aleph_3, \aleph_1) \twoheadrightarrow (\aleph_2, \aleph_0)$).

Open Problem 2.14. Is it consistent to have for all $m > n$ $(\aleph_{m+2}, \aleph_m) \twoheadrightarrow (\aleph_{n+2}, \aleph_n)$?

A general solution to this problem would probably solve many problems in this area.

Another problem is:

Open Problem 2.15. Does "for all $m > n$ $(\aleph_{m+1}, \aleph_m) \twoheadrightarrow (\aleph_{n+1}, \aleph_n)$" imply "$\aleph_\omega$ is Jonsson"?

Finally, since the method of proof for Theorems 1.2 & 1.3 was the same it would be interesting to establish a direct relationship between saturated ideals and transfer properties. Shelah has shown that "there is an $\aleph_2$-complete ideal $\mathcal{I}$ on $\aleph_2$ such that $\wp(\aleph_2)/\mathcal{I}$ has a dense ω-closed subset" implies $(\aleph_2, \aleph_1) \twoheadrightarrow (\aleph_1, \aleph_0)$. The proof however is tied very closely to the cardinals $\aleph_2$, $\aleph_1$ and $\aleph_0$. What is desired is some more general statement for cardinals bigger than $\aleph_2$. For the sake of concreteness we mention a possibility.

Open Problem 2.16. Does "$\aleph_2$ carries an $\aleph_3$-saturated ideal and $\aleph_3$ carries precipitous ideal" imply $(\aleph_3, \aleph_2) \twoheadrightarrow (\aleph_2, \aleph_1)$?

§3. We now turn our attention to proving Theorem 1.4. For this section a "saturated ideal" on κ will mean a normal, κ-complete, κ^+-saturated ideal on κ.

Proof of Theorem 1.4. Let κ be a huge cardinal. Then there is a set $A \subseteq \kappa$ with measure one with respect to the measure on κ induced by the huge embedding, such that for all $\alpha, \beta \in A$, $\beta > \alpha$ implies: There is an almost huge embedding $j_{\alpha,\beta}$ with $\mathrm{crit}(j_{\alpha,\beta}) = \alpha$ and $j_{\alpha,\beta} = \beta$. Fix such a set A. If $\alpha \in A$, let α' be the successor of $\alpha \in A$.

We need to redefine our $\mathbb{R}(\kappa, \lambda)$'s. Assume by induction that we have defined $\mathbb{R}(\kappa, \lambda')$ for all regular $\kappa > \omega$ and all measurable $\lambda' < \lambda$, and that $\mathbb{R}(\kappa, \lambda')$ is closed, has cardinality λ' and is λ'-c.c.

If $X \subseteq \lambda$ is a set for measurable cardinals we define the **standard iteration** I_X as follows:

(a) I_X is an iteration of length $\leq \lambda$ with Easton supports. We shall define the notion of an ordinal in X being mentioned in $(I_X)_\alpha$ by induction on α. To start with, no ordinals are mentioned in $(I_X)_0$.

(b) At stage $\alpha + 1$: If there are fewer than five elements of X not mentioned in $(I_X)_\alpha$, then $I_X = (I_X)_\alpha$.

Otherwise, let $\gamma_0 < \gamma_1 < \cdots < \gamma_4$ be the first five elements of X not mentioned in $(I_X)_\alpha$. If $\alpha \geq \gamma_0$ let,

$$(I_X)_{\alpha+1} = (I_X)_\alpha * [\bar{\mathbb{R}}(\gamma_0, \gamma_1) * \bar{\mathbb{R}}(\gamma_1, \gamma_2) * \bar{\mathbb{R}}(\gamma_2, \gamma_3) * \bar{\mathbb{R}}(\gamma_3, \gamma_4)].$$

The set of elements of X mentioned in $(I_X)_{\alpha+1}$ is defined to be elements of X mentioned in $(I_X)_\alpha$ together with $\gamma_0, \ldots, \gamma_4$.

If $\alpha < \gamma_0$ then $(I_X)_{\alpha+1} = (I_X)_\alpha * 1$, and the elements of X mentioned in $(I_X)_{\alpha+1}$ are the same as the elements of X mentioned in $(I_X)_\alpha$.

Informally, this iteration is the result of iterating, five at a time, the elements of X strung together by the $\mathbb{R}$'s.

Proposition 3.1. For every X, I_X is λ-c.c. and $I_X/(I_X)_\alpha$ is γ_0-closed in $V^{(I_X)_\alpha}$, where γ_0 is the least ordinal mentioned in $(I_X)_{\alpha+1}$ not mentioned in $(I_X)_\alpha$.

Proof. This uses standard facts about iterations with Easton support.
$\dashv$ (Proposition 3.1)

We can now define our new version of $\mathbb{R}(\kappa, \lambda)$ for all regular $\kappa > \omega$. Again $\mathbb{R}(\kappa, \lambda) = \prod_{n \in \omega} S^n(\kappa, \lambda)$ where $S^n(\alpha, \lambda)$ are defined by induction on n, simultaneously for all regular α, $\omega < \alpha < \lambda$.

Let $S^0(\alpha, \lambda) =$ the Silver collapse of λ to α^+

$$S^{n+1}(\alpha, \lambda) = \prod_{\alpha\text{-supports}} \{A(\mathbb{R}(\alpha, \beta); C(\beta, \gamma) \times [I_X * \bar{S}^n(\gamma, \lambda)]) : \\ \alpha < \beta < \gamma < \lambda, \beta, \gamma \text{ are measurable and} \\ X \subseteq \gamma \backslash \beta \text{ is a set of measurable cardinals.}\}$$

Let $\mathbb{R}(\alpha, \lambda) = \prod_{n \in \omega} S^n(\alpha, \lambda)$.

Using the techniques of §2 it is now easy to verify that if $\omega < \alpha < \gamma < \lambda$, α is regular, β, γ and λ are measurable and $X \subseteq \gamma \backslash \beta$ is a set of measurable cardinals then there is a map ψ

$$\psi \colon \prod_{n \in \omega} A(\mathbb{R}(\alpha, \beta)\ ;\ C(\beta, \gamma) \times [I_X * \bar{S}^n(\gamma, \lambda)]) \hookrightarrow \mathbb{R}(\alpha, \lambda).$$

From this (as before) we conclude that if $\mathrm{id} \colon \mathbb{R}(\alpha, \beta) \hookrightarrow \mathbb{R}(\alpha, \lambda)$ then there is a map φ

$$\varphi \colon \mathbb{R}(\alpha, \beta) \times \prod_{n \in \omega} A(\mathbb{R}(\alpha, \beta)\ ;\ C(\beta, \gamma) \times [I_X * \bar{S}^n(\gamma, \lambda)]) \hookrightarrow \mathbb{R}(\alpha, \lambda).$$

Again this implies that: if $\mathrm{id} \colon \mathbb{R}(\alpha, \beta) \hookrightarrow \mathbb{R}(\alpha, \lambda)$ then there is a map φ

$$\varphi \colon \mathcal{B}(\mathbb{R}(\alpha, \beta) * [\bar{C}(\beta, \gamma) \times (I_X * \bar{\mathbb{R}}(\gamma, \lambda))]) \hookrightarrow \mathcal{B}(\mathbb{R}(\kappa, \lambda))$$

LEMMA 3.2. If $\kappa > \omega$ is regular and $\lambda > \kappa$ is measurable

(a) $\mathbb{R}(\kappa, \lambda)$ has the λ-c.c.
(b) $\mathbb{R}(\kappa, \lambda)$ is κ-closed.
(c) In $V^{\mathbb{R}(\kappa, \lambda)}$, $\lambda = \kappa^+$.
(d) $|\mathbb{R}(\kappa, \lambda)| = \lambda$.
(e) $\prod_{n \in \omega, n \geq 1} S^n(\kappa, \lambda)$ is κ^+-closed.
(f) $\prod_{n \in \omega} S^n(\kappa, \lambda)$ is κ-closed.

PROOF. Essentially the same as in Section 2. ⊣ (Lemma 3.2)

To build our model of Theorem 1.3 we start by first doing the following preparatory forcing. (Recall that κ is our huge cardinal, j our huge embedding, and the definition of A.)

Let

$$\mathbb{P}_{\leq \kappa} = I_A * \bar{\mathbb{R}}(\kappa, \kappa') * \bar{\mathbb{R}}(\kappa', \kappa'') * \bar{\mathbb{R}}(\kappa'', \kappa''') * \bar{\mathbb{R}}(\kappa''', \kappa^{\mathrm{iv}}),$$

where I_A is the standard iteration of A and $\kappa', \kappa'', \kappa''', \kappa^{\mathrm{iv}}$ are the first four elements of $j(A)$ above κ.

PROPOSITION 3.3. In $\mathbb{P}_{\leq \kappa}$, we have that

(a) $\kappa' = \kappa^+, \kappa'' = \kappa^{++}, \kappa''' = \kappa^{+3}, \kappa^{\text{iv}} = \kappa^{+4}$,
(b) κ remains κ^{+5} supercompact by a measure μ_5 such that there is a set of μ_5-measure-one worth of $x \in \mathbb{P}_\kappa(\kappa^{+5})$ such that if $\alpha = \text{crit}(x)$, then $\alpha^+, \alpha^{+2}, \alpha^{+3}$ carry saturated ideals, and
(c) $2^\kappa = \kappa^+$.

PROOF. (a) I_A is κ-c.c. by earlier comments. In V^{I_A}, $\mathbb{R}(\kappa, \kappa')$ makes κ' into κ^+ and is κ-closed. In the $V^{I_A * \bar{\mathbb{R}}(\kappa,\kappa')}$, $\bar{\mathbb{R}}(\kappa', \kappa'') * \bar{\mathbb{R}}(\kappa'', \kappa''') * \bar{\mathbb{R}}(\kappa''', \kappa^{\text{iv}})$ is κ'-closed. Hence, in $V^{\mathbb{P}_{\leq\kappa}}$, $\kappa' = \kappa^+$. In $V^{I_A * \mathbb{R}(\kappa,\kappa')}$, $\mathbb{R}(\kappa', \kappa'')$ makes κ'' into $(\kappa')^+$. Since $\mathbb{R}(\kappa'', \kappa''') * \bar{\mathbb{R}}(\kappa''', \kappa^{\text{iv}})$ is $< \kappa''$-closed in $V^{I_A * \mathbb{R}(\kappa,\kappa') * \mathbb{R}(\kappa',\kappa'')}$, $\kappa'' = (\kappa')^+$ in $V^{\mathbb{P}_{\leq\kappa}}$. Hence $\kappa'' = \kappa^{+2}$ in $V^{\mathbb{P}_{\leq\kappa}}$. Similarly we see that in $V^{\mathbb{P}_{\leq\kappa}}$, $\kappa''' = \kappa^{+3}$ and $\kappa^{\text{iv}} = \kappa^{+4}$.

(b) This is a standard argument using the fact that $\mathbb{P}_{\leq\kappa}$ is initial segment of $j(\mathbb{P}_{\leq\kappa})$ and $j(\mathbb{P}_{\leq\kappa})/\mathbb{P}_{\leq\kappa}$ is $(\beth_3(\kappa^{+5}))^{V^{\mathbb{P}_{\leq\kappa}}}$-closed in $V^{\mathbb{P}_{\leq\kappa}}$. (*Cf.* [Bau83]). In fact we can require that the measure μ_5 induces on κ extends the measure j induces on κ. Hence, $\{\alpha$: at some stage γ in the iteration I_A, α is the least ordinal not mentioned in $(I_A)_\gamma\}$ has measure one with respect to the measure μ_5 induces on κ.

At some stage γ, if α is the least ordinal not mentioned in $(I_A)_\gamma$ and $\gamma \geq \alpha$,

$$I_A = (I_A)_\gamma * \bar{\mathbb{R}}(\alpha, \alpha') * \bar{\mathbb{R}}(\alpha', \alpha'') * \bar{\mathbb{R}}(\alpha'', \alpha''') * \bar{\mathbb{R}}(\alpha''', \alpha^{\text{iv}}) * (I_A/(I_A)_{\gamma+1}).$$

By Lemma 2.1, it is enough to see that in $V^{(I_A)_{\gamma+1}}$, $\alpha^+, \alpha^{++}, \alpha^{+3}$ carry saturated ideals.

Since, $\alpha', \alpha'' \in A$, there is an almost huge embedding in V, $j_{\alpha,\alpha'}$ such that $j_{\alpha'}(\alpha') = \alpha''$ and $\text{crit}(j_\alpha) = \alpha'$. By Magidor's modification of the Kunen theorem (with $\mathbb{P}' = (I_A)_\alpha * \mathbb{R}(\alpha, \alpha')$ and $\varphi = \bar{\mathbb{R}}(\alpha', \alpha'')$ and $j_{\alpha',\alpha''}$ the almost huge embedding). There is a saturated ideal on α' in $V^{(I_A)_\alpha * \bar{\mathbb{R}}(\alpha,\alpha') * \bar{\mathbb{R}}(\alpha',\alpha'')}$. Since $(I_A)_{\alpha+1}/((I)_\alpha * \mathbb{R}(\alpha, \alpha') * \mathbb{R}(\alpha', \alpha''))$ is α''-closed we can apply Lemma 2.1 to get that there is a saturated ideal on α^+ in $V^{(I_A)_{\alpha+1}}$. Entirely similar arguments work for α^{+2} and α^{+3}.

(c) is standard. ⊣ (Proposition 3.3)

In our final model, κ will become $\aleph_\omega$ using a technique of Magidor [Mag78]. Let $V' = V^{\mathbb{P}_{\leq\kappa}}$.

LEMMA 3.4. Let $\gamma < \beta$ be inaccessible elements of A in V'. In $(V')^{C(\gamma^{+4},\beta)}$, γ^{+4} carries a $(\gamma^{+4})^+$-saturated ideal.

PROOF. In $(V')^{C(\gamma^{+4},\beta)}$, $(\gamma^{+4})^+ = (\beta^+)^{V'}$. Let α_1 and α_2 be the stages where γ and β are the least elements of A not mentioned in $(I_A)_{\alpha_1}$ and $(I_A)_{\alpha_2}$ respectively.

By Proposition 3.1, $\mathbb{P}_{\leq\kappa}/(\mathbb{P}_{\alpha_2} * \bar{\mathbb{R}}(\beta, \beta'))$ is β'-closed. Further $C(\gamma^{+4}, \beta)^{V'} = C(\gamma^{+4}, \beta)^{V^{\mathbb{P}_{\alpha_1+1}}}$ and has cardinality β. Thus we are in position to apply

Lemma 2.1 in $V^{\mathbb{P}_{\alpha_2}*\mathbb{R}(\beta,\beta')}$. Note $(V')^{C(\gamma^{+4},\beta)}$ is an extension of $V^{\mathbb{P}_{\alpha_2}*\mathbb{P}(\beta,\beta')}$ of the form

$$(\beta'\text{-c.c.}) \times (\beta'\text{-closed})$$
$$(i.e., \ C(\gamma^{+4}, \beta) * \mathbb{P}_{\leq\kappa}/(\mathbb{P}_{\alpha_2} * \mathbb{P}(\beta, \beta')))$$

Thus by Lemma 2.1, it is enough to see that γ^{+4} carries a β'-saturated ideal in $V^{\mathbb{P}_{\alpha_2}*\bar{\mathbb{R}}(\beta,\beta')*\bar{C}(\gamma^{+4},\beta)}$. Let $\mathbb{Q}' = \mathbb{P}_{\alpha_1} * \bar{\mathbb{R}}(\gamma, \gamma') * \bar{\mathbb{R}}(\gamma', \gamma'') * \bar{\mathbb{R}}(\gamma'', \gamma''')$. Then $\mathbb{P}_{\alpha_1+1} = \mathbb{Q}' * \mathbb{R}(\gamma''', \gamma^{\text{iv}})$. Let $j_{\gamma^{\text{iv}},\beta'}$ be the almost huge embedding in V with $\text{crit}(j_{\gamma^{\text{iv}},\beta'}) = \gamma^{\text{iv}}$ and $j_{\gamma^{\text{iv}},\beta'}(\gamma^{\text{iv}}) = \beta'$. Then

$$j_{\gamma^{\text{iv}},\beta'}(\mathbb{Q}' * \bar{\mathbb{R}}(\gamma''', \gamma^{\text{iv}})) = \mathbb{Q}' * \mathbb{R}(\gamma''', \beta').$$

Since $\mathbb{Q}' * \bar{\mathbb{R}}(\gamma''', \gamma^{\text{iv}}) \subseteq V_{\gamma^{\text{iv}}}$ and $\mathbb{Q}' * \bar{\mathbb{R}}(\gamma''', \gamma^{\text{iv}})$ is γ^{iv}-c.c., $j \restriction \mathbb{R}(\gamma''', \gamma^{\text{iv}}) = \text{id}$ and hence

$$\text{id}\colon \mathbb{Q}' * \bar{\mathbb{R}}(\gamma''', \gamma^{\text{iv}}) \hookrightarrow \mathbb{Q}' * \bar{\mathbb{R}}(\gamma''', \beta').$$

Thus by our construction of $\bar{\mathbb{R}}(\gamma''', \beta')$, there is an embedding

$$i\colon \mathbb{Q}' * \bar{\mathbb{R}}(\gamma''', \gamma^{\text{iv}}) * [\bar{C}(\gamma^{\text{iv}}, \beta) * \{(\mathbb{P}_{\alpha_2}/\mathbb{P}_{\alpha_1+1}) * \bar{\mathbb{R}}(\beta, \beta')\}] \hookrightarrow \mathbb{Q}' * \bar{\mathbb{R}}(\gamma''', \beta').$$

Since $\bar{C}(\gamma^{\text{iv}}, \beta) \times \{(\mathbb{P}_{\alpha_2}/\mathbb{P}_{\alpha_1+1}) * \bar{\mathbb{R}}(\beta, \beta')\}$ is γ^{iv}-closed in $V^{\mathbb{Q}'*\mathbb{R}(\gamma''',\gamma^{\text{iv}})}$, Magidor's remarks about Kunen's theorem (in Section 2) apply with $\mathbb{P}' = \mathbb{Q}' * \bar{\mathbb{R}}(\gamma''', \gamma^{\text{iv}})$ and $\mathbb{Q} = \bar{C}(\gamma^{\text{iv}}, \beta) \times \{(\mathbb{P}_{\alpha_2}/\mathbb{P}_{\alpha_1+1}) * \bar{\mathbb{R}}(\beta, \beta')\}$ and we conclude that in the model $V^{\mathbb{Q}'*\bar{\mathbb{R}}(\gamma''',\gamma^{\text{iv}})*[\bar{C}(\gamma^{\text{iv}},\beta)\times\{(\mathbb{P}_{\alpha_2}/\mathbb{P}_{\alpha_1+1})*\mathbb{R}(\beta,\beta')\}]}$, γ^{iv} carries a $(\gamma^{\text{iv}})^+$-saturated ideal. Pictorially:

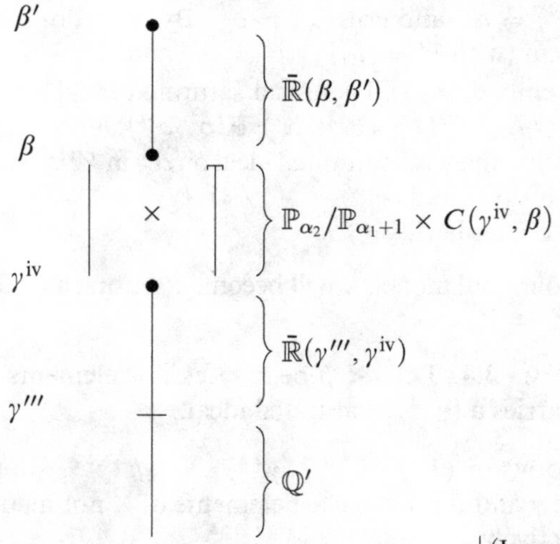

⊣ (Lemma 3.4)

We are now ready to define our forcing conditions for the model in Theorem 1.4.

This forcing is a direct variant of the Magidor forcing [Mag78]. However, in our forcing we collapse the points on the Prikry sequence. This necessitates a slightly different argument which we show below.

For the moment, we shall regard V' as our ground model. Note that in V', κ^+ carries a saturated ideal. We shall define two forcings, $\mathbb{P}$ and $\mathbb{P}^\pi$ with a projection map

$$\pi\colon \mathbb{P} \to \mathbb{P}^\pi.$$

Our final model will be $(V')^{\mathbb{P}^\pi}$. We shall use $\mathbb{P}$ to show that $\mathbb{P}^\pi$ has nice Prikry-type properties.

Let μ_3 be the κ^{+3}-supercompact measure on $\wp_\kappa(\kappa^{+3})$ in V' that is induced by μ_5. For $x \in \wp_\kappa(\kappa^{+3})$, let $x \cap \kappa = \kappa_x$. Let

$$N = \{x : x \in \wp_\kappa(\kappa^{+3}) \text{ and } x \cap \kappa \text{ is an inaccessible cardinal, } x \cap \kappa \in A\}.$$

Standard arguments show that N has μ_3-measure one and that μ_3 is closed under the following kind of diagonal intersection:

Suppose $\rho\colon \wp_\kappa(\kappa^{+3}) \times V_\kappa \to \wp(\wp_\kappa(\kappa^{+3}))$ is such that for all (x, y), $\rho(x, y)$ has the μ_3-measure one. Define $\Delta\rho = \{z :$ for all $x \prec z$ and for all $y \in V_{\text{crit}(z)}, z \in \rho(x, y)\}$. Then $\Delta\rho$ has μ_3-measure one.

A condition $p \in \mathbb{P}$ will be of the form $p = (x_0, p_0, x_1, p_1, \dots, x_n, p_n, f, B)$ where

(a) $x_i \in N$, $x_i \prec x_{i+1}$,
(b) $p_i \in C(\kappa_{x_i}^{\text{iv}}, \kappa_{x_{i+1}})$,
(c) $B \subseteq \wp_\kappa(\kappa^{+3}) \cap N$ and B has μ_3-measure one, and
(d) $f\colon B \to V_\kappa$ and for all $x \in B$, $f(x) \in C(\kappa_x^{\text{iv}}, \kappa)$.

If $q = (x_0, q_0, x_1, q_1, \dots, x_n, q_n, y_1, r_1, y_2, r_2, \dots, y_m, r_m, g, C)$ then $q \le_{\mathbb{P}} p$ if and only if

(a) $y_i \in B$ for all $i \le m$,
(b) $q_i \le p_i$ for all $i \le n$,
(c) $r_i \le f(y_i)$ for all $i \le m$,
(d) $C \subseteq B$, and
(e) for all $x \in C$, $g(x) \le f(x)$.

$\mathbb{P}$ has the Prikry property in the following sense:

LEMMA 3.5. Let $b \in \mathcal{B}(\mathbb{P})$.

(a) There is a condition (g, D) that decides b (by which we mean that either $(g, D) \le b$ or (g, D) is incompatible with b. In symbols: $(g, D) \parallel b$).
(b) If $p = (x_0, p_0, \dots, x_n, p_n, g, D) \in \mathbb{P}$ there is a condition $q \le p$ such that $q = (x_0, p_0, \dots, x_n, p_n, g', D') \parallel b$ and the length of q is equal to the length of p.

A proof of this appears in Magidor [Mag78]; also cf. [FW91]. The fact that the conditions have a slightly different form does not change the proof.

If $G \subseteq \mathbb{P}$ is generic and $p = (x_0, p_0, \ldots, x_n, p_n, g, D) \in G$, then G_p is defined to be $\{(x_0, q_0, \ldots, q_{n-1}, x_n, q_n) :$ there is a $q \in G, \mathrm{lh}(q) = \mathrm{lh}(p), q = (x_0, q_0, \ldots, q_{n-1}, x_n, q_n, g', D')\}$. If $\lambda < \kappa$ then $G \restriction \lambda$ is defined to be $\{(x_0, q_0, \ldots, x_n, q_n) :$ there is a $q \in G, q = (x_0, q_0, \ldots, x_n, q_n, g, D)$, and $\sup q_n < \lambda\}$.

Using Lemma 3.5 we can easily establish the following corollary:

COROLLARY 3.6.
(a) Suppose $p \in \mathbb{P}$ and $p \Vdash \dot{\tau} \subseteq \alpha$ and $\alpha < \kappa_x$ for some $x \in \wp_\kappa(\kappa^{+3})$ occurring in p. Then $p \Vdash \dot{\tau} \in V[G_p]$.
(b) Let $p = (x_0, p_0, \ldots, x_n, p_n, g, D)$ and $H \subseteq C(\omega, \kappa_{x_0})$ be generic. If $\alpha < \kappa_{x_n}, \tau \in V'[H]^{\mathbb{P}}$ and $p \Vdash_{\mathbb{P}} \dot{\tau} \subseteq \alpha$, then $p \Vdash \dot{\tau} \in V'[\dot{G}_p \times H]$. ($\dot{G}_p$ is a canonical term for the generic object $G \subseteq \mathbb{P}$ restricted to p.)

We now make the following claim.

CLAIM 3.7. If $G \times H \subseteq \mathbb{P} \times C(\aleph_0, \kappa_{x_0})$ is a generic over V', then $V'[G \times H] \models$ for all $n \in \omega$, $\aleph_n$ carries an $\aleph_n$ carries $\aleph_{n+1}$-saturated ideal.

PROOF. Let $\langle \kappa_i : i \in \omega \rangle$ be the sequence of κ_{x_i} such that x_i occurs in some condition in G. By Magidor's arguments, κ becomes $\aleph_\omega$. By the corollary, it is enough to see that for each i,

$V'[G \restriction \kappa_i \times H] \Vdash$ for all $n \in \omega$ ($\aleph_n < \kappa_i$ implies $\aleph_n$ carries a saturated ideal).

Our sequence of $\aleph_n$'s in $V[G \times H]$ is

$$\begin{aligned} \aleph_0 &= \omega, \\ \aleph_1 &= \kappa_0', \\ \aleph_2 &= \kappa_0'', \\ \aleph_3 &= \kappa_0''', \\ \aleph_4 &= \kappa_0^{\mathrm{iv}}, \\ \aleph_5 &= \kappa_1', \\ &\vdots \\ \aleph_{4n+k} &= \kappa_n^{+k} \quad 1 \le k \le 4. \end{aligned}$$

If $p = (x_1, p_1, \ldots, x_{n-1}, p_{n-1}, g, D)$, then in V', $G \restriction \kappa_n \times H$ is generic over

$$[C(\kappa_1^{\mathrm{iv}}, \kappa_2) \times \cdots \times C(\kappa_{n-1}^{\mathrm{iv}}, \kappa_n)] \times C(\omega, \kappa_0).$$

Hence, if $\aleph_m = \kappa_i^k$, $1 \le k \le 3$, we claim $\aleph_m$ carries a saturated ideal in $V'[G \restriction \kappa_n \times H]$ since it does in V' and our forcing is of the form

$$(\text{cardinality} < \aleph_m) \times (\aleph_{m+1}\text{-closed}).$$

Any forcing of cardinality$< \aleph_m$ preserves a saturated ideal on $\aleph_m$ by Lemma 2.2. Thus by Lemma 2.1 we can conclude that $\aleph_m$ has a saturated ideal in $V'[G \restriction \kappa_n \times$

$H]$. If $\aleph_m = \kappa_i^{\text{iv}}$ for some $i < n$, by Lemma 3.4, $\aleph_m$ carries a saturated ideal in $(V')^{C(\kappa_i^{\text{iv}},\kappa_{i+1})}$. To pick up $V'[G \restriction \kappa_n \times H]$ from $(V')^{C(\kappa_i^{\text{iv}},\kappa_{i+1})}$ we need forcing of the form

$$(\text{cardinality} < \aleph_m) \times ((\aleph_{\kappa_{i+1}}^{\text{iv}}, \infty)\text{-distributive})$$

(namely:

$$[C(\omega, \kappa_0) \times C(\kappa_0^{\text{iv}}, \kappa_1) \times \cdots \times C(\kappa_{i-1}^{\text{iv}}, \kappa_i)] \times [C(\kappa_{i+1}^{\text{iv}}, \kappa_{i+2}) \times \cdots \times C(\kappa_{n-1}^{\text{iv}}, \kappa_n)]).$$

Thus by Lemma 2.2, $\aleph_m$ carries a saturated ideal in $V'[G \restriction \kappa_n \times H]$. This establishes the claim. ⊣ (Claim 3.7)

Unfortunately, in $V'[G \times H]$, $(\kappa^+)^{V'}$ has cardinality κ, since $\bigcup\{x_i : x_i \text{ occurs in some } p \in G\} = (\kappa^{+3})^{V'}$. We now define a κ^+-c.c. partial ordering $\mathbb{P}^\pi$, a condition $p_0 \in \mathbb{P}$ and a projection map

$$\pi\colon \mathbb{P}/p_0 \to \mathbb{P}^\pi.$$

If $G \times H \subseteq \mathbb{P} \times (\omega, \kappa_{x_0})$ then $V'[G \times H]$ and $V'[\pi'' G \times H]$ will have the same bounded subsets of κ.

Let U be the measure on κ induced by μ_3. Let $\mathbb{Q}$ be the following partial ordering: $p \in \mathbb{Q}$ if and only if $p\colon X \to V_\kappa$ for a set $X \subseteq \kappa$, $X \in U$ and if $\alpha \in X$, $p(\alpha) \in \mathcal{B}(C(\alpha^{+4}, \kappa))$.

If p and q are elements of $\mathbb{Q}$, $p \leq_{\mathbb{Q}} q$ if and only if $\{\alpha : p(\alpha) \leq q(\alpha)\} \in U$.

If $B \subseteq \wp_\kappa(\kappa^{+3})$ is a set of measure one (for μ_3) and $g\colon B \to V_\kappa$ is a function such that for all $x \in \mathcal{B}$, $g(x) \in C(\kappa_x^{+4}, \kappa)$, we can associate an element $\bar{g}_B$ of $\mathbb{Q}$ with domain $\pi(B) = \langle \alpha : \exists x \in B, \kappa_x = \alpha \text{ and } \alpha \in A\rangle$ by

$$\bar{g}_B(\alpha) = \bigvee_{\substack{\kappa_x = \alpha \\ x \in B}} g(x)$$

Thus if $q \in C(\alpha^{+4}, \kappa)$ for some $\alpha \in \pi(B)$ and $q \wedge \bar{g}_B(\alpha) \neq 0$, we can find an $x \in B$, $\kappa_x = \alpha$ and $q \wedge g(x) \neq 0$.

A function g as above naturally gives rise to a filter $\mathcal{F}_g$, namely

$$\mathcal{F}_g = \{p \in \mathbb{Q} : \text{ there is a set } B \in \mu_3, g \text{ is defined on } B \text{ and } p \geq \bar{g}_B.\}$$

If $g\colon B \to R_\kappa$, $h\colon C \to R_\kappa$ are functions as above with $\mu_3(B) = \mu_3(C) = 1$ and $g \leq h$ almost everywhere with respect to μ_3, then $\mathcal{F}_g \supseteq \mathcal{F}_h$. (If $g \leq h$ almost everywhere we shall say $g \leq_{\mu_3} h$.)

Claim 3.8. If $\langle g_\alpha : \alpha < \beta\rangle \subseteq \{h : \text{there is a } B \subseteq N,\ \mu_3(B) = 1,\ h\colon B \to R_\kappa, \text{ and for all } x \in B,\ h(x) \in C(\kappa_x^{+4}, \kappa)\}$ is a $\leq_{\mu_3}$-descending sequence of length $\leq \kappa^{+3}$ then there is a g_β and a set B_β, $g_\beta\colon B_\beta \to \mathbb{R}_\kappa$ and for all $x \in B_\beta$, $g(x) \in C(\kappa_x^{+4}, \kappa)$ and for all $\alpha < \beta$, $g_\beta \leq_{\mu_3} g_\alpha$.

PROOF. We view each g_α as $[g_\alpha] \in V'^{\wp_\kappa(\kappa^{+3})}/\mu_3 = M$. Then $g_\alpha \in C(\kappa^{+4}, j(\kappa))^M$. If $\alpha < \alpha' < \beta$ then $[g_\alpha] \geq [g_{\alpha'}]$. $C(\kappa^{+4}, j(\kappa))$ is κ^{+4}-closed in M, hence, using κ^{+3}-supercompactness, there is a $q \in C(\kappa^{+4}, j(\kappa))^M$. such that $q \leq [g_\alpha]$ for all $\alpha < \beta$. Let $g_\beta \colon \wp_\kappa(\kappa^{+3}) \to R_\kappa$ be such that $[g_\beta] = q$. Let $B_\beta = \{x : g(x) \in C(\kappa_x^{+4}, \kappa)\}$. Then it is easy to verify that q_β and B_β satisfy the conclusions of the claim. ⊣ (Claim 3.8)

We now build a sequence $\langle g_\alpha : \alpha < \kappa^{++}\rangle$ such that if $g \leq_{\mu_3} g_\alpha$ for all α and $h \leq_{\mu_3} g$, then $\mathcal{F}_g = \mathcal{F}_h$.

To do this: At successor stages $\alpha+1$ pick a $g_{\alpha+1} \leq_{\mu_3} g_\alpha$ such that $\mathcal{F}_{g_{\alpha+1}} \supsetneq \mathcal{F}_{g_\alpha}$ if possible. At limit stages β, pick a $g_\beta \leq_{\mu_3} g_\alpha$ for all $\alpha < \beta$. Let $g \leq_{\mu_3} g_\alpha$ for all $\alpha < \kappa^{++}$. Since $2^\kappa = \kappa^+$ in V', we cannot have a strictly increasing chain of filters of order type κ^{++}. Hence for a tail of κ^{++}, $\mathcal{F}_{g_\alpha}$ is constant. From this we conclude that if $h \leq g$, $\mathcal{F}_g = \mathcal{F}_h$. (In fact it is possible to show that $\mathcal{F}_g$ is an ultrafilter.) Fix such a g.

A condition in the projected forcing $\mathbb{P}^\pi$ will be of the form

$$p = (\alpha_0, p_0, \alpha_1, p_1, \ldots, \alpha_n, p_n, b, B)$$

where

(a) $\alpha_0 < \alpha_1 < \alpha_2 < \cdots < \alpha_n \in A$,
(b) $p_i \in C(\alpha_i^{+4}, \alpha_{i+1})$,
(c) $B \subseteq \kappa$, $B \in U$, and
(d) $b \in \mathcal{F}_g$ and b is defined on B.

We let

$$(\alpha_0, p_0, \alpha_1, p_1, \ldots, \alpha_n, p_n, b, B) \geq (\alpha_0, p_0', \alpha_1, p_1', \ldots, \alpha_n, p_n', \beta_1, q_1, \ldots, \beta_m, q_m, c, C)$$

if and only if

(a) $p_i' \leq p_i$ for all $i \leq n$,
(b) for all $j \leq m$, $\beta_j \in B$,
(c) for all j, $q_j \leq b(\beta_j)$,
(d) $C \subseteq B$, and
(e) $c \leq b$ in $\mathbb{Q}$.

If $D \in \mu_3$, let $\pi(D) = \{\alpha : \text{there is an } x \in D, \kappa_x = \alpha\}$. Then $\pi(D) \in U$. Let $p_0 = (g, \operatorname{dom} g)$, then $p_0 \in \mathbb{P}$. If $p \in \mathbb{P}$, $p \leq p_0$, $p = (x_0, p_0, x_1, p_1, \ldots, x_n, p_n, h, B)$, let $\pi(p) = (\kappa_{x_0}, p_0, \kappa_{x_1}, p_1, \ldots, \kappa_{x_n}, p_n, \bar{h}_B, \pi(B))$. Then π is clearly order preserving. By Lemma 2.4, to see that π is a projection map we must show that for all $p \in \mathbb{P}/p_0$ there is a $q \in \mathbb{P}^\pi$, $q \leq \pi(p)$ such that for all $q' \leq q$ there is a $p' \leq p$ with $\pi(p') \leq q'$. (It is here that this argument differs slightly from earlier arguments of this type, hence we include it.)

Let $p \in \mathbb{P}/p_0$, $p = (x_0, p_0, \ldots, x_n, p_n, h, B)$, and $\pi(p) = (\kappa_{x_0}, p_0, \ldots, \kappa_{x_n}, p_n, \bar{h}, \pi(B))$. We want to define $q \leq \pi(p)$. We assume that for all $y \in B$,

$x_n \prec y$. We shall build a descending sequence $\langle A_i : i \in \omega\rangle \subseteq \mu_3$ with the property that for all $x \in A_{i+1}$ and for all $\alpha < \kappa_x$, if for some p there is $y \in A_i$, $\alpha = \kappa_y$,

$$p \in C(\alpha^{+4}, \kappa_x)$$
$$p \wedge \bar{h}_{A_i}(\alpha) \neq 0 \quad (\text{in } \mathcal{B}(C(\alpha^{+4}, \kappa)))$$

then there is a $y \in A_i$, $\alpha = \kappa_y$ such that

(a) $p \wedge h(y) \neq 0$,
(b) $y \prec x$, and
(c) $h(y) \in C(\alpha^{+4}, \kappa_x)$.

(It is reasonable to hope that $h(y) \in Col(\alpha^{+4}, \kappa_x)$ because $C(\alpha^{+4}, \kappa_x) \subseteq C(\alpha^{+4}, \kappa)$.)

Let $A_0 = B \cap \{x : \kappa_x > \sup p_n \text{ and } x_n \prec x\}$. Assume that A_i has been defined. For each $\alpha \in \pi(A_i)$ and each regular $\beta > \alpha$, pick a set $Y_\beta \subseteq A_i$ of cardinality β such that

$$\{h(y) \cap C(\alpha^{+4}, \beta) : y \in A_i\} = \{h(y) \cap C(\alpha^{+4}, \beta) : y \in Y_\beta\}.$$

Let $X_\beta = \{x \in A_i : \text{for all } y \in Y_\beta, y \prec x \text{ and } \kappa_x > \sup h(y)\}$. Let $Z_\alpha = \triangle_{\beta<\kappa} X_\beta = \{x : \text{for all } \beta < \kappa_x,\ x \in X_\beta\}$. Then Z_α has μ_3-measure one and if $x \in Z_\alpha$, $p \in C(\alpha^{+4}, \kappa_x)$ and p is compatible with the $\bigvee\{h(y) : y \in A_i \text{ and } \kappa_y = \alpha\}$, there is a $y \in A_i$, $\kappa_y = \alpha$ and p is compatible with $h(y)$. But $h(y) \cap (\alpha^{+4}, \kappa_x)$ is bounded in κ_x hence there is a $\beta < \kappa_x$, $\beta > \sup p$, $h(y) \cap (\alpha^{+4}, \kappa) = h(y) \cap (\alpha^{+4}, \beta)$. Pick $y' \in Y_\beta$ such that

$$h(y) \cap (\alpha^{+4}, \beta) = h(y') \cap (\alpha^{+4}, \beta).$$

Then $y' \prec x$, $\sup h(y') < x$ and $h(y')$ is compatible with p.

Let $A_{i+1} = \triangle_{\alpha<\kappa} Z_\alpha = \{x : \text{for all } \alpha < \kappa_x,\ x \in Z_\alpha\}$. Then A_{i+1} has the desired properties (a)–(c).

Let $q = (\kappa_{x_0}, p_0, \kappa_{x_1}, p_1, \ldots, \kappa_{x_n}, p_n, \bar{h}_{\bigcap A_i}, \pi(\bigcap A_i))$. Let $q' \leq q$, and suppose that $q' = (\kappa_{x_0}, q_0, \ldots, \kappa_{x_n}, q_n, \alpha_1, q_{n+1}, \ldots, \alpha_m, q_{n+m}, c, C)$. We must define $p' \in \mathbb{P}$ with $\pi(p') \leq q'$. Pick $y_m \in A_m$, $h(y_m) \wedge q_{n+m} \neq 0$ and $\kappa_{y_m} = \alpha_m$. Pick $p'_{n+m} \leq h(y_m) \wedge q_{n+m}$. Since $y_m \in A_m$ there is a $y_{m-1} \in A_{m-1}$, $\kappa_{y_{m-1}} = \alpha_{m-1}$, $\sup h(y_{m-1}) < \kappa_{y_m}$ and $h(y_{m-1}) \wedge q_{n+m-1} \neq 0$. Pick such a y_{m-1} and $p_{n+m-1} \leq q_{n+m-1} \wedge h(y_{m-1})$. Continue in this way for $0 \leq k \leq m$ to pick $y_{m-(k+1)} \in A_{m-(k+1)}$ with $y_{m-(k+1)} \prec y_{m-k}$, $\kappa_{y_{m-(k+1)}} = \alpha_{m-(k+1)}$, $\sup h(y_{m-(k+1)}) < \kappa_{y_{m-k}}$ and $q_{n+m-(k-1)}$ is compatible with $h(y_{m-(k+1)})$. Pick $p'_{m-(k+1)} \leq q_{n+m-(k+1)} \wedge h(y_{m-(k+1)})$. Let $p' = (x_1, q_1, x_2, q_2, \ldots, x_n, q_n, y_1, p'_{n+1}, y_2, p'_{n+2}, \ldots, y_m, p'_{n+m}, g, D)$ where $g \leq h$, $D \subseteq \cap A_i$ and $\bar{g}_D \leq c$ and $\pi(D) \subseteq C$.

It is clear that $p' \le p$ and $\pi(p') \le q'$. Thus $\pi\colon \mathbb{P}/p_0 \to \mathbb{P}^\pi$ is a projection map.

Let $G \times H \subseteq \mathbb{P} \times C(\aleph_0, \kappa_0)$ be V'-generic. Let G^π be $\pi''G$. Then $G^\pi \times H$ is V'-generic for $\mathbb{P}^\pi \times C(\aleph_0, \kappa_0)$. If $\tau \subseteq \alpha < \kappa$, $\tau \in V'[G \times H]$ then $\tau \in V'[G_p \times H]$ for some $p \in G$ and from this we conclude $\tau \in V'[G^\pi \times H]$. Hence $V'[G^\pi \times H] \models$"for all $n \in \omega$, $\aleph_n$ carries a saturated ideal".

To see that $\aleph_{\omega+1}$ carries a saturated ideal it suffices to show that $\mathbb{P}^\pi \times C(\aleph_0, \kappa_0)$ is κ-centered.

To see this we define an equivalence relation on $\mathbb{P}^\pi \times C(\aleph_0, \kappa_0)$ with κ equivalence classes and show that if $p_1, \ldots, p_n$ lie in the same equivalence class then $\bigwedge_{i=1}^n p_i \ne 0$.

If we are given $p = ((\alpha_1, p_1, \ldots, \alpha_n, p_n, c, C), p') \in \mathbb{P}^\pi \times C(\aleph_0, \kappa_0)$ and $q = ((\beta_1, q_1, \ldots, \beta_n, q_n, b, B), q') \in \mathbb{P}^\pi \times C(\aleph_0, \kappa_0)$ set $p \sim q$ if and only if

(a) $n = m$ and for all $k \le n$ $\alpha_k = \beta_k$, $p_k = q_k$, and
(b) $p' = q'$.

There are only $|\kappa^\omega = \kappa|$ equivalence classes. If we have equivalent elements $p_1, \ldots, p_n$ then to find a $p \le \bigwedge_{i=1}^n p_i$ we merely take the intersection of the boolean values in q (this is non-zero since they all lie in the filter $\mathcal{F}_g$) and the intersection of the sets of measure one. This establishes Theorem 1.4.

⊣ (Theorem 1.4)

With a little more work, starting from a 2-huge cardinal we can also get $(\aleph_{n+1}, \aleph_n) \twoheadrightarrow (\aleph_{m+1}, \aleph_m)$ for all $m > n$ to hold in the model satisfying Theorem 1.4.

To prove Con(ZFC+"there is a huge cardinal") implies Con(ZFC+"all regular cardinals κ carry a saturated ideal") we do the same preparatory forcing $\mathbb{P}_{\le\kappa}$. We then do the modified Radin forcing that appears in [FW91].

The auxiliary function again takes its values in the appropriate collapse algebra. There are no essential new ideas needed to do this that do not appear in this paper or in [FW91].

REFERENCES

JAMES E. BAUMGARTNER

[Bau83] *Iterated forcing*, ***Surveys in set theory*** (A. R. D. Mathias, editor), London Mathematical Society Lecture Note Series, vol. 87, Cambridge University Press, 1983, pp. 1–59.

MATTHEW FOREMAN

[For82] *Large cardinals and strong model theoretic transfer properties*, ***Transactions of the American Mathematical Society***, vol. 272 (1982), no. 2, pp. 427–463.

MATTHEW FOREMAN AND RICHARD LAVER

[FL88] *Some downwards transfer properties for* $\aleph_2$, ***Advances in Mathematics***, vol. 67 (1988), no. 2, pp. 230–238.

Matthew Foreman and W. Hugh Woodin

[FW91] *The generalized continuum hypothesis can fail everywhere*, ***Annals of Mathematics***, vol. 133 (1991), no. 1, pp. 1–35.

Akihiro Kanamori and Menachem Magidor

[KM78] *The evolution of large cardinal axioms in set theory*, ***Higher Set Theory. Proceedings, Oberwolfach, Germany, April 13–23, 1977*** (Gert H. Müller and Dana S. Scott, editors), Lecture Notes in Mathematics, vol. 669, Springer-Verlag, 1978, pp. 99–275.

Kenneth Kunen

[Kun78] *Saturated ideals*, ***The Journal of Symbolic Logic***, vol. 43 (1978), no. 1, pp. 65–76.

Richard Laver

[Lav82A] *An* $(\aleph_2, \aleph_2, \aleph_0)$*-saturated ideal on* ω_1, ***Logic Colloquium '80. Papers intended for the European Summer Meeting of the Association for Symbolic Logic to have been held in Prague, August 24–30, 1980*** (Dirk van Dalen, D. Lascar, and T. J. Smiley, editors), Studies in Logic and the Foundations of Mathematics, vol. 108, North-Holland, 1982, pp. 173–180.

[Lav82B] *Saturated ideals and nonregular ultrafilters*, ***Patras Logic Symposion (Patras, 1980)***, Studies in Logic and the Foundations of Mathematics, vol. 109, North-Holland, Amsterdam, 1982, pp. 297–305.

Menachem Magidor

[Mag78] *On the singular cardinals problem. I*, ***Israel Journal of Mathematics***, vol. 28 (1978), no. 1–2, pp. 1–31.

DEPARTMENT OF MATHEMATICS
UNIVERSITY OF CALIFORNIA, IRVINE
265 MULTIPURPOSE SCIENCE & TECHNOLOGY BUILDING
IRVINE, CALIFORNIA 92697
UNITED STATES OF AMERICA
E-mail: mdforema@uci.edu

THE FOURTEEN VICTORIA DELFINO PROBLEMS AND THEIR STATUS IN THE YEAR 2020

ANDRÉS EDUARDO CAICEDO, BENEDIKT LÖWE

§1. Introduction. The Victoria Delfino problems played an important role in the development of descriptive set theory in the context of the Cabal. The first set of problems (# 1 to # 5) were announced during one of the *Very Informal Gatherings of Logicians* (VIG) at UCLA in 1978. They were subsequently published as an Appendix [KM78A] in [Cabal i] with the following explanations and rules:

> The following list of problems was distributed during a very informal gathering of logicians at UCLA in January 1978. We are reproducing it here because of its obvious relevance to the contents of this volume.
>
> A cash prize of $100 is offered by the logicians in the Los Angeles area for the solution of each of the following five problems. This competition is financed by the Victoria Delfino Fund for the Advancement of Logic which was established by a generous contribution from Miss Victoria Delfino.
>
> Employees of UCLA and Caltech and their immediate families (other than students) are ineligible for these prizes; competition is open to everyone else. All decisions by the judges are final. Multiple entries are allowed.

1.1. Victoria Delfino. Victoria Delfino was a realtor in the Los Angeles area who helped Yiannis Moschovakis buy his house.[1] When Tony Martin moved to UCLA, Moschovakis referred him to Delfino, who also became Martin's realtor and found the house where Martin still lives. Two weeks after the sale was finalised, Delfino gave Moschovakis an amount of money as commission for the referral, and did not accept his attempts to reject it.

As a result, Moschovakis decided instead to use the money to help fund the series of Very Informal Gatherings, the first of which had taken place in

The first author thanks the National Science Foundation for partial support through grant DMS-0801189.

[1] Most of this section is based on recollections shared by Moschovakis with the first author during a telephone conversation.

Large Cardinals, Determinacy, and Other Topics: The Cabal Seminar, Volume IV
Edited by A. S. Kechris, B. Löwe, J. R. Steel
Lecture Notes in Logic, 49

the fall of 1975.[2] The second Very Informal Gathering, in 1978, started a new tradition: with a single exception, all subsequent VIGs have taken place on Super Bowl weekend, in late January or early February. Moschovakis comments:

> The time of the only exception, there was an earthquake! A clear sign that moving the date was a mistake.

Together with the funding of the Very Informal Gatherings, the money was also set to cover the prizes for the solutions of the five original Delfino Problems. (Contrary to popular belief, no monetary prize was attached to further problems.)

When Moschovakis introduced these five problems (in what he described as one of the most significant Very Informal Gatherings to date), and mentioned the *Victoria Delfino Fund*, Martin, taken by surprise, exclaimed "That's my broker!" Not all in attendance heard this, and Moschovakis offered no further explanation for the name of the fund. This led for a short while to a variety of conjectures trying to find appropriate interpretations to explain the name.

Originally, the fund was kept in a joint account by Alexander Kechris, Martin, and Moschovakis. It was supplemented by occasional donations from other logicians in the area. Martin reports (personal communication) that by 1998, "all the money in the fund had been used and we had stopped asking people to contribute to it." Eventually, it became so low that it made sense to use it all and close the account. Nowadays, the Very Informal Gatherings are typically funded through support of the NSF.

As for Delfino, she eventually retired, moved out of state to take care of an ill relative, and her trail disappears there. It is unknown whether she ever found out that her name was associated with the problems or with the Cabal.

1.2. The problems. After the first announcement of the Victoria Delfino Problems, progress reports were published in [Cabal ii] and [Cabal iii]. In 1985, three of the original problems had been solved, and seven new problems (# 6 to # 12) were added and published as [KMS88A] in [Cabal iv], preceded by the following comment:

> At the "Very Informal Gathering" of January 1984, the Cabal announced the addition of seven problems to the Victoria Delfino list. We are happy (and not at all embarrassed) to report that since then four of these problems have been solved. Below we list the new problems, beginning with # 6 since there were five problems on the original list. For each we describe briefly what was known when it was added to the list, and what has been its fate since.

[2]There is some uncertainty about the date of the first VIG; in preparation for the twentieth VIG in February 2019, the original organisers discussed this question and concluded that "our best recollection now is that the first VIG was in the fall of 1975" (Kechris, personal communication, 2018).

In the years following the publication of the final original Cabal volume, there were two more problems announced at one of the Very Informal Gatherings in the late 1980s or early 1990s (the precise date could not be identified), but they were never published as Victoria Delfino problems. We include these problems as # 13 and # 14.

Today, two of the problems remain open. The first one is better known under the name of *Martin's Conjecture* (# 5), the other one has now been embedded into Woodin's theory AD^+ (# 14). In Table 1, the reader can find a synoptic list of the problems with their current status.

	Published in	Status
# 1	[KM78A]	Solved by Steve Jackson (1983)
# 2	[KM78A]	Solved by Yiannis Moschovakis (1981)
# 3	[KM78A]	Solved by Howard Becker & Alexander Kechris (1983)
# 4	[KM78A]	Solved by John Steel (1993)
# 5	[KM78A]	Open
# 6	[KMS88A]	Solved by John Steel (1984)
# 7	[KMS88A]	Solved by Steve Jackson (1985)
# 8	[KMS88A]	Solved by John Steel (1994)
# 9	[KMS88A]	Solved by Tony Martin & John Steel (1985)
# 10	[KMS88A]	Solved by W. Hugh Woodin & Saharon Shelah (1985)
# 11	[KMS88A]	Solved by John Steel (1994)
# 12	[KMS88A]	Solved by John Steel (1997)
# 13	unpublished	Solved by W. Hugh Woodin (1999)
# 14	unpublished	Open

TABLE 1. List of the Victoria Delfino problems and their current status

This paper is organised as follows: Each problem is presented in its own section that, except for the last two problems, starts with a quote from the original Cabal volumes under the headline ***Original problem***. The quotation is essentially literal, although we have followed the general practice of modernising and homogenising notation and writing style. For the first five problems, the original formulation is followed by one or several subsections entitled ***Progress report*** where we reproduce the text from subsequent Cabal volumes providing updates on the problem. We then proceed with a brief discussion of the current state of knowledge in a subsection entitled ***2020 comments***.

1.3. Acknowledgements. We should like to thank Kai Hauser, Daisuke Ikegami, Antonio Montalbán, Jan Reimann, Ralf Schindler, Ted Slaman, John Steel, Simon Thomas, Hugh Woodin, and Yizheng Zhu for detailed remarks and comments. Particular thanks are due to Yiannis Moschovakis and Tony

Martin for their recollections on Victoria Delfino, the Delfino Fund, and the early history of the Very Informal Gatherings.

 # 1. Projective Ordinals.

Original problem [KM78A]. For each positive integer n, let $\underset{\sim}{\delta}^1_n$ be the least nonzero ordinal not the length of a $\underset{\sim}{\Delta}^1_n$ prewellordering of the reals. *Assume* AD + DC. It is known that $\underset{\sim}{\delta}^1_1 = \omega_1$, $\underset{\sim}{\delta}^1_2 = \omega_2$, $\underset{\sim}{\delta}^1_3 = \omega_{\omega+1}$, $\underset{\sim}{\delta}^1_4 = \omega_{\omega+2}$, $\underset{\sim}{\delta}^1_{2n+2} = (\underset{\sim}{\delta}^1_{2n+1})^+$, and $\underset{\sim}{\delta}^1_{2n+1}$ is always the successor (cardinal) of a cardinal of cofinality ω.

PROBLEM # 1. *Compute* $\underset{\sim}{\delta}^1_5$.

Kunen has some partial results on this problem, results which suggest the answer ω_{ω^3+1}.

The problem is related to that of whether $\underset{\sim}{\delta}^1_3 \to (\underset{\sim}{\delta}^1_3)^{\underset{\sim}{\delta}^1_3}$. Kunen has shown that $\underset{\sim}{\delta}^1_3 \to (\underset{\sim}{\delta}^1_3)^\alpha$ for each $\alpha < \underset{\sim}{\delta}^1_3$. Results of Kleinberg imply that $\underset{\sim}{\delta}^1_3$ has exactly three normal measures. It is likely that the regular cardinals between $\underset{\sim}{\delta}^1_3$ and $\underset{\sim}{\delta}^1_5$ are exactly the ultrapowers of $\underset{\sim}{\delta}^1_3$ with respect to these normal measures. This would be important in getting an upper bound on $\underset{\sim}{\delta}^1_5$ from Choice plus $\mathsf{AD}^{\mathbf{L}(\mathbb{R})}$, the hypothesis that every set of reals in $\mathbf{L}(\mathbb{R})$ is determined.

(Needless to say, the decision of the judges as to what constitutes a "computation" of $\underset{\sim}{\delta}^1_5$ will be final.)

Progress report [KMM81A]. Martin has established the conjectured lower bound for $\underset{\sim}{\delta}^1_5$ by proving (from AD + DC) that

$$\underset{\sim}{\delta}^1_5 \geq \aleph_{\omega^3+1};$$

moreover Martin showed (from AD) that the ultrapowers of $\underset{\sim}{\delta}^1_3 = \aleph_{\omega+1}$ under the three normal measures on $\underset{\sim}{\delta}^1_3$ are exactly $\underset{\sim}{\delta}^1_4 = \aleph_{\omega+2}$ (this was known to Kunen), $\aleph_{\omega\cdot 2+1}$ and $\aleph_{\omega^2+1}$ and that these three cardinals are measurable (and hence regular), so that (in particular), $\underset{\sim}{\delta}^1_5$ is not the first regular cardinal after $\underset{\sim}{\delta}^1_4$. We still have no upper bounds for $\underset{\sim}{\delta}^1_5$ from AD.

Progress report [KMM83A]. It was announced in [KMM81A] that Martin had shown $\underset{\sim}{\delta}^1_5 \geq \aleph_{\omega^3+1}$ and that the ultrapowers of $\underset{\sim}{\delta}^1_3$ with respect to the three normal measures on $\underset{\sim}{\delta}^1_3$ are $\aleph_{\omega+2}$, $\aleph_{\omega\cdot 2+1}$ and $\aleph_{\omega^2+1}$. The proof of part of the last assertion, that the ultrapower by the ω_2-cofinal measure is $\leq \aleph_{\omega^2+1}$, was incorrect. Actually this ultrapower is larger ($\aleph_{\omega^\omega+1}$).

Steve Jackson has completely solved the first problem. He first proved that $\underset{\sim}{\delta}^1_5 \leq \aleph_{\omega^{(\omega^\omega)}+1}$. This result will appear in his UCLA Ph.D. Thesis. He next used the machinery for getting this upper bound to analyze all measures on $\underset{\sim}{\delta}^1_3$ and to get a good representation of functions with respect to these measures. Martin observed that this representation and ideas of Kunen allow one to show

$\underset{\sim}{\delta}^1_3 \to (\underset{\sim}{\delta}^1_3)^{\underset{\sim}{\delta}^1_3}$. From this it follows by a result of Martin that the ultrapower of $\underset{\sim}{\delta}^1_3$ with respect to any of its measures is a cardinal. Jackson's analysis then gives $\underset{\sim}{\delta}^1_5 \geq \aleph_{\omega^{(\omega^\omega)}+1}$ so $\underset{\sim}{\delta}^1_5 = \aleph_{\omega^{(\omega^\omega)}+1}$.

2020 comments. Steve Jackson not only solved Problem # 1, but also solved the problem in general for all projective ordinals. He computed $\underset{\sim}{\delta}^1_{2n+1}$ to be $\aleph_{\mathbf{e}_n+1}$ where $\mathbf{e}_0 := 0$ and $\mathbf{e}_{i+1} := \omega^{(\omega^{\mathbf{e}_i})}$ (*i.e.*, $\mathbf{e}_n$ is an exponential ω-tower of height $2n-1$).

However, Jackson's paper [Jac88] where the inequality $\underset{\sim}{\delta}^1_{2n+1} \leq \aleph_{\mathbf{e}_n+1}$ is established, is notoriously hard to read, and so in the decades following his solution of the problem, Jackson produced various expositions of the results. As the title "*A computation of* $\underset{\sim}{\delta}^1_5$" suggests, his book [Jac99] focuses on the (complete) computation of $\underset{\sim}{\delta}^1_5$ as asked in the original problem, and explains how to proceed to compute all projective ordinals via an inductive analysis. His survey paper [Jac10] also discusses extensions of these results beyond the projective ordinals:

> In the early 1980s, Martin [Mar] obtained a result on the ultrapowers of $\underset{\sim}{\delta}^1_3$ by the normal measures on $\underset{\sim}{\delta}^1_3$. Building on this and some joint work with Martin, [Jackson] computed $\underset{\sim}{\delta}^1_5$. In the mid-1980s, this was extended to compute all the $\underset{\sim}{\delta}^1_n$, and to develop the combinatorics of the cardinal structure of the cardinals up to that point. The analysis, naturally, proceeded by induction. The complete "first-step" of the induction appears in [Jac99]. The analysis revealed a rich combinatorial structure to these cardinals. [. . .] A goal, then, is to extend some version of this "very-fine" structure theory to the entire model $\mathbf{L}(\mathbb{R})$. In the late 1980s, [Jackson] extended the analysis further, up to the least inaccessible cardinal in $\mathbf{L}(\mathbb{R})$, although this lengthy analysis has never been written up. It was clear, however, that new, serious problems were being encountered shortly past the least inaccessible. In [Jac91], for example, results were given that show that the theory fell far short of $\kappa^{\mathbb{R}}$, the ordinal of the inductive sets (the Wadge ordinal of the least non-selfdual pointclass closed under real quantification). [Jac10, p. 1755]

Part of the extended results is what is known as *Kechris's theorem*:

THEOREM 1. *Assume* $\mathsf{AD} + \mathbf{V}{=}\mathbf{L}(\mathbb{R})$. *If* κ *is an inaccessible Suslin cardinal,*[3] *then* κ, κ^+, *and* κ^{++} *are measurable.*

Kechris's theorem remained unpublished for many years; a proof and generalisations to polarised partition properties for κ, κ^+, and κ^{++} can be found in [AJL13].

[3] *E.g.*, the Kleene ordinal discussed in Problem # 7 or $\kappa^{\mathbb{R}}$, the least non-hyperprojective ordinal; *cf.* [AJL13, Proposition 5].

Further projects to make the proof and its generalisations more accessible are what Jackson calls "description theory" (cf. [JK16]) and the "simple inductive arguments" given in [Löw02, BL07, Bol09]. On the basis of the inductive analysis, Jackson and the second author developed an abstract theory of *canonical measure assignments* that allows blackboxing the proofs of partition properties and deriving consequences (such as the behaviour of the cofinality function or the calculation of the measurable cardinals) directly by induction (cf. [JL13]).

The ZFC *context.* At the end of the original formulation of Problem # 1, the question of calculating bounds for the projective ordinals in ZFC + $\mathsf{AD}^{\mathbf{L}(\mathbb{R})}$ is mentioned. The original AD-results listed in the original formulation of Problem # 1 yield upper bounds in this context: $\underset{\sim}{\boldsymbol{\delta}}^1_1 = \aleph_1$, $\underset{\sim}{\boldsymbol{\delta}}^1_2 \leq \aleph_2$, $\underset{\sim}{\boldsymbol{\delta}}^1_3 \leq \aleph_3$, $\underset{\sim}{\boldsymbol{\delta}}^1_4 \leq \aleph_4$, $\underset{\sim}{\boldsymbol{\delta}}^1_5 \leq \aleph_7$. In general, Jackson's analysis shows that under AD, there are exactly $2^n - 1$ regular cardinals below $\underset{\sim}{\boldsymbol{\delta}}^1_{2n+1}$; thus ZFC + $\mathsf{AD}^{\mathbf{L}(\mathbb{R})}$ proves that $\underset{\sim}{\boldsymbol{\delta}}^1_n < \aleph_\omega$ for every natural number n. Martin conjectured that "for all n, $\underset{\sim}{\boldsymbol{\delta}}^1_n = \aleph_n$" should follow from ZFC + $\mathsf{AD}^{\mathbf{L}(\mathbb{R})}$ plus reasonable additional assumptions (*cf.*, *e.g.*, [Woo99, p. 5]).

In [Woo99], Woodin develops a very powerful technique to produce models of ZFC as forcing extensions of models of determinacy, the analysis of which provides a solution to Martin's conjecture in the case $n = 2$:

THEOREM 2 (Woodin; [Woo99, Theorem 1.1 & § 3.1]). *If the nonstationary ideal on ω_1 is ω_2-saturated and $\wp(\omega_1)^\#$ exists, then $\underset{\sim}{\boldsymbol{\delta}}^1_2 = \aleph_2$.*

On the other hand, Woodin points out that current techniques produce models where $\underset{\sim}{\boldsymbol{\delta}}^1_3 < \Theta^{\mathbf{L}(\mathbb{R})} = \aleph_3$, and asks—in contrast to Martin's conjecture—whether it is a theorem of ZFC+$\mathsf{AD}^{\mathbf{L}(\mathbb{R})}$ that $\Theta^{\mathbf{L}(\mathbb{R})} \leq \aleph_3$ [Woo99, § 1.5]. Martin's conjecture for $n > 2$ and the competing question by Woodin remain open.

2. The extent of definable scales.

Original problem [KM78A]. A **semiscale** on a set $P \subseteq \mathbb{R}^k$ ($\mathbb{R} = {}^\omega\omega$) is a sequence $\vec{\varphi} = \{\varphi_n : n \in \omega\}$ of **norms** on P, where each $\varphi_n : P \to \lambda$ maps P into some ordinal λ and the following convergence condition holds: If $x_0, x_1, \dots \in P$ and for each n the sequence $\varphi_n(x_0), \varphi_n(x_1), \varphi_n(x_2), \dots$ is ultimately constant, then $x \in P$. We call $\vec{\varphi}$ a **scale** if, under the same hypotheses, we can infer that

$$\varphi_n(x) \leq \varphi_n(x_i) \text{ for all large } i.$$

A semiscale $\vec{\varphi}$ is in a class of relations Γ if both relations

$$U(n, x, y) \iff x \in P \wedge [y \notin P \vee \varphi_n(x) \leq \varphi_n(y)]$$
$$V(n, x, y) \iff x \in P \wedge [y \notin P \vee \varphi_n(x) < \varphi_n(y)]$$

are in Γ.

It is easy to check that a set P admits a semiscale $\vec{\varphi}$ into λ if and only if P is λ-Suslin, *i.e.*, P is the projection of some tree T on $\omega^k \times \lambda$; moreover, T is definable exactly when $\vec{\varphi}$ is definable. Sets which admit definable scales are well-behaved in many ways, e.g. we can use a scale on $P \subseteq \mathbb{R} \times \mathbb{R}$ to uniformise P.

Granting projective determinacy, we can prove that every projective set admits a projective scale (Moschovakis); on the other hand it is easy to check that $\{(x, y) : x \text{ is not ordinal definable from } y\}$ does not admit a scale which is OD in a real, granting only that for each y there is some x which is not OD in y. Thus not every "definable" set admits a "definable" scale.

The strongest result we can get with current methods is that inductive sets admit inductive scales, granting inductive determinacy; here P is inductive if P is Σ_1 over the smallest admissible set M which contains the reals, $\mathbb{R} \in M$.

Problem #2. *Assume* $\mathsf{ZF} + \mathsf{DC} + \mathsf{AD} + \mathbf{V}{=}\mathbf{L}(\mathbb{R})$; *prove or disprove that every coinductive set of reals is λ-Suslin for some λ.*

Progress report [KMM81A]. The problem was solved by Moschovakis who showed (from AD+DC) that every coinductive pointset admits a scale. If we let Σ_0^* be the set of all Boolean combinations of inductive and coinductive sets and then define Σ_n^* by counting quantifiers over $\mathbb{R}$ in front of a Σ_0^* matrix in the usual way, then the proof shows that every coinductive set admits a scale $\{\varphi_n : n \in \omega\}$, where each φ_n is a Σ_{n+1}^*-norm, uniformly in n.

Martin and Steel extended the method used by Moschovakis in this proof and showed that $\mathsf{ZF} + \mathsf{DC} + \mathsf{AD} + \mathbf{V}{=}\mathbf{L}(\mathbb{R})$ implies that every Σ_1^2 set admits a Σ_1^2-scale; this combines with an earlier result of Kechris and Solovay to show that $\mathsf{ZF} + \mathsf{DC} + \mathsf{AD} + \mathbf{V}{=}\mathbf{L}(\mathbb{R})$ implies that a pointset admits a scale if and only if it is $\undertilde{\Sigma}_1^2$.

Martin then combined these ideas with the technique of the Third Periodicity Theorem [Mos80, Theorem 6E.1] and showed that under reasonable hypotheses of determinacy for games on $\mathbb{R}$, (namely, $\mathsf{AD}_{\mathbb{R}}$), the scale property is preserved by the game quantifier $\Game^2$ on $\mathbb{R}$, where

$$(\Game^2\alpha)P(x,\alpha) \iff (\exists\alpha_0)(\forall\alpha_1)(\exists\alpha_2)(\forall\alpha_3)\ldots P(x,\langle\alpha_0,\alpha_1,\ldots\rangle).$$

This result produces scales for sets that are not $\undertilde{\Sigma}_1^2$ in $\mathbf{L}(\mathbb{R})$ and leaves open the general question of the extent of scales in the presence of axioms stronger than AD.

2020 comments. Moschovakis's paper appeared as [Mos83]. The result of Martin and Steel appears in [MS83]. Martin's theorem on preservation of scales under the game quantifier $\Game^2$ is in [Mar83]. Related results by Steel are in [Ste83B] and [Ste83A]. The latter paper introduces the key fine structural analysis of $\mathbf{L}(\mathbb{R})$ via gaps that is now used in the core model induction. These results have been further extended by Steel (under appropriate large cardinals or stronger determinacy assumptions), *cf.*, *e.g.*, [Ste08E, Ste08D], and the introduction [Ste08B]. These extensions are needed for core model inductions whose goal is to reach models of strong determinacy assumptions.

 # 3. The invariance of $\mathbf{L}[T^3]$.

Original problem [KM78A]. Let n be an *odd* integer. Let P be a complete Π^1_n set of reals and assuming PD let $\vec{\varphi} = \{\varphi_m : m \in \omega\}$ be a Π^1_n-scale on P. (It is understood here that each φ_m maps P onto an initial segment of the ordinals.) The tree $T^n = T^n(\vec{\varphi})$ associated with this scale is defined by

$$T^n = \{\langle \alpha(0), \varphi_0(\alpha), \ldots, \alpha(k), \varphi_k(\alpha)\rangle : \alpha \in P\}.$$

Let $\mathsf{AD}^{\mathbf{L}(\mathbb{R})}$ be the hypothesis that every set of reals in $\mathbf{L}(\mathbb{R})$ is determined.

Problem # 3. *Assume* $\mathsf{ZF} + \mathsf{DC} + \mathsf{AD}^{\mathbf{L}(\mathbb{R})}$. *Prove or disprove that* $\mathbf{L}[T^3] = \mathbf{L}[T^3(\vec{\varphi})]$ *is independent of the choice of the complete* Π^1_3 *set* P *and the particular* Π^1_3*-scale* $\vec{\varphi}$ *on* P.

It is known that $\mathbf{L}[T^1] = \mathbf{L}$ (Moschovakis). Also under the above hypothesis it is known that for all odd n and all $T^n = T^n(\vec{\varphi})$, $\mathbf{L}[T^n] \cap \mathbb{R} = \mathrm{C}_{n+1}$, where C_{n+1} is the largest countable Σ^1_{n+1} set of reals (Harrington-Kechris), so that $\mathbb{R} \cap \mathbf{L}[T^n]$ does not depend on the choice of T^n.

In many ways, the model $\mathbf{L}[T^n]$ is an excellent analog of $\mathbf{L}$ for the $(n+1)$-st level of the analytical hierarchy.

Progress report [KMM81A]. Kechris showed in [Kec81] that if $T^3 = T^3(\vec{\varphi})$ is the tree associated with some Π^1_3-scale $\vec{\varphi}$ on a Π^1_3-complete set P and if

$$\widetilde{\mathbf{L}}[T^3] = \bigcup_{\alpha \in \mathbb{R}} \mathbf{L}[T^3, \alpha],$$

then $\mathsf{ZF} + \mathsf{AD} + \mathsf{DC} + \underset{\sim}{\delta}^1_3 \to (\underset{\sim}{\delta}^1_3)^{\underset{\sim}{\delta}^1_3}$ implies that $\widetilde{\mathbf{L}}[T^3]$ is independent of the choice of P and $\vec{\varphi}$.

This partial result emphasises the importance of the question of *the strong partition property* for $\underset{\sim}{\delta}^1_3$ which is still open.

Progress report [KMM83A]. The problem was solved by Becker and Kechris who showed that $L[T^3]$ is independent of the choice of T^3. This is a consequence of the following fact, which is a theorem of $\mathsf{ZF} + \mathsf{DC}$.

Theorem 3. *Let* Γ *be an* ω*-parametrised pointclass closed under* $\wedge$ *and recursive substitution and containing all recursive sets. Let* $P \subset \mathbb{R}$ *be a complete* Γ *set,* $\vec{\varphi} = \{\varphi_i : i \in \omega\}$ *be an* $\exists^{\mathbb{R}}\Gamma$*-scale on* P *such that all norms* φ_i *are regular, and* $\kappa = \sup\{\varphi_i(x) : i \in \omega, x \in P\}$. *Let* $T(\vec{\varphi})$ *be the tree on* $\omega \times \kappa$ *associated with* $\vec{\varphi}$. *For any set* $A \subset \kappa$, *if* A *is* $\exists^{\mathbb{R}}\Gamma$*-in-the-codes with respect to* $\vec{\varphi}$ *(that is, if the set* $\{\langle i, x\rangle \in \omega \times \mathbb{R} : x \in P \wedge \varphi_i(x) \in A\}$ *is* $\exists^{\mathbb{R}}\Gamma$*), then* $A \in \mathbf{L}[T(\vec{\varphi})]$.

In general, given two such scales $\vec{\varphi}$, $\vec{\psi}$, it is not known that $T(\vec{\psi})$ is $\exists^{\mathbb{R}}\Gamma$-in-the-codes with respect to $\vec{\varphi}$, so the invariance of $\mathbf{L}[T(\vec{\varphi})]$ has not been shown in this generality. However there are special cases where invariance can be proved. Henceforth, assume AD.

In Moschovakis [Mos80, p. 562], a model H_Γ is defined for every pointclass Γ which resembles Π^1_1; this includes the pointclasses Π^1_n for odd n. It follows from Theorem 3, together with known results about the H_Γ's [Mos80, 8G], that for any Γ, P, $\vec{\varphi}$ such that Γ resembles Π^1_1 and Γ, P, $\vec{\varphi}$ satisfy the assumptions of Theorem 3, $L[T(\vec{\varphi})] = H_\Gamma$, and hence $\mathbf{L}[T(\vec{\varphi})]$ is independent of the choice of P and $\vec{\varphi}$. For $\Gamma = \Pi^1_3$ this solves the third problem.

While the invariance problem for $\mathbf{L}[T^n]$ is thus solved for odd n, for even n the situation is still unclear. Call a Σ^1_3-scale on a Π^1_2 set **good** if it satisfies the ordinal quantification property of Kechris-Martin [KM78]. It follows from the above theorem that $\mathbf{L}[T^2]$ is independent of the choice of a complete Π^1_2 set and of the choice of a good scale. Whether or not it is independent of the choice of an arbitrary scale is unknown. For even $n > 2$, it is not known whether there exist any good scales.

2020 comments. The result by Becker and Kechris for odd n appears in [BK84]. In [Hjo96B], Hjorth shows that, under $\mathsf{Det}(\undertilde{\mathbf{\Pi}}^1_2)$, the model $\mathbf{L}[T^2]$ is independent of the exact choice of T^2. His argument uses forcing to analyze Π^1_3 equivalence relations. In [Hjo95], he uses properties of $\mathbf{L}[T^2]$ to draw descriptive set theoretic consequences of the assumption that all reals have sharps, in particular showing that if all reals have sharps and MA_{ω_1} holds, then all $\undertilde{\mathbf{\Sigma}}^1_3$ sets are Lebesgue measurable. Further work on $\mathbf{L}[T^2]$ using fine-structural techniques has been carried out by Hauser [Hau99].

In [Atm19], Atmai shows that $\mathbf{L}[T^{2n}]$ is independent of the choice of T^{2n}, assuming $\mathsf{Det}(\undertilde{\mathbf{\Pi}}^1_{2n})$. His proof involves an appropriate generalisation of the Kechris-Martin theorem to the odd levels of the projective hierarchy. Atmai also shows that the $\mathbf{L}[T^{2n}]$ are not extender models, but satisfy some of their properties, such as GCH.

Meanwhile, developments in inner model theory have provided us both with new methods for analyzing the models $\mathbf{L}[T^n]$, and with the proper analogues of $\mathbf{L}$ for higher levels of the analytic hierarchy, the fine structural models $\mathbf{M}_n$. Recall that (under appropriate large cardinal assumptions) $\mathbf{M}_n$ is the canonical minimal inner model for the assumption that there are n Woodin cardinals. In [Ste95B], Steel gives a precise definition of $\mathbf{M}_n$ in terms of n-smallness and shows that $\mathbf{M}_n$ is Σ^1_n-correct, and that $\mathbb{R} \cap \mathbf{M}_n = \mathrm{C}_n$ for n even, and $\mathbb{R} \cap \mathbf{M}_n = \mathrm{Q}_n$ for odd n. For odd n, it is unknown whether the sets C_n in general have an inner model theoretic characterisation. For $n = 1$, Guaspari, Kechris, and Sacks independently showed that $\mathrm{C}_1 = \{x \in \mathbb{R} : x \in \mathbf{L}_{\omega_1^x}\}$ [Gua73, Kec75, Sac76]. For $k > 0$, the analogous statement "C_{2k+1} is the set of reals Δ^1_{2k+1}-equivalent to the first order theory of some level of $\mathbf{M}_{2k}$

projecting to ω" is open and known as the C_3 **conjecture** [Ste08B, p. 13] (*cf.* also [GH76, Cra85, Zhu17]).

 # 4. The strength of $\mathsf{Sep}(\Sigma^1_3)$ in the presence of sharps.

Original problem [KM78A]. Let (#) stand for "for all $x \subseteq \omega$, $x^{\#}$ exists" and let $\mathsf{Sep}(\Sigma^1_3)$ denote "for every $x \subseteq \omega$, every two disjoint $\Sigma^1_3(x)$ sets of reals can be separated by a $\Delta^1_3(x)$ set."

PROBLEM # 4. *Prove or disprove that*

$$\mathsf{ZFC} + \mathsf{Sep}(\Sigma^1_3) + (\#) \textit{ implies } \mathsf{Det}(\underset{\sim}{\mathbf{\Delta}}^1_2).$$

Harrington has shown that $\mathsf{ZFC}+\mathsf{Sep}(\Sigma^1_3)$ is consistent relative to ZFC. However, using Jensen's Absoluteness Theorem for the core model $\mathbf{K}$ (which states that if (#) holds and Σ^1_3 formulas are not absolute for $\mathbf{K}$, then $0^\dagger$ exists) one can see that

$$\mathsf{ZF}+\mathsf{DC}+\mathsf{Sep}(\Sigma^1_3)+(\#) \text{ implies that } x^\dagger \text{ exists for all } x \subseteq \omega.$$

2020 comments. Problem # 4 was solved with core model techniques by John Steel, following the approach mentioned in the last paragraph of the original problem. The result appears as [Ste96, Corollary 7.14]. The key result is that if there are no inner models with Woodin cardinals and there exists a measurable cardinal, then $\mathbf{K}$ is Σ^1_3-correct [Ste96, Theorem 7.9]. In the setting of that book, an additional larger measurable Ω is assumed in the background and a set sized $\mathbf{K}$ is built of height Ω; this additional assumption is now known not to be necessary; cf. [JS13].

To solve Problem # 4 affirmatively, Steel argues that $\mathsf{Sep}(\Sigma^1_3) + (\#)$ implies that for every real x there is a proper class model M with $x \in M$, and an ordinal δ such that $\mathbf{V}^M_{\delta+1}$ is countable, and δ is Woodin in M. By results of Woodin, this implies $\mathsf{Det}(\underset{\sim}{\mathbf{\Delta}}^1_2)$ (*cf.* the ***2020 comments*** on Problem # 9 and [Nee10, Corollary 6.12]). To see that such a model M exists, one first uses the core model argument mentioned in the original problem: for any real y, the Σ^1_3-correctness of the Mitchell core model gives a proper class model N with $y \in N$ and two measurable cardinals.[4]

Once we have N, Steel argues that if y is chosen carefully to ensure that $\mathsf{Sep}(\Sigma^1_3)$ relativises down from $\mathbf{V}$ to N, the $\mathbf{K}_x$ construction inside N must fail: Assuming that $(\mathbf{K}_x)^N$ exists, then it is Σ^1_3-correct inside N. But there is a $\Delta^1_3(x)$ well-ordering of the reals of $(\mathbf{K}_x)^N$, which implies the failure of $\mathsf{Sep}(\Sigma^1_3)$ inside $(\mathbf{K}_x)^N$. But by our choice of y, $\mathsf{Sep}(\Sigma^1_3)$ relativises down from $\mathbf{V}$ to N and the correctness of $\mathbf{K}_x$ inside N implies that it further relativises down from N to $(\mathbf{K}_x)^N$, which is impossible.

---

[4] If we make use of [JS13], we only need one measurable cardinal and could use the Dodd-Jensen core model here.

Thus, the $\mathbf{K}_x$ construction inside N fails and and therefore $(\mathbf{K}_x^c)^N$ reaches a Woodin cardinal, and an iterate of an appropriate hull of $(\mathbf{K}_x^c)^N$ is the model M as needed.

It is still open whether there is a Σ_3^1-correctness theorem for $\mathbf{K}$ (in the absence of Woodin cardinals) without additional assumptions beyond the existence of sharps.

5. A classification of functions on the Turing degrees.

Original problem [KM78A]. We write $\mathcal{D}$ for the set of Turing degrees. A property P of degrees holds almost everywhere (a.e.) if and only if there is a $\mathbf{c}$ such that for all $\mathbf{d} \geq \mathbf{c}$, we have $P(\mathbf{d})$. For $f, g : \mathcal{D} \to \mathcal{D}$, let $f \leq_{\mathrm{m}} g$ if and only if $f(\mathbf{d}) \leq g(\mathbf{d})$ a.e. A function $f : \mathcal{D} \to \mathcal{D}$ is **representable** if and only if there is some $F : {}^{\omega}\omega \to {}^{\omega}\omega$ such that for all x, $\deg(F(x)) = f(\deg(x))$.

PROBLEM # 5. *Working in* $\mathsf{ZF} + \mathsf{AD} + \mathsf{DC}$, *settle the following conjectures of D. Martin*:

(*a*) *If* $f : \mathcal{D} \to \mathcal{D}$ *is representable and* $\mathbf{d} \not\leq f(\mathbf{d})$ *a.e., then there is a* $\mathbf{c}$ *such that* $f(\mathbf{d}) = \mathbf{c}$ *a.e.*
(*b*) *The relation* $\leq_{\mathrm{m}}$ *is a prewellorder of* $\{f : f$ *is representable and* $\mathbf{d} \leq f(\mathbf{d})$ *a.e.*$\}$.

Further, if f *has rank* α *in* $\leq_{\mathrm{m}}$, *then* f' *has rank* $\alpha + 1$, *where* $f'(\mathbf{d}) = f(\mathbf{d})'$, *the Turing jump of* $f(\mathbf{d})$.

REMARKS. With regard to (a), it is well known that if $f(\mathbf{d}) \leq \mathbf{d}$ and $\forall \mathbf{c}\,(\mathbf{c} \leq f(\mathbf{d})$ a.e.), then $f(\mathbf{d}) = \mathbf{d}$ a.e. It is known that conjecture (b) is true when restricted to uniformly representable f so that $\mathbf{d} \leq f(\mathbf{d})$ a.e. (A function f is **uniformly representable** if there is an $F : {}^{\omega}\omega \to {}^{\omega}\omega$ such that for all x, we have $\deg(F(x)) = f(\deg(x))$ and, moreover, there is a $t : \omega \to \omega$ such that for all x and y, if $x \equiv_{\mathrm{T}} y$ via e then $F(x) \equiv_{\mathrm{T}} F(y)$ via $t(e)$.) It is conjectured that every representable $f : \mathcal{D} \to \mathcal{D}$ is uniformly representable.

A proof of conjecture (b) would yield a strong negative answer to a question of Sacks: is there a degree invariant solution to Post's problem?

Progress report [KMM81A]. It follows from unpublished results of Kechris and Solovay that $\mathsf{ZF} + \mathsf{AD} + \mathsf{DC} + \mathbf{V}{=}\mathbf{L}(\mathbb{R})$ implies that every function $f : \mathcal{D} \to \mathcal{D}$ on the degrees is representable. Although this has no direct bearing on a possible solution of the fifth problem, it underscores the generality of the question.

Progress report [KMM83A]. Slaman and Steel have proved two theorems relevant to Problem # 5. The first verifies a special case of conjecture (a):

THEOREM 4. $(\mathsf{ZF} + \mathsf{AD} + \mathsf{DC})$. *Let* $f : \mathcal{D} \to \mathcal{D}$ *be such that* $f(\mathbf{d}) < \mathbf{d}$ *a.e.*; *then for some* c, $f(\mathbf{d}) = \mathbf{c}$ *a.e.*

The second verifies a special case of conjecture (b). Call $f : \mathcal{D} \to \mathcal{D}$ **order-preserving a.e.** if and only if there is a $\mathbf{c}$ such that for all $\mathbf{a}, \mathbf{b} \geq \mathbf{c}$, we have that $\mathbf{a} \leq \mathbf{b}$ implies $f(\mathbf{a}) \leq f(\mathbf{b})$.

THEOREM 5. $(\mathsf{ZF} + \mathsf{AD} + \mathsf{DC})$. *Let* $f : \mathcal{D} \to \mathcal{D}$ *be order-preserving a.e. and such that* $\mathbf{d} < f(\mathbf{d})$ *a.e. Then either*

(*i*) $\exists \alpha < \omega_1 \, (f(\mathbf{d}) = \mathbf{d}^{\alpha}$ *a.e.*), *or*

(*ii*) *For a.e.* $\mathbf{d}$, $\forall \alpha < \omega_1^{\mathbf{d}} \, (f(\mathbf{d}) > \mathbf{d}^{\alpha})$.

(Here ω_1 is the least uncountable ordinal, and $\omega_1^{\mathbf{d}}$ is the least $\mathbf{d}$-admissible ordinal greater than ω.)

2020 comments. Problem # 5 is commonly known as **Martin's Conjecture**. We shall refer to its restriction to uniformly representable functions as the **Uniform Martin's Conjecture**. Steel had proved part (b) of the Uniform Martin's Conjecture [Ste82A]; for further work in this direction, *cf.* [Bec86]. The partial results by Slaman and Steel listed in the 1983 progress report appear in [SS88] and constitute a proof of the Uniform Martin's Conjecture. Theorem 5 can be improved using results of Woodin [Woo08], so that part (ii) of the conclusion can be strengthened to $f(\mathbf{d}) > \mathcal{O}^{\mathbf{d}}$. Kihara and Montalbán recently refined the Uniform Martin's Conjecture to functions from the Turing degrees to the many-one degrees [KM18].

A competing conjecture from the theory of Borel equivalence relations is in conflict with Martin's conjecture: For Polish spaces X and Y and equivalence relations $\equiv$ and $\equiv'$ on X and Y, respectively, we say that $\equiv$ is **Borel reducible** to $\equiv'$ if and only if there is a Borel function $f : X \to Y$ such that for all $x, x' \in X$ we have

$$x \equiv x' \iff f(x) \equiv' f(x').$$

An equivalence relation on X is **Borel** if and only if it is a Borel subset of $X \times X$, and it is **countable** if and only if all its equivalence classes are countable. A countable Borel equivalence relation is **universal** if and only if all countable Borel equivalence relations are Borel reducible to it. Kechris asked (*cf.* [Kec92, Problem 17, p. 99]):[5]

QUESTION 6. *Is Turing equivalence* $\equiv_{\mathrm{T}}$ *universal*?

Slaman and Steel have also shown that *arithmetic* equivalence is universal [MSS16, § 2], but the question remains open for Turing equivalence. A positive answer to Kechris's question would contradict Martin's conjecture: if there is a Borel reduction of two disjoint copies of $\equiv_{\mathrm{T}}$ to $\equiv_{\mathrm{T}}$, then the range of one of the copies under the reduction would be a set disjoint from a cone.[6] A detailed discussion of the current state of knowledge, including a proof of the Slaman-Steel result on arithmetic equivalence, can be found in [MSS16].

[5] This question is sometimes stated as a conjecture; *cf.* [DK00, Conjecture, p. 86].

[6] Details can be found in [DK00, second Fact on p. 86].

Montalbán, Reimann, and Slaman, have shown (in unpublished work) that Turing equivalence is not *uniformly* universal [Sla09].

 # 6. The extent of definable scales.

Original problem [KMS88A].

PROBLEM # 6. *Assume* $\undertilde{\Pi}^1_1$-AD^{Σ_3}. *Do all* $\Game^{\Sigma_2}\undertilde{\Pi}^1_1$ *sets admit* $\mathbf{HOD}(\mathbb{R})$ *scales?*

The terminology is explained in Steel's paper [Ste88]. The strongest result in this direction has been Martin's theorem that for $\lambda < \omega_1$ a limit ordinal, $\undertilde{\Pi}^1_1$-AD^{λ} implies all $\Game^{\lambda}\undertilde{\Pi}^1_1$ sets admit $\Game^{\lambda}\undertilde{\Pi}^1_1$ scales [Mar83]. Work of Woodin and Steel had shown that a positive answer to # 6 implies that some form of definable determinacy (*i.e.*, $\undertilde{\Pi}^1_1$-AD^{Σ_3}) yields an inner model of $\mathsf{AD}_{\mathbb{R}}$.

Steel obtained a positive answer to # 6 in February 1984; his results in this area are described in [Ste88].

2020 comments. There has been a significant amount of additional work on determinacy of long games and regularity of associated sets. In [Ste08C], Steel extends the work published in [Ste88]:

Say that T is an ω_1-tree if and only if $T \subseteq \omega^{<\omega_1}$ and T is closed under initial segments.[7] For an ω_1-tree T, the game $\mathcal{G}(T)$ is the following (closed) game on ω of length ω_1: For any countable α, at stage α, player I plays an integer m_α and player II replies an integer n_α. Letting $\langle\cdot,\cdot\rangle$ denote a (natural) pairing function, let $f : \omega_1 \to \omega$ be the function defined at any α by $f(\alpha) = \langle m_\alpha, n_\alpha\rangle$. Player II wins this run of the game if and only if $f \in [T]$, the set of length-ω_1 branches through T.

We say that **ω_1-open-projective determinacy** holds if for all ω_1-trees T definable over $\mathbf{H}(\omega_1)$ from parameters, the game $\mathcal{G}(T)$ is determined. We let $\Game^{\omega_1}$(open-analytical) be the pointclass of all sets of the form $\Game^{\omega_1}(T)$ for such a tree.

THEOREM 7 (Steel). *If* ω_1*-open-projective determinacy holds, then the pointclass* $\Game^{\omega_1}$*(open-analytical) has the scale property.*

The determinacy property is called "ω_1-open-projective" because an ω_1-tree is definable over $\mathbf{H}(\omega_1)$ from parameters if and only if it can be coded by a projective set of reals. In [Nee04], Neeman proved that ω_1-open-projective determinacy follows from a traditional large cardinal assumption, *viz.*, that for every real x there is a countable, $\omega_1 + 1$-iterable (coarse) mouse M with $x \in M$ and $M \models \mathsf{ZFC} - \mathsf{P}$ + "there is a measurable Woodin cardinal", where $\mathsf{ZFC} - \mathsf{P}$ denotes ZFC without the power set axiom. The monograph [Nee04] describes the state of the art in the theory of determinacy of long games around

[7] The usual definition implies that ω_1-trees have height ω_1 and that each level be countable. The present form weakens both requirements but keeps that each node has at most countably many immediate successors, while simultaneously providing a uniform way of ensuring the countability of each of these sets of successors.

2004, although a few results of Woodin in the area remain unpublished and the field has further developed since then.

7. The Kleene ordinal.

Original problem [KMS88A].

PROBLEM #7. *Let κ be the least ordinal not the order type of a prewellordering of $\mathbb{R}$ recursive in Kleene's ^3E and a real. Assume $\mathsf{AD}^{\mathbf{L}(\mathbb{R})}$. Is κ the least weakly inaccessible cardinal?*

That the answer is positive is an old conjecture of Moschovakis, who had shown that κ is a regular limit of Suslin cardinals [Mos70, Mos78]. Steel showed in [Ste81A] that κ is the least regular limit of Suslin cardinals. Thus the problem amounted to bounding the growth of the Suslin cardinals below κ. Building on work of Kunen and Martin, Jackson had done this for the first ω Suslin cardinals; this work is described in his long paper [Jac88].

In the fall of 1985, Jackson obtained a positive answer to # 7. His new work extends the theory presented in [Jac88]. Because of its length and complexity, as of now no one but Jackson has been through this new work.

2020 comments. Jackson's result remains unpublished; he comments:

> Steel has developed a "fine structure theory" for $\mathbf{L}(\mathbb{R})$ assuming $\mathsf{ZF} + \mathsf{AD}$. This suffices to answer certain questions about $\mathbf{L}(\mathbb{R})$, for example, it gives a complete description of the scale property in $\mathbf{L}(\mathbb{R})$. Other problems, however, such as whether every regular cardinal is measurable seem to require a more detailed understanding of $\mathbf{L}(\mathbb{R})$.
>
> Our results provide such a detailed analysis for an initial segment of the $\mathbf{L}_\alpha(\mathbb{R})$ hierarchy. Exactly how far this enables one to go is not clear, and is the subject of current investigation. However, the author has verified that the theory extends through the Kleene ordinal $\kappa = \mathrm{o}(^3E)$, and in fact, considerably beyond. This analysis is quite involved, however, and has not yet been written up. One consequence is the solution to a problem of Moschovakis, who conjectured in $\mathsf{ZF} + \mathsf{AD} + \mathsf{DC}$ that the Kleene ordinal should be the least inaccessible cardinal (this is the seventh Victoria Delfino problem). [Jac89, p. 80]

As already quoted in our comments to Problem # 1, Jackson reports in 2010 that he had

> extended the analysis further, up to the least inaccessible cardinal in $\mathbf{L}(\mathbb{R})$, although this lengthy analysis has never been written up. [Jac10, p. 1755]

To the best of our knowledge, no alternative approaches (via the **HOD** analysis or otherwise) have been suggested. Portions of the analysis have appeared in [Jac91, Jac92, Jac10].

8. Regular cardinals in $\mathbf{L}(\mathbb{R})$.

Original problem [KMS88A].

PROBLEM # 8. *Assume* $\mathsf{AD} + \mathbf{V}{=}\mathbf{L}(\mathbb{R})$. *Are all regular cardinals below* Θ *measurable*?

Moschovakis and Kechris had shown, in $\mathsf{ZFC} + \mathsf{AD}^{\mathbf{L}(\mathbb{R})}$, that if κ is regular (in $\mathbf{V}$, where AC holds!) and $\kappa < \Theta^{\mathbf{L}(\mathbb{R})}$, then $\mathbf{L}(\mathbb{R}) \models$ "κ is measurable". This led them to conjecture a positive answer to # 8. Jackson's detailed analysis of cardinals and measures had verified the conjecture for κ below the supremum of the first ω Suslin cardinals (cf. [Jac88]).

The only progress on this problem since its addition to the list is Jackson's new work cited above, which presumably yields a positive answer to # 8 for κ below the Kleene ordinal.

2020 comments. Problem # 8 was solved by John Steel using core model techniques, specifically through the beginning of what we now call the **HOD** analysis. The proof is published as [Ste10, Theorem 8.27].

Steel realised, under the assumption of determinacy, the fragment of the model $\mathbf{HOD}^{\mathbf{L}(\mathbb{R})}$ below Θ as a fine structural mouse, specifically as the direct limit of a system whose objects are certain countable mice and whose commuting maps are appropriate iterations. Analysis of this system allows us to conclude (combinatorial or descriptive set theoretic) properties of its direct limit from (fine structural) properties of the mice, and many different results have been established this way. In particular:

THEOREM 8 (Steel; [Ste95A]). *Assume* $\mathsf{AD}^{\mathbf{L}(\mathbb{R})}$ *and work in* $\mathbf{L}(\mathbb{R})$. *Then for every* $x \in \mathbb{R}$ *and* $\kappa < \underset{\sim}{\delta}^2_1$ *such that* κ *is regular in* $\mathbf{HOD}(x)$, *the following implication holds*:

$$\mathrm{cf}(\kappa) > \omega \text{ implies } \mathbf{HOD}(x) \models \text{“}\kappa \text{ is measurable”}.$$

These measures on κ in $\mathbf{HOD}(x)$ for different x can be amalgamated via the directed system that guides the iterations mentioned above; the result now follows via reflection; *cf.* also [Ste10, Lemma 8.25]

This analysis of $\mathbf{V}_\Theta \cap \mathbf{HOD}^{\mathbf{L}(\mathbb{R})}$ has been extended by Woodin to a full analysis of **HOD** via a longer directed system, while identifying the correct *hybrid* rather than purely fine structural mice that make up **HOD**; *cf.* [SW16]. A similar analysis of the **HOD** of larger models than $\mathbf{L}(\mathbb{R})$ has become a key tool in recent work in determinacy, in particular, in the proofs of partial versions of the *mouse set conjecture* (*cf.* the ***2020 comments*** on Problem # 11).

To illustrate the reach of the **HOD** analysis, we mention some further applications (the list is not exhaustive): Recall that, assuming determinacy, κ^1_{2n+1} is the cardinal predecessor of the projective ordinal $\underset{\sim}{\delta}^1_{2n+1}$. In [Sar13B, Theorem 5.2.2], Sargsyan proves Woodin's theorem that, under $\mathsf{AD} + \mathbf{V}{=}\mathbf{L}(\mathbb{R})$, for all $n \in \omega$, κ^1_{2n+3} is the least cardinal δ of **HOD** such that $\mathbf{M}_{2n}(\mathbf{HOD}|\delta) \models$ "δ is Woodin." This identifies a purely descriptive set theoretic characterisation of cardinals with a fine structural characterisation, and provides us with precise information of how much large cardinal strength the relevant cardinals retain when passing from $\mathbf{V}$ to nice inner models. In [Sar14], Sargsyan uses the **HOD** analysis to prove the strong partition property of $\underset{\sim}{\delta}^2_1$, a result first established in [KKMW81]. In [Nee07B], Neeman uses the analysis to provide a characterisation of supercompactness measures for ω_1 in $\mathbf{L}(\mathbb{R})$. In [JKSW14], the authors use the analysis to prove Woodin's result that, under $\mathsf{AD} + \mathbf{V}{=}\mathbf{L}(\mathbb{R})$, every uncountable cardinal below Θ is Jónsson and, if its cofinality is ω, then it is even Rowbottom. This drastically extends previous results of Kleinberg [Kle77] and their generalisation by the second author [Löw02].

9. Large cardinals implying determinacy.

Original problem [KMS88A].

PROBLEM # 9. *Does the existence of a nontrivial, elementary* $j : \mathbf{V}_{\lambda+1} \to \mathbf{V}_{\lambda+1}$ *imply* $\underset{\sim}{\mathbf{\Pi}}^1_3$ *determinacy*?

The world view embodied in the statements of this and the succeeding problem was seriously mistaken. That view was inspired by Martin's result that the existence of a nontrivial, Σ_1-elementary $j : \mathbf{V}_{\lambda+1} \to \mathbf{V}_{\lambda+1}$ implies $\underset{\sim}{\mathbf{\Pi}}^1_2$ determinacy [Mar80], together with work of Mitchell [Mit79] which promised to lead to a proof that nothing much weaker than the existence of such an embedding would imply $\underset{\sim}{\mathbf{\Pi}}^1_2$ determinacy. Martin naturally conjectured that a nontrivial, fully elementary $j : \mathbf{V}_{\lambda+1} \to \mathbf{V}_{\lambda+1}$ would yield PD; hence the inclusion of # 9 on our list.

Partly because this view was so mistaken, progress in this area since 1984 has been dramatic. From February to April of 1984, Woodin showed that the existence of a nontrivial, elementary $j : \mathbf{L}(\mathbf{V}_{\lambda+1}) \to \mathbf{L}(\mathbf{V}_{\lambda+1})$ implies PD and in fact $\mathsf{AD}^{\mathbf{L}(\mathbb{R})}$. This was still consistent with the view underlying # 9, and in spirit was a positive answer, although even for $\underset{\sim}{\mathbf{\Pi}}^1_3$ determinacy Woodin's result required a hypothesis slightly stronger than allowed in # 9. However, at about the same time Foreman, Magidor and Shelah [FMS88] developed a powerful new technique for producing generic elementary embeddings under relatively "weak" large cardinal hypotheses such as the existence of supercompact cardinals. Woodin realised at once the potential in their technique and used it to show, in May 1984, that the existence of a supercompact cardinal implies all projective sets of reals are Lebesgue measurable. Immediately thereafter, Shelah and Woodin improved this to include all sets in $\mathbf{L}(\mathbb{R})$.

If the relationship between large cardinals and determinacy were to exhibit anything like the pattern it had previously, supercompact cardinals had to imply $\mathsf{AD}^{\mathbf{L}(\mathbb{R})}$. In September 1985, Martin and Steel showed that in fact they do (thereby answering # 9 positively). (Their proof of PD is self-contained. Their proof of $\mathsf{AD}^{\mathbf{L}(\mathbb{R})}$ requires work done by Woodin using the generic embedding techniques.) The Martin-Steel theorem required much less than supercompactness; e.g., for $\boldsymbol{\Pi}^1_{n+1}$ determinacy it required the existence of n "Woodin cardinals" with a measurable above them all. [The notion of a "Woodin cardinal" had been isolated by Woodin in his work on generic embeddings; it is a refinement of a notion due to Shelah.] In May–July of 1986, Martin and Steel pushed the theory of inner models for large cardinals far enough to show that the hypothesis of their theorem was best possible: the existence of n Woodin cardinals does not imply $\boldsymbol{\Pi}^1_{n+1}$ determinacy. More recently, Woodin has obtained relative consistency results in this direction by a different method; cf. Problem # 10 below.

Unfortunately, with the exception of [FMS88], none of this recent work has been published.

2020 comments. The relationship between determinacy and large cardinals is now well documented. Since this relationship is of fundamental importance to the field and the Cabal, we use this opportunity to give a brief exposition of the developments and the current topics of research.

Woodin cardinals. The mentioned Shelah-Woodin results on Lebesgue measurability of all sets of reals in $\mathbf{L}(\mathbb{R})$ in the presence of large cardinals appear in [SW90]. The paper defines the notions now known as Shelah and Woodin cardinals, although the notation it uses is different, cf. [SW90, p. 384 and Definitions 3.5 & 4.1]. Since the paper was not published until after the importance of Woodin cardinals had become apparent, the name *Woodin cardinal* appears in this paper:

> We define here two large cardinals: $\mathrm{Pr}_{\mathrm{a}}(\lambda, f)$, $\mathrm{Pr}_{\mathrm{a}}(\lambda)$ by Shelah (Definition 3.5) and $\mathrm{Pr}_{\mathrm{b}}(\lambda)$ by Woodin—now called a Woodin cardinal. [SW90, p. 384]

The mentioned Martin-Steel results appear in [MS88] and [MS89], which also mark the first appearance of the term *Woodin cardinal* in the literature. The definition of Woodin cardinals given in [SW90] is easily seen to be equivalent to the modern definition: suppose that δ is an infinite ordinal and that $A \subseteq \mathbf{V}_\delta$. A cardinal $\lambda < \delta$ is **$<\delta$-A-strong** if and only if for any $\mu < \delta$ there is a nontrivial elementary embedding $j : \mathbf{V} \to M$ with critical point λ and such that $j(\lambda) > \mu$, $\mathbf{V}_\mu \subset M$, and $j(A) \cap \mathbf{V}_\mu = A \cap \mathbf{V}_\mu$. The ordinal δ is a **Woodin cardinal** if and only if it is an inaccessible cardinal and for all $A \subseteq \mathbf{V}_\delta$ there is a $<\delta$-A-strong cardinal.

Woodinness was instantly recognised as a pivotal large cardinal notion, and its properties were immediately studied in detail. The realisation that Woodin cardinals form a key step in the development of the inner model program confirmed their importance for the field: comparison of mice is central to the theory of fine structural models; comparisons at the level of cardinals that could be reached by the techniques of the early 1980s were linear and this imposed serious limitations on the nature of the corresponding models, *e.g.*, all of them admitted $\underset{\sim}{\mathbf{\Delta}}^1_3$ well-orderings of their set of reals. As a consequence, none of them could be models of projective determinacy. Thus, if inner model theory had any hope of reaching supercompact cardinals, essential changes were needed.

The crucial change connected to Woodin cardinals was the increase in the complexity of the comparison process from linear iterations to what are now called *iteration trees*. The development of the appropriate fine structure followed shortly thereafter [MaS94, MiS94] and led to the precise determination of the effect of Woodin cardinals on the complexity of the reals present in canonical inner models, on the amount of determinacy outright provable or provably consistent, and on the amount of correctness that a model would satisfy or that could be forced of an iterate of the model. As a consequence, the set theoretic landscape transformed significantly thanks to the introduction of Woodin cardinals.

Determinacy from large cardinals. The Martin-Steel theorem mentioned in the original problem, "n Woodin cardinals and a measurable above imply $\mathsf{Det}(\underset{\sim}{\mathbf{\Pi}}^1_{n+1})$", is published in [MS89]. The optimal result is that if for every real x there is a suitable model M that is iterable and contains x and n Woodin cardinals, then $\mathsf{Det}(\underset{\sim}{\mathbf{\Pi}}^1_{n+1})$ holds [Nee95].

That assuming just n Woodin cardinals does not suffice follows from inner model theory: from the existence of n Woodin cardinals, a fine structural model with n Woodin cardinals can be obtained, in which the reals admit a Δ^1_{n+2} well-ordering [MaS94, Ste95B], and therefore $\mathsf{Det}(\Pi^1_{n+1})$ fails in the model by [Kan94, Exercise 27.14].

> $\mathsf{AD}^{\mathbf{L}(\mathbb{R})}$ from infinitely many Woodin cardinals and a measurable cardinal above them is due to Woodin, proved using the methods of stationary tower forcing and an appeal to the main theorem, Theorem 5.11, in Martin-Steel [MS89]. A proof using Woodin's genericity iterations and fine structure instead of stationary tower forcing is due to Steel, and the proof reached in this chapter (using a second form of genericity iterations and no fine structure) is due to Neeman. [Nee10, p. 1880]

These arguments can be pushed much further, and the determinacy of stronger pointclasses than $\wp(\mathbb{R}) \cap \mathbf{L}(\mathbb{R})$ is provable by similar methods from large cardinals still in the region of Woodin cardinals (in particular, well before reaching the level of rank-to-rank embeddings or even supercompactness).

The consistency strength of the Axiom of Determinacy. Woodin's derived model theorem shows that infinitely many Woodin cardinals without a measurable above suffice to establish the consistency of determinacy in $\mathbf{L}(\mathbb{R})$:[8]

THEOREM 9. *If λ is a limit of Woodin cardinals, G is $\mathrm{Col}(\omega, {<}\lambda)$-generic over $\mathbf{V}$, and $\mathbb{R}^* = \bigcup_{\alpha<\lambda} \mathbb{R} \cap \mathbf{V}[G|\alpha]$, then $\mathbf{L}(\mathbb{R}^*)$ is a model of determinacy.*

In fact, we have that $\mathbb{R}^* = \mathbb{R} \cap \mathbf{V}(\mathbb{R}^*)$ and, letting Γ denote the collection of all sets of reals $A \subseteq \mathbb{R}^*$ in $\mathbf{V}(\mathbb{R}^*)$ such that $\mathbf{L}(A, \mathbb{R}^*) \models \mathsf{AD}^+$, then we have that $\mathbf{L}(\Gamma, \mathbb{R}^*)$ is also a model of determinacy.

Here, AD^+ denotes Woodin's strengthening of the axiom of determinacy; *cf.* the ***2020 comments*** on Problem # 14 below.

Conversely, if $\mathsf{AD}^{\mathbf{L}(\mathbb{R})}$ holds, then in a forcing extension there is a model of choice with ω Woodin cardinals (*cf.* [Ste09, KW10, ST10, Zhu10, Zhu15]).

One of the important early results concerning proofs of the existence of large cardinals in inner models from determinacy is the following theorem (*cf.* [KW10, Theorem 5.1]):

THEOREM 10 (Woodin). *Assume $\mathsf{AD} + \mathbf{V}{=}\mathbf{L}(\mathbb{R})$. Then Θ is a Woodin cardinal in* **HOD**.

Lightface determinacy. Harrington's results on getting sharps from analytic determinacy are lightface: if x is a real and $\mathsf{Det}(\Pi^1_1(x))$ holds, then $x^\#$ exists [Har78].

Moving up to Π^1_2, we get that if both pointclasses $\boldsymbol{\Pi}^1_1$ and Π^1_2 are determined, then $\mathbf{M}_1^\#$ exists and is ω_1-iterable [SW16, Corollary 4.17]. In 1995, Woodin claimed the following boldface generalisations of this result:[9]

THEOREM 11. *If $\mathsf{Det}(\boldsymbol{\Pi}^1_{n+1})$ holds, then $\mathbf{M}_n^\#(x)$ exists and is ω_1-iterable for all reals x.*

The result remained unpublished until [MSW16], where the following strengthening is established:

THEOREM 12 (Woodin). *Assume $\mathsf{Det}(\Pi^1_{n+1})$ and $\mathsf{Det}(\boldsymbol{\Pi}^1_n)$. If there is no $\boldsymbol{\Sigma}^1_{n+2}$ sequence of length ω_1 of distinct reals, then $\mathbf{M}_n^\#$ exists and is ω_1-iterable.*

The proof uses inner model theory and relativizes to give Theorem 11.

In [MSW16, § 4.2], the authors further conjecture the strengthening of Theorem 12 where the assumption about the existence of uncountable sequences of reals is removed. For $n = 1$, this is [SW16, Corollary 4.17] mentioned above. The conjecture remains open in general, but was settled affirmatively by Zhu for odd numbers in [Zhu16].

[8] Appropriate weakenings hold for finitely many Woodin cardinals; *e.g.*, if δ is Woodin and G is $\mathrm{Col}(\omega, \delta)$-generic over $\mathbf{V}$, then $\mathsf{Det}(\Delta^1_2)$ holds in $\mathbf{V}[G]$ [Nee95, Corollary 2.3].

[9] *Cf.* [Nee95, p. 328] and [Nee04, p. 9].

The Solovay sequence. Theorem 10 has been significantly generalised and is part of the **HOD** analysis mentioned in connection with Problem # 8. In [Sol78B], Solovay introduced the **Solovay sequence** $\langle \Theta_\alpha \mid \alpha \leq \Omega \rangle$ as a way of measuring the strength of determinacy models: We assume determinacy and let Θ_0 is the supremum of all ordinals α for which there is an ordinal definable pre-wellordering of a subset of $\mathbb{R}$ of length α. If Θ_α is defined for all $\alpha < \beta$, and β is limit, then Θ_β is defined as their supremum. Finally, if Θ_α is defined and is less than Θ, then $\Theta_{\alpha+1}$ is the supremum of the lengths of all pre-wellorderings of subsets of $\mathbb{R}$ that are definable from ordinals and a set of reals of Wadge rank Θ_α. The sequence ends once an ordinal Ω is reached such that $\Theta_\Omega = \Theta$.

In $\mathbf{L}(\mathbb{R})$, we have that $\Theta = \Theta_0$, but longer sequences are possible and correspond to models of stronger versions of determinacy. It turns out that all $\Theta_{\alpha+1}$ are Woodin cardinals in **HOD**. The situation at limit ordinals is more delicate and still being explored; *cf.* [Sar15]. Conversely, starting with models with many Woodin cardinals, the derived model construction provides us with models of strong versions of determinacy; *cf.*, *e.g.*, [Ste08A].

Very large cardinals. Although no longer relevant to the goal of deriving determinacy from large cardinals, Woodin's original approach led to the development of the theory of large cardinals past the level of rank-to-rank embeddings. The motivation was the realisation that there was a strong analogy between the theory of $\mathbf{L}(\mathbb{R})$ in the presence of determinacy, and the theory of $\mathbf{L}(\mathbf{V}_{\lambda+1})$ in the presence of nontrivial embeddings $j\colon \mathbf{L}(\mathbf{V}_{\lambda+1}) \to \mathbf{L}(\mathbf{V}_{\lambda+1})$ with λ being the supremum of the associated critical sequence. Some results illustrating this can be found in [Kaf04], where versions of the coding lemma are established. For more recent developments, *cf.*, *e.g.*, [Dim11, BKW17].

10. Supercompacts in $\mathbf{HOD}^{\mathbf{L}(\mathbb{R})}$.

Original problem [KMS88A].

PROBLEM # 10. *Assume* $\mathsf{AD}^{\mathbf{L}(\mathbb{R})}$. *Does* $\mathbf{HOD}^{\mathbf{L}(\mathbb{R})}$ *satisfy "there is a* κ *such that* κ *is* 2^κ*-supercompact"?*

Becker and Moschovakis [BM81] had shown that $\mathbf{HOD}^{\mathbf{L}(\mathbb{R})} \models$ "there is a κ such that $o(\kappa) = \kappa^+$". Martin (unpublished) then showed $\mathbf{HOD}^{\mathbf{L}(\mathbb{R})} \models$ "there is a κ such that κ is μ-measurable". Steel (unpublished) then showed $\mathbf{HOD}^{\mathbf{L}(\mathbb{R})} \models$ "there is a κ such that κ is λ-strong, where $\lambda > \kappa$ is measurable". Inspired by these results, the Cabal conjectured that the model $\mathbf{HOD}^{\mathbf{L}(\mathbb{R})}$ satisfies all large cardinal hypotheses weaker than that which implies $\mathsf{AD}^{\mathbf{L}(\mathbb{R})}$ (which is false in $\mathbf{HOD}^{\mathbf{L}(\mathbb{R})}$). Problem # 10 resulted from our mistaken guess as to what these hypotheses are.

The Woodin-Shelah Theorem that the existence of supercompacts implies all sets in $\mathbf{L}(\mathbb{R})$ are Lebesgue measurable settles # 10 negatively, since, assuming

$\mathsf{AD}^{\mathbf{L}(\mathbb{R})}$, $\mathbf{HOD}^{\mathbf{L}(\mathbb{R})} \models$ "there is a wellorder of $\mathbb{R}$ in $\mathbf{L}(\mathbb{R})$". However, except for the mistake about the cardinals involved, the answer to # 10 is positive. Woodin has recently (February 1987) shown that, assuming $\mathsf{AD}^{\mathbf{L}(\mathbb{R})}$, $\mathbf{HOD}^{\mathbf{L}(\mathbb{R})} \models$ "there is a κ such that κ is a Woodin cardinal", and under the same assumption found a natural submodel of $\mathbf{HOD}^{\mathbf{L}(\mathbb{R})}$ satisfying "there are ω Woodin cardinals". The work of Martin, Steel and Woodin referred to in the discussion of # 9, together with further work of Woodin reducing its large cardinal hypothesis, shows that $\mathsf{AD}^{\mathbf{L}(\mathbb{R})}$ follows from the existence of ω Woodin cardinals with a measurable above them all, so that Woodin's recent work is in spirit a positive answer to # 10.

2020 comments. The remarks we gave on Problem # 9 apply here as well. The paper [KW10] shows how to find Woodin cardinals in **HOD**. Assuming strong forms of determinacy, the question of precisely which large cardinals can be present in **HOD** remains open, with modern research in descriptive inner model theory motivated by the expectation that at least a very large initial segment of the large cardinal hierarchy should be realised within the **HOD** models of strong models of determinacy [Sar13A].

11. The GCH in $\mathbf{HOD}^{\mathbf{L}(\mathbb{R})}$.

Original problem [KMS88A].

PROBLEM # 11. *Assume* $\mathsf{AD}^{\mathbf{L}(\mathbb{R})}$. *Does* $\mathbf{HOD}^{\mathbf{L}(\mathbb{R})}$ *satisfy the* GCH?

Becker [Bec80] has shown that, assuming $\mathsf{AD}^{\mathbf{L}(\mathbb{R})}$, $\mathbf{HOD}^{\mathbf{L}(\mathbb{R})} \models 2^\kappa = \kappa^+$ for many cardinals κ. There has been little progress on this question since January 1984. Woodin's recent work on large cardinals in $\mathbf{HOD}^{\mathbf{L}(\mathbb{R})}$ does show that

$$\mathbf{HOD}^{\mathbf{L}(\mathbb{R})} \models \text{"}(\underset{\sim}{\delta}^2_1)^{\mathbf{L}(\mathbb{R})} \text{ is } \Theta^{\mathbf{L}(\mathbb{R})}\text{-strong"}.$$

It follows by an easy reflection argument that if $\mathbf{HOD}^{\mathbf{L}(\mathbb{R})}$ satisfies the GCH below $(\underset{\sim}{\delta}^2_1)^{\mathbf{L}(\mathbb{R})}$, then it satisfies the GCH.

2020 comments. Steel's analysis of **HOD** below $\underset{\sim}{\delta}^2_1$, mentioned in the solution to Problem # 8, also solves # 11, *cf.* [Ste10, Corollary 8.22]. Beyond the fine structural analysis, Steel's argument uses the result mentioned in the original wording of the problem, that under AD, $\underset{\sim}{\delta}^2_1$ is strong up to Θ in $\mathbf{HOD}^{\mathbf{L}(\mathbb{R})}$; *cf.* [KW10]. It also uses that there is a set $P \subseteq \Theta$ in $\mathbf{L}(\mathbb{R})$ such that $\mathbf{HOD}^{\mathbf{L}(\mathbb{R})} = \mathbf{L}(P)$. Both these results are due to Woodin. The second follows from the analysis of the Vopěnka algebra; *cf.* [SW16].

The argument generalises to the **HOD** of larger models of determinacy, as long as the models allow a version of the **HOD** analysis. At the moment, this falls within the region below a Woodin limit of Woodin cardinals or, in terms of determinacy assumptions, somewhere in the neighbourhood of $\mathsf{AD}_{\mathbb{R}}$+"Θ is regular" [Sar15, Tra14, AS19, ST16]. The expectation is that the result should hold in general.

 # 12. Projective uniformisation, measure, and category.

Original problem [KMS88A].

PROBLEM # 12. *Does the theory* ZFC+*"Every projective relation can be uniformised by a projective function"* + *"Every projective set is Lebesgue measurable and has the property of Baire" prove* PD?

Woodin [Woo82] showed that the theory in question proves $\forall x \subseteq \omega(x^\dagger$ exists) and more in this direction, together with some other consequences of PD, and conjectured a positive answer to # 12.

There has been no direct progress on this problem since 1984.

2020 comments. Although the expectation was a positive answer, Problem # 12 was solved negatively by Steel in 1997. The precise strength of the theory in question is that of ZFC together with the existence of a cardinal δ with countable cofinality that is the limit of cardinals that are δ-strong.

Details can be found as handwritten notes by Schindler [Sch99], and in Philipp Doebler's Master's thesis [Doe06]. Steel showed that the large cardinal mentioned above suffices to produce a model of the theory under consideration, and Schindler proved that this is indeed an equiconsistency, *cf.* [Sch02, Theorem 9.1].

We sketch Steel's argument. If there is a cardinal δ as required, then there is a minimal, fully iterable, fine structural inner model $\mathbf{L}[E]$ witnessing that there is such a cardinal δ; this model admits a $\undertilde{\mathbf{\Sigma}}^1_3$ well-ordering of its reals and this means that $\mathsf{Det}(\undertilde{\mathbf{\Delta}}^1_2)$ must fail.

Steel argues by forcing with $\mathrm{Col}(\omega, \delta)$ over $\mathbf{L}[E]$. In the resulting model, all projective sets are Lebesgue measurable, and have the Baire property and we have projective uniformisation. Furthermore, $\mathbf{L}[E]$ is the core model of any of its forcing extensions, and thus $\mathbf{L}[E] \prec_{\Sigma^1_3} \mathbf{L}[E][G]$. Since $\mathsf{Det}(\undertilde{\mathbf{\Delta}}^1_2)$ is a $\undertilde{\mathbf{\Sigma}}^1_3$-statement, we obtain that $\mathsf{Det}(\undertilde{\mathbf{\Delta}}^1_2)$ must fail in $\mathbf{L}[E][G]$.

Using the additional assumption that $\mathbb{R}^\#$ exists (in order to implement the core model theory of [Ste96]) and results of Schindler on the complexity of $K \cap \mathbf{H}(\omega_1)$,[10] Hauser and Schindler showed that the theory in Problem # 12 gives us an inner model with a cardinal δ and an ω-sequence of cardinals cofinal in δ and δ-strong [HS00]. Finally, in [Sch02], Schindler shows that, at the level of the theories under consideration, core model theory works without this additional assumption and therefore provides us with a genuine equiconsistency.

---

[10] *Cf.* [HS00, Theorems 3.4 & 3.6] which in turn relied on earlier work by Hauser and Hjorth [HH97].

From further results in [HS00] and the same argument from [Sch02], we also have that the theory ZF+"Every projective relation can be uniformised by a projective function" + "Every projective set is Lebesgue measurable and has the property of Baire" (*i.e.*, the theory considered in Problem # 12 without the Axiom of Choice) gives us an inner model with a cardinal δ of cofinality ω that is the limit of cardinals that are λ-strong for all $\lambda < \delta$. This is also an equiconsistency, as can be verified by starting with the corresponding minimal $\mathbf{L}[E]$ model for this large cardinal assumption, and forcing now with the symmetric collapse of δ.

Two variations of Problem # 12 remain open:[11]

In the first variation, we strengthen the theory by changing the assumption of projective uniformisation with its level-by-level version, namely, that for each n, any $\boldsymbol{\Pi}^1_{2n+1}$ subset of $\mathbb{R}^2$ can be uniformised by a function with a $\boldsymbol{\Pi}^1_{2n+1}$ graph. Steel has shown that this version implies $\mathsf{Det}(\boldsymbol{\Delta}^1_2)$; *cf.* [Ste96, Corollary 7.14].

In the second variation, we replace the assumption with its lightface version, *i.e.*, that all lightface projective subsets of $\mathbb{R}^2$ can be uniformised by a function with a lightface projective graph.

 # 13. The cofinal branches hypothesis.

The **cofinal branches hypothesis**, introduced by Martin and Steel [MaS94, pp. 50–53], is the statement that every countable iteration tree on $\mathbf{V}$ has at least one cofinal well-founded branch; we write CBH for this statement.

PROBLEM # 13. *Does* CBH *hold*?

The **unique branches hypothesis**, UBH, also introduced by Martin-Steel [MaS94], is the statement that every countable iteration tree on $\mathbf{V}$ has at most one cofinal well-founded branch. As long as the iteration tree $\mathcal{T}$ under consideration is sufficiently closed, UBH for $\mathcal{T}$ implies CBH for $\mathcal{T}$.

2020 comments. A few years after the problem was formulated, Woodin refuted UBH using large cardinals at the level of embeddings $j : \mathbf{V}_\lambda \to \mathbf{V}_\lambda$. Later, in 1999, he also refuted CBH, from the existence of a supercompact with a Woodin above, showing from these assumptions that there is an iteration tree of length ω^2 with no cofinal well-founded branch. The tree is formed by an ultrapower by an extender, followed by an ω-sequence of alternating chains on the ultrapower model.

The argument also refutes UBH from the same assumptions, the counterexample being a single ultrapower, now followed by an alternating chain on the ultrapower model, both of whose branches are well-founded.

Details for the case of UBH were presented by Woodin at a meeting at the *American Institute of Mathematics* (AIM) in December 2004. Later, Neeman

[11] *Cf.* [Hau00] for more information on both of them.

and Steel significantly lowered the large cardinal assumption needed for both results, to something weaker than the existence of a cardinal strong past a Woodin. More precisely, Neeman and Steel obtained their counterexamples (using the same tree structure as in Woodin's results) from the assumption that there exists a cardinal δ and an extender F such that F has critical point below δ, support δ, and is δ-strong, and δ is Woodin in the smallest admissible set containing $\mathbf{V}_\delta \cup \{F\}$.

Details, including a discussion of revised versions of both hypotheses that remain open, together with partial positive results, can be found in [NS06].

 # 14. ∞-Borel sets.

Informally, a set is ∞-Borel if it can be generated from open sets by closing under the operations of complementation and well-ordered union. Since we are in a choiceless context, we need to give the formal definition in terms of ∞-Borel codes. In analogy to standard Borel codes, we define the class of ∞-Borel codes by recursion as follows: a tree T is an **∞-Borel code** if and only if

(i) either $T = \{\langle n \rangle\}$ for some $n \in \omega$,
(ii) or $T = \bigvee_\alpha T_\alpha := \{\langle \bigvee, \alpha \rangle^\frown t : \alpha < \tau \text{ and } t \in T_\alpha\}$, where τ is an ordinal, and each $T_\alpha \in \mathsf{BC}$,
(iii) or $T = \neg T' := \{\langle \neg \rangle^\frown t : t \in T'\}$, where $T' \in \mathsf{BC}$.

Now fix a bijection $\ulcorner \cdot, \cdot \urcorner : \omega^2 \to \omega$; given $T \in \mathsf{BC}$, we define its interpretation by recursion via

(i) $A_T = \{x \in {}^\omega\omega : x(k) = \ell\}$ if $T = \{\langle \ulcorner k, \ell \urcorner \rangle\}$,
(ii) $A_T = \bigcup_{\alpha<\tau} A_{T_\alpha}$ if $T = \bigvee_\alpha T_\alpha$, and
(iii) $A_T = {}^\omega\omega \setminus A_{T'}$ if $T = \neg T'$.

Then we say that a set A is **∞-Borel** if and only if there is an ∞-Borel code T such that $A = A_T$.

PROBLEM # 14. *Does* AD *imply that all sets of reals are ∞-Borel?*

A possibly weaker version of the problem is: Does $\mathsf{AD} + \mathsf{DC}_\mathbb{R}$ imply that all sets of reals are ∞-Borel?[12]

2020 comments. Both versions of Problem # 14 are open. The problem is now considered part of the question whether Woodin's AD^+ is equivalent to AD.

In order to define AD^+, we first need to formulate the concept of *ordinal determinacy*: if $\lambda < \Theta$, we endow λ with the discrete topology, and consider the product topology on ${}^\omega\lambda$. Given a set $A \subseteq {}^\omega\omega$ and a function $f : {}^\omega\lambda \to {}^\omega\omega$, we consider the game $\mathrm{G}(f, A)$ to be the game of length ω on λ with payoff set

[12]The axiom $\mathsf{DC}_\mathbb{R}$, or (more precisely) $\mathsf{DC}_\omega(\mathbb{R})$, is the statement that whenever $R \subseteq \mathbb{R}^2$ satisfies that for any real x there is a y with $x \mathrel{R} y$, then there is a function $f : \omega \to \mathbb{R}$ such that $f(n) \mathrel{R} f(n+1)$ for all n. Equivalently, any tree T on a subset of $\mathbb{R}$ with no end nodes has an infinite branch.

$f^{-1}[A]$. We say that **ordinal determinacy** holds if for any $\lambda < \Theta$, any continuous $f : {}^{\omega}\lambda \to {}^{\omega}\omega$, and any set of reals A, the game $\mathrm{G}(f, A)$ is determined. Now AD^+ is the conjunction of "All sets are ∞-Borel", $\mathsf{DC}_{\mathbb{R}}$, and ordinal determinacy.[13] It is not known whether any of the three components of AD^+ follows from AD.

It is known that AD^+ holds in natural models of determinacy, such as models of the form $\mathbf{L}(\wp(\mathbb{R}))$ obtained through the derived model construction. Woodin has shown that $\mathsf{AD}_{\mathbb{R}}$ (in fact, $\mathsf{AD} + \mathsf{Uniformisation}$) implies that all sets of reals are ∞-Borel (*cf.*, *e.g.*, [IW09, Theorem 4.10]).

The problem is closely connected to a number of other famous open problems in the area:

If every set of reals is ∞-Borel and there is no uncountable sequence of distinct reals, then all sets of reals are Ramsey, Lebesgue measurable, have the Baire property, and the perfect set property (*cf.* [CK11]); therefore a positive answer for Problem # 14 would imply that AD implies that every set of reals is Ramsey (*cf.* [Kan94, Question 27.18]).

In unpublished work, Woodin has shown that from the consistency of $\mathsf{ZF} + \mathsf{DC} + \mathsf{AD} +$ "not every set of reals is ∞-Borel" one can prove the consistency of $\mathsf{ZF} + \mathsf{DC} + \mathsf{AD} +$ "there exists $\kappa > \Theta$ with the strong partition property". This connects the problem with the open problem whether it is consistent to have a strong partition cardinal above Θ.

REFERENCES

ARTHUR W. APTER, STEPHEN C. JACKSON, AND BENEDIKT LÖWE

[AJL13] *Cofinality and measurability of the first three uncountable cardinals*, ***Transactions of the American Mathematical Society***, vol. 365 (2013), no. 1, pp. 59–98.

RACHID ATMAI

[Atm19] *An analysis of the models* $\mathbf{L}[T^{2n}]$, ***The Journal of Symbolic Logic***, vol. 84 (2019), pp. 1–26.

RACHID ATMAI AND GRIGOR SARGSYAN

[AS19] **HOD** *up to* $\mathsf{AD}_{\mathbb{R}} + \Theta$ *is measurable*, ***Annals of Pure and Applied Logic***, vol. 170 (2019), no. 1, pp. 95–108.

JOAN BAGARIA, PETER KOELLNER, AND W. HUGH WOODIN

[BKW17] *Large cardinals beyond choice*, 2017, preprint.

JAMES E. BAUMGARTNER, DONALD A. MARTIN, AND SAHARON SHELAH

[BMS84] ***Axiomatic Set Theory. Proceedings of the AMS-IMS-SIAM joint summer research conference held in Boulder, Colo., June 19–25, 1983***, Contemporary Mathematics, vol. 31, American Mathematical Society, 1984.

HOWARD S. BECKER

[Bec80] *Thin collections of sets of projective ordinals and analogs of* $\mathbf{L}$, ***Annals of Mathematical Logic***, vol. 19 (1980), pp. 205–241.

[Bec86] *Inner model operators and the continuum hypothesis*, ***Proceedings of the American Mathematical Society***, vol. 96 (1986), no. 1, pp. 126–129.

[13] *Cf.* [CK11, § 2] for an introduction to AD^+.

HOWARD S. BECKER AND ALEXANDER S. KECHRIS
[BK84] *Sets of ordinals constructible from trees and the third Victoria Delfino problem*, in Baumgartner et al. [BMS84], pp. 13–29.

HOWARD S. BECKER AND YIANNIS N. MOSCHOVAKIS
[BM81] *Measurable cardinals in playful models*, in Kechris et al. [CABAL ii], pp. 203–214, reprinted in [CABAL III], pp. 115–125.

STEFAN BOLD
[Bol09] *Cardinals as Ultrapowers. A Canonical Measure Analysis under the Axiom of Determinacy*, ***Ph.D. thesis***, Rheinische Friedrich-Wilhelms-Universität Bonn, 2009.

STEFAN BOLD AND BENEDIKT LÖWE
[BL07] *A simple inductive measure analysis for cardinals under the Axiom of Determinacy*, ***Advances in logic. Papers from the North Texas Logic Conference held at the University of North Texas, Denton, TX, October 8–10, 2004*** (Su Gao, Steve Jackson, and Yi Zhang, editors), Contemporary Mathematics, vol. 425, American Mathematical Society, Providence, RI, 2007, pp. 23–41.

ANDRÉS EDUARDO CAICEDO AND RICHARD KETCHERSID
[CK11] *A trichotomy theorem in natural models of* AD^+, ***Set theory and its applications. Papers from the Annual Boise Extravaganzas (BEST) held in Boise, ID, 1995–2010*** (Liljana Babinkostova, Andrés E. Caicedo, Stefan Geschke, and Marion Scheepers, editors), Contemporary Mathematics, vol. 533, American Mathematical Society, 2011, pp. 227–258.

MARK CRAWSHAW
[Cra85] *Explicit Formulas for the Jump of* Q*-Degrees*, ***Ph.D. thesis***, California Institute of Technology, 1985.

VINCENZO DIMONTE
[Dim11] *Totally non-proper ordinals beyond* $\mathbf{L}(\mathbf{V}_{\lambda+1})$, ***Archive for Mathematical Logic***, vol. 50 (2011), no. 5-6, pp. 565–584.

PHILIPP DOEBLER
[Doe06] *The* 12*th Delfino Problem and universally Baire sets of reals*, ***Master's thesis***, Westfälische Wilhelms-Universität Münster, 2006.

RANDALL DOUGHERTY AND ALEXANDER S. KECHRIS
[DK00] *How many Turing degrees are there?*, ***Computability Theory and its Applications. Current Trends and Open Problems. Proceedings of the AMS-IMS-SIAM Joint Summer Research Conference held at the University of Colorado, Boulder, CO, June 13–17, 1999*** (Peter A. Cholak, Steffen Lempp, Manuel Lerman, and Richard A. Shore, editors), Contemporary Mathematics, vol. 257, American Mathematical Society, 2000, pp. 83–94.

MATTHEW FOREMAN, MENACHEM MAGIDOR, AND SAHARON SHELAH
[FMS88] *Martin's maximum, saturated ideals and nonregular ultrafilters. I*, ***Annals of Mathematics***, vol. 127 (1988), no. 1, pp. 1–47.

DAVID GUASPARI
[Gua73] *Thin and wellordered analytical sets*, ***Ph.D. thesis***, University of Cambridge, 1973.

DAVID GUASPARI AND LEO HARRINGTON
[GH76] *Characterizing* C_3 *(the largest countable* Π^1_3 *set)*, ***Proceedings of the American Mathematical Society***, vol. 57 (1976), no. 1, pp. 127–129.

LEO A. HARRINGTON

[Har78] *Analytic determinacy and* $0^{\#}$, ***The Journal of Symbolic Logic***, vol. 43 (1978), no. 4, pp. 685–693.

KAI HAUSER

[Hau99] *Towards a fine structural representation of the Martin-Solovay tree*, 1999, preprint.

[Hau00] *Reflections on the last Delfino problem*, ***Logic Colloquium '98. Proceedings of the Annual European Summer Meeting of the Association of Symbolic Logic, held in Prague, Czech Republic, August 9–15, 1998*** (Samuel R. Buss, Petr Hájek, and Pavel Pudlák, editors), Lecture Notes in Logic, vol. 13, Cambridge University Press, 2000, pp. 206–225.

KAI HAUSER AND GREG HJORTH

[HH97] *Strong cardinal in the core model*, ***Annals of Pure and Applied Logic***, vol. 83 (1997), no. 2, pp. 165–198.

KAI HAUSER AND RALF SCHINDLER

[HS00] *Projective uniformization revisited*, ***Annals of Pure and Applied Logic***, vol. 103 (2000), no. 1-3, pp. 109–153.

GREGORY HJORTH

[Hjo95] *The size of the ordinal* u_2, ***Journal of the London Mathematical Society***, vol. 52 (1995), no. 3, pp. 417–433.

[Hjo96B] *Variations of the Martin-Solovay tree*, ***The Journal of Symbolic Logic***, vol. 61 (1996), no. 1, pp. 40–51.

DAISUKE IKEGAMI AND W. HUGH WOODIN

[IW09] *Real determinacy and real Blackwell determinacy*, Institut Mittag-Leffler Report No. 32, 2009/2010, Fall, 2009, available online.

STEPHEN JACKSON

[Jac88] AD *and the projective ordinals*, in Kechris et al. [CABAL iv], pp. 117–220, reprinted in [CABAL II], pp. 364–483.

[Jac89] AD *and the very fine structure of* $\mathbf{L}(\mathbb{R})$, ***Bulletin of the American Mathematical Society***, vol. 21 (1989), no. 1, pp. 77–81.

[Jac91] *Admissible Suslin cardinals in* $\mathbf{L}(\mathbb{R})$, ***The Journal of Symbolic Logic***, vol. 56 (1991), no. 1, pp. 260–275.

[Jac92] *Admissibility and Mahloness in* $\mathbf{L}(\mathbb{R})$, ***Set theory of the continuum. Papers from the workshop held in Berkeley, California, October 16–20, 1989*** (Haim Judah, Winfried Just, and Hugh Woodin, editors), Mathematical Sciences Research Institute Publications, vol. 26, Springer-Verlag, 1992, pp. 63–74.

[Jac99] ***A Computation of*** $\underset{\sim}{\delta}^1_5$, vol. 140, Memoirs of the AMS, no. 670, American Mathematical Society, July 1999.

[Jac10] *Structural consequences of* AD, in Kanamori and Foreman [KF10], pp. 1753–1876.

STEPHEN JACKSON, RICHARD KETCHERSID, FARMER SCHLUTZENBERG, AND W. HUGH WOODIN

[JKSW14] *Determinacy and Jónsson cardinals in* $\mathbf{L}(\mathbb{R})$, ***The Journal of Symbolic Logic***, vol. 79 (2014), no. 4, pp. 1184–1198.

STEPHEN JACKSON AND FARID KHAFIZOV

[JK16] *Descriptions and cardinals below* $\underset{\sim}{\delta}^1_5$, ***The Journal of Symbolic Logic***, vol. 81 (2016), no. 4, pp. 1177–1224.

STEVE JACKSON AND BENEDIKT LÖWE

[JL13] *Canonical measure assignments*, ***The Journal of Symbolic Logic***, vol. 78 (2013), no. 2, pp. 403–424.

RONALD B. JENSEN AND JOHN R. STEEL

[JS13] **K** *without the measurable*, ***The Journal of Symbolic Logic***, vol. 78 (2013), no. 3, pp. 708–734.

GEORGE KAFKOULIS

[Kaf04] *Coding lemmata in* $\mathbf{L}(\mathbf{V}_{\lambda+1})$, ***Archive for Mathematical Logic***, vol. 43 (2004), no. 2, pp. 193–213.

AKIHIRO KANAMORI

[Kan94] ***The Higher Infinite. Large Cardinals in Set Theory from Their Beginnings***, Perspectives in Mathematical Logic, Springer-Verlag, Berlin, 1994.

AKIHIRO KANAMORI AND MATTHEW FOREMAN

[KF10] ***Handbook of Set Theory***, Springer-Verlag, 2010.

ALEXANDER S. KECHRIS

[Kec75] *The theory of countable analytical sets*, ***Transactions of the American Mathematical Society***, vol. 202 (1975), pp. 259–297.

[Kec81] *Homogeneous trees and projective scales*, in Kechris et al. [CABAL ii], pp. 33–74, reprinted in [CABAL II], pp. 270–303.

[Kec92] *The structure of Borel equivalence relations in polish spaces*, ***Set Theory of the Continuum. Papers from the workshop held in Berkeley, California, October 16–20, 1989*** (Haim Judah, Winfried Just, and Hugh Woodin, editors), Mathematical Sciences Research Institute Publications, vol. 26, Springer-Verlag, 1992, pp. 89–102.

ALEXANDER S. KECHRIS, EUGENE M. KLEINBERG, YIANNIS N. MOSCHOVAKIS, AND W. H. WOODIN

[KKMW81] *The axiom of determinacy, strong partition properties, and nonsingular measures*, in Kechris et al. [CABAL ii], pp. 75–99, reprinted in [CABAL I], pp. 333–354.

ALEXANDER S. KECHRIS, BENEDIKT LÖWE, AND JOHN R. STEEL

[CABAL I] *Games, Scales, and Suslin cardinals: the Cabal Seminar, volume I*, Lecture Notes in Logic, vol. 31, Cambridge University Press, 2008.

[CABAL II] *Wadge Degrees and Projective Ordinals: the Cabal Seminar, volume II*, Lecture Notes in Logic, vol. 37, Cambridge University Press, 2012.

[CABAL III] *Ordinal Definability and Recursion Theory: the Cabal Seminar, volume III*, Lecture Notes in Logic, vol. 43, Cambridge University Press, 2016.

ALEXANDER S. KECHRIS AND DONALD A. MARTIN

[KM78] *On the theory of* Π^1_3 *sets of reals*, ***Bulletin of the American Mathematical Society***, vol. 84 (1978), no. 1, pp. 149–151.

ALEXANDER S. KECHRIS, DONALD A. MARTIN, AND YIANNIS N. MOSCHOVAKIS

[KMM81A] *Appendix: Progress report on the Victoria Delfino problems*, in *Cabal Seminar* 77–79 [CABAL ii], pp. 273–274.

[CABAL ii] *Cabal Seminar* 77–79, Lecture Notes in Mathematics, vol. 839, Berlin, Springer-Verlag, 1981.

[KMM83A] *Appendix: Progress report on the Victoria Delfino problems*, in *Cabal Seminar* 79–81 [CABAL iii], pp. 283–284.

[CABAL iii] *Cabal Seminar* 79–81, Lecture Notes in Mathematics, vol. 1019, Berlin, Springer-Verlag, 1983.

ALEXANDER S. KECHRIS, DONALD A. MARTIN, AND JOHN R. STEEL

[KMS88A] *Appendix: Victoria Delfino problems II*, in *Cabal Seminar* 81–85 [CABAL iv], pp. 221–224.

[CABAL iv] *Cabal Seminar* 81–85, Lecture Notes in Mathematics, vol. 1333, Berlin, Springer-Verlag, 1988.

ALEXANDER S. KECHRIS AND YIANNIS N. MOSCHOVAKIS

[KM78A] *Appendix. The Victoria Delfino problems*, in *Cabal Seminar* 76–77 [CABAL i], pp. 279–282.

[CABAL i] *Cabal Seminar* 76–77, Lecture Notes in Mathematics, vol. 689, Berlin, Springer-Verlag, 1978.

TAKAYUKI KIHARA AND ANTONIO MONTALBÁN

[KM18] *The uniform Martin's conjecture for many-one degrees*, ***Transactions of the American Mathematical Society***, vol. 370 (2018), no. 12, pp. 9025–9044.

EUGENE M. KLEINBERG

[Kle77] ***Infinitary Combinatorics and the Axiom of Determinateness***, Lecture Notes in Mathematics, vol. 612, Springer-Verlag, 1977.

PETER KOELLNER AND W. HUGH WOODIN

[KW10] *Large cardinals from determinacy*, in Kanamori and Foreman [KF10], pp. 1951–2119.

BENEDIKT LÖWE

[Löw02] *Kleinberg sequences and partition cardinals below* $\underset{\sim}{\delta}^1_5$, ***Fundamenta Mathematicae***, vol. 171 (2002), no. 1, pp. 69–76.

ANDREW MARKS, THEODORE A. SLAMAN, AND JOHN R. STEEL

[MSS16] *Martin's conjecture, arithmetic equivalence, and countable Borel equivalence relations*, in Kechris et al. [CABAL III], pp. 493–520.

DONALD A. MARTIN

[Mar80] *Infinite games*, ***Proceedings of the International Congress of Mathematicatians, Helsinki 1978*** (Olli Lehto, editor), Academia Scientiarum Fennica, 1980, pp. 269–273.

[Mar83] *The real game quantifier propagates scales*, in Kechris et al. [CABAL iii], pp. 157–171, reprinted in [CABAL I], pp. 209–222.

[Mar] AD *and the normal measures on* $\underset{\sim}{\delta}^1_3$, unpublished, undated.

DONALD A. MARTIN AND JOHN R. STEEL

[MS83] *The extent of scales in* $\mathbf{L}(\mathbb{R})$, in Kechris et al. [CABAL iii], pp. 86–96, reprinted in [CABAL I], pp. 110–120.

[MS88] *Projective determinacy*, ***Proceedings of the National Academy of Sciences of the United States of America***, vol. 85 (1988), no. 18, pp. 6582–6586.

[MS89] *A proof of projective determinacy*, ***Journal of the American Mathematical Society***, vol. 2 (1989), no. 1, pp. 71–125.

[MaS94] *Iteration trees*, ***Journal of the American Mathematical Society***, vol. 7 (1994), no. 1, pp. 1–73.

WILLIAM J. MITCHELL

[Mit79] *Hypermeasurable cardinals*, ***Logic Colloquium '78. Proceedings of the Colloquium held in Mons, August 24–September 1, 1978*** (Maurice Boffa, Dirk van Dalen, and Kenneth McAloon, editors), Studies in Logic and the Foundations of Mathematics, vol. 97, North-Holland, Amsterdam, 1979, pp. 303–316.

WILLIAM J. MITCHELL AND JOHN R. STEEL

[MiS94] ***Fine Structure and Iteration Trees***, Lecture Notes in Logic, vol. 3, Springer-Verlag, Berlin, 1994.

YIANNIS N. MOSCHOVAKIS

[Mos70] *Determinacy and prewellorderings of the continuum*, ***Mathematical Logic and Foundations of Set Theory. Proceedings of an international colloquium held under the auspices of the Israel Academy of Sciences and Humanities, Jerusalem, 11–14 November 1968*** (Y. Bar-Hillel, editor), Studies in Logic and the Foundations of Mathematics, North-Holland, Amsterdam-London, 1970, pp. 24–62.

[Mos78] *Inductive scales on inductive sets*, in Kechris and Moschovakis [CABAL i], pp. 185–192, reprinted in [CABAL I], pp. 94–101.

[Mos80] ***Descriptive Set Theory***, Studies in Logic and the Foundations of Mathematics, vol. 100, North-Holland, Amsterdam, 1980.

[Mos83] *Scales on coinductive sets*, in Kechris et al. [CABAL iii], pp. 77–85, reprinted in [CABAL I], pp. 102–109.

SANDRA MÜLLER, RALF SCHINDLER, AND W. HUGH WOODIN

[MSW16] *Mice with finitely many Woodin cardinals from optimal determinacy hypotheses*, submitted, 2016.

ITAY NEEMAN

[Nee95] *Optimal proofs of determinacy*, ***The Bulletin of Symbolic Logic***, vol. 1 (1995), no. 3, pp. 327–339.

[Nee04] ***The Determinacy of Long Games***, de Gruyter Series in Logic and its Applications, vol. 7, Walter de Gruyter, Berlin, 2004.

[Nee07B] *Inner models and ultrafilters in* $\mathrm{L}(\mathbb{R})$, ***The Bulletin of Symbolic Logic***, vol. 13 (2007), no. 1, pp. 31–53.

[Nee10] *Determinacy in* $\mathrm{L}(\mathbb{R})$, in Kanamori and Foreman [KF10], pp. 1877–1950.

ITAY NEEMAN AND JOHN R. STEEL

[NS06] *Counterexamples to the unique and cofinal branches hypotheses*, ***The Journal of Symbolic Logic***, vol. 71 (2006), no. 3, pp. 977–988.

GERALD E. SACKS

[Sac76] *Countable admissible ordinals and hyperdegrees*, ***Advances in Mathematics***, vol. 20 (1976), no. 2, pp. 213–262.

GRIGOR SARGSYAN

[Sar13A] *Descriptive inner model theory*, ***The Bulletin of Symbolic Logic***, vol. 19 (2013), no. 1, pp. 1–55.

[Sar13B] *On the prewellorderings associated with the directed systems of mice*, ***The Journal of Symbolic Logic***, vol. 78 (2013), no. 3, pp. 735–763.

[Sar14] *An inner model proof of the strong partition property for* $\underset{\sim}{\delta}^2_1$, ***Notre Dame Journal of Formal Logic***, vol. 55 (2014), no. 4, pp. 563–568.

[Sar15] ***Hod Mice and the Mouse Set Conjecture***, vol. 236, Memoirs of the American Mathematical Society, no. 1111, American Mathematical Society, 2015.

GRIGOR SARGSYAN AND NAM TRANG

[ST16] ***The Largest Suslin Axiom***, 2016, book manuscript, submitted.

RALF SCHINDLER

[Sch99] *The Delfino problem # 12, Talks in Bonn*, 4/ 21/ 99–4/ 23/ 99, handwritten manuscript, 50 pages, available online, 1999.

[Sch02] *The core model for almost linear iterations*, ***Annals of Pure and Applied Logic***, vol. 116 (2002), no. 1-3, pp. 205–272.

SAHARON SHELAH AND W. HUGH WOODIN

[SW90] *Large cardinals imply that every reasonably definable set of reals is Lebesgue measurable*, ***Israel Journal of Mathematics***, vol. 70 (1990), no. 3, pp. 381–394.

THEODORE A. SLAMAN

[Sla09] *Degree invariant functions*, slides of a talk at the 2009 VIG, available online, 2009.

THEODORE A. SLAMAN AND JOHN R. STEEL

[SS88] *Definable functions on degrees*, in Kechris et al. [CABAL iv], pp. 37–55, reprinted in [CABAL III], pp. 458–475.

Robert M. Solovay

[Sol78B] *The independence of* DC *from* AD, in Kechris and Moschovakis [Cabal i], pp. 171–184, reprinted in this volume.

John R. Steel

[Ste81A] *Closure properties of pointclasses*, in Kechris et al. [Cabal ii], pp. 147–163, reprinted in [Cabal II], pp. 102–117.

[Ste82A] *A classification of jump operators*, ***The Journal of Symbolic Logic***, vol. 47 (1982), no. 2, pp. 347–358.

[Ste83A] *Scales in* $\mathbf{L}(\mathbb{R})$, in Kechris et al. [Cabal iii], pp. 107–156, reprinted in [Cabal I], pp. 130–175.

[Ste83B] *Scales on* Σ^1_1 *sets*, in Kechris et al. [Cabal iii], pp. 72–76, reprinted in [Cabal I], pp. 90–93.

[Ste88] *Long games*, in Kechris et al. [Cabal iv], pp. 56–97, reprinted in [Cabal I], pp. 223–259.

[Ste95A] $\mathbf{HOD}^{\mathbf{L}(\mathbb{R})}$ *is a core model below* Θ, ***The Bulletin of Symbolic Logic***, vol. 1 (1995), no. 1, pp. 75–84.

[Ste95B] *Projectively wellordered inner models*, ***Annals of Pure and Applied Logic***, vol. 74 (1995), no. 1, pp. 77–104.

[Ste96] ***The Core Model Iterability Problem***, Lecture Notes in Logic, no. 8, Springer-Verlag, Berlin, 1996.

[Ste08A] *Derived models associated to mice*, ***Computational Prospects of Infinity. Part II. Presented Talks*** (Chitat Chong, Qi Feng, Theodore A. Slaman, W. Hugh Woodin, and Yue Yang, editors), World Scientific, 2008, pp. 105–193.

[Ste08B] *Games and scales. Introduction to Part I*, in Kechris et al. [Cabal I], pp. 3–27.

[Ste08C] *The length-*ω_1 *open game quantifier propagates scales*, in Kechris et al. [Cabal I], pp. 260–269.

[Ste08D] *Scales in* $\mathbf{K}(\mathbb{R})$ *at the end of a weak gap*, ***The Journal of Symbolic Logic***, vol. 73 (2008), no. 2, pp. 369–390.

[Ste08E] *Scales in* $\mathbf{K}(\mathbb{R})$, in Kechris et al. [Cabal I], pp. 176–208.

[Ste09] *The derived model theorem*, ***Logic Colloquium '06. Proc. of the Annual European Conference on Logic of the Association for Symbolic Logic held at the Radboud University, Nijmegen, July 27–August 2, 2006*** (S. Barry Cooper, Herman Geuvers, Anand Pillay, and Jouko Väänänen, editors), Lecture Notes in Logic, vol. 19, Association for Symbolic Logic, 2009, pp. 280–327.

[Ste10] *An outline of inner model theory*, in Kanamori and Foreman [KF10], pp. 1595–1684.

John R. Steel and Nam Trang

[ST10] AD^+, *derived models, and* Σ_1 *reflection*, Notes from the first Münster conference on the core model induction and hod mice, available online, 2010.

John R. Steel and W. Hugh Woodin

[SW16] **HOD** *as a core model*, in Kechris et al. [Cabal III], pp. 257–345.

Nam Trang

[Tra14] **HOD** *in natural models of* AD^+, ***Annals of Pure and Applied Logic***, vol. 165 (2014), no. 10, pp. 1533–1556.

W. Hugh Woodin

[Woo82] *On the consistency strength of projective uniformization*, ***Proceedings of the Herbrand Symposium. Logic Colloquium '81. Held in Marseille, July 16–24, 1981*** (Jacques Stern, editor), Studies in Logic and the Foundations of Mathematics, vol. 107, North-Holland, Amsterdam, 1982, pp. 365–384.

[Woo99] ***The Axiom of Determinacy, Forcing Axioms, and the Nonstationary Ideal***, de Gruyter Series in Logic and its Applications, vol. 1, Walter de Gruyter, Berlin, 1999.

[Woo08] *A tt version of the Posner-Robinson theorem*, ***Computational Prospects of Infinity. Part II. Presented Talks*** (Chitat Chong, Qi Feng, Theodore A. Slaman, W. Hugh Woodin, and Yue Yang, editors), World Scientific, 2008, pp. 355–392.

YIZHENG ZHU

[Zhu10] *The derived model theorem II*, Notes on lectures given by H. Woodin at the first Münster conference on the core model induction and hod mice, available online, 2010.

[Zhu15] *Realizing an* AD^+ *model as a derived model of a premouse*, ***Annals of Pure and Applied Logic***, vol. 166 (2015), no. 12, pp. 1275–1364.

[Zhu16] *Lightface mice with finitely many Woodin cardinals from optimal determinacy hypotheses*, 2016, preprint, arXiv 1610.02352v1.

[Zhu17] *The higher sharp I: on* $\mathbf{M}_1^{\#}$, 2017, preprint, arXiv:1604.00481v4.

MATHEMATICAL REVIEWS
416 FOURTH STREET
ANN ARBOR, MI 48103-4820
UNITED STATES OF AMERICA
E-mail: aec@ams.org

INSTITUTE FOR LOGIC, LANGUAGE AND COMPUTATION
UNIVERSITEIT VAN AMSTERDAM
POSTBUS 94242
1090 GE AMSTERDAM
THE NETHERLANDS

FACHBEREICH MATHEMATIK
UNIVERSITÄT HAMBURG
BUNDESSTRASSE 55
20146 HAMBURG
GERMANY

CHURCHILL COLLEGE
UNIVERSITY OF CAMBRIDGE
STOREY'S WAY
CAMBRIDGE CB3 0DS
ENGLAND
E-mail: bloewe@science.uva.nl

BIBLIOGRAPHY

JOHN W. ADDISON

[Add58A] *Separation principles in the hierarchies of classical and effective descriptive set theory*, ***Fundamenta Mathematicae***, vol. 46 (1958–9), pp. 123–135.

[Add58B] *Some consequences of the axiom of constructibility*, ***Fundamenta Mathematicae***, vol. 46 (1958–9), pp. 337–357.

JOHN W. ADDISON AND YIANNIS N. MOSCHOVAKIS

[AM68] *Some consequences of the axiom of definable determinateness*, ***Proceedings of the National Academy of Sciences of the United States of America***, vol. 59 (1968), pp. 708–712.

DONALD J. ALBERS AND GERALD L. ALEXANDERSON

[AA85] ***Mathematical People. Profiles and Interviews***, Birkhäuser, Boston, MA, 1985.

ALESSANDRO ANDRETTA, ITAY NEEMAN, AND JOHN R. STEEL

[ANS01] *The domestic levels of* $\mathbf{K}^c$ *are iterable*, ***Israel Journal of Mathematics***, vol. 125 (2001), pp. 157–201.

ARTHUR W. APTER, STEPHEN C. JACKSON, AND BENEDIKT LÖWE

[AJL13] *Cofinality and measurability of the first three uncountable cardinals*, ***Transactions of the American Mathematical Society***, vol. 365 (2013), no. 1, pp. 59–98.

RACHID ATMAI

[Atm19] *An analysis of the models* $\mathbf{L}[T^{2n}]$, ***The Journal of Symbolic Logic***, vol. 84 (2019), pp. 1–26.

RACHID ATMAI AND GRIGOR SARGSYAN

[AS19] **HOD** *up to* $\mathsf{AD}_{\mathbb{R}} + \Theta$ *is measurable*, ***Annals of Pure and Applied Logic***, vol. 170 (2019), no. 1, pp. 95–108.

JOAN BAGARIA, NEUS CASTELLS, AND PAUL B. LARSON

[BCL06] *An* Ω*-logic primer*, ***Set theory*** (Joan Bagaria and Stevo Todorčević, editors), Trends in Mathematics, Birkhäuser, Basel, 2006, pp. 1–28.

JOAN BAGARIA, PETER KOELLNER, AND W. HUGH WOODIN

[BKW17] *Large cardinals beyond choice*, 2017, preprint.

STEFAN BANACH AND ALFRED TARSKI

[BT24] *Sur la décomposition des ensembles de points en parties respectivement congruentes*, ***Fundamenta Mathematicae***, vol. 6 (1924), pp. 244–277.

JON BARWISE

[Bar76] ***Admissible Sets and Structures. An approach to definability theory***, Perspectives in Mathematical Logic, Springer-Verlag, 1976.

Large Cardinals, Determinacy, and Other Topics: The Cabal Seminar, Volume IV
Edited by A. S. Kechris, B. Löwe, J. R. Steel
Lecture Notes in Logic, 49

James E. Baumgartner

[Bau83] *Iterated forcing*, ***Surveys in set theory*** (A. R. D. Mathias, editor), London Mathematical Society Lecture Note Series, vol. 87, Cambridge University Press, 1983, pp. 1–59.

James E. Baumgartner, Donald A. Martin, and Saharon Shelah

[BMS84] ***Axiomatic Set Theory. Proceedings of the AMS-IMS-SIAM joint summer research conference held in Boulder, Colo., June 19–25, 1983***, Contemporary Mathematics, vol. 31, American Mathematical Society, 1984.

Howard S. Becker

[Bec78] *Partially playful universes*, in Kechris and Moschovakis [Cabal i], pp. 55–90, reprinted in [Cabal III], pp. 49–85.

[Bec80] *Thin collections of sets of projective ordinals and analogs of* L, ***Annals of Mathematical Logic***, vol. 19 (1980), pp. 205–241.

[Bec81] AD *and the supercompactness of* $\aleph_1$, ***The Journal of Symbolic Logic***, vol. 46 (1981), pp. 822–841.

[Bec85] *A property equivalent to the existence of scales*, ***Transactions of the American Mathematical Society***, vol. 287 (1985), no. 2, pp. 591–612.

[Bec86] *Inner model operators and the continuum hypothesis*, ***Proceedings of the American Mathematical Society***, vol. 96 (1986), no. 1, pp. 126–129.

Howard S. Becker and Alexander S. Kechris

[BK84] *Sets of ordinals constructible from trees and the third Victoria Delfino problem*, in Baumgartner et al. [BMS84], pp. 13–29.

Howard S. Becker and Yiannis N. Moschovakis

[BM81] *Measurable cardinals in playful models*, in Kechris et al. [Cabal ii], pp. 203–214, reprinted in [Cabal III], pp. 115–125.

Mohamed Bekkali

[Bek91] ***Topics in Set Theory: Lebesgue measurability, large cardinals, forcing axioms, rho-functions. Notes on lectures by Stevo Todorčević***, Lecture Notes in Mathematics, vol. 1476, Springer-Verlag, Berlin, 1991.

David Blackwell

[Bla67] *Infinite games and analytic sets*, ***Proceedings of the National Academy of Sciences of the United States of America***, vol. 58 (1967), pp. 1836–1837.

[Bla69] *Infinite* G_δ*-games with imperfect information*, ***Polska Akademia Nauk. Instytut Matematyczny. Zastosowania Matematyki***, vol. 10 (1969), pp. 99–101.

Andreas Blass

[Bla75] *Equivalence of two strong forms of determinacy*, ***Proceedings of the American Mathematical Society***, vol. 52 (1975), pp. 373–376.

Stefan Bold

[Bol09] *Cardinals as Ultrapowers. A Canonical Measure Analysis under the Axiom of Determinacy*, ***Ph.D. thesis***, Rheinische Friedrich-Wilhelms-Universität Bonn, 2009.

Stefan Bold and Benedikt Löwe

[BL07] *A simple inductive measure analysis for cardinals under the Axiom of Determinacy*, ***Advances in logic. Papers from the North Texas Logic Conference held at the University of North Texas, Denton, TX, October 8–10, 2004*** (Su Gao, Steve Jackson, and Yi Zhang, editors), Contemporary Mathematics, vol. 425, American Mathematical Society, Providence, RI, 2007, pp. 23–41.

NICOLAS BOURBAKI

[Bou75] *Eléments de mathématique.* ***Topologie générale***, 3ème ed., Hermann, Paris, 1975.

L. E. J. BROUWER

[Bro24] *Beweis dass jede volle Funktion gleichmässig stetig ist*, ***Koninklijke Akademie van Wetenschappen te Amsterdam. Proceedings of the Section of Sciences***, vol. 27 (1924), pp. 189–193.

JOHN P. BURGESS

[Bur74] *Infinitary Languages and Descriptive Set Theory*, ***Ph.D. thesis***, University of California at Berkeley, 1974.

[Bur78] *Equivalences generated by families of Borel sets*, ***Proceedings of the American Mathematical Society***, vol. 69 (1978), no. 2, pp. 323–326.

DOUGLAS R. BUSCH

[Bus73] *Some Problems Connected with the Axiom of Determinacy*, ***Ph.D. thesis***, Rockefeller University, 1973.

DANIEL BUSCHE AND RALF SCHINDLER

[BS09] *The strength of choiceless patterns of singular and weakly compact cardinals*, ***Annals of Pure and Applied Logic***, vol. 159 (2009), no. 1-2, pp. 198–248.

ANDRÉS EDUARDO CAICEDO AND RICHARD KETCHERSID

[CK11] *A trichotomy theorem in natural models of* AD^+, ***Set theory and its applications. Papers from the Annual Boise Extravaganzas (BEST) held in Boise, ID, 1995–2010*** (Liljana Babinkostova, Andrés E. Caicedo, Stefan Geschke, and Marion Scheepers, editors), Contemporary Mathematics, vol. 533, American Mathematical Society, 2011, pp. 227–258.

ANDRÉS EDUARDO CAICEDO, PAUL B. LARSON, GRIGOR SARGSYAN, RALF SCHINDLER, JOHN R. STEEL, AND MARTIN ZEMAN

[CLS17] *Square principles in* $\mathbb{P}_{\max}$ *extensions*, ***Israel Journal of Mathematics***, vol. 217 (2017), no. 1, pp. 231–261.

GUSTAVE CHOQUET

[Cho55] *Theory of capacities*, ***Annales de l'Institut Fourier***, vol. 5 (1955), pp. 131–295.

[Cho59] *Forme abstraite du théorème de capacitabilité*, ***Annales de l'Institut Fourier***, vol. 9 (1959), pp. 83–89.

S. BARRY COOPER

[Coo04] ***Computability Theory***, Chapman & Hall/CRC, Boca Raton, FL, 2004.

MARK CRAWSHAW

[Cra85] *Explicit Formulas for the Jump of Q-Degrees*, ***Ph.D. thesis***, California Institute of Technology, 1985.

JAMES CUMMINGS AND MATTHEW FOREMAN

[CF98] *The tree property*, ***Advances in Mathematics***, vol. 133 (1998), no. 1, pp. 1–32.

ROY O. DAVIES AND C. AMBROSE ROGERS

[DR69] *The problem of subsets of finite positive measure*, ***Bulletin of the London Mathematical Society***, vol. 1 (1969), pp. 47–54.

MORTON DAVIS

[Dav64] *Infinite games of perfect information*, ***Advances in Game Theory*** (Melvin Dresher, Lloyd S. Shapley, and Alan W. Tucker, editors), Annals of Mathematical Studies, vol. 52, Princeton University Press, 1964, pp. 85–101.

CLAUDE DELLACHERIE

[Del72] ***Ensembles analytiques, capacités, mesures de Hausdorff***, Lecture Notes in Mathematics, vol. 295, Springer-Verlag, Heidelberg, 1972.

[Del80] *Un cours sur les ensembles analytiques*, ***Analytic sets (Developed from lectures given at the London Mathematical Society Instructional Conference on Analytic Sets, University College London, July 1978)*** (C. A. Rogers, J. E. Jayne, Claude Dellacherie, Flemming Topsøe, Jørgen Hoffmann-Jørgensen, Donald A. Martin, Alexander S. Kechris, and A. H. Stone, editors), Academic Press, 1980, pp. 183–316.

KEITH J. DEVLIN

[Dev84] ***Constructibility***, Perspectives in Mathematical Logic, Springer-Verlag, Berlin, 1984.

VINCENZO DIMONTE

[Dim11] *Totally non-proper ordinals beyond* $\mathbf{L}(\mathbf{V}_{\lambda+1})$, ***Archive for Mathematical Logic***, vol. 50 (2011), no. 5-6, pp. 565–584.

ANTHONY DODD

[Dod82] ***The Core Model***, London Mathematical Society Lecture Note Series, vol. 61, Cambridge University Press, 1982.

PHILIPP DOEBLER

[Doe06] *The* 12*th Delfino Problem and universally Baire sets of reals*, ***Master's thesis***, Westfälische Wilhelms-Universität Münster, 2006.

[Doe10] *Stationary set preserving* $\mathcal{L}$*-forcings and their applications*, ***Ph.D. thesis***, Westfälische Wilhelms-Universität Münster, 2010.

PHILIPP DOEBLER AND RALF SCHINDLER

[DS13] *The extender algebra and vagaries of* Σ^2_1 *absoluteness*, ***Münster Journal of Mathematics***, vol. 6 (2013), pp. 117–166.

RANDALL DOUGHERTY AND ALEXANDER S. KECHRIS

[DK00] *How many Turing degrees are there?*, ***Computability Theory and its Applications. Current Trends and Open Problems. Proceedings of the AMS-IMS-SIAM Joint Summer Research Conference held at the University of Colorado, Boulder, CO, June 13–17, 1999*** (Peter A. Cholak, Steffen Lempp, Manuel Lerman, and Richard A. Shore, editors), Contemporary Mathematics, vol. 257, American Mathematical Society, 2000, pp. 83–94.

DERRICK ALBERT DUBOSE

[DuB90] *The equivalence of determinacy and iterated sharps*, ***The Journal of Symbolic Logic***, vol. 55 (1990), no. 2, pp. 502–525.

PAUL ERDŐS AND ANDRÁS HAJNAL

[EH58] *On the structure of set mappings*, ***Acta Mathematica Academiae Scientiarum Hungaricae***, vol. 9 (1958), pp. 111–131.

[EH66] *On a problem of B. Jónsson*, ***Bulletin de l'Académie Polonaise des Sciences***, vol. 14 (1966), pp. 19–23.

ILIJAS FARAH

[Far07] *A proof of the* Σ^2_1*-absoluteness theorem*, ***Advances in logic. Papers from the North Texas Logic Conference held at the University of North Texas, Denton, TX, October 8–10, 2004*** (Yi Zhang Su Gao, Steve Jackson, editor), Contemporary Mathematics, vol. 425, American Mathematical Society, 2007, pp. 9–22.

ILIJAS FARAH, RICHARD O. KETCHERSID, PAUL B. LARSON, AND MENACHEM MAGIDOR

[FKLM08] *Absoluteness for universally Baire sets and the uncountable II*, ***Computational Prospects of Infinity. Part II. Presented Talks*** (Chitat Chong, Qi Feng, Theodore A. Slaman, W. Hugh Woodin, and Yue Yang, editors), World Scientific, 2008, pp. 163–192.

SOLOMAN FEFERMAN AND AZRIEL LÉVY

[FL63] *Independence results in set theory by Cohen's method II*, ***Notices of the American Mathematical Society***, vol. 10 (1963), p. 593.

QI FENG, MENACHEM MAGIDOR, AND W. HUGH WOODIN

[FMW92] *Universally Baire sets of reals*, in Judah et al. [JJW92], pp. 203–242.

MATTHEW FOREMAN

[For82] *Large cardinals and strong model theoretic transfer properties*, ***Transactions of the American Mathematical Society***, vol. 272 (1982), no. 2, pp. 427–463.

[For86] *Potent axioms*, ***Transactions of the American Mathematical Society***, vol. 294 (1986), no. 1, pp. 1–28.

MATTHEW FOREMAN AND RICHARD LAVER

[FL88] *Some downwards transfer properties for* $\aleph_2$, ***Advances in Mathematics***, vol. 67 (1988), no. 2, pp. 230–238.

MATTHEW FOREMAN, MENACHEM MAGIDOR, AND RALF-DIETER SCHINDLER

[FMS01] *The consistency strength of successive cardinals with the tree property*, ***The Journal of Symbolic Logic***, vol. 66 (2001), no. 4, pp. 1837–1847.

MATTHEW FOREMAN, MENACHEM MAGIDOR, AND SAHARON SHELAH

[FMS88] *Martin's maximum, saturated ideals and nonregular ultrafilters. I*, ***Annals of Mathematics***, vol. 127 (1988), no. 1, pp. 1–47.

MATTHEW FOREMAN AND W. HUGH WOODIN

[FW91] *The generalized continuum hypothesis can fail everywhere*, ***Annals of Mathematics***, vol. 133 (1991), no. 1, pp. 1–35.

HARVEY FRIEDMAN

[Fri71A] *Determinateness in the low projective hierarchy*, ***Fundamenta Mathematicae***, vol. 72 (1971), no. 1, pp. 79–95. (errata insert).

[Fri71B] *Higher set theory and mathematical practice*, ***Annals of Mathematical Logic***, vol. 2 (1971), no. 3, pp. 325–357.

[Fri73] *Countable models of set theories*, ***Cambridge Summer School in Mathematical Logic (held in Cambridge, England, August 1–21, 1971)*** (A. R. D. Mathias and H. Rogers, editors), Lecture Notes in Mathematics, vol. 337, Springer-Verlag, 1973, pp. 539–573.

DAVID GALE AND FRANK M. STEWART

[GS53] *Infinite games with perfect information*, ***Contributions to the theory of games, vol. 2***, Annals of Mathematics Studies, no. 28, Princeton University Press, 1953, pp. 245–266.

MOTI GITIK

[Git80] *All uncountable cardinals can be singular*, ***Israel Journal of Mathematics***, vol. 35 (1980), no. 1-2, pp. 61–88.

MOTI GITIK, RALF SCHINDLER, AND SAHARON SHELAH

[GSS06] *PCF theory and Woodin cardinals*, ***Logic Colloquium '02. Joint proceedings of the Annual European Summer Meeting of the Association for Symbolic Logic and the Biannual Meeting of the German Association for Mathematical Logic and the Foundations of Exact Sciences (the Colloquium Logicum) held in Münster, August 3–11, 2002*** (Zoé Chatzidakis, Peter Koepke, and Wolfram Pohlers, editors), Lecture Notes in Logic, vol. 27, Association for Symbolic Logic, 2006, pp. 172–205.

JOHN TOWNSEND GREEN

[Gre78] *Determinacy and the Existence of Large Measurable Cardinals*, ***Ph.D. thesis***, University of California at Berkeley, 1978.

DAVID GUASPARI

[Gua73] *Thin and wellordered analytical sets*, ***Ph.D. thesis***, University of Cambridge, 1973.

DAVID GUASPARI AND LEO HARRINGTON

[GH76] *Characterizing* C_3 *(the largest countable* Π^1_3 *set)*, ***Proceedings of the American Mathematical Society***, vol. 57 (1976), no. 1, pp. 127–129.

JACQUES HADAMARD

[Had05] *Cinq letters sur la théorie des ensembles*, ***Bulletin de la Societé mathématique de France***, vol. 33 (1905), pp. 261–273.

ANDRÁS HAJNAL

[Haj56] *On a consistency theorem connected with the generalized continuum problem*, ***Zeitschrift für Mathematische Logik und Grundlagen der Mathematik***, vol. 2 (1956), pp. 131–136.

[Haj61] *On a consistency theorem connected with the generalized continuum problem*, ***Acta Mathematica Academiae Scientiarum Hungaricae***, vol. 12 (1961), pp. 321–376.

PAUL R. HALMOS

[Hal50] ***Measure Theory***, D. Van Nostrand Company, Inc., New York, 1950.

JOEL DAVID HAMKINS AND W. HUGH WOODIN

[HW00] *Small forcing creates neither strong nor Woodin cardinals*, ***Proceedings of the American Mathematical Society***, vol. 128 (2000), no. 10, pp. 3025–3029.

VICTOR HARNIK AND MICHAEL MAKKAI

[HM77] *A tree argument in infinitary model theory*, ***Proceedings of the American Mathematical Society***, vol. 67 (1977), no. 1, pp. 309–314.

LEO A. HARRINGTON

[Har78] *Analytic determinacy and* $0^\#$, ***The Journal of Symbolic Logic***, vol. 43 (1978), no. 4, pp. 685–693.

[Har] *A powerless proof of a theorem of Silver*, unpublished notes, undated.

LEO A. HARRINGTON AND ALEXANDER S. KECHRIS

[HK81] *On the determinacy of games on ordinals*, ***Annals of Mathematical Logic***, vol. 20 (1981), pp. 109–154.

FELIX HAUSDORFF

[Hau08] *Grundzüge einer Theorie der geordneten Mengen*, ***Mathematische Annalen***, vol. 65 (1908), pp. 435–505.

[Hau14] *Bemerkung über den Inhalt von Punktmengen*, ***Mathematische Annalen***, vol. 75 (1914), pp. 428–434.

KAI HAUSER

[Hau99] *Towards a fine structural representation of the Martin-Solovay tree*, 1999, preprint.

[Hau00] *Reflections on the last Delfino problem*, ***Logic Colloquium '98. Proceedings of the Annual European Summer Meeting of the Association of Symbolic Logic, held in Prague, Czech Republic, August 9–15, 1998*** (Samuel R. Buss, Petr Hájek, and Pavel Pudlák, editors), Lecture Notes in Logic, vol. 13, Cambridge University Press, 2000, pp. 206–225.

KAI HAUSER AND GREG HJORTH

[HH97] *Strong cardinal in the core model*, ***Annals of Pure and Applied Logic***, vol. 83 (1997), no. 2, pp. 165–198.

KAI HAUSER AND RALF SCHINDLER

[HS00] *Projective uniformization revisited*, ***Annals of Pure and Applied Logic***, vol. 103 (2000), no. 1-3, pp. 109–153.

JAMES HENLE, A. R. D. MATHIAS, AND W. HUGH WOODIN

[HMW85] *A barren extension*, ***Methods in mathematical logic. Proceedings of the sixth Latin American symposium on mathematical logic held in Caracas, August 1–6, 1983*** (Carlos A. Di Prisco, editor), Lecture Notes in Mathematics, vol. 1130, Springer-Verlag, Berlin, 1985, pp. 195–207.

GREGORY HJORTH

[Hjo95] *The size of the ordinal* u_2, ***Journal of the London Mathematical Society***, vol. 52 (1995), no. 3, pp. 417–433.

[Hjo96A] Π^1_2 *Wadge degrees*, ***Annals of Pure and Applied Logic***, vol. 77 (1996), no. 1, pp. 53–74.

[Hjo96B] *Variations of the Martin-Solovay tree*, ***The Journal of Symbolic Logic***, vol. 61 (1996), no. 1, pp. 40–51.

[Hjo97] *Some applications of coarse inner model theory*, ***The Journal of Symbolic Logic***, vol. 62 (1997), no. 2, pp. 337–365.

PAUL HOWARD AND JEAN E. RUBIN

[HR98] ***Consequences of the Axiom of Choice***, Mathematical Surveys and Monographs, vol. 59, American Mathematical Society, 1998.

DAISUKE IKEGAMI AND W. HUGH WOODIN

[IW09] *Real determinacy and real Blackwell determinacy*, Institut Mittag-Leffler Report No. 32, 2009/2010, Fall, 2009, available online.

STEPHEN JACKSON

[Jac88] AD *and the projective ordinals*, in Kechris et al. [CABAL iv], pp. 117–220, reprinted in [CABAL II], pp. 364–483.

[Jac89] AD *and the very fine structure of* $\mathrm{L}(\mathbb{R})$, ***Bulletin of the American Mathematical Society***, vol. 21 (1989), no. 1, pp. 77–81.

[Jac91] *Admissible Suslin cardinals in* $\mathrm{L}(\mathbb{R})$, ***The Journal of Symbolic Logic***, vol. 56 (1991), no. 1, pp. 260–275.

[Jac92] *Admissibility and Mahloness in* $\mathrm{L}(\mathbb{R})$, ***Set theory of the continuum. Papers from the workshop held in Berkeley, California, October 16–20, 1989*** (Haim Judah, Winfried Just, and Hugh Woodin, editors), Mathematical Sciences Research Institute Publications, vol. 26, Springer-Verlag, 1992, pp. 63–74.

[Jac99] ***A Computation of*** $\underset{\sim}{\delta}^1_5$, vol. 140, Memoirs of the AMS, no. 670, American Mathematical Society, July 1999.

[Jac08] *Suslin cardinals, partition properties, homogeneity. Introduction to Part II*, in Kechris et al. [CABAL I], pp. 273–313.

[Jac10] *Structural consequences of* AD, in Kanamori and Foreman [KF10], pp. 1753–1876.

STEPHEN JACKSON, RICHARD KETCHERSID, FARMER SCHLUTZENBERG, AND W. HUGH WOODIN

[JKSW14] *Determinacy and Jónsson cardinals in* $\mathrm{L}(\mathbb{R})$, ***The Journal of Symbolic Logic***, vol. 79 (2014), no. 4, pp. 1184–1198.

STEPHEN JACKSON AND FARID KHAFIZOV

[JK16] *Descriptions and cardinals below* $\underset{\sim}{\delta}^1_5$, ***The Journal of Symbolic Logic***, vol. 81 (2016), no. 4, pp. 1177–1224.

STEVE JACKSON AND BENEDIKT LÖWE

[JL13] *Canonical measure assignments*, ***The Journal of Symbolic Logic***, vol. 78 (2013), no. 2, pp. 403–424.

JOHN E. JAYNE

[Jay76] *Structure of analytic Hausdorff spaces*, ***Mathematika***, vol. 23 (1976), pp. 208–211.

THOMAS JECH

[Jec03] ***Set Theory***, Springer Monographs in Mathematics, Springer-Verlag, Berlin, 2003, the third millennium edition, revised and expanded.

RONALD B. JENSEN

[Jen72] *The fine structure of the constructible hierarchy*, ***Annals of Mathematical Logic***, vol. 4 (1972), pp. 229–308; erratum, p. 443.

RONALD B. JENSEN AND JOHN R. STEEL

[JS13] **K** *without the measurable*, ***The Journal of Symbolic Logic***, vol. 78 (2013), no. 3, pp. 708–734.

H. JUDAH, W. JUST, AND W. HUGH WOODIN

[JJW92] ***Set Theory of the Continuum***, MSRI publications, vol. 26, Springer-Verlag, 1992.

GEORGE KAFKOULIS

[Kaf04] *Coding lemmata in* $\mathbf{L}(\mathbf{V}_{\lambda+1})$, ***Archive for Mathematical Logic***, vol. 43 (2004), no. 2, pp. 193–213.

LÁSZLÓ KALMÁR

[Kal28] *Zur Theorie der abstrakten Spiele*, ***Acta Scientiarum Mathematicarum (Szeged)***, vol. 4 (1928–29), no. 1–2, pp. 65–85.

AKIHIRO KANAMORI

[Kan94] ***The Higher Infinite. Large Cardinals in Set Theory from Their Beginnings***, Perspectives in Mathematical Logic, Springer-Verlag, Berlin, 1994.

[Kan95] *The emergence of descriptive set theory*, ***From Dedekind to Gödel. Essays on the Development of the Foundations of Mathematics. Proceedings of a conference held at Boston University, Boston, MA, April 5–7, 1992*** (Jaakko Hintikka, editor), Synthese Library, vol. 251, Kluwer Academic Publishers, Dordrecht, 1995, pp. 241–262.

[Kan03] ***The Higher Infinite. Large Cardinals in Set Theory from Their Beginnings***, second ed., Springer Monographs in Mathematics, Springer-Verlag, Berlin, 2003.

AKIHIRO KANAMORI AND MATTHEW FOREMAN

[KF10] ***Handbook of Set Theory***, Springer-Verlag, 2010.

AKIHIRO KANAMORI AND MENACHEM MAGIDOR

[KM78] *The evolution of large cardinal axioms in set theory*, ***Higher Set Theory. Proceedings, Oberwolfach, Germany, April 13–23, 1977*** (Gert H. Müller and Dana S. Scott, editors), Lecture Notes in Mathematics, vol. 669, Springer-Verlag, 1978, pp. 99–275.

ALEXANDER S. KECHRIS

[Kec74] *On projective ordinals*, ***The Journal of Symbolic Logic***, vol. 39 (1974), pp. 269–282.

[Kec75] *The theory of countable analytical sets*, ***Transactions of the American Mathematical Society***, vol. 202 (1975), pp. 259–297.

[Kec78] AD *and projective ordinals*, in Kechris and Moschovakis [CABAL i], pp. 91–132, reprinted in [CABAL II], pp. 304–345.

[Kec81] *Homogeneous trees and projective scales*, in Kechris et al. [CABAL ii], pp. 33–74, reprinted in [CABAL II], pp. 270–303.

[Kec84] *The axiom of determinacy implies dependent choices in* $\mathbf{L}(\mathbb{R})$, ***The Journal of Symbolic Logic***, vol. 49 (1984), no. 1, pp. 161–173.

[Kec88] *A coding theorem for measures*, in Kechris et al. [CABAL iv], pp. 103–109, reprinted in [CABAL I], pp. 398–403.

[Kec92] *The structure of Borel equivalence relations in polish spaces*, ***Set Theory of the Continuum. Papers from the workshop held in Berkeley, California, October 16–20, 1989*** (Haim Judah, Winfried Just, and Hugh Woodin, editors), Mathematical Sciences Research Institute Publications, vol. 26, Springer-Verlag, 1992, pp. 89–102.

[Kec95] ***Classical Descriptive Set Theory***, Graduate Texts in Mathematics, vol. 156, Springer-Verlag, 1995.

ALEXANDER S. KECHRIS, EUGENE M. KLEINBERG, YIANNIS N. MOSCHOVAKIS, AND W. H. WOODIN

[KKMW81] *The axiom of determinacy, strong partition properties, and nonsingular measures*, in Kechris et al. [CABAL ii], pp. 75–99, reprinted in [CABAL I], pp. 333–354.

ALEXANDER S. KECHRIS, BENEDIKT LÖWE, AND JOHN R. STEEL

[CABAL I] *Games, Scales, and Suslin cardinals: the Cabal Seminar, volume I*, Lecture Notes in Logic, vol. 31, Cambridge University Press, 2008.

[CABAL II] *Wadge Degrees and Projective Ordinals: the Cabal Seminar, volume II*, Lecture Notes in Logic, vol. 37, Cambridge University Press, 2012.

[CABAL III] *Ordinal Definability and Recursion Theory: the Cabal Seminar, volume III*, Lecture Notes in Logic, vol. 43, Cambridge University Press, 2016.

ALEXANDER S. KECHRIS AND DONALD A. MARTIN

[KM78] *On the theory of Π^1_3 sets of reals*, ***Bulletin of the American Mathematical Society***, vol. 84 (1978), no. 1, pp. 149–151.

ALEXANDER S. KECHRIS, DONALD A. MARTIN, AND YIANNIS N. MOSCHOVAKIS

[KMM81A] *Appendix: Progress report on the Victoria Delfino problems*, in *Cabal Seminar* 77–79 [CABAL ii], pp. 273–274.

[CABAL ii] *Cabal Seminar* 77–79, Lecture Notes in Mathematics, vol. 839, Berlin, Springer-Verlag, 1981.

[KMM83A] *Appendix: Progress report on the Victoria Delfino problems*, in *Cabal Seminar* 79–81 [CABAL iii], pp. 283–284.

[CABAL iii] *Cabal Seminar* 79–81, Lecture Notes in Mathematics, vol. 1019, Berlin, Springer-Verlag, 1983.

ALEXANDER S. KECHRIS, DONALD A. MARTIN, AND JOHN R. STEEL

[KMS88A] *Appendix: Victoria Delfino problems II*, in *Cabal Seminar* 81–85 [CABAL iv], pp. 221–224.

[CABAL iv] *Cabal Seminar* 81–85, Lecture Notes in Mathematics, vol. 1333, Berlin, Springer-Verlag, 1988.

ALEXANDER S. KECHRIS AND YIANNIS N. MOSCHOVAKIS

[KM72] *Two theorems about projective sets*, ***Israel Journal of Mathematics***, vol. 12 (1972), pp. 391–399.

[KM78A] *Appendix. The Victoria Delfino problems*, in *Cabal Seminar* 76–77 [CABAL i], pp. 279–282.

[KM78B] *Notes on the theory of scales*, in *Cabal Seminar* 76–77 [CABAL i], pp. 1–53, reprinted in [CABAL I], pp. 28–74.

[CABAL i] *Cabal Seminar* 76–77, Lecture Notes in Mathematics, vol. 689, Berlin, Springer-Verlag, 1978.

ALEXANDER S. KECHRIS AND ROBERT M. SOLOVAY

[KS85] *On the relative consistency strength of determinacy hypotheses*, ***Transactions of the American Mathematical Society***, vol. 290 (1985), no. 1, pp. 179–211.

ALEXANDER S. KECHRIS, ROBERT M. SOLOVAY, AND JOHN R. STEEL

[KSS81] *The axiom of determinacy and the prewellordering property*, in Kechris et al. [CABAL ii], pp. 101–125, reprinted in [CABAL II], pp. 118–140.

ALEXANDER S. KECHRIS AND W. HUGH WOODIN

[KW83] *Equivalence of partition properties and determinacy*, ***Proceedings of the National Academy of Sciences of the United States of America***, vol. 80 (1983), no. 6 i., pp. 1783–1786.

H. Jerome Keisler

[Kei71] ***Model Theory for Infinitary Logic***, Studies in Logic and the Foundations of Mathematics, vol. 62, North-Holland, 1971.

Richard O. Ketchersid, Paul B. Larson, and Jindřich Zapletal

[KLZ10] *Regular embeddings of the stationary tower and Woodin's Σ^2_2 maximality theorem*, ***The Journal of Symbolic Logic***, vol. 75 (2010), pp. 711–727.

Richard O. Ketchersid and Stuart Zoble

[KZ06] *On the extender algebra being complete*, ***Mathematical Logic Quarterly***, vol. 52 (2006), no. 6, pp. 531–533.

Takayuki Kihara and Antonio Montalbán

[KM18] *The uniform Martin's conjecture for many-one degrees*, ***Transactions of the American Mathematical Society***, vol. 370 (2018), no. 12, pp. 9025–9044.

Stephen C. Kleene

[Kle38] *On notation for ordinal numbers*, ***The Journal of Symbolic Logic***, vol. 3 (1938), pp. 150–155.

[Kle43] *Recursive predicates and quantifiers*, ***Transactions of the American Mathematical Society***, vol. 53 (1943), pp. 41–73.

[Kle55A] *Arithmetical predicates and function quantifiers*, ***Transactions of the American Mathematical Society***, vol. 79 (1955), pp. 312–340.

[Kle55B] *Hierarchies of number-theoretic predicates*, ***Bulletin of the American Mathematical Society***, vol. 61 (1955), pp. 193–213.

[Kle55C] *On the forms of the predicates in the theory of constructive ordinals. II*, ***American Journal of Mathematics***, vol. 77 (1955), pp. 405–428.

Eugene M. Kleinberg

[Kle70] *Strong partition properties for infinite cardinals*, ***The Journal of Symbolic Logic***, vol. 35 (1970), pp. 410–428.

[Kle77] ***Infinitary Combinatorics and the Axiom of Determinateness***, Lecture Notes in Mathematics, vol. 612, Springer-Verlag, 1977.

Peter Koellner

[Koe] *Incompatible* AD^+ *models*, preprint.

Peter Koellner and W. Hugh Woodin

[KW10] *Large cardinals from determinacy*, in Kanamori and Foreman [KF10], pp. 1951–2119.

[KW] *Foundations of set theory: The search for new axioms*, in preparation.

Motokiti Kondô

[Kon38] *Sur l'uniformization des complementaires analytiques et les ensembles projectifs de la seconde classe*, ***Japanese Journal of Mathematics***, vol. 15 (1938), pp. 197–230.

Dénes Kőnig

[Kőn27] *Über eine Schlussweise aus dem Endlichen ins Unendliche*, ***Acta Scientiarum Mathematicarum (Szeged)***, vol. 3 (1927), no. 2–3, pp. 121–130.

Kenneth Kunen

[Kun70] *Some applications of iterated ultrapowers in set theory*, ***Annals of Mathematical Logic***, vol. 1 (1970), pp. 179–227.

[Kun71A] *Elementary embeddings and infinitary combinatorics*, ***The Journal of Symbolic Logic***, vol. 36 (1971), pp. 407–413.

[Kun71B] *A remark on Moschovakis' uniformization theorem*, circulated note, March 1971.

[Kun71C] *Some singular cardinals*, circulated note, September 1971.

[Kun71D] *Some more singular cardinals*, circulated note, September 1971.

[Kun78] *Saturated ideals*, ***The Journal of Symbolic Logic***, vol. 43 (1978), no. 1, pp. 65–76.

[Kun80] ***Set Theory: An Introduction to Independence Proofs***, North-Holland, 1980.

[Kun11] ***Set Theory***, Studies in Logic: Mathematical Logic and Foundations, vol. 34, College Publications, 2011.

KAZIMIERZ KURATOWSKI

[Kur36] *Sur les théorèmes de séparation dans las théorie des ensembles*, ***Fundamenta Mathematicae***, vol. 26 (1936), pp. 183–191.

[Kur58] ***Topologie. Vol. I***, 4ème ed., Monografie Matematyczne, vol. 20, Państwowe Wydawnictwo Naukowe, Warsaw, 1958.

PAUL B. LARSON

[Lar04] ***The Stationary Tower: Notes on a Course by W. Hugh Woodin***, University Lecture Series, vol. 32, American Mathematical Society, Providence, RI, 2004.

[Lar05] *The canonical function game*, ***Archive for Mathematical Logic***, vol. 44 (2005), no. 7, pp. 817–827.

[Lar11] *Three days of Ω-logic*, ***Annals of the Japan Association for Philosophy of Science***, vol. 19 (2011), pp. 57–86.

[Lar12] *A brief history of determinacy*, ***Sets and Extensions in the Twentieth Century*** (Dov M. Gabbay, Akihiro Kanamori, and John Woods, editors), Handbook of the History of Logic, vol. 6, Elsevier, 2012, pp. 457–507.

RICHARD LAVER

[Lav82A] *An $(\aleph_2, \aleph_2, \aleph_0)$-saturated ideal on ω_1*, ***Logic Colloquium '80. Papers intended for the European Summer Meeting of the Association for Symbolic Logic to have been held in Prague, August 24–30, 1980*** (Dirk van Dalen, D. Lascar, and T. J. Smiley, editors), Studies in Logic and the Foundations of Mathematics, vol. 108, North-Holland, 1982, pp. 173–180.

[Lav82B] *Saturated ideals and nonregular ultrafilters*, ***Patras Logic Symposion (Patras, 1980)***, Studies in Logic and the Foundations of Mathematics, vol. 109, North-Holland, Amsterdam, 1982, pp. 297–305.

HENRI LEBESGUE

[Leb05] *Sur les fonctions représentables analytiquement*, ***Journal de Mathématiques Pures et Appliquées***, vol. 1 (1905), pp. 139–216.

[Leb18] *Remarques sur les théories de le mesure et de l'intégration*, ***Annales de l'École Normale supérieure***, vol. 35 (1918), pp. 191–250.

AZRIEL LÉVY

[Lév57] *Indépendance conditionnelle de* **V=L** *et d'axiomes qui se rattachent au système de M. Gödel*, ***Comptes rendus hebdomadaires des séances de l'Académie des Sciences***, vol. 245 (1957), pp. 1582–1583.

[Lév60] *A generalization of Gödel's notion of constructibility*, ***The Journal of Symbolic Logic***, vol. 25 (1960), pp. 147–155.

[Lév65A] *Definability in axiomatic set theory. I*, ***Logic, Methodology and Philosophy of Science. Proceedings of the 1964 International Congress*** (Yehoshua Bar-Hillel, editor), Studies in Logic and the Foundations of Mathematics, North-Holland, 1965, pp. 127–151.

[Lév65B] *A hierarchy of formulas in set theory*, ***Memoirs of the American Mathematical Society***, vol. 57 (1965), p. 76.

[Lév79] ***Basic Set Theory***, Springer-Verlag, Berlin, 1979.

ALAIN LOUVEAU

[Lou79] *Familles séparantes pour les ensembles analytiques*, ***Comptes Rendus de l'Académie des Sciences***, vol. 288 (1979), pp. 391–394.

ALAIN LOUVEAU AND JEAN SAINT-RAYMOND

[LSR87] *Borel classes and closed games: Wadge-type and Hurewicz-type results*, ***Transactions of the American Mathematical Society***, vol. 304 (1987), no. 2, pp. 431–467.

[LSR88] *The strength of Borel Wadge determinacy*, in Kechris et al. [CABAL iv], pp. 1–30, reprinted in [CABAL II], pp. 74–101.

BENEDIKT LÖWE

[Löw02] *Kleinberg sequences and partition cardinals below $\underset{\sim}{\delta}^1_5$*, ***Fundamenta Mathematicae***, vol. 171 (2002), no. 1, pp. 69–76.

NIKOLAI LUZIN

[Luz25A] *Les proprietes des ensembles projectifs*, ***Comptes rendus hebdomadaires des séances de l'Académie des Sciences***, vol. 180 (1925), pp. 1817–1819.

[Luz25B] *Sur les ensembles projectifs de M. Henri Lebesgue*, ***Comptes rendus hebdomadaires des séances de l'Académie des Sciences***, vol. 180 (1925), pp. 1318–1320.

[Luz25C] *Sur un problème de M. Emil Borel et les ensembles projectifs de M. Henri Lebesgue: les ensembles analytiques*, ***Comptes rendus hebdomadaires des séances de l'Académie des Sciences***, vol. 164 (1925), pp. 91–94.

[Luz27] *Sur les ensembles analytiques*, ***Fundamenta Mathematicae***, vol. 10 (1927), pp. 1–95.

[Luz30A] *Analogies entre les ensembles mesurables B et les ensembles analytiques*, ***Fundamenta Mathematicae***, vol. 16 (1930), pp. 48–76.

[Luz30B] *Sur le problème de M. J. Hadamard d'uniformisation des ensembles*, ***Comptes rendus hebdomadaires des séances de l'Académie des Sciences***, vol. 190 (1930), pp. 349–351.

NIKOLAI LUZIN AND PETR NOVIKOV

[LN35] *Choix effectif d'un point dans un complemetaire analytique arbitraire, donne par un crible*, ***Fundamenta Mathematicae***, vol. 25 (1935), pp. 559–560.

NIKOLAI LUZIN AND WACŁAW SIERPIŃSKI

[LS18] *Sur quelques propriétés des ensembles (A)*, ***Bulletin de l'Académie des Sciences Cracovie, Classe des Sciences Mathématiques, Série A***, (1918), pp. 35–48.

[LS23] *Sur un ensemble non measurable B*, ***Journal de Mathématiques Pures et Appliqueées***, vol. 2 (1923), no. 9, pp. 53–72.

MENACHEM MAGIDOR

[Mag78] *On the singular cardinals problem. I*, ***Israel Journal of Mathematics***, vol. 28 (1978), no. 1–2, pp. 1–31.

[Mag80] *Precipitous ideals and $\underset{\sim}{\Sigma}^1_4$ sets*, ***Israel Journal of Mathematics***, vol. 35 (1980), no. 1-2, pp. 109–134.

MICHAEL MAKKAI

[Mak77] *An "admissible" generalization of a theorem on countable Σ^1_1 sets of reals with applications*, ***Annals of Mathematical Logic***, vol. 11 (1977), no. 1, pp. 1–30.

RICHARD MANSFIELD

[Man70] *Perfect subsets of definable sets of real numbers*, ***Pacific Journal of Mathematics***, vol. 35 (1970), no. 2, pp. 451–457.

[Man71] *A Souslin operation on Π^1_2*, ***Israel Journal of Mathematics***, vol. 9 (1971), no. 3, pp. 367–379.

LEO MARCUS

[Mar80] *The number of countable models of a theory of one unary function*, ***Fundamenta Mathematicae***, vol. 108 (1980), no. 3, pp. 171–181.

ANDREW MARKS, THEODORE A. SLAMAN, AND JOHN R. STEEL

[MSS16] *Martin's conjecture, arithmetic equivalence, and countable Borel equivalence relations*, in Kechris et al. [CABAL III], pp. 493–520.

DONALD A. MARTIN

[Mar68] *The axiom of determinateness and reduction principles in the analytical hierarchy*, ***Bulletin of the American Mathematical Society***, vol. 74 (1968), pp. 687–689.

[Mar70] *Measurable cardinals and analytic games*, ***Fundamenta Mathematicae***, vol. 66 (1970), pp. 287–291.

[Mar75] *Borel determinacy*, ***Annals of Mathematics***, vol. 102 (1975), no. 2, pp. 363–371.

[Mar76] *Proof of a conjecture of Friedman*, ***Proceedings of the American Mathematical Society***, vol. 55 (1976), no. 1, p. 129.

[Mar80] *Infinite games*, ***Proceedings of the International Congress of Mathematicatians, Helsinki 1978*** (Olli Lehto, editor), Academia Scientiarum Fennica, 1980, pp. 269–273.

[Mar83] *The real game quantifier propagates scales*, in Kechris et al. [CABAL iii], pp. 157–171, reprinted in [CABAL I], pp. 209–222.

[Mar85] *A purely inductive proof of Borel determinacy*, ***Recursion Theory***, Proceedings of Symposia in Pure Mathematics, vol. 42, AMS, Providence, RI, 1985, pp. 303–308.

[Mar90] *An extension of Borel determinacy*, ***Annals of Pure and Applied Logic***, vol. 49 (1990), no. 3, pp. 279–293.

[Mar98] *The determinacy of Blackwell games*, ***The Journal of Symbolic Logic***, vol. 63 (1998), no. 4, pp. 1565–1581.

[Mar03] *A simple proof that determinacy implies Lebesgue measurability*, ***Università e Politecnico di Torino. Seminario Matematico. Rendiconti***, vol. 61 (2003), no. 4, pp. 393–397.

[Mar20] *Games of countable length*, 2020, this volume.

[Mar] AD *and the normal measures on* $\underset{\sim}{\delta}^1_3$, unpublished, undated.

DONALD A. MARTIN, YIANNIS N. MOSCHOVAKIS, AND JOHN R. STEEL

[MMS82] *The extent of definable scales*, ***Bulletin of the American Mathematical Society***, vol. 6 (1982), pp. 435–440.

DONALD A. MARTIN, ITAY NEEMAN, AND MARCO VERVOORT

[MNV03] *The strength of Blackwell determinacy*, ***The Journal of Symbolic Logic***, vol. 68 (2003), no. 2, pp. 615–636.

DONALD A. MARTIN AND JEFF B. PARIS

[MP71] AD $\Rightarrow \exists$ *exactly* 2 *normal measures on* ω_2, circulated note, March 1971.

DONALD A. MARTIN AND ROBERT M. SOLOVAY

[MS69] *A basis theorem for* Σ^1_3 *sets of reals*, ***Annals of Mathematics***, vol. 89 (1969), pp. 138–160.

DONALD A. MARTIN AND JOHN R. STEEL

[MS83] *The extent of scales in* $\mathbf{L}(\mathbb{R})$, in Kechris et al. [CABAL iii], pp. 86–96, reprinted in [CABAL I], pp. 110–120.

[MS88] *Projective determinacy*, ***Proceedings of the National Academy of Sciences of the United States of America***, vol. 85 (1988), no. 18, pp. 6582–6586.

[MS89] *A proof of projective determinacy*, ***Journal of the American Mathematical Society***, vol. 2 (1989), no. 1, pp. 71–125.

[MaS94] *Iteration trees*, ***Journal of the American Mathematical Society***, vol. 7 (1994), no. 1, pp. 1–73.

[MS08] *The tree of a Moschovakis scale is homogeneous*, in Kechris et al. [CABAL I], pp. 404–420.

A. R. D. MATHIAS

[Mat68] *On a generalization of Ramsey's theorem*, ***Notices of the American Mathematical Society***, vol. 15 (1968), p. 931.

[Mat77] *Happy families*, ***Annals of Mathematical Logic***, vol. 12 (1977), no. 1, pp. 59–111.

R. Daniel Mauldin

[Mau81] ***The Scottish Book: Mathematics from the Scottish Café***, Birkhäuser, Boston, MA, 1981.

Arnold W. Miller

[Mil77] *Some Problems in Set Theory and Model Theory*, ***Ph.D. thesis***, University of California at Berkeley, 1977.

[Mil95] ***Descriptive Set Theory and Forcing: How to prove theorems about Borel sets the hard way***, Lecture Notes in Logic, vol. 4, Springer-Verlag, Berlin, 1995.

William J. Mitchell

[Mit79] *Hypermeasurable cardinals*, ***Logic Colloquium '78. Proceedings of the Colloquium held in Mons, August 24–September 1, 1978*** (Maurice Boffa, Dirk van Dalen, and Kenneth McAloon, editors), Studies in Logic and the Foundations of Mathematics, vol. 97, North-Holland, Amsterdam, 1979, pp. 303–316.

William J. Mitchell and John R. Steel

[MiS94] ***Fine Structure and Iteration Trees***, Lecture Notes in Logic, vol. 3, Springer-Verlag, Berlin, 1994.

Gabriel Mokobodzki

[Mok78] *Ensembles à coupes dénombrables et capacités dominées par une mesure*, ***Séminaire de Probabilités XII***, Lecture Notes in Mathematics, vol. 649, Springer-Verlag, Heidelberg, 1978, pp. 492–511.

Michael Morley

[Mor70] *The number of countable models*, ***The Journal of Symbolic Logic***, vol. 35 (1970), pp. 14–18.

Yiannis N. Moschovakis

[Mos67] *Hyperanalytic predicates*, ***Transactions of the American Mathematical Society***, vol. 129 (1967), pp. 249–282.

[Mos69A] *Abstract first order computability I*, ***Transactions of the American Mathematical Society***, vol. 138 (1969), pp. 427–463.

[Mos69B] *Abstract first order computability II*, ***Transactions of the American Mathematical Society***, vol. 138 (1969), pp. 464–504.

[Mos70] *Determinacy and prewellorderings of the continuum*, ***Mathematical Logic and Foundations of Set Theory. Proceedings of an international colloquium held under the auspices of the Israel Academy of Sciences and Humanities, Jerusalem, 11–14 November 1968*** (Y. Bar-Hillel, editor), Studies in Logic and the Foundations of Mathematics, North-Holland, Amsterdam-London, 1970, pp. 24–62.

[Mos71] *Uniformization in a playful universe*, ***Bulletin of the American Mathematical Society***, vol. 77 (1971), pp. 731–736.

[Mos73] *Analytical definability in a playful universe*, ***Logic, Methodology, and Philosophy of Science IV. Proceedings of the Fourth International Congress for Logic, Methodology and Philosophy of Science, Bucharest, 29 August–4 September, 1971*** (Patrick Suppes, Leon Henkin, Athanase Joja, and Gr. C. Moisil, editors), North-Holland, 1973, pp. 77–83.

[Mos78] *Inductive scales on inductive sets*, in Kechris and Moschovakis [Cabal i], pp. 185–192, reprinted in [Cabal I], pp. 94–101.

[Mos80] ***Descriptive Set Theory***, Studies in Logic and the Foundations of Mathematics, vol. 100, North-Holland, Amsterdam, 1980.

[Mos81] *Ordinal games and playful models*, in Kechris et al. [Cabal ii], pp. 169–201, reprinted in [Cabal III], pp. 86–114.

[Mos83] *Scales on coinductive sets*, in Kechris et al. [Cabal iii], pp. 77–85, reprinted in [Cabal I], pp. 102–109.

[Mos09] ***Descriptive Set Theory***, second ed., Mathematical Surveys and Monographs, vol. 155, American Mathematical Society, 2009.

SANDRA MÜLLER, RALF SCHINDLER, AND W. HUGH WOODIN

[MSW16] *Mice with finitely many Woodin cardinals from optimal determinacy hypotheses*, submitted, 2016.

JAN MYCIELSKI

[Myc64] *On the axiom of determinateness*, ***Fundamenta Mathematicae***, vol. 53 (1964), pp. 205–224.

[Myc66] *On the axiom of determinateness. II*, ***Fundamenta Mathematicae***, vol. 59 (1966), pp. 203–212.

JAN MYCIELSKI AND HUGO STEINHAUS

[MS62] *A mathematical axiom contradicting the axiom of choice*, ***Bulletin de l'Académie Polonaise des Sciences***, vol. 10 (1962), pp. 1–3.

JAN MYCIELSKI AND STANISŁAW ŚWIERCZKOWSKI

[MŚ64] *On the Lebesgue measurability and the axiom of determinateness*, ***Fundamenta Mathematicae***, vol. 54 (1964), pp. 67–71.

MARK E. NADEL

[Nad71] *Model Theory in Admissible Sets*, ***Ph.D. thesis***, University of Wisconsin, 1971.

ITAY NEEMAN

[Nee95] *Optimal proofs of determinacy*, ***The Bulletin of Symbolic Logic***, vol. 1 (1995), no. 3, pp. 327–339.

[Nee00] *Unraveling $\utilde{\Pi}^1_1$ sets*, ***Annals of Pure and Applied Logic***, vol. 106 (2000), no. 1-3, pp. 151–205.

[Nee02A] *Inner models in the region of a Woodin limit of Woodin cardinals*, ***Annals of Pure and Applied Logic***, vol. 116 (2002), no. 1-3, pp. 67–155.

[Nee02B] *Optimal proofs of determinacy II*, ***Journal of Mathematical Logic***, vol. 2 (2002), no. 2, pp. 227–258.

[Nee04] ***The Determinacy of Long Games***, de Gruyter Series in Logic and its Applications, vol. 7, Walter de Gruyter, Berlin, 2004.

[Nee05] *An introduction to proofs of determinacy of long games*, ***Logic Colloquium '01. Proceedings of the Annual European Summer Meeting of the Association for Symbolic Logic held in Vienna, August 6–11, 2001*** (Matthias Baaz, Sy-David Friedman, and Jan Krajíček, editors), Lecture Notes in Logic, vol. 20, Association for Symbolic Logic, 2005, pp. 43–86.

[Nee06A] *Determinacy for games ending at the first admissible relative to the play*, ***The Journal of Symbolic Logic***, vol. 71 (2006), no. 2, pp. 425–459.

[Nee06B] *Unraveling $\utilde{\Pi}^1_1$ sets, revisited*, ***Israel Journal of Mathematics***, vol. 152 (2006), pp. 181–203.

[Nee07A] *Games of length ω_1*, ***Journal of Mathematical Logic***, vol. 7 (2007), no. 1, pp. 83–124.

[Nee07B] *Inner models and ultrafilters in* $\mathbf{L}(\mathbb{R})$, ***The Bulletin of Symbolic Logic***, vol. 13 (2007), no. 1, pp. 31–53.

[Nee10] *Determinacy in* $\mathbf{L}(\mathbb{R})$, in Kanamori and Foreman [KF10], pp. 1877–1950.

ITAY NEEMAN AND JOHN R. STEEL

[NS06] *Counterexamples to the unique and cofinal branches hypotheses*, ***The Journal of Symbolic Logic***, vol. 71 (2006), no. 3, pp. 977–988.

ITAY NEEMAN AND JINDŘICH ZAPLETAL

[NZ00] *Proper forcing and* $\mathbf{L}(\mathbb{R})$, preprint, arXiv:0003027v1, 2000.

[NZ01] *Proper forcing and* $\mathbf{L}(\mathbb{R})$, ***The Journal of Symbolic Logic***, vol. 66 (2001), no. 2, pp. 801–810.

PETR NOVIKOV

[Nov35] *Sur la séparabilité des ensembles projectifs de seconde class*, ***Fundamenta Mathematicae***, vol. 25 (1935), pp. 459–466.

JOHN C. OXTOBY

[Oxt80] ***Measure and Category***, second ed., Graduate Texts in Mathematics, vol. 2, Springer-Verlag, New York, 1980.

JEFF B. PARIS

[Par72] $\mathsf{ZF} \vdash \Sigma^0_4$ *determinateness*, ***The Journal of Symbolic Logic***, vol. 37 (1972), pp. 661–667.

DAVID PREISS

[Pre73] *Metric spaces in which Prohorov's theorem is not valid*, ***Zeitschrift für Wahrscheinlichkeitstheorie und Verwandte Gebiete***, vol. 27 (1973), pp. 109–116.

KAREL PŘÍKRÝ

[Pří76] *Determinateness and partitions*, ***Proceedings of the American Mathematical Society***, vol. 54 (1976), pp. 303–306.

FRANK RAMSEY

[Ram30] *On a problem of formal logic*, ***Proceedings of the London Mathematical Society***, vol. 30 (1930), no. 2, pp. 2–24.

MATI RUBIN

[Rub74] *Thoeries of linear order*, ***Israel Journal of Mathematics***, vol. 17 (1974), pp. 392–443.

[Rub77] *Vaught's conjecture for linear orderings*, ***Notices of the American Mathematical Society***, vol. 24 (1977), p. A 390.

GERALD E. SACKS

[Sac76] *Countable admissible ordinals and hyperdegrees*, ***Advances in Mathematics***, vol. 20 (1976), no. 2, pp. 213–262.

GRIGOR SARGSYAN

[Sar13A] *Descriptive inner model theory*, ***The Bulletin of Symbolic Logic***, vol. 19 (2013), no. 1, pp. 1–55.

[Sar13B] *On the prewellorderings associated with the directed systems of mice*, ***The Journal of Symbolic Logic***, vol. 78 (2013), no. 3, pp. 735–763.

[Sar14] *An inner model proof of the strong partition property for* $\boldsymbol{\delta}^2_1$, ***Notre Dame Journal of Formal Logic***, vol. 55 (2014), no. 4, pp. 563–568.

[Sar15] ***Hod Mice and the Mouse Set Conjecture***, vol. 236, Memoirs of the American Mathematical Society, no. 1111, American Mathematical Society, 2015.

GRIGOR SARGSYAN AND NAM TRANG

[ST16] ***The Largest Suslin Axiom***, 2016, book manuscript, submitted.

ERNEST SCHIMMERLING

[Sch95] *Combinatorial principles in the core model for one Woodin cardinal*, ***Annals of Pure and Applied Logic***, vol. 74 (1995), no. 2, pp. 153–201.

[Sch01] *The ABC's of mice*, ***The Bulletin of Symbolic Logic***, vol. 7 (2001), no. 4, pp. 485–503.

[Sch07] *Coherent sequences and threads*, ***Advances in Mathematics***, vol. 216 (2007), no. 1, pp. 89–117.

[Sch10] *A core model toolbox and guide*, in Kanamori and Foreman [KF10], pp. 1685–1752.

[Sch] *Notes on Woodin's extender algebra*, preprint, undated.

ERNEST SCHIMMERLING AND MARTIN ZEMAN

[SZ01] *Square in core models*, ***The Bulletin of Symbolic Logic***, vol. 7 (2001), no. 3, pp. 305–314.

RALF SCHINDLER

[Sch99] *The Delfino problem # 12, Talks in Bonn*, 4/ 21/ 99–4/ 23/ 99, handwritten manuscript, 50 pages, available online, 1999.

[Sch02] *The core model for almost linear iterations*, ***Annals of Pure and Applied Logic***, vol. 116 (2002), no. 1-3, pp. 205–272.

ULRICH SCHWALBE AND PAUL WALKER

[SW01] *Zermelo and the early history of game theory*, ***Games and Economic Behavior***, vol. 34 (2001), no. 1, pp. 123–137.

SAHARON SHELAH

[She84] *Can you take Solovay's inaccessible away?*, ***Israel Journal of Mathematics***, vol. 48 (1984), no. 1, pp. 1–47.

[She98] ***Proper and Improper Forcing***, second ed., Perspectives in Mathematical Logic, Springer-Verlag, Berlin, 1998.

SAHARON SHELAH AND W. HUGH WOODIN

[SW90] *Large cardinals imply that every reasonably definable set of reals is Lebesgue measurable*, ***Israel Journal of Mathematics***, vol. 70 (1990), no. 3, pp. 381–394.

DAVID DAWSON SHOCHAT

[Sho72] *Capacitability of* $\underset{\sim}{\Sigma}^1_2$ *sets*, ***Ph.D. thesis***, University of California at Los Angeles, 1972.

JOSEPH R. SHOENFIELD

[Sho61] *The problem of predicativity*, ***Essays on the Foundations of Mathematics. Dedicated to A. A. Fraenkel on his Seventieth Anniversary*** (Y. Bar-Hillel, E. I. J. Poznanski, M. O. Rabin, and A. Robinson, editors), Magnes Press, Jerusalem, 1961, pp. 132–139.

[Sho67] ***Mathematical Logic***, Addison-Wesley, 1967.

WACŁAW SIERPIŃSKI

[Sie24] *Sur une propriété des ensembles ambigus*, ***Fundamenta Mathematicae***, vol. 6 (1924), pp. 1–5.

[Sie25] *Sur une class d'ensembles*, ***Fundamenta Mathematicae***, vol. 7 (1925), pp. 237–243.

[Sie38] *Fonctions additives non complètement additives et fonctions non mesurables*, ***Fundamenta Mathematicae***, vol. 30 (1938), pp. 96–99.

JACK H. SILVER

[Sil71] *Some applications of model theory in set theory*, ***Annals of Mathematical Logic***, vol. 3 (1971), no. 1, pp. 45–110.

[Sil75] *On the singular cardinals problem*, ***Proceedings of the International Congress of Mathematicians (Vancouver, B.C., 1974), Vol. 1***, Canadian Mathematical Congress, 1975, pp. 265–268.

[Sil] $\underset{\sim}{\Pi}^1_1$ *equivalence relations*, unpublished notes, undated.

MAURICE SION

[Sio63] *On capacitability and measurability*, ***Annales de l'Institut Fourier***, vol. 13 (1963), pp. 88–99.

THEODORE A. SLAMAN

[Sla09] *Degree invariant functions*, slides of a talk at the 2009 VIG, available online, 2009.

THEODORE A. SLAMAN AND JOHN R. STEEL

[SS88] *Definable functions on degrees*, in Kechris et al. [CABAL iv], pp. 37–55, reprinted in [CABAL III], pp. 458–475.

ROBERT I. SOARE

[Soa87] ***Recursively Enumerable Sets and Degrees***, Perspectives in Mathematical Logic, Springer-Verlag, Berlin, 1987.

ROBERT M. SOLOVAY

[Sol66] *On the cardinality of* Σ^1_2 *set of reals*, ***Foundations of Mathematics: Symposium papers commemorating the 60th birthday of Kurt Gödel*** (Jack J. Bulloff, Thomas C. Holyoke, and S. W. Hahn, editors), Springer-Verlag, 1966, pp. 58–73.

[Sol67A] *Measurable cardinals and the axiom of determinateness*, lecture notes prepared in connection with the Summer Institute of Axiomatic Set Theory held at UCLA, Summer 1967.

[Sol67B] *A nonconstructible Δ^1_3 set of integers*, ***Transactions of the American Mathematical Society***, vol. 127 (1967), no. 1, pp. 50–75.

[Sol70] *A model of set-theory in which every set of reals is Lebesgue measurable*, ***Annals of Mathematics***, vol. 92 (1970), pp. 1–56.

[Sol78A] *A Δ^1_3 coding of the subsets of ω_ω*, in Kechris and Moschovakis [Cabal i], pp. 133–150, reprinted in [Cabal II], pp. 346–363.

[Sol78B] *The independence of* DC *from* AD, in Kechris and Moschovakis [Cabal i], pp. 171–184, reprinted in this volume.

John R. Steel

[Ste81A] *Closure properties of pointclasses*, in Kechris et al. [Cabal ii], pp. 147–163, reprinted in [Cabal II], pp. 102–117.

[Ste81B] *Determinateness and the separation property*, ***The Journal of Symbolic Logic***, vol. 46 (1981), no. 1, pp. 41–44.

[Ste82A] *A classification of jump operators*, ***The Journal of Symbolic Logic***, vol. 47 (1982), no. 2, pp. 347–358.

[Ste82B] *Determinacy in the Mitchell models*, ***Annals of Mathematical Logic***, vol. 22 (1982), no. 2, pp. 109–125.

[Ste83A] *Scales in* $\mathbf{L}(\mathbb{R})$, in Kechris et al. [Cabal iii], pp. 107–156, reprinted in [Cabal I], pp. 130–175.

[Ste83B] *Scales on* Σ^1_1 *sets*, in Kechris et al. [Cabal iii], pp. 72–76, reprinted in [Cabal I], pp. 90–93.

[Ste88] *Long games*, in Kechris et al. [Cabal iv], pp. 56–97, reprinted in [Cabal I], pp. 223–259.

[Ste95A] $\mathbf{HOD}^{\mathbf{L}(\mathbb{R})}$ *is a core model below* Θ, ***The Bulletin of Symbolic Logic***, vol. 1 (1995), no. 1, pp. 75–84.

[Ste95B] *Projectively wellordered inner models*, ***Annals of Pure and Applied Logic***, vol. 74 (1995), no. 1, pp. 77–104.

[Ste96] ***The Core Model Iterability Problem***, Lecture Notes in Logic, no. 8, Springer-Verlag, Berlin, 1996.

[Ste02] *Core models with more Woodin cardinals*, ***The Journal of Symbolic Logic***, vol. 67 (2002), no. 3, pp. 1197–1226.

[Ste05] PFA *implies* $\mathsf{AD}^{\mathbf{L}(\mathbb{R})}$, ***The Journal of Symbolic Logic***, vol. 70 (2005), no. 4, pp. 1255–1296.

[Ste08A] *Derived models associated to mice*, ***Computational Prospects of Infinity. Part II. Presented Talks*** (Chitat Chong, Qi Feng, Theodore A. Slaman, W. Hugh Woodin, and Yue Yang, editors), World Scientific, 2008, pp. 105–193.

[Ste08B] *Games and scales. Introduction to Part I*, in Kechris et al. [Cabal I], pp. 3–27.

[Ste08C] *The length-ω_1 open game quantifier propagates scales*, in Kechris et al. [Cabal I], pp. 260–269.

[Ste08D] *Scales in* $\mathbf{K}(\mathbb{R})$ *at the end of a weak gap*, ***The Journal of Symbolic Logic***, vol. 73 (2008), no. 2, pp. 369–390.

[Ste08E] *Scales in* $\mathbf{K}(\mathbb{R})$, in Kechris et al. [Cabal I], pp. 176–208.

[Ste09] *The derived model theorem*, ***Logic Colloquium '06. Proc. of the Annual European Conference on Logic of the Association for Symbolic Logic held at the Radboud University, Nijmegen, July 27–August 2, 2006*** (S. Barry Cooper, Herman Geuvers, Anand Pillay, and Jouko Väänänen, editors), Lecture Notes in Logic, vol. 19, Association for Symbolic Logic, 2009, pp. 280–327.

[Ste10] *An outline of inner model theory*, in Kanamori and Foreman [KF10], pp. 1595–1684.

John R. Steel and Nam Trang

[ST10] AD^+, *derived models, and* Σ_1 *reflection*, Notes from the first Münster conference on the core model induction and hod mice, available online, 2010.

John R. Steel and Robert Van Wesep

[SVW82] *Two consequences of determinacy consistent with choice*, ***Transactions of the American Mathematical Society***, vol. 272 (1982), no. 1, pp. 67–85.

John R. Steel and W. Hugh Woodin

[SW16] **HOD** *as a core model*, in Kechris et al. [Cabal III], pp. 257–345.

John R. Steel and Stuart Zoble

[SZ] *Determinacy from strong reflection*, in preparation.

Mikhail Ya. Suslin

[Sus17] *Sur une définition des ensembles mesurables* B *sans nombres transfinis*, ***Comptes rendus hebdomadaires des séances de l'Académie des Sciences***, vol. 164 (1917), pp. 88–91.

Stevo Todorčević

[Tod84] *A note on the proper forcing axiom*, in Baumgartner et al. [BMS84], pp. 209–218.

Nam Trang

[Tra14] **HOD** *in natural models of* AD^+, ***Annals of Pure and Applied Logic***, vol. 165 (2014), no. 10, pp. 1533–1556.

Stanisław Ulam

[Ula60] ***A Collection of Mathematical Problems***, Interscience Tracts in Pure and Applied Mathematics, vol. 8, Interscience Publishers, New York–London, 1960.

Robert Van Wesep

[Van78A] *Separation principles and the axiom of determinateness*, ***The Journal of Symbolic Logic***, vol. 43 (1978), no. 1, pp. 77–81.

[Van78B] *Wadge degrees and descriptive set theory*, in Kechris and Moschovakis [Cabal i], pp. 151–170, reprinted in [Cabal II], pp. 24–42.

Robert L. Vaught

[Vau74] *Invariant sets in topology and logic*, ***Fundamenta Mathematicae***, vol. 82 (1974), pp. 269–293.

Boban Veličković

[Vel92] *Forcing axioms and stationary sets*, ***Advances in Mathematics***, vol. 94 (1992), no. 2, pp. 256–284.

Giuseppe Vitali

[Vit05] *Sul problema della misura dei gruppi di punti di una retta*, ***Tipografia Gamberini e Parmeggiani***, (1905), pp. 231–235.

John von Neumann and Oskar Morgenstern

[vNM04] ***Theory of Games and Economic Behavior***, Princeton University Press, 2004, Reprint of the 1980 edition.

Stan Wagon

[Wag93] ***The Banach-Tarski paradox***, Cambridge University Press, 1993, corrected reprint of the 1985 original.

Philip Wolfe

[Wol55] *The strict determinateness of certain infinite games*, ***Pacific Journal of Mathematics***, vol. 5 (1955), pp. 841–847.

W. Hugh Woodin

[Woo82] *On the consistency strength of projective uniformization*, ***Proceedings of the Herbrand Symposium. Logic Colloquium '81. Held in Marseille, July 16–24, 1981*** (Jacques Stern, editor), Studies in Logic and the Foundations of Mathematics, vol. 107, North-Holland, Amsterdam, 1982, pp. 365–384.

[Woo83] *Some consistency results in* ZFC *using* AD, in Kechris et al. [Cabal iii], pp. 172–198, reprinted in this volume.

[Woo85] Σ^2_1-*absoluteness*, handwritten note, May 1985.

[Woo86] *Aspects of determinacy*, ***Logic, Methodology and Philosophy of Science. VII. Proceedings of the Seventh International Congress held at the University of Salzburg, Salzburg, July 11–16, 1983*** (Ruth Barcan Marcus, Georg J. W. Dorn, and Paul Weingartner, editors), Studies in Logic and the Foundations of Mathematics, vol. 114, North-Holland, Amsterdam, 1986, pp. 171–181.

[Woo88] *Supercompact cardinals, sets of reals, and weakly homogeneous trees*, ***Proceedings of the National Academy of Sciences of the United States of America***, vol. 85 (1988), no. 18, pp. 6587–6591.

[Woo99] ***The Axiom of Determinacy, Forcing Axioms, and the Nonstationary Ideal***, de Gruyter Series in Logic and its Applications, vol. 1, Walter de Gruyter, Berlin, 1999.

[Woo01] *The* Ω *conjecture*, ***Aspects of Complexity. Minicourses in Algorithmics, Complexity and Computational Algebra. Proceedings of the Workshop on Computability, Complexity, and Computational Algebra held in Kaikoura, January 7–15, 2000*** (Rod Downey and Denis Hirschfeldt, editors), de Gruyter Series in Logic and its Applications, vol. 4, de Gruyter, Berlin, 2001, pp. 155–169.

[Woo02] *Beyond* Σ^2_1 *absoluteness*, ***Proceedings of the International Congress of Mathematicians, Vol. I. Plenary Lectures and Ceremonies. Held in Beijing, August 20–28, 2002*** (Beijing) (Tatsien Li, editor), Higher Education Press, 2002, pp. 515–524.

[Woo08] *A tt version of the Posner-Robinson theorem*, ***Computational Prospects of Infinity. Part II. Presented Talks*** (Chitat Chong, Qi Feng, Theodore A. Slaman, W. Hugh Woodin, and Yue Yang, editors), World Scientific, 2008, pp. 355–392.

[Woo10] ***The Axiom of Determinacy, Forcing Axioms, and the Nonstationary Ideal***, revised ed., de Gruyter Series in Logic and its Applications, vol. 1, Walter de Gruyter, Berlin, 2010.

[Woo] *An* $\aleph_1$ *dense ideal on* $\aleph_1$, in preparation.

Ernst Zermelo

[Zer04] *Beweis, daß jede Menge wohlgeordnet werden kann*, ***Mathematische Annalen***, vol. 59 (1904), pp. 514–516.

[Zer13] *Über eine Anwendung der Mengenlehre auf die Theorie des Schachspiels*, ***Proceedings of the Fifth International Congress of Mathematicians***, vol. 2, 1913, pp. 501–504.

Yizheng Zhu

[Zhu10] *The derived model theorem II*, Notes on lectures given by H. Woodin at the first Münster conference on the core model induction and hod mice, available online, 2010.

[Zhu15] *Realizing an* AD^+ *model as a derived model of a premouse*, ***Annals of Pure and Applied Logic***, vol. 166 (2015), no. 12, pp. 1275–1364.

[Zhu16] *Lightface mice with finitely many Woodin cardinals from optimal determinacy hypotheses*, 2016, preprint, arXiv 1610.02352v1.

[Zhu17] *The higher sharp I: on* $\mathbf{M}_1^{\#}$, 2017, preprint, arXiv:1604.00481v4.

CPSIA information can be obtained
at www.ICGtesting.com
Printed in the USA
LVHW041205091220
673703LV00003B/10